Handbook and Atlas of CURVES

Handbook and Atlas of CURVES

Eugene V. Shikin

CRC Press
Boca Raton New York London Tokyo

Library of Congress Cataloging-in-Publication Data

Shikin, E. V.
Handbook and atlas of curves / author, Eugene V. Shikin.
p. cm.
Includes bibliographical references and index.
ISBN 0-8493-8963-1
1. Curves--Handbooks, Manuals, etc. I. Title.
QA565.S55 1995
516'.15--dc20 94-35135
CIP

Direct all inquiries to CRC Press, Inc., 2000 Corporate Blvd., N.W., Boca Raton, Florida 33431.

International Standard Book Number 0-8493-8963-1
Library of Congress Card Number 94-35135
Printed in the United States of America 1 2 3 4 5 6 7 8 9 0
Printed on acid-free paper

To my wife Clara
and
our daughter Guzel

The Author

Eugene V. Shikin, D.Sc., is a Professor at Moscow State University, Russia. Dr. Shikin received his Master's Degree in 1964, Candidate of Science Degree in 1970 at Moscow State University, and Doctor of Science Degree at the Mathematics Institute of the USSR Academy of Sciences, Moscow, in 1977.

Dr. Shikin has worked in various fields of geometry and its applications. Originally, he was interested in the regular isometric imbeddings of Riemannian manifolds with negative curvature into Euclidean spaces. This area of study led to investigations of geometric properties of the nonlinear partial equations.

Dr. Shikin's main interest at present is development of geometric approaches to various problems in mathematics and informatics, from dynamic searching games to computer graphics. He has written five books, various reviews, and scientific papers.

During fifteen years he was an Editorial Board member of the *Mathematical Notes* journal. Currently he is an Editorial Board member of the *Fundamental and Applied Mathematics* journal.

Preface

The known statement of Oscar Wilde "Every good storyteller nowadays starts with the end, and then goes on to the beginning, and concludes with the middle" concerns above all the authors of handbooks, and this *Handbook* is not an exception to the rule.

A great deal of time elapsed from the first conceptions to the moment the *Handbook* was accomplished. Time was required to come to a decision and time was required to carry out the decision. Now all is in the past and I hope that soon the *Handbook* will meet its readers and, if I am lucky, its users.

Two main considerations had to be taken into account during the preparation of the *Handbook*. First, it is the availability of the fundamental books containing a great body of data on plane curves such as the two-volume edition *Spezielle algebraische and transcendente ebene Kurven* by G. Loria. Second, it is the increasing use of personal computers possessing wide dynamic and graphic possibilities for the preparation of the new types of handbooks — computer handbooks. I have attempted to take a middle course between these two approaches by connecting the differentially geometric apparatus and computer-aided figures.

Although I have done my best for the *Handbook* to be free of shortcomings, I suspect that the book has both strengths and weaknesses. And I hope that the readers will multiply the first, and rectify the latter.

I feel indebted to many good friends for help and advice in the preparation of this book.

But most of all I want to express my gratitude to Anna Kirin and Michail Frank-Kamenetskii for very valuable and helpful suggestions.

Moscow
January, 1995

EUGENE V. SHIKIN

Contents

Introduction

At present, there is no way of telling when the interest for curve investigation came into existence for the first time. Clearly, it goes back a long way in history. We can see the evidence of those times in the names of many curves. Some of them are connected with the names of the investigators who have drawn attention to the curves, and at times we even have several names for one curve. Poetic names of the others have obvious Greek roots. Some of the curves are especially rich in helpful properties, and the history of each of them deserves special attention.

Though the initial research has not been of a regular nature, many interesting curve classes were disclosed as a result. Descartes' coordinate method gave a substantial impetus for researching curves. The possibility of using powerful mathematical apparatus has alleviated the problem of their analysis. But some difficulties remain.

One of the essential difficulties which any investigator comes across inevitably consists of the following: the word "curve" is among a small set of mathematical terms which refer to a number of different, but closely related, notions.

Let us consider two major approaches to the notion of the plane curve.

Approach A. Plane curve is considered to be a point set.

Approach B. Plane curve is considered to be a point set with a fixed sequence of points.

Before discussing the forms of precise definitions, let us explain the difference between these two approaches.

Example 1. Let γ be a plane point set generated by two congruent circles which are tangent to one another as shown in Figure 1. This point set may be followed sequentially in a number of ways (Figure 2). The first approach provides us with only one curve, whereas the second approach gives us at least four distinct curves.

The two most frequently used methods of curve definition are connected with the two above-mentioned approaches to the curve notion. Staying with the intuitive point of view, let us consider these two ways more closely.

We shall suggest a Euclidean plane referred to Cartesian coordinates Oxy.

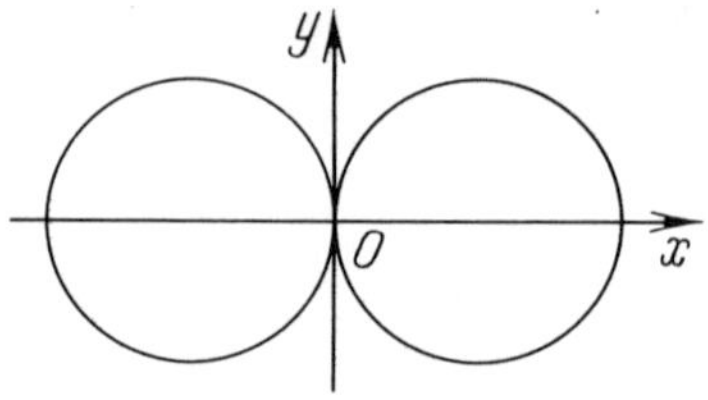

Figure 1.

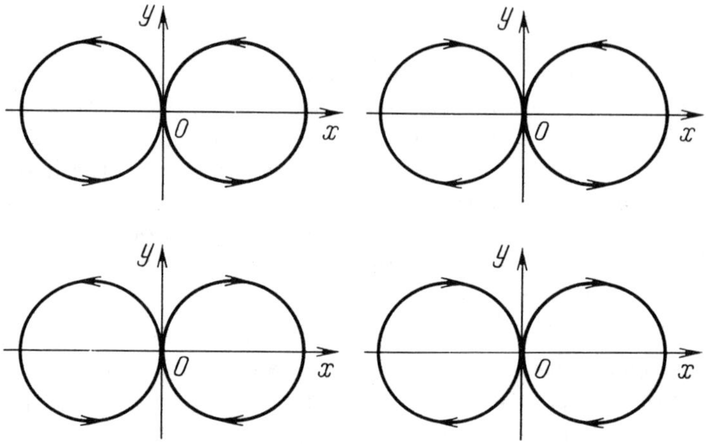

Figure 2.

With *an implicit description*, the plane curve may be thought of as a set of points $M(x, y)$, so that

$$F(x, y) = 0 \tag{A}$$

where $F(x, y)$ is a function of two variables x and y.

With *a parametric description*

$$x = x(t), \quad y = y(t) \tag{B}$$

the curve may be considered as a trace of a moving point (parameter t is time).

All the curves in this handbook are given either by the implicit or parametric way.

In actual practice, the problem of the plane curve construction is often reduced either to one equation (A) or to two equations (B). For this reason, we must constantly keep in mind that the respective plane set, *geometric image*, may have a very unusual appearance.

In the following examples we illustrate some such cases.

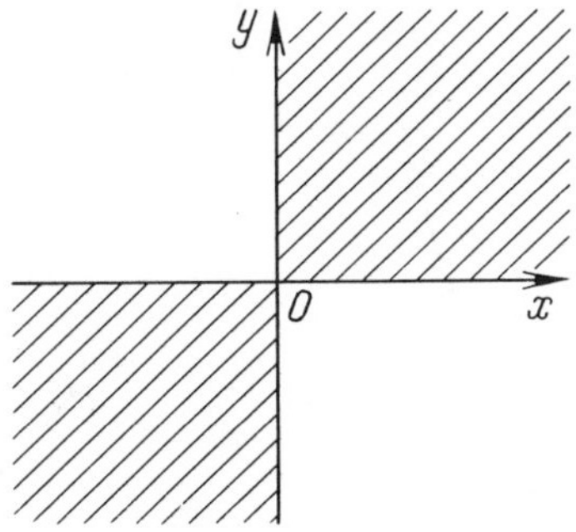

Figure 3.

Description way (A).

$$F(x, y) = 0$$

Example 2. Let us consider a function

$$F(x, y) = x^2 + y^2 + 1$$

It is easy to see that the set of points $M(x, y)$ where

$$x^2 + y^2 + 1 = 0$$

is empty.

Let function $F(x, y)$ have the form

$$F(x, y) = x^3|x|^3 - y^3|y|^3$$

The set of points $M(x, y)$ on a plane, the coordinates of which are connected by the equality

$$x^3|x|^3 - y^3|y|^3 = 0$$

is the union of the points of the first and third quadrants (Figure 3).

Description way (B).

$$x = x(t), \quad y = y(t)$$

It is natural to assume that ranges of definition for both functions $x(t)$ and $y(t)$ are the same.

Example 3. Let us take

$$x = t\cos t, \quad y = (t\cos t)^2$$

It is easy to prove that every point of this curve lies on parabola

$$y = x^2$$

(Figure 4) but under the arbitrary (admissible) changes of parameter, we are obligated to return to each point of the parabola repeatedly.

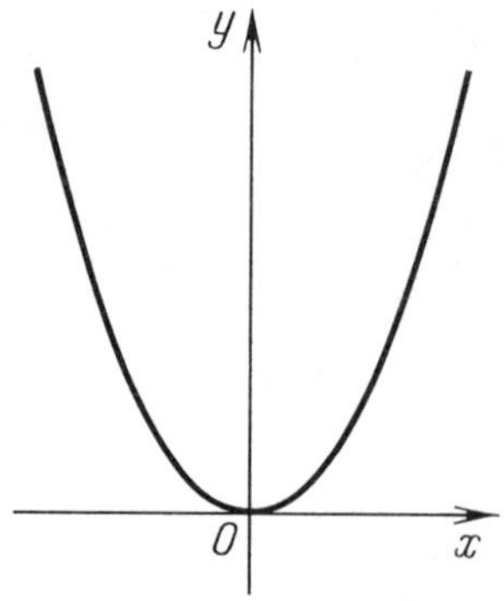

Figure 4.

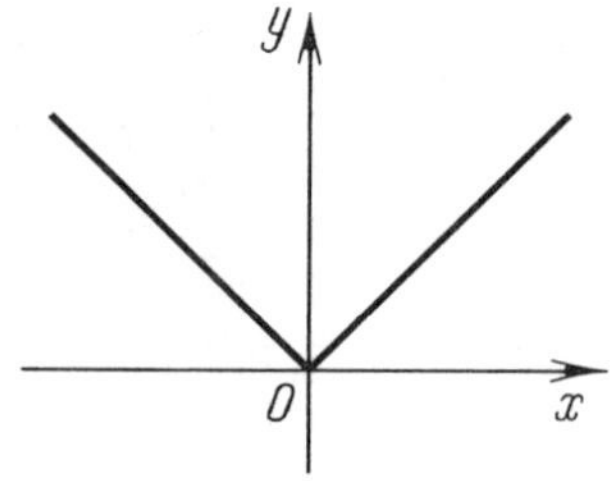

Figure 5.

The plane set described by the equations

$$x = t^5, \quad y = |t|^5$$

has a singularity (break point) (Figure 5); therefore, both functions $x(t)$ and $y(t)$ have continuous derivatives of the fourth order.

We have seen that the smoothness of the functions describing the curves is not necessarily safe against the singularities both in the parametric way of description and in the implicit way. For this reason, we must place added constraints on them if we want to receive a smoothly varying curve.

In conclusion, we want to give equations which describe the curves in Figure 1 and Figure 2.

(A)

$$(x^2 + y^2)^2 - x^2 = 0$$

(B)

$$\begin{aligned} x &= \begin{cases} \pm\cos^2 t, & -\pi/2 \le t \le \pi/2 \\ \mp\cos^2 t, & \pi/2 < t \le 3\pi/2 \end{cases} \\ y &= \begin{cases} \pm\cos t \sin t, & -\pi/2 \le t \le \pi/2 \\ \mp\cos t \sin t, & \pi/2 < t \le 3\pi/2 \end{cases} \end{aligned} \tag{1}$$

$$\begin{aligned} x &= \begin{cases} \pm\cos^2 t, & -\pi/2 \le t \le \pi/2 \\ \mp\cos^2 t, & \pi/2 < t \le 3\pi/2 \end{cases} \\ y &= \pm\cos t \sin t, \quad -\pi/2 \le t \le 3\pi/2 \end{aligned} \tag{2}$$

There are many other ways of curve definitions (with the help of differential equations, tangent families, mechanical or geometric properties, natural equations, etc.). But here we use just the implicit and parametric approaches. Of course, the essential features of curves do not have a coordinate, but a geometric nature. Nevertheless, the coordinate method of curve description permits the use of powerful analytic means for their investigation. The capabilities of the differential geometry methods for the examination of the plane and spatial curves are presented in detail in this handbook (Chapters 1 and 3).

An important part of the book is the Atlas of Plane Curves (Chapter 2), which contains detailed and illustrated descriptions of about 200 specific plane curve classes. The classes have been chosen for their interesting and important properties because investigators have at different times drawn attention to them, because they can be described by elementary functions, and because they have their own names. A brief definition of each class includes its name, equations, singular and characteristic points, asymptotes and their form, symmetry (if any), admissible values of the parameters specifying a given class, and some geometric properties of the class. The dynamics of each class is visually represented by a series of specially arranged drawings showing the qualitative change of the curve's behavior as the parameters defining the class vary. The pictures in the Atlas were constructed using an IBM PC/AT and are precise, which makes the variations due to parameter changes more interesting and distinct.

In Chapter 4 we consider at some length the geometric and cinematic properties of the specific curve classes and we also touch slightly upon the geometric splines.

Chapter 1

Plane Curves

In this chapter we introduce key notions and principal facts unveiling the potential of the differential geometric approach to investigations of properties of plane curves, and provide descriptions of some interesting classes of the plane curves defined by diverse geometric conditions.

1.1 Main definitions and facts

A compromise between a visual conception of plane curves and simple receptions of calculus gives us an efficient and valuable tool for researching their various properties. In this section we present and elucidate principal concepts — notions and facts — of the differential geometric apparatus of the smooth plane curve theory.

1.1.1 Parametric equations of plane curves

A plane set γ of points $M(x, y)$ whose coordinates x and y are defined by formulae

$$x = x(t), \quad y = y(t), \quad a \leq t \leq b \tag{$*$}$$

where functions $x(t)$ and $y(t)$ are continuous on segment $[a, b]$, or in vector form,

$$\mathbf{r} = \mathbf{r}(t), \quad a \leq t \leq b \tag{$*$}$$

where $\mathbf{r}(t) = (x(t), y(t))$ is called *a parametrically defined (parametrized) plane curve.*

The relations $(*)$ are called *parametric equations of plane curve* γ, or simply *a parametrization of curve* γ.

Point $M(x, y)$ where coordinates x and y are calculated by the formulae $(*)$ is called *a point of curve* γ.

Point $M(x(t), y(t))$, or $M(t)$, is called *a point of curve γ corresponding to parameter value t.* In this case curve γ is said *to go through point M* (Figure 1).

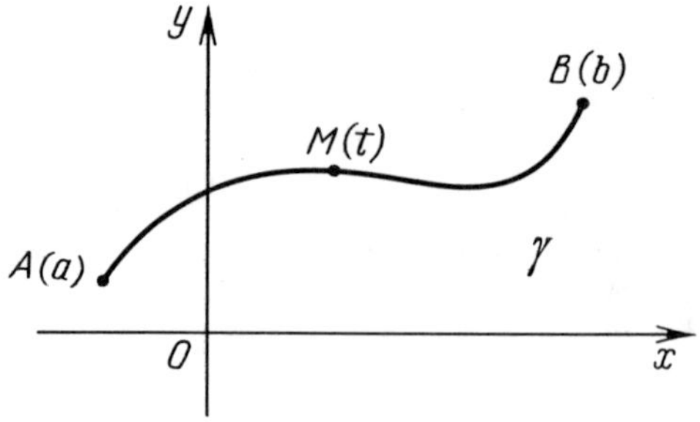

Figure 1.

Point $A(x(a), y(a))$ is called *an initial point*, and point $B(x(b), y(b))$ — *a terminal point* of the curve. Points A and B are named *end-points* of curve γ; curve γ *connects* point A with the point B.

Curve γ is called a *simple*, or *Jordan* curve when both equalities

$$x(t_1) = x(t_2), \quad y(t_1) = y(t_2)$$

hold, if and only if

$$t_1 = t_2$$

A simple curve has no self-intersection points.

A curve is named *a closed curve*, or *a contour*, if its initial and terminal points coincide.

Closed curve γ is named *a simple closed* curve when both equalities

$$x(t_1) = x(t_2), \quad y(t_1) = y(t_2)$$

hold, if and only if

$$t_1 = t_2$$

or

$$t_1 = a, \quad t_2 = b$$

Important note. The relations $(*)$ define not only a plane point-set, but indicate the order of its points under changes of parameter as well. It is most clearly seen when parameter t is interpreted to be a physical quantity — *time*. We may then consider a parametrically defined curve to be a trajectory of running point $M(x, y)$ moved according to rule $(*)$.

Example. Relations

$$x = t, \quad y = t, \quad -1 \le t \le 1$$

and

$$x = -t, \quad y = -t, \quad -1 \le t \le 1$$

describe the same plane point-sets — a segment of bisectrix of the first and the third coordinate angles. However, under changes of parameter t from

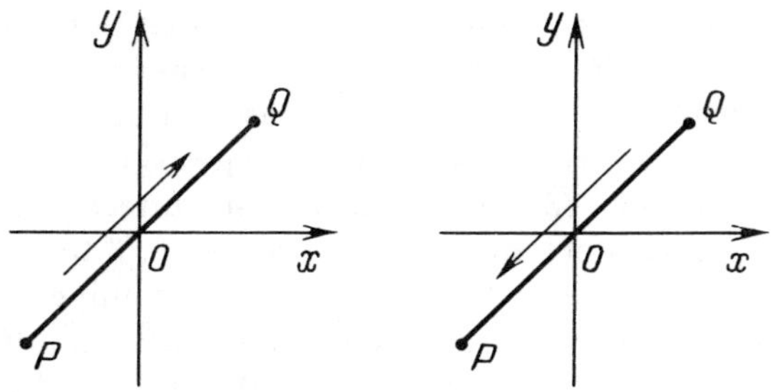

Figure 2.

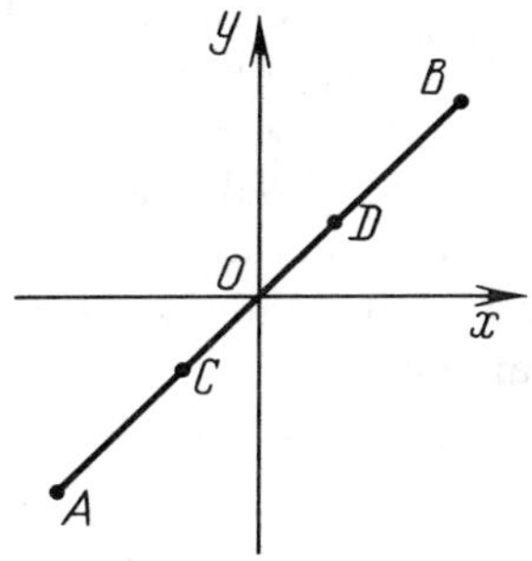

Figure 3.

-1 to 1, running point $M(t)$ moves from position P to position Q in the first case, and in the opposite direction, from point Q to point P, in the second (Figure 2).

It is significant that the same plane point-sets may be described by various parametric equations. As this takes place, the curve orientation may coincide or be opposite.

Let us consider one more example.

Example. Every pair of the following relations describes one plane point-set — the segment of bisectrix of the first and the third coordinate angles (Figure 3):

$$x = t^5, \quad y = t^5, \quad -1 \le t \le 1 \tag{a}$$

$$x = \begin{cases} t^3, & -1 \le t < 0 \\ 8t^3, & 0 \le t \le 1/2 \end{cases} \quad y = \begin{cases} t^3, & -1 \le t < 0 \\ 8t^3, & 0 \le t \le 1/2 \end{cases} \tag{b}$$

$$x = \frac{1}{6}(t^3 - t), \quad y = \frac{1}{6}(t^3 - t), \quad -2 \le t \le 2 \tag{c}$$

$$x = \begin{cases} (t+1)^3, & 2 \le t < -1 \\ 0, & -1 \le t \le 1 \\ (t-1)^3, & 1 < t \le 2 \end{cases} \quad y = \begin{cases} (t+1)^3, & 2 \le t < -1 \\ 0, & -1 \le t \le 1 \\ (t-1)^3, & 1 < t \le 2 \end{cases} \tag{d}$$

In cases (a) and (b), the running point $M(t)$ moves along each curve from position A to position B with parameter t increasing from -1 to 1, even though the velocities are different in each case. In case (c), running point $M(t)$ moves along the curve from position A to position D when parameter t increases from -2 to $-1/\sqrt{3}$, then in the opposite direction from position D to position C when parameter t increases from $-1/\sqrt{3}$ to $1/\sqrt{3}$, and finally when parameter t increases from $1/\sqrt{3}$ to 2, the running point once more changes direction of movement and completes it at point B (line segment CD is the superposition section of the curve). In case (d), running point $M(t)$ moves from position A to position B keeping the direction of motion, but point $M(t)$ rests at origin $O(0,0)$ when parameter t lies within segment $[-1,1]$.

Thus, in all cases discussed above we have the same plane point-sets. But can we think that we have only one curve?

There is more than a single answer to this question. But, ultimately we consider a curve in this handbook, nevertheless, to be a point-set. For this reason, all parametric equations defining the same set of points may be considered, under certain conditions (described below), as *various* parametrizations of the same curve.

Parametric equations of plane curves:

in the vector form $\mathbf{r} = \mathbf{r}(t), \quad a \le t \le b$

in Cartesian coordinates $x = x(t),\ y = y(t), \quad a \le t \le b$

in polar coordinates $\rho = \rho(t),\ \theta = \theta(t), \quad a \le t \le b$

1.1.2 Reparametrization

Let us consider two parametrized curves: γ with parametric equations

$$x = x(t), \quad y = y(t), \quad a \le t \le b \tag{$*$}$$

and $\tilde{\gamma}$ with parametric equations

$$x = \tilde{x}(\tau), \quad y = \tilde{y}(\tau), \quad \alpha \le \tau \le \beta \tag{$\tilde{*}$}$$

Let

$$\tau = \tau(t)$$

be a continuous monotonically increasing function with segment $[a,b]$ as a range of definition and with segment $[\alpha,\beta]$ as a range of values, so that

$$x(t) = \tilde{x}(\tau(t)), \quad y(t) = \tilde{y}(\tau(t))$$

for all t from segment $[a,b]$ (Figure 4).

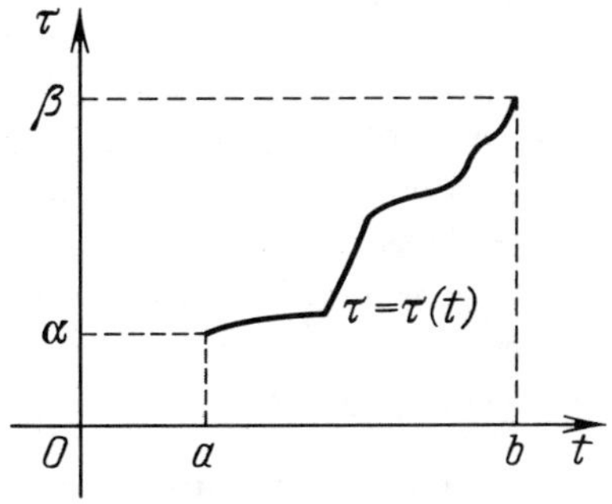

Figure 4.

In this case, the set of points defined by equations $(*)$ coincides with the set of points defined by equations $(\tilde{*})$,

$$\gamma = \tilde{\gamma}$$

Their velocity vectors point in the same direction but are distinguished from one another. Relations $(*)$ and $(\tilde{*})$ are *distinct parametrizations* of curve γ.

We say that

$$\tau = \tau(t), \quad a \le t \le b$$

is *a continuous transformation of the parameter* on curve γ if

a) $\tau(t)$ is a continuous function on line segment $[a, b]$,

b) $\tau(t)$ is a strictly increasing function on line segment $[a, b]$, and

c) its range of values is segment $[\alpha, \beta]$.

The equations

$$x = \tilde{x}(\tau), \quad y = \tilde{y}(\tau)$$

are *another parametrization* of curve γ.

It is easy to see that any curve may be parametrized in many ways.

Example. The curve

$$x = t, \quad y = t, \quad -1 \le t \le 1$$

has parametrizations

$$x = \ln\tau, \quad y = \ln\tau, \quad e^{-1} \le \tau \le e; \tag{a}$$

$$x = \tau^3, \quad y = \tau^3, \quad -1 \le \tau \le 1; \tag{b}$$

$$x = \tau\,|\,\tau\,|, \quad y = \tau\,|\,\tau\,|, \quad -1 \le \tau \le 1; \tag{c}$$

$$x = \begin{cases} \tau, & -1 \le \tau < 0 \\ 2\tau, & 0 \le \tau \le 1/2 \end{cases} \quad y = \begin{cases} \tau, & -1 \le \tau < 0 \\ 2\tau, & 0 \le \tau \le 1/2 \end{cases} \tag{d}$$

etc.

1.1.3 Smooth plane curves

To use the apparatus and the methods of mathematical analysis for examination of curve properties, it is necessary to place some additional restrictions on the functions through which the curves may be described. One of such restrictions is the smoothness of curve.

We say that plane curve γ is *an s-smooth curve with respect to given parametrization*

$$x = x(t), \quad y = y(t), \quad a \le t \le b \tag{$*$}$$

if functions $x(t)$ and $y(t)$ are s-smooth functions on segment $[a, b]$ (see Appendix A). A plane curve is called a *∞-smooth curve with respect to parametrization* $(*)$ if functions $x(t)$ and $y(t)$ have the derivatives of every order on segment $[a, b]$. We say simply that curve γ and the associated parametrization $(*)$ are *smooth* when order s of smoothness is not essential.

Let γ be a closed curve. We can construct the periodic extension of functions $x(t)$ and $y(t)$ onto the whole t-axis (with the period equal to $b - a$). Closed curve γ is called *s-smooth* (*∞-smooth*, *analytic*), if $(b - a)$-periodic extensions of functions $x(t)$ and $y(t)$ onto the t-axis are s-smooth (∞-smooth, analytic) functions; and s-smoothness of a closed curve with equations

$$x = x(t), \quad y = y(t), \quad a \le t \le b$$

is equivalent to the following conditions:

1) functions $x(t)$ and $y(t)$ are s-smooth on segment $[a, b]$, and

2) one-sided derivatives of the functions at $t = a$ and at $t = b$ are respectively equal:

$$\begin{array}{ll} x'(a+0) = x'(a-0), & y'(b+0) = y'(b-0), \\ x''(a+0) = x''(a-0), & y''(b+0) = y''(b-0), \\ \ldots & \\ x^{(s)}(a+0) = x^{(s)}(a-0), & y^{(s)}(b+0) = y^{(s)}(b-0) \end{array}$$

The next example shows that a plane curve has both smooth and nonsmooth parametrizations.

Example. We can parametrize the curve in Figure 5 in several ways:

$$x = \begin{cases} t, & -1 \le t < 0 \\ 2t, & 0 \le t \le 1/2 \end{cases} \qquad y = \begin{cases} t, & -1 \le t < 0 \\ 2t, & 0 \le t \le 1/2 \end{cases}$$

(nonsmooth parametrization: derivatives of functions $x(t)$ and $y(t)$ have discontinuity of the first kind at $t = 0$);

$$x = \begin{cases} t^3, & -1 \le t < 0 \\ 8t^3, & 0 \le t \le 1/2 \end{cases} \qquad y = \begin{cases} t^3, & -1 \le t < 0 \\ 8t^3, & 0 \le t \le 1/2 \end{cases}$$

(2-smooth parametrization); and

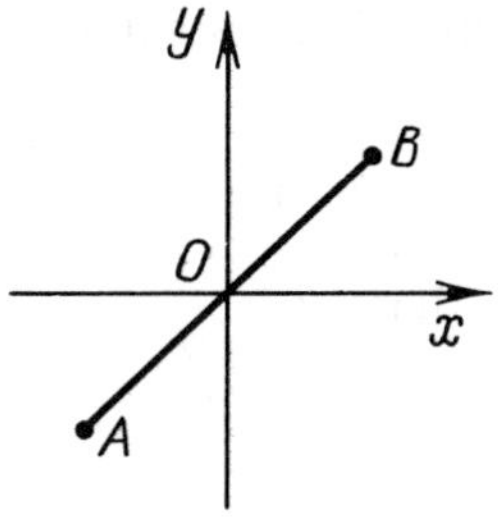

Figure 5.

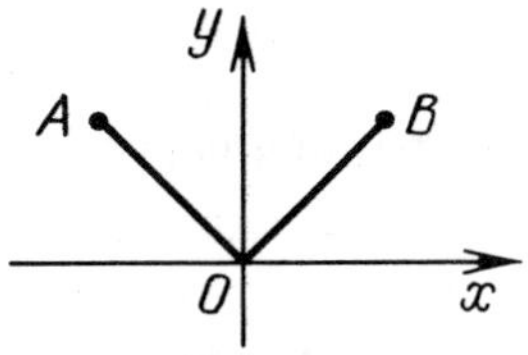

Figure 6.

$$x = t, \quad y = t, \quad -1 \le t \le 1/2$$

(analytic parametrization).

Let us consider an example of a different type.

Example. Let γ be a curve defined by equations

$$x = t^3, \quad y = t^2 \mid t \mid, \quad -1 \le t \le 1$$

With respect to this parametrization, curve γ is 2-smooth. But the correspondent point-set (geometric image) described by the given formulae has a singularity — *an angular point, break* (Figure 6). Remark that both derivatives

$$x'(t) = 3t^2, \quad y'(t) = 3t \mid t \mid$$

are equal to zero at $t = 0$.

This example shows that the requirements of smoothness on parametrization may not be sufficient for the absence of curve singularities.

Let γ be an s-smooth curve and

$$x = x(t), \quad y = y(t), \quad a \le t \le b$$

are the corresponding equations.

We say that point $M(t_0), a < t_0 < b$, is *a singular (with respect to given parametrization) point* of curve γ if

$$x'(t_0) = 0, \quad y'(t_0) = 0$$

Point $M(t), a < t < b$, of curve γ is called *regular (nonsingular)* if

$$[x'(t_0)]^2 + [y'(t_0)]^2 > 0$$

The singularities of a smooth parametrized plane curve may be of two kinds.

A. *Parametrization singularities (remediable)* disappear with the corresponding reparametrization of the curve.

Example. It is easy to see that point $O(0,0)$ is a singular point of the parametrized curve

$$x = t^3, \quad y = t^3, \quad -1 \le t \le 1$$

Let τ be a new parameter for this curve so that

$$\tau = t^3$$

Then every point of the parametrized curve

$$x = \tau, \quad y = \tau, \quad -1 \le \tau \le 1$$

is nonsingular (regular).

B. *Geometric singularities (nonremediable)* are impossible to remedy by reparametrization (see break point of graph $y = |x|$ (Figure 6)).

Ranges of the parameter may not only be segments, but also intervals (finite or infinite) and half-intervals. It is easy to reformulate the definitions above for each of these cases. There are only two notions (of a simple curve and of an analytic curve) which deserve special attention.

Let, for example, the range of definition be an interval.

1. Parametrized curve

$$x = x(t), \quad y = y(t), \quad a < t < b$$

is said to be *a simple Jordan curve* if equalities

$$x(t_1) = x(t_2), \quad y(t_1) = y(t_2)$$

hold for $t_1 = t_2$ only, and close points of the curve correspond to close values of the parameter.

Example. We can explain the necessity of the second requirement in the definition above with the help of the curve described by equations

$$x = \frac{t^2 - 1}{t^2 + 1}, \quad y = t\frac{t^2 - 1}{t^2 + 1}, \quad -1 < t < 2$$

(arc of strophoide (Figure 7)). Let us consider $t = -1 + \varepsilon$ ($\varepsilon > 0$ is small number) and $t_0 = 1$, with the corresponding points

$$M\left(-\frac{(2-\varepsilon)\varepsilon}{(1-\varepsilon)^2 + 1}, \varepsilon\frac{(1-\varepsilon)(2-\varepsilon)}{(1-\varepsilon)^2 + 1}\right) \quad \text{and} \quad O(0,0)$$

on the curve. The smaller is ε (consequently, the greater is the distinction between t and t_0), the closer are points M and O. The curve on Figure 7 is not said to be a simple curve.

2. We say that a plane curve is *an analytic curve with respect to parametrization*

$$x = x(t), \quad y = y(t), \quad a < t < b$$

if $x(t)$ and $y(t)$ are analytic functions on interval (a, b) (see Appendix A).

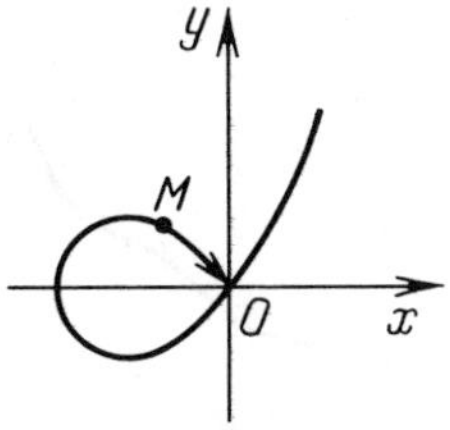

Figure 7.

1.1.4 Regular plane curves

Plane curve γ is named *an s-regular curve with respect to parametrization*

$$x = x(t), \quad y = y(t), \quad a \leq t \leq b \tag{$*$}$$

if

1) this curve is s-smooth with respect to parametrization $(*)$, and
2) $[x'(t)]^2 + [y'(t)]^2 > 0$ at every point of segment $[a, b]$.

The last condition implies that the *scalar velocity*

$$\sqrt{[x'(t)]^2 + [y'(t)]^2}$$

of curve point $M(t)$ is positive. As the parameter t changes, running point $M(t)$ moves along the regular curve γ without stopping and returning back because the curve velocity (with respect to given parametrization) is distinct from zero for any value of parameter.

Reparametrization of regular curves. Regular transformation of parameter. Let function

$$t = t(\tau)$$

be defined on segment $[\alpha, \beta]$ and possess the following properties:

1) function $t = t(\tau)$ is s-smooth on segment $[\alpha, \beta]$,
2) its set of values coincides with segment $[a, b]$, and
3) $t'(\tau) > 0$ at every point of segment $[\alpha, \beta]$ (Figure 8).

Note. Under the stated conditions there exists function

$$\tau = \tau(t)$$

inverse to the function $t = t(\tau)$, and with analogous properties:

1′) the function $\tau(t)$ is s-smooth on the segment $[a, b]$,
2′) the set of values of the function $\tau(t)$ coincides with the segment $[\alpha, \beta]$, and
3′) $\tau'(t) > 0$ at every point of the segment $[a, b]$.

By replacing parameter t in the formulae $(*)$ which define regular curve γ, with function $t(\tau)$ possessing properties 1–3, we obtain new parametric

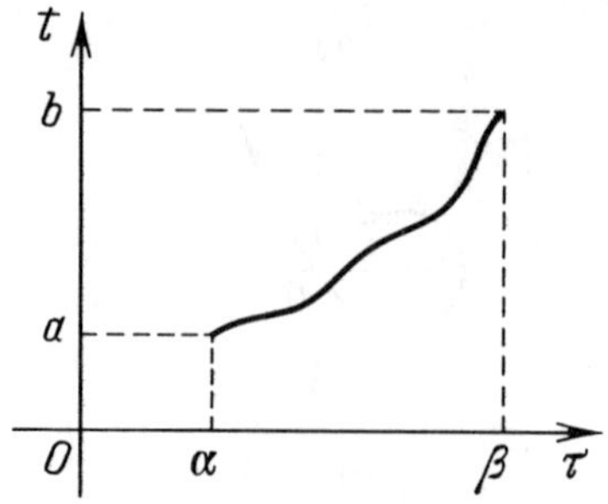

Figure 8.

equations of curve γ

$$x = \tilde{x}(\tau) = x(t(\tau)), \quad y = \tilde{y}(\tau) = y(t(\tau)), \qquad (\tilde{*})$$
$$\alpha \leq \tau \leq \beta$$

Curve γ with respect to this new parametrization is also s-regular and

$$[\tilde{x}'(\tau)]^2 + [\tilde{y}'(\tau)]^2 = ([x'(t)]^2 + [y'(t)]^2)(t'(\tau))^2 > 0$$

The parameter transformation

$$t = t(\tau)$$

with properties 1–3 is called *s-regular* or simply *regular* (when the value of s is immaterial). Curve γ is said to be *s-regular* if there exists for it such a parametrization

$$x = x(t), \quad y = y(t), \quad a \leq t \leq b$$

with respect to which curve γ is s-regular, that is, s-smooth, and the inequality

$$[x'(t)]^2 + [y'(t)]^2 > 0$$

holds at every point of segment $[a, b]$.

Advantages of regular parametrization. Let curve γ with parametric equations

$$x = x(t), \quad y = y(t), \quad a \leq t \leq b$$

be s-regular and $M_0(t_0)$ be a point on the curve (for the sake of definition, $x'(t_0) \neq 0$).

According to the well-known implicit-function theorem, equation

$$x = x(t)$$

is solvable with respect to t in the neighborhood of the point t_0:

$$t = t(x)$$

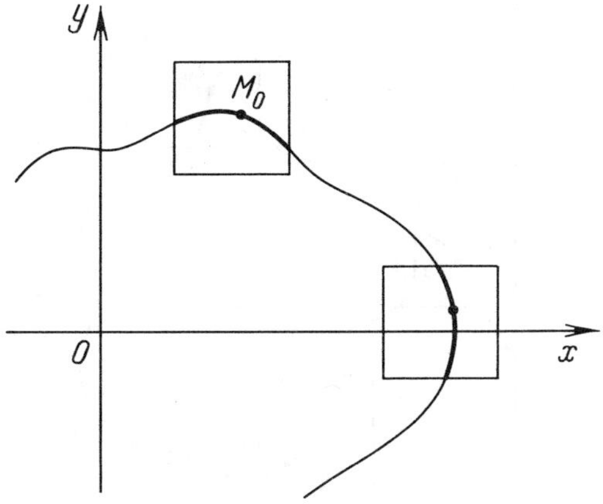

Figure 9.

and, thus, we can describe curve γ in the vicinity of point M_0 by equation

$$y = y(x)$$

that is, we may consider it as a graph of function $y = y(x)$ (Figure 9).

If $y'(t_0) \neq 0$, then we may describe curve γ in the neighborhood of point M_0 by equation

$$x = x(y)$$

In a similar manner, the notions of a regular curve and a regular transformation of parameter may be used in cases where domain T of the parameter variability is an interval or a half-interval.

Let us formulate, for example, the definitions of a ∞-regular curve and a regular analytic curve (in the case that T is interval).

Curve γ is called a ∞-*regular* curve (*regular analytic* curve, respectively) if there exists for it such a parametrization

$$x = x(t), \quad y = y(t), \quad a \leq t \leq b$$

with respect to which curve γ is ∞-smooth (analytic, respectively), and the inequality

$$[x'(t)]^2 + [y'(t)]^2 > 0$$

holds at every point of interval (a, b).

1.1.5 Implicit definition of a plane curve

Another method for defining a plane curve in use is the *implicit way* — when the curve is described by equation in the form

$$F(x, y) = 0 \qquad (*)$$

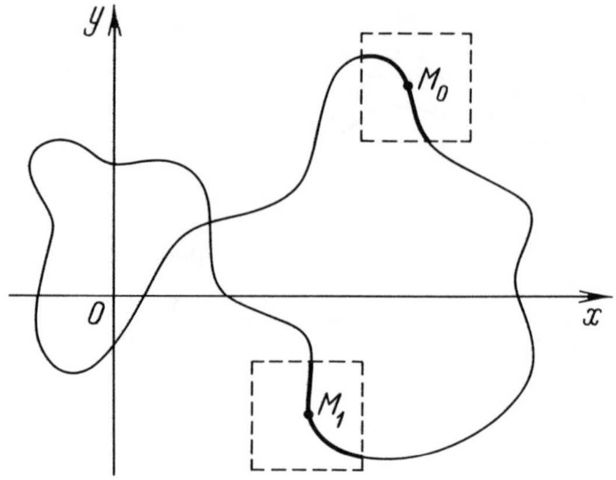

Figure 10.

It is apparent that function $F(x,y)$ is nonarbitrary. To receive a point-set corresponding to the intuitive view on plane curve, we must subject this function $F(x,y)$ to definite conditions. The requirements imposed on these conditions must be such that we have a reason to call a plane curve the set of points $M(x,y)$ whose coordinates x and y transform the equation $(*)$ to identity (or each connected component of this point-set).

Let us point out an important special case which we shall use further.

Let $F(x,y)$ be a smooth function and

$$F(x_0,y_0)=0, \quad F_x^2(x_0,y_0)+F_y^2(x_0,y_0)>0 \tag{$**$}$$

at some point $M_0(x_0,y_0)$. It is well known (the implicit-function theorem) that the fulfillment of these conditions is sufficient for expressing one variable as a function of another. In particular, the sufficient condition for the existence of a solution of the equation $(*)$ in the form

$$y=y(x)$$

is the fulfillment of the inequality

$$F_y(x_0,y_0)\neq 0$$

In this case, the equation $(*)$ describes a simple Jordan curve in some vicinity of point $M_0(x_0,y_0)$. Moreover, the curve is regular (Figure 10) and we can choose the coordinate x as a parameter for this curve.

A condition

$$F_x(x_0,y_0)\neq 0$$

provides a solvability of the equation $(*)$ in the form

$$x=x(y)$$

Point $M_0(x_0, y_0)$ of the implicitly defined curve is called *a simple (non-singular, regular) point* if the conditions (**) are fulfilled.

Point $M_0(x_0, y_0)$ of the implicitly defined curve is called its *singular* point if

$$F(x_0, y_0) = 0, \quad F_x(x_0, y_0) = 0, \quad F_y(x_0, y_0) = 0$$

1.1.6 Comparison of two ways for defining a plane curve

The implicit definition.

Advantages.

A. The curve definition with the help of equation

$$F(x, y) = 0$$

is very important for classification of plane curves. Under this classification, all plane curves are divided into two classes: algebraic and transcendental curves.

Algebraic curves (the function $F(x, y)$ is an algebraic function of variables x and y, in particular, polynomial

$$F(x, y) = \sum_{p+q=0}^{m} a_{pq} x^p y^q$$

where p and q are nonnegative integers and a_{pq} is numerical coefficient);

Transcendental curves (the function $F(x, y)$ is nonalgebraic).

In its turn, the class of algebraic curves is divided into subclasses depending on the power of polynomial $F(x, y)$: algebraic curves of the second order, algebraic curves of the third order, algebraic curves of the fourth order, and algebraic curves of the higher orders.

B. If function $F(x, y)$ is a product of two functions

$$F(x, y) = G(x, y)H(x, y)$$

then a curve defined by equation

$$F(x, y) = 0$$

is a point set union of two sets described by equations

$$G(x, y) = 0$$

and

$$H(x, y) = 0$$

respectively.

C. Singular points of the implicitly defined curves and the character of their singularity are examined with the unified method.

Disadvantages.

The construction of the implicitly defined curve in practice (in particular, its visualization on display) ever so often seems to be difficult. For this reason, we need to resolve this equation with respect to x and/or y.

Parametric definition $x = x(t), \quad y = y(t)$.

Advantages.

A. Curve definition with the help of equations

$$x = x(t), \quad y = y(t)$$

is convenient to use for curve construction. A parametrically defined curve may be visualized on the display of a personal computer with relative ease.

B. The formulae describing the main notions and facts take a rather simple form and clear geometric contents (especially for the curves with unit velocity).

C. Parametric equations describe not only the geometric image of the curve (as is the case in implicit way), but also the movement of the running point along it under changes of parameter.

Disadvantages.

A. A unified way to investigate the type of any singular point is absent (we are forced to seek self-intersection points for the parametrically defined curve by the special method differing from the one used for other singular points).

1.1.7 Tangents and normals to plane curves

Let γ be a smooth curve with parametric equations

$$x = x(t), \quad y = y(t) \tag{$*$}$$

and M_0 and M — points on it defined by values t_0 and t (the last from a neighborhood of point t_0) of the parameter, respectively (Figure 11). It is obvious that $M \to M_0$ when $t \to t_0$. That is, the length of straight-line segment MM_0 tends to zero.

A straight line M_0T is said to be ***a tangent*** to curve γ at point M_0 if the lesser angle between this line and variable straight line M_0M (angle $\Delta\Theta$ in Figure 12) tends to zero when $M \to M_0$.

Let P be an orthogonal projection of point M on line M_0T, and h — a length of line segment MP (Figure 13). Straight line M_0T is tangent to curve γ at point M_0 if

$$\frac{h}{d} \to 0$$

for $M \to M_0$.

If M_0 is a regular point of curve γ, that is,

$$[x'(t_0)]^2 + [y'(t_0)]^2 > 0$$

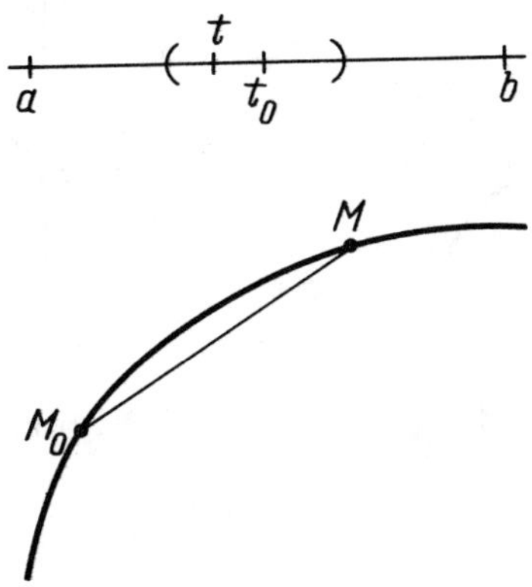

Figure 11.

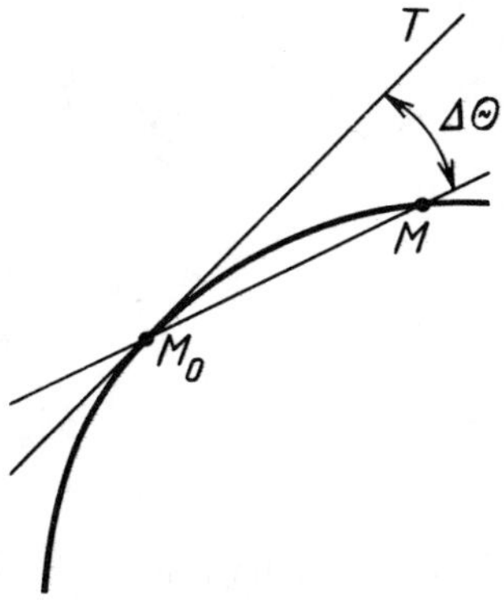

Figure 12.

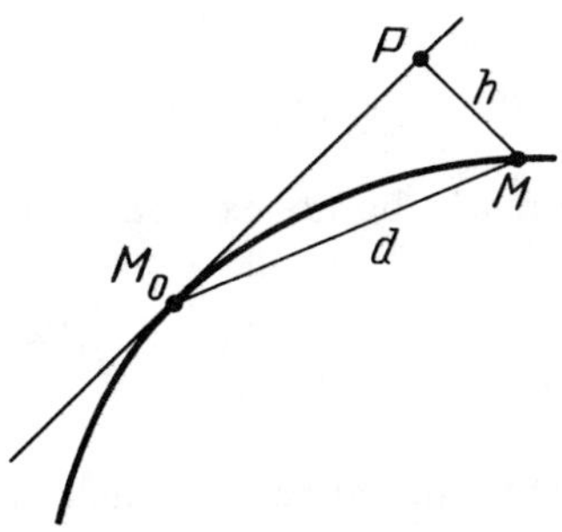

Figure 13.

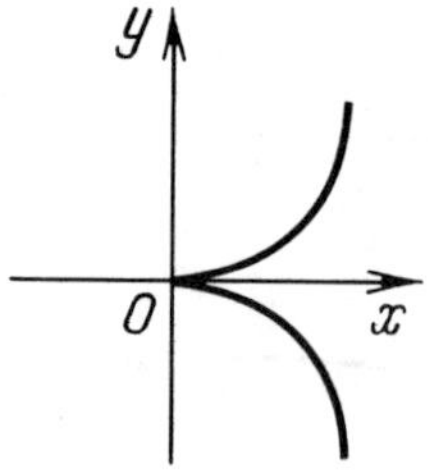

Figure 14.

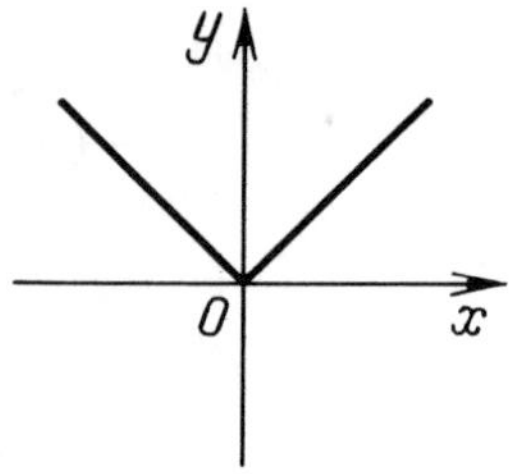

Figure 15.

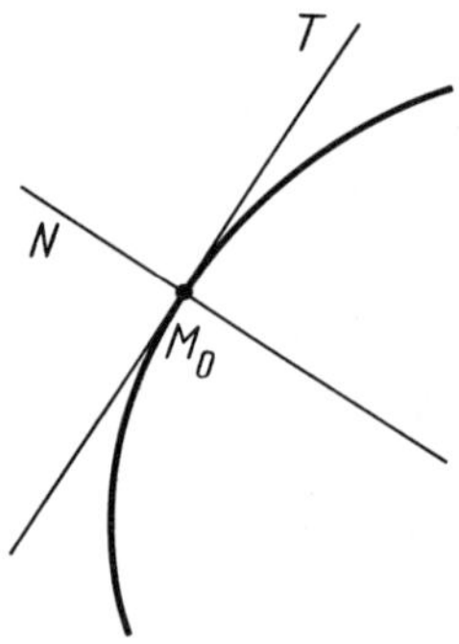

Figure 16.

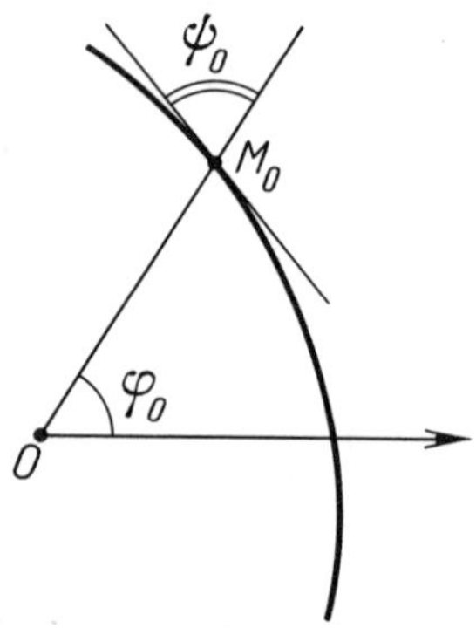

Figure 17.

then there exists a tangent to the curve at this point and the tangent is uniquely determined by point M_0.

In case

$$x'(t_0) = 0, \quad y'(t_0) = 0$$

the tangent to the curve at point M_0 may exist, for example, for the curve in Figure 14:

$$x = t^3, \quad y = t^3, \quad -1 < t < 1$$

($t_0 = 0$) or it just as well might be nonexistent, for example, for the curve in Figure 15:

$$x = t^3, \quad y = t^2|t|, \quad -1 < t < 1$$

($t_0 = 0$).

A regular curve has a unique tangent at each of its points.

The normal of plane curve γ at point M_0 is said to be a straight line which passes through this point perpendicularly to the tangent of curve γ at point M_0 (Figure 16).

Let curve γ be given in polar coordinates (r, φ) by equation of the form

$$r = r(\varphi)$$

where $r(\varphi_0) > 0$. Then the equation of the tangent to the curve at point $M_0(r(\varphi_0), \varphi_0)$ takes the next form

$$\rho(\cos(\Theta - \psi_0)) = r(\varphi_0)$$

where ψ_0 is an angle between the tangent to the curve at point M_0 (Figure 17) and radius vector of this point; the angle ψ_0 can be found from relationship

$$\tan \psi_0 = \frac{r(\varphi_0)}{r'(\varphi_0)}$$

Equations of a tangent and a normal for a plane curve at regular point M_0.

If the curve is defined *in a parametric form*

$$x = x(t), \quad y = y(t),$$

$$x_0 = x(t_0), \quad y_0 = y(t_0), \quad [x'(t_0)]^2 + [y'(t_0)]^2 > 0$$

then equations of the tangent have the form

$$\frac{x - x_0}{x'(t_0)} = \frac{y - y_0}{y'(t_0)}$$

or

$$x = x_0 + ux'(t_0), \quad y = y_0 + uy'(t_0),$$

$$-\infty < u < \infty$$

and

$$x'(t_0)(x - x_0) + y'(t_0)(y - y_0) = 0$$

is an equation of the normal.

If the curve is defined *in a vector form*

$$\mathbf{r} = \mathbf{r}(t),$$

$$\mathbf{r}_0 = \mathbf{r}(t_0), \quad \mathbf{r}'(t_0) \neq \mathbf{0}$$

then an equation of the tangent has the form

$$\mathbf{r} = \mathbf{r}(t_0) + u\mathbf{r}'(t_0), \quad -\infty < u < \infty$$

and

$$(\mathbf{r} - \mathbf{r}_0, \mathbf{r}'(t_0)) = 0$$

is a vector equation of the normal.

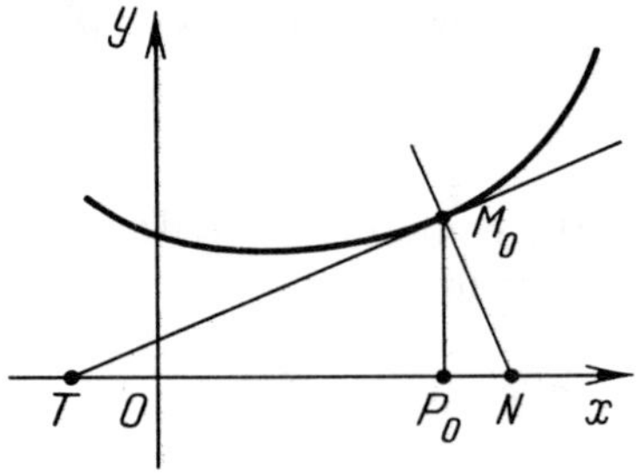

Figure 18.

If the curve is defined *in an implicit form*

$$F(x, y) = 0,$$

$$F_x^2(x_0, y_0) + F_y^2(x_0, y_0) > 0$$

then an equation of the tangent has the form

$$F_x(x_0, y_0)(x - x_0) + F_y(x_0, y_0)(y - y_0) = 0$$

and

$$\frac{x - x_0}{F_x(x_0, y_0)} = \frac{y - y_0}{F_y(x_0, y_0)}$$

is an equation of the normal.

If the curve is defined *in an explicit form*

$$y = y(x), \quad y_0 = y(x_0)$$

then

$$y - y_0 = y'(x_0)(x - x_0)$$

is an equation of the tangent and

$$x - x_0 + y'(x_0)(y - y_0) = 0$$

is an equation of the normal.

Let γ be a regular curve, M_0 — a point on curve γ, and P_0 — a projection of the point on the x-axis. Let tangent M_0T to curve γ at point M_0 intersect the x-axis at point T (Figure 18). Projection P_0T of line segment M_0T on the axis of abscissae (taken, if necessary, with a sign appropriated to the orientation of the x-axis) is called *a subtangent*. Projection P_0N of line segment M_0T on the axis of abscissae (from point N of its intersection with the x-axis to point M_0) is called *a subnormal*.

For the projection axis we may take any (not just the coordinate) axis. In particular, for a polar axis the subtangent and the subnormal are said to be *polar*.

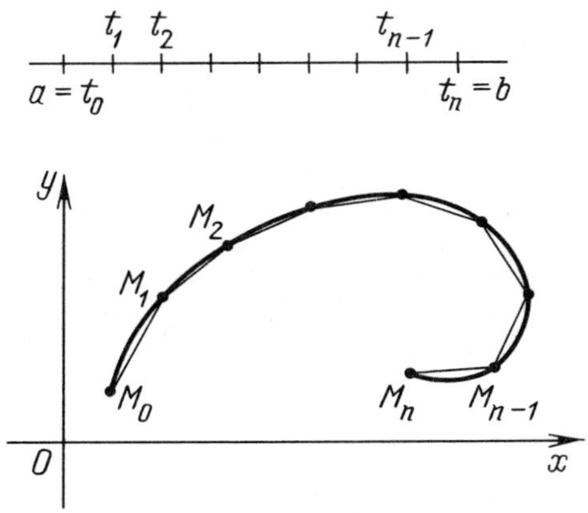

Figure 19.

1.1.8 Arc lengths of plane curves

Let curve γ be defined by equations

$$x = x(t), \quad y = y(t), \quad a \leq t \leq b$$

Divide segment $[a, b]$ into parts by points

$$t_0, t_1, \ldots, t_{n-1}, t_n$$

so that

$$a = t_0 < t_1 < \ldots < t_{n-1} < t_n = b$$

Mark points M_i on it, corresponding to values t_i of parameter, and construct broken line $M_0 M_1 \ldots M_{n-1} M_n$ inscribed in curve γ (Figure 19).

Curve γ is said to be *rectifiable* if the set of lengths of all broken lines inscribed in curve γ as above is bounded. Least upper bound of this length set is called *an arc length of curve* γ, or simply *a length of curve* γ.

Notation: $s(\gamma)$, S.

Every smooth plane curve is rectifiable; its length may be calculated by one of the following formulae:

$$s = \int_a^b \sqrt{[x'(t)]^2 + [y'(t)]^2} dt$$

for parametric description $(x = x(t),\ y = y(t))$,

$$s = \int_a^b \mid \mathbf{r}'(t) \mid \, dt$$

for vector description $(\mathbf{r} = \mathbf{r}(t))$,

$$s = \int_a^b \sqrt{1 + [y'(x)]^2}\, dx$$

for explicit description ($y = y(x)$),

$$s = \int_{\theta_1}^{\theta_2} \sqrt{\rho^2(\theta) + [\rho'(\theta)]^2}\, d\theta$$

for polar description ($\rho = \rho(\theta)$).

1.1.9 Natural parametrization of regular plane curves

Let γ be an s-regular curve with parametric equations

$$x = x(t), \quad y = y(t), \quad a \le t \le b$$

and S — a length of the curve.

A function

$$s(t) = \int_a^t \sqrt{[x'(\xi)]^2 + [y'(\xi)]^2}\, d\xi$$

strictly increases on segment $[a, b]$ and its derivative

$$s'(t) = \sqrt{[x'(t)]^2 + [y'(t)]^2}$$

is positive. In addition, function $s(t)$ is s-smooth on segment $[a, b]$ and its set of values coincides with segment $[0, S]$.

Thus, the transformation of parameter

$$t = t(s)$$

is s-regular and we may take the arc length as a new *natural* parameter on curve γ.

A curve parametrization with an arc length as a parameter is called *a natural parametrization*. The natural parametrization of an s-regular curve is s-regular.

If

$$x = x(s), \quad y = y(s), \quad 0 \le t \le S$$

is the natural parametrization of the curve, then

$$|\, \mathbf{r}'(s)\, | = \sqrt{[x'(s)]^2 + [y'(s)]^2} = 1$$

For this reason, a naturally parametrized curve is said to be *a unit velocity curve*.

For the parametrization

$$x = x(t), \quad y = y(t), \quad a \le t \le b$$

of a smooth curve to be natural, it is necessary and sufficient that

$$[x'(t)]^2 + [y'(t)]^2 = 1$$

Let t_0 be an inner point of segment $[a, b]$. Then function

$$s(t) = \int_{t_0}^{t} \sqrt{[x'(\xi)]^2 + [y'(\xi)]^2}\, d\xi$$

takes both negative ($a \le t < t_0$) and positive ($t_0 < t \le b$) values. The function keeps the name of *a natural parameter.*

Note. The above permits the use of natural parametrization in cases where domain T of the parameter definition is an interval or a half-interval (finite or infinite).

The natural parametrization of an ∞-regular (regular analytic, respectively) curve is ∞-regular (regular analytic, respectively).

1.1.10 Curvature of plane curves

Let γ be a regular curve and M_0 — a point on it. *Curvature* k of the curve is defined as a limit of fraction

$$\frac{\Delta\Theta}{\Delta s}$$

as $M \to M_0$, where $\Delta\Theta$ is the lesser of the two angles between tangents to curve γ at points M and M_0, and Δs is a length of arc M_0M (Figure 20); 2-regular curve has definite curvature k at every point.

If

$$\mathbf{r} = \mathbf{r}(s)$$

is a natural parametrization of curve γ, then its curvature

$$k(s) = | \mathbf{r}''(s) |$$

For an arbitrary 2-smooth parametrization

$$\mathbf{r} = \mathbf{r}(t)$$

the curvature of a 2-regular plane curve is calculated by formula

$$k(t) = \frac{| \mathbf{r}'(t) \times \mathbf{r}''(t) |}{| \mathbf{r}'(t) |^3}$$

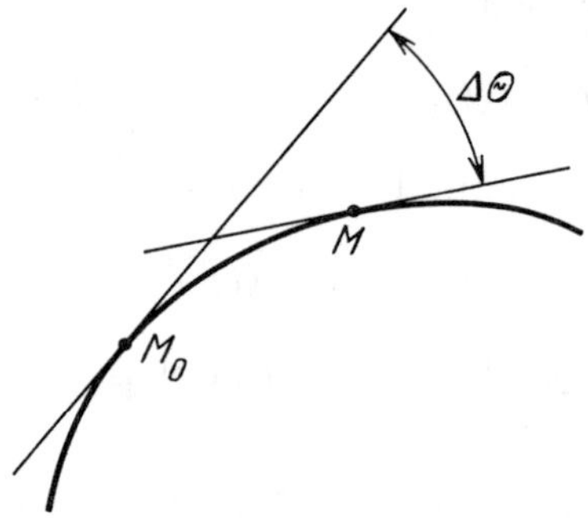

Figure 20.

If the curve is defined by parametric equations

$$x = x(t), \quad y = y(t)$$

then its curvature

$$k(t) = \frac{\mid x'(t)y''(t) - x''(t)y'(t) \mid}{([x'(t)]^2 + [y'(t)]^2)^{3/2}}$$

For the explicit way of curve description, $y = y(x)$, we have

$$k(x) = \frac{\mid y''(x) \mid}{(1 + [y'(t)]^2)^{3/2}}$$

And for the implicit way, $F(x, y) = 0$, the curvature of the regular curve may be calculated by either of the two formulae

$$k = \frac{\mid F_{xx}F_y^2 - 2F_{xy}F_xF_y + F_{yy}F_x^2 \mid}{(F_x^2 + F_y^2)^{3/2}}$$

or

$$k = \left| \left(\frac{F_x}{\sqrt{F_x^2 + F_y^2}} \right)_x + \left(\frac{F_y}{\sqrt{F_x^2 + F_y^2}} \right)_y \right|$$

If

$$\rho = \rho(\theta)$$

is the curve equation in polar coordinates (ρ, θ) then

$$k = \frac{\mid \rho^2(\theta) + 2[\rho'(\theta)]^2 - \rho(\theta)\rho''(\theta) \mid}{(\rho^2(\theta) + [\rho'(\theta)]^2)^{3/2}}$$

or

$$k = (1/\rho + (1/\rho)'')(1 + (\rho'/\rho)^2)^{3/2}$$

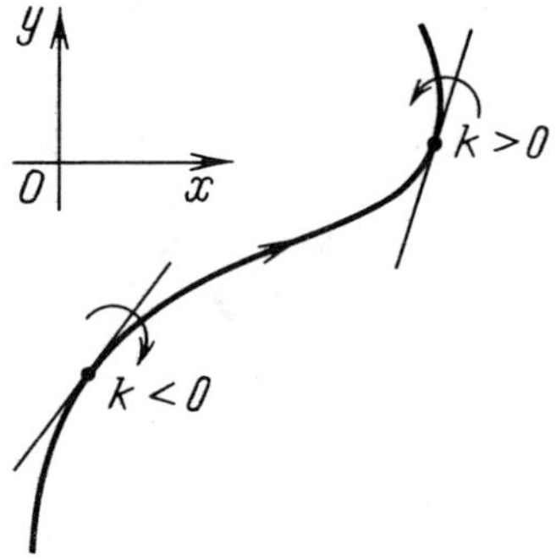

Figure 21.

By definition, the curvature of a plane curve is nonnegative. However, in many cases it is useful to ascribe a sign to the curve. The choosing of the sign is commonly connected with the sense of the tangent rotation (in moving along the curve in the direction of the increasing parameter):

the curvature of the curve is positive when its tangent rotates counterclockwise (in the positive direction),

the curvature of the curve is negative when its tangent rotates clockwise (in the negative direction) (Figure 21).

In particular, for the explicit way of curve definition $y = y(x)$, the sign of curvature

$$k(x) = \frac{| y''(x) |}{(1 + [y'(t)]^2)^{3/2}}$$

coincides with the sign of the second derivative $y''(x)$.

The curvature of a straight line is equal to zero at every point. The reverse is also true: if the curvature of a curve is equal to zero at every point then this curve is a straight line or a connected part of a straight line.

The curvature of a circle with radius R is a constant and equal to $1/R$ at every point. The reverse is also true: if the curvature of a curve is equal to a positive constant a, then the curve is an arc of a circle with radius $1/a$.

Let M_0 be a point of regular curve γ so that

$$k(M_0) \neq 0$$

Then a number

$$R = \frac{1}{k(M_0)}$$

is defined. The number is known as *a curvature radius* of the curve at point M_0.

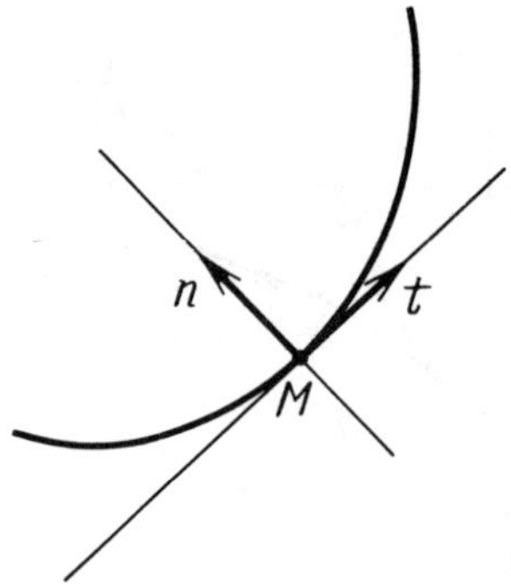

Figure 22.

A point of a curve is said to be *a point of straightness* if its curvature at this point is equal to zero. By definition, a radius of curvature at this point is infinite.

Let α be an angle between a tangent of curve γ and a fixed direction on a plane. Then curvature k of the curve may be calculated by formulae

$$k = \frac{d\alpha}{ds}$$

or

$$k = -\frac{d\alpha}{ds}$$

where s is a natural parameter.

1.1.11 Frénet formulae for plane curves

Let γ be a regular curve.

The following two mutually orthogonal unit vectors,
$\mathbf{t}$ — *a unit vector of tangent* and
$\mathbf{n}$ — *a unit vector of normal,*
are defined at every point of the curve (Figure 22).

Let

$$\mathbf{r} = \mathbf{r}(s), \quad 0 \leq s \leq S$$

be a natural parametrization of a regular curve. Then

$$\mathbf{t}(s) = \mathbf{r}'(s), \quad \mathbf{n}(s) = \frac{\mathbf{r}''(s)}{k(s)}$$

(the last formula is true only in cases when $k(s) \neq 0$).

For the parametric description of the curve

$$x = x(t), \quad y = y(t)$$

we have

$$\mathbf{t}(t) = \frac{1}{\sqrt{[x'(t)]^2 + [y'(t)]^2}}(x'(t), y'(t))$$

$$\mathbf{n}(t) = \frac{1}{\sqrt{[x'(t)]^2 + [y'(t)]^2}}(-y'(t), x'(t))$$

Vectors $\mathbf{t}$ and $\mathbf{n}$ are connected by *Frénet (Serret-Frénet) formulae*:

$$\frac{d\mathbf{t}}{ds} = k\mathbf{n}, \quad \frac{d\mathbf{n}}{ds} = -k\mathbf{t}$$

Let

$$x = x(t), \quad y = y(t), \quad 0 \le t \le 1$$

be parametric equations of a regular curve γ. There exist some geometrically substantiated ways for introducing parameters.

1. *Natural parametrization*:

$$s = \int_0^t \sqrt{[x'(\xi)]^2 + [y'(\xi)]^2}\, d\xi$$

2. *Normalized natural parametrization*:

$$\tau = \frac{\int_0^t \sqrt{[x'(\xi)]^2 + [y'(\xi)]^2}\, d\xi}{\int_0^1 \sqrt{[x'(\xi)]^2 + [y'(\xi)]^2}\, d\xi}$$

3. *Angular parametrization*:

$$l = \int_0^t |\, k(\xi) \,| \sqrt{[x'(\xi)]^2 + [y'(\xi)]^2}\, d\xi$$

4. *Normalized angular parametrization*:

$$\sigma = \frac{\int_0^t |\, k(\xi) \,| \sqrt{[x'(\xi)]^2 + [y'(\xi)]^2}\, d\xi}{\int_0^1 |\, k(\xi) \,| \sqrt{[x'(\xi)]^2 + [y'(\xi)]^2}\, d\xi}$$

Let us explain the meaning of the last parametrization.

An angle between tangent vector $(x'(t), y'(t))$ of a given curve and a unit vector of x-axis is equal to

$$\arctan \frac{y'(t)}{x'(t)}$$

Its derivative

$$(\arctan \frac{y'(t)}{x'(t)})' = k(t)\sqrt{[x'(t)]^2 + [y'(t)]^2}$$

shows that the curvature of the curve measures the range of the tangent direction change when we move along the curve with constant velocity.

Example. It is easy to verify that for parabola

$$x = t, \quad y = t^2$$

the relationship

$$k(t)\sqrt{[x'(t)]^2 + [y'(t)]^2} = \frac{1}{1+t^2}$$

holds; whence it follows that

$$\tau = (4/\pi)\arctan t$$

1.1.12 Natural equations of plane curves

Equations of the form

$$\begin{gathered} k = k(s) \\ F(k, s) = 0 \\ k = k(t), \quad s = s(t) \end{gathered}$$

where k is a curvature and s is a natural parameter of a curve, are said to be *the natural equations* of the plane curve.

If any of the natural equations are given, then its natural parametrization may be described in the form

$$x = x_0 + \int_0^s \cos\alpha(\xi)\,d\xi, \quad y = y_0 + \int_0^s \sin\alpha(\xi)\,d\xi$$

or, more generally, in the form

$$x = x_0 + \int_0^s \cos\left(\alpha_0 + \int_0^\xi k(\eta)\,d\eta\right) d\xi,$$

$$y = y_0 + \int_0^s \sin\left(\alpha_0 + \int_0^\xi k(\eta)\,d\eta\right) d\xi,$$

where α is an angle the curve tangent makes with the x-axis (Figure 23). We may find this angle from equation

$$\frac{d\alpha}{ds} = k$$

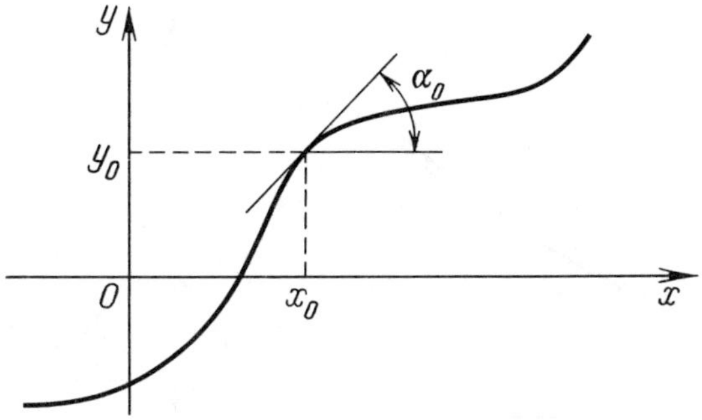

Figure 23.

A plane curve is uniquely defined by its natural equation $k = k(s)$; that is to say that for equation $k = k(s)$, given point $M_0(x_0, y_0)$, and direction α_0, there exists a unique curve where M_0 is a point on it and α_0 is an angle of the slope of its tangent at point M_0 to x-axis.

Connection between the natural and the parametric descriptions of regular curves.

Let γ be a regular curve, R — a radius of its curvature, s — an arc length of the curve, and α — an angle between the tangent to the curve and a fixed direction. Then

$$x = x_0 + \int_{\alpha_0}^{\alpha} f(\beta) \cos \beta \, d\beta$$

where $R = f(\alpha)$

$$y = y_0 + \int_{\alpha_0}^{\alpha} f(\beta) \sin \beta \, d\beta$$

$$x = x_0 + \int_{R_0}^{R} \rho f'(\rho) \cos[f(\rho)] \, d\rho$$

where $\alpha = f(R)$

$$y = y_0 + \int_{R_0}^{R} \rho f'(\rho) \sin[f(\rho)] \, d\rho$$

$$x = x_0 + \int_{\alpha_0}^{\alpha} f'(\beta) \cos \beta \, d\beta$$

where $s = f(\alpha)$

$$y = y_0 + \int_{\alpha_0}^{\alpha} f'(\beta) \sin \beta \, d\beta$$

$$x = x_0 + \int_{s_0}^{s} \cos f(\sigma) \, d\sigma$$

where $\alpha = f(s)$

$$y = y_0 + \int_{s_0}^{s} \sin f(\sigma) \, d\sigma$$

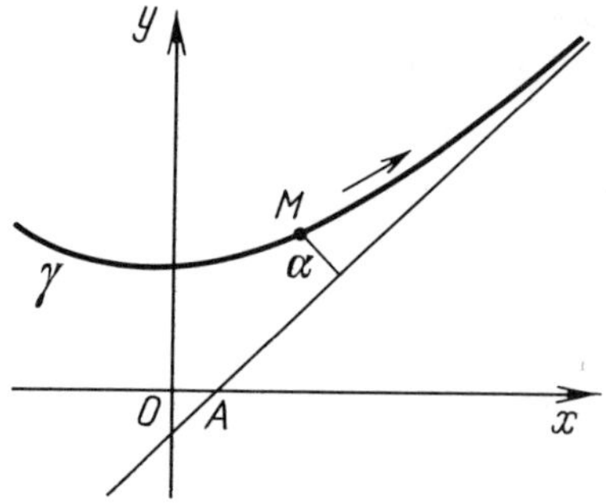

Figure 24.

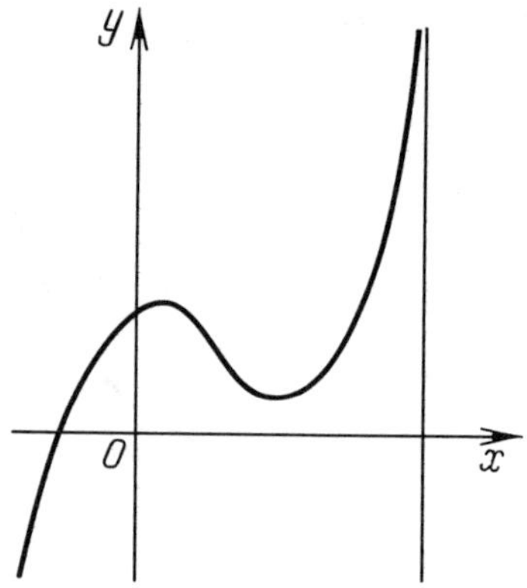

Figure 25.

1.1.13 Asymptotes of plane curves

Curve γ defined by equations

$$x = x(t), \quad y = y(t), \quad a < t < b \qquad (*)$$

is said *to go to infinity* if

$$x^2(t) + y^2(t) \to \infty$$

when $t \to a$ and/or $t \to b$.

This property is independent of parametrization.

Straight line A is called *an asymptote* for curve γ going to infinity if distance $d(t)$ from curve point $M(x(t), y(t))$ to line A tends to zero when $t \to a$ and/or $t \to b$.

Let curve γ go to infinity (for the sake of definition when $t \to a$).

The existence criterion for inclined and horizontal asyptotes. In order for straight line $y = kx + b$ to be an (*inclined* — $k \neq 0$, *horizontal* — $k = 0$) asymptote of curve γ with reference to a given Cartesian coordinate system (Figure 24), it is necessary and sufficient that the next two limits

$$\lim_{t\to a} \frac{y(t)}{x(t)} = k,$$

$$\lim_{t\to a} (y(t) - kx(t)) = b$$

exist concurrently.

The existence criterion for vertical asymptotes. The necessary and sufficient condition for straight line $x = A$ to be a *vertical* asymptote of curve γ (Figure 25) is the concurrent existence of the next two limits

$$\lim_{t\to a} x(t) = A, \quad \lim_{t\to a} y(t) = \pm\infty$$

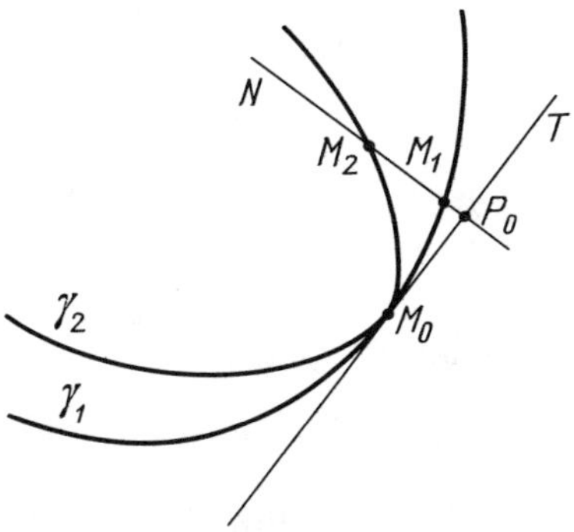

Figure 26.

1.1.14 Osculation of plane curves

Two curves are said *to be tangent to one another at a common point*, or *to be contacting curves* if both curves have their tangents at the point and the tangents coincide.

Let γ_1 and γ_2 be regular curves, M_0 — their common point, and M_0T — their common tangent at the point (Figure 26). Let us take point P on the tangent close to point M_0 and pass straight line PN through point P perpendicularly to straight line M_0T. Line PN intersects curves γ_1 and γ_2 at points M_1 and M_2, respectively. Denote the length of line segment M_0P by d and the length of line segment M_1M_2 by h.

We say that *an order of osculation* (*an order of contact*) of curves γ_1 and γ_2 at the common point M_0 is not below $n \in \mathbf{N}$ if

$$\lim_{d \to 0} \frac{h}{d^n} = 0 \qquad (*)$$

If, in addition,

$$\lim_{d \to 0} \frac{h}{d^n} \neq 0$$

or a ratio

$$\frac{h}{d^{n+1}}$$

has no limit under $d \longrightarrow 0$, then *the order of osculation* (*the order of contact*) of curves γ_1 and γ_2 at point M_0 is said to *be equal to* n.

When the limiting relation $(*)$ holds for every natural number n, curves γ_1 and γ_2 have at the common point M_0 *an infinite order of osculation* (*contact*) by definition.

Example. Calculate the order of osculation at origin $O(0,0)$ for parabola $y = x^2$ and one of the following three curves described by explicit equations:

$$y = x^2 + x^3 \qquad \text{(a)}$$

$$y = \begin{cases} x^2 \sin \frac{1}{x}, & x \neq 0 \\ 0, & x = 0 \end{cases} \tag{b}$$

$$y = \begin{cases} x^2 - x^2 e^{-1/x^2}, & x \neq 0 \\ 0, & x = 0 \end{cases} \tag{c}$$

Solutions.

(a) The order of osculation is equal to 2:

$$h = |\, x \,|^3, \quad d = |\, x \,|,$$

$$\lim_{x \to 0} \frac{h}{d^2} = 0, \quad \lim_{x \to 0} \frac{h}{d^3} = 1 \neq 0$$

(b) The order of osculation is equal to 1:

$$h = x^2 \left(1 - \sin \frac{1}{x} \right), d = |\, x \,|,$$

$$\lim_{x \to 0} \frac{h}{d^1} = 0$$

the ratio

$$\frac{h}{d^2} = 1 - \sin \frac{1}{x}$$

has no limit under $x \to 0$.

(c) The order of osculation is infinite:

$$h = x^2 e^{-1/x^2}, \quad d = |\, x \,|,$$

$$\lim_{x \to 0} \frac{h}{d^n} = 0$$

for every $n \in \mathbf{N}$.

Let γ_1 be a regular curve defined by the implicit way

$$F(x, y) = 0$$

γ_2 — a regular curve with parametric equations

$$x = x(t), \quad y = y(t)$$

and $M_0(x_0, y_0)$ where $x_0 = x(t_0), y_0 = y(t_0)$ be their common point so that

$$F_x^2(x_0, y_0) + F_y^2(x_0, y_0) > 0$$

Curves γ_1 and γ_2 have an osculation of order n at point M_0 if, and only if, function

$$\varphi(t) = F(x(t), y(t))$$

satisfies the next conditions:

$$\varphi(t_0) = \varphi'(t_0) = \varphi''(t_0) = \ldots = \varphi^n(t_0) = 0$$

but

$$\varphi^{n+1}(t_0) \neq 0$$

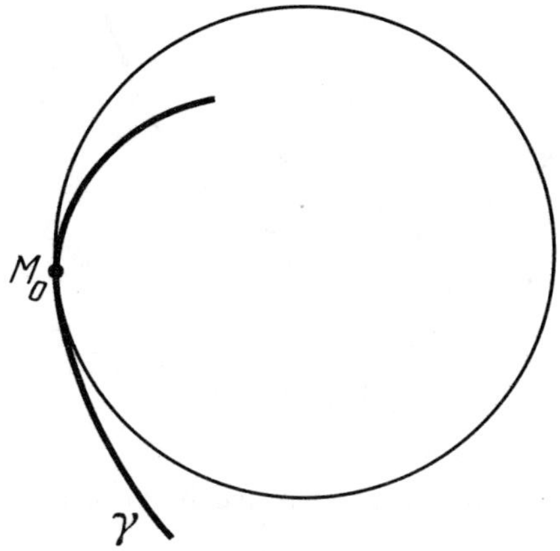

Figure 27.

1.1.15 Osculating circle of a plane curve

A circle passing through point M_0 of regular curve γ is said to be *an osculating circle* of the curve at point M_0 if their order of osculation is not below 2 (Figure 27).

Let

$$x = x(t), \quad y = y(t), \quad t \in T$$

be a regular parametrization of curve γ. Take point M_0 on it corresponding to parameter value t_0 so that $k(t_0) \neq 0$. That is, the point is not the point of straightness of the curve. Then there exists a unique osculating circle of curve γ at point M_0. Radius R of the osculating circle is defined by equality

$$R = \frac{1}{k(t_0)}$$

and its center has coordinates

$$a = x(t_0) - \frac{y'(t_0)}{k(t_0)},$$

$$b = y(t_0) + \frac{x'(t_0)}{k(t_0)}$$

The center of an osculating circle is called *a center of curvature* of the curve at the given point and its radius is called *a radius of curvature.*

An osculating circle is often called *a circle of curvature* (see above 1.1.10).

If a regular curve is described by vector equation

$$\mathbf{r} = \mathbf{r}(t), \quad t \in T$$

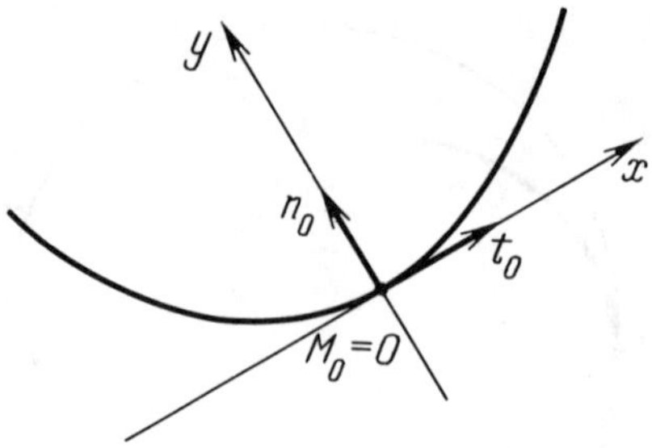

Figure 28.

then its center of curvature is determined by vector

$$\mathbf{r}(t_0) + \frac{1}{k(t_0)}\mathbf{n}(t_0)$$

where $\mathbf{n}(t_0)$ is a unit vector of the normal to the curve at point M_0.

If

$$\mathbf{r} = \mathbf{r}(s)$$

is a natural parametrization of the curve, then its radius of curvature may be calculated by formula

$$R = \frac{1}{|\,\mathbf{r}''(s_0)\,|}$$

and its center of curvature is determined by vector

$$\mathbf{r}(s_0) + \frac{\mathbf{r}''(s_0)}{|\,\mathbf{r}''(s_0)\,|}$$

Circle

$$(x-a)^2 + (y-b)^2 = R^2$$

is an osculating circle of plane curve γ defined by parametric equations

$$x = x(t), \quad y = y(t)$$

at point M_0 corresponding to parameter value t_0 if, and only if, the next conditions hold:

$$\varphi(t_0) = 0, \quad \varphi'(t_0) = 0, \quad \varphi''(t_0) = 0$$

where

$$\varphi(t) = (x-a)^2 + (y-b)^2 - R^2$$

Take coordinate system Oxy so that origin $O(0,0)$ coincides with point M_0, x-axis with a tangent of curve γ at this point, and y-axis with its normal at point M_0 (Figure 28). Then we have

$$a = x(t_0) = 0, \quad b = y(t_0) + \frac{1}{k(t_0)} = \frac{1}{k(t_0)}$$

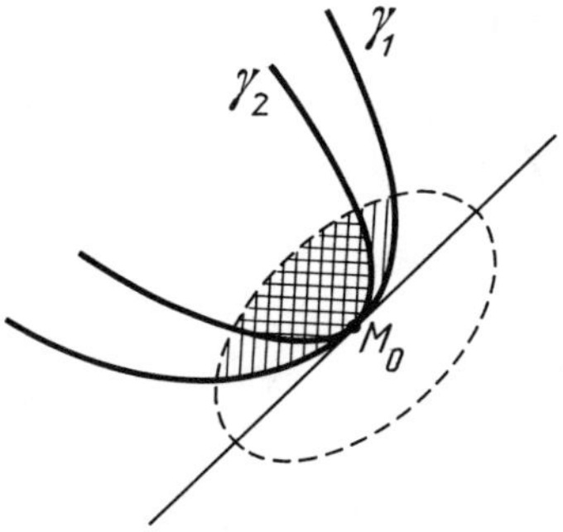

Figure 29.

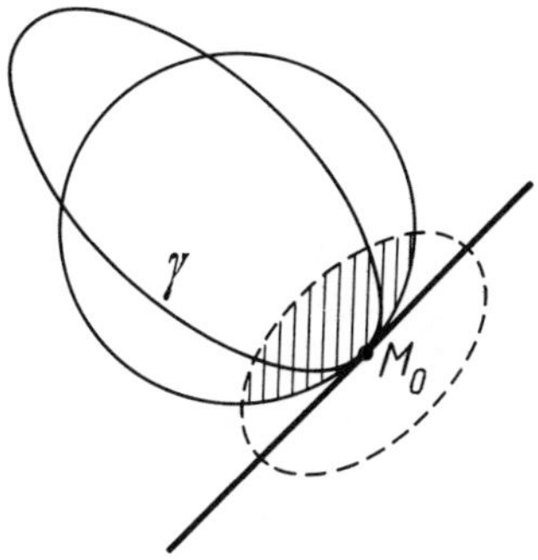

Figure 30.

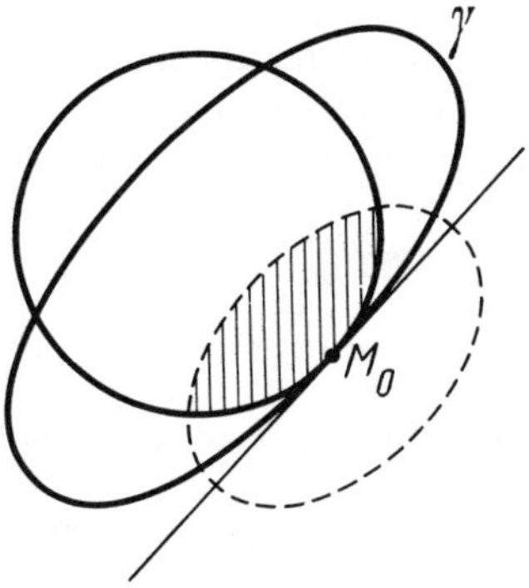

Figure 31.

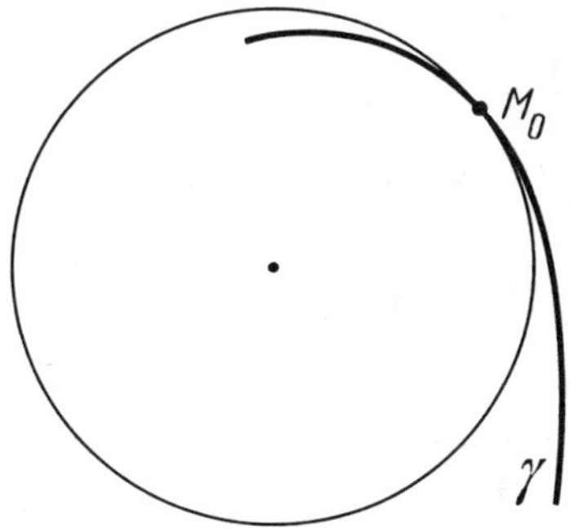

Figure 32.

That is, the equation of the osculating circle in the coordinate system takes the form

$$x^2 + \left(y - \frac{1}{k(t_0)}\right)^2 = \frac{1}{k^2(t_0)}$$

When M_0 is a point of straightness (straightening point) of the curve (curvature k vanishes at the point), an osculating circle does not exist. In this case, we say that the osculating circle degenerates to a straight line or the radius of the osculating circle is infinite.

Mutual arrangement of two contacting curves. Let regular plane curves γ_1 and γ_2 be tangent to one another at point M_0. Let their curvatures k_1 and k_2, respectively, be subject to relationship

$$0 < k_1 < k_2$$

at the point. Then in the neighborhood of point M_0, curve γ_1 wraps around curve γ_2 (Figure 29).

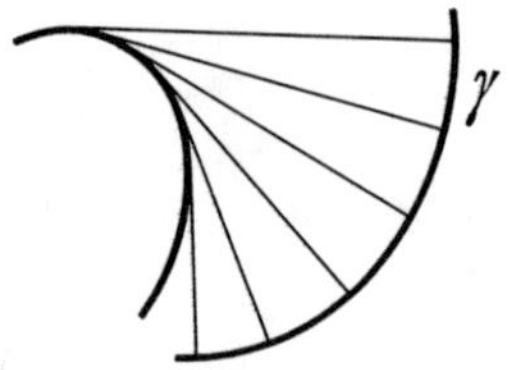

Figure 33.

The curve and circle of curvature. If the curvature radius of plane curve γ has a (local) maximum at its point M_0, then in the neighborhood of the point curve γ lies inside the circle of curvature (Figure 30).

If the curvature radius of plane curve γ has a (local) minimum at its point M_0, then in the neighborhood of the point curve γ lies outside the circle of curvature (Figure 31).

If in the neighborhood of point M_0 the curvature of regular plane curve γ changes monotonically, then the curve leaves its circle of curvature in the direction of the increasing radius of curvature, and enters it in the direction of the decreasing radius of curvature (Figure 32).

1.1.16 Evolute and evolvent of a plane curve

Let γ be a regular plane curve. A set of its centers of curvature is called *an evolute* of the curve.

Let

$$x = x(t), \quad y = y(t)$$

be parametric equations of curve γ. Then the equations of its evolute have the form

$$x = x(t) - y'(t)\frac{[x'(t)]^2 + [y'(t)]^2}{x'(t)y''(t) - x''(t)y'(t)}$$

$$y = y(t) + x'(t)\frac{[x'(t)]^2 + [y'(t)]^2}{x'(t)y''(t) - x''(t)y'(t)}$$

If

$$\mathbf{r} = \mathbf{r}(t)$$

is a vector equation of curve γ, then a vector equation of its evolute may be written in the form

$$\mathbf{r} = \mathbf{r}(t) + \frac{1}{k(t)}\mathbf{n}(t)$$

where $\mathbf{n}(t)$ is a unit vector of the normal to the curve.

If curvature $k(s)$ of regular curve γ is nonzero at every point and derivative $k'(s)$ keeps its sign along the curve, then the evolute of the curve is a regular curve (Figure 33).

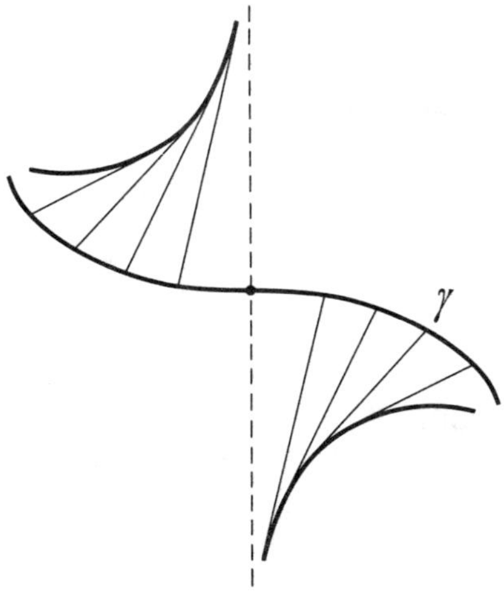

Figure 34.

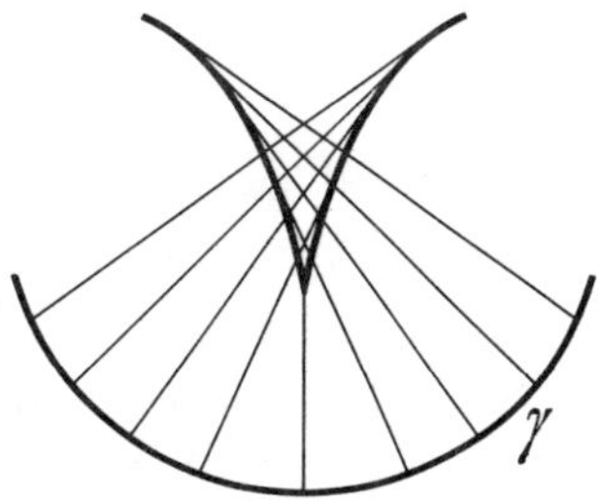

Figure 35.

If curvature $k(s)$ of regular curve γ is equal to zero at point M_0 corresponding to value s_0 of the natural parameter, and derivative $k'(s)$ keeps its sign along the curve, then the evolute of the curve falls into two regular curves so that each of them is an evolute of its parts, for $s < s_0$ and $s > s_0$, respectively, and goes to infinity (Figure 34).

If curvature $k(s)$ of regular curve γ is nonzero at every point and

$$k'(s_0) = 0, \quad k''(s_0) \neq 0$$

at point M_0 corresponding to value s_0 of the natural parameter, then the point of the evolute of the curve corresponding to point M_0 is a cuspidal point of the first kind (Figure 35).

A normal to regular curve γ coincides with a tangent to its evolute at a corresponding point.

Let M_1 and M_2 be points of regular curve γ, and R_1 and R_2 be radii of curvature at these points. The length of the evolute arc corresponding to arc M_1M_2 of the curve is equal to the absolute value of difference $R_1 - R_2$.

An evolute of a plane curve is an envelope of its normal family (see 1.1.19).

Curve γ_* is called *an evolvent* of given curve γ if curve γ is an evolute of curve γ_*.

The evolvent of curve γ is a set of end-points of its tangent segments with the following property: the lengths of the segments decrease along the curve in the magnitude equal to the increment of the curve's arc length.

Let

$$\mathbf{r} = \mathbf{r}(t)$$

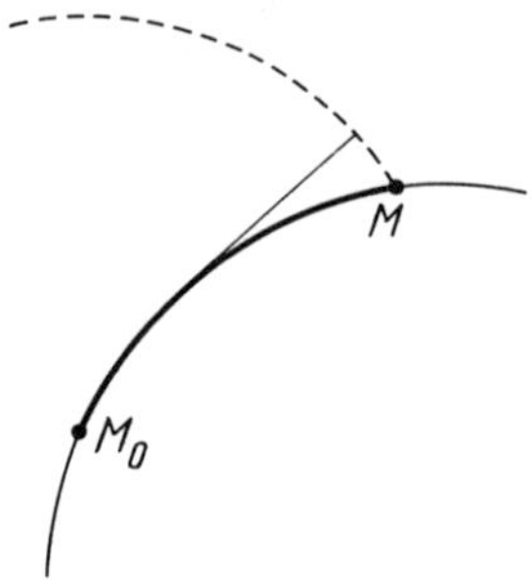

Figure 36.

be a natural parametrization of curve γ. Then a vector equation of the evolvent has the form

$$\rho(s) = \mathbf{r}(s) + (c - s)\mathbf{t}(s)$$

where $\mathbf{t}(s)$ is a unit tangent vector of the curve, and c — an arbitrary constant.

A regular curve has an infinite number of evolvents, but one, and only one, evolvent passes through every point of any tangent to an evolute.

Let

$$x = x(t), \quad y = y(t)$$

be parametric equations of reqular plane curve γ. Then the evolvent equations take the form

$$x = x(t) - \frac{x'(t)}{\sqrt{[x'(t)]^2 + [y'(t)]^2}} \int \sqrt{[x'(t)]^2 + [y'(t)]^2}\, dt$$

$$y = y(t) - \frac{y'(t)}{\sqrt{[x'(t)]^2 + [y'(t)]^2}} \int \sqrt{[x'(t)]^2 + [y'(t)]^2}\, dt$$

For vector equation

$$\mathbf{r} = \mathbf{r}(t)$$

of regular curve γ, we have

$$\rho = \mathbf{r}(t) - \frac{\mathbf{r}'(t)}{|\,\mathbf{r}'(t)\,|} \int |\,\mathbf{r}'(t)\,|\, dt$$

A visual way of an evolvent construction. Let us plot on curve γ, from its arbitrary point M_0, an arc of length c and denote the other end-point as M. Let a flexible, nonstreching thread be superimposed on arc M_0M and one of its ends be fixed at point M_0. Leading the stretched thread from the curve as though it were a template, the other end-point M describes the evolvent of curve γ (Figure 36).

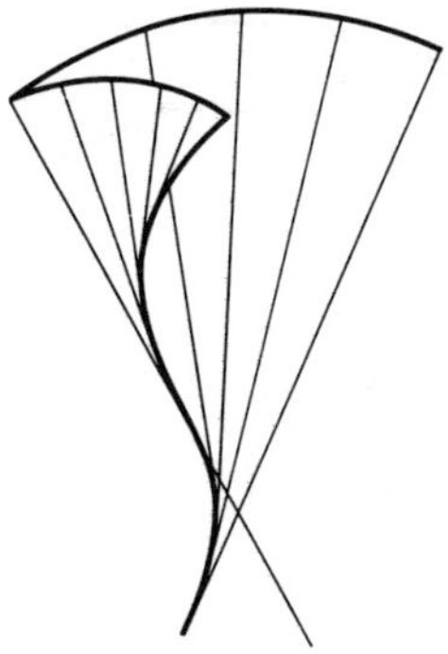

Figure 37.

The radius of the evolvent curvature changes monotonically: it decreases as the arc length increases.

The evolvent of a regular curve with a nonvanished curvature is also a regular curve.

If curvature k vanishes at just one point M_0 corresponding to value s_0 of the natural parameter and

$$k(s_0) = 0, \quad k'(s_0) \neq 0$$

then the evolvent point corresponding to point M_0 is a cuspidal point of the second kind (Figure 37).

Let curve γ be defined by a natural equation

$$R = \varphi(s)$$

Then

$$\rho = R\frac{dR}{ds}, \quad \rho = \varphi(s)\varphi'(s)$$

$$\sigma = R + c, \quad \sigma = \varphi(s) + c$$

(here c is a constant) are the natural equations of the evolute, and

$$\rho = \sigma$$

$$\sigma = \int \frac{s}{\varphi(s)}\,ds$$

are the natural equations of the evolvent.

1. The length of an evolute arc is equal to the difference of the lengths of the tangent segments to the evolute at the end-points of the arc (the segments are plotted from the points of tangency to meeting with the evolvent).

2. A tangent to the evolute is a normal to the evolvent.

3. A normal to the evolvent is a tangent to the evolute.

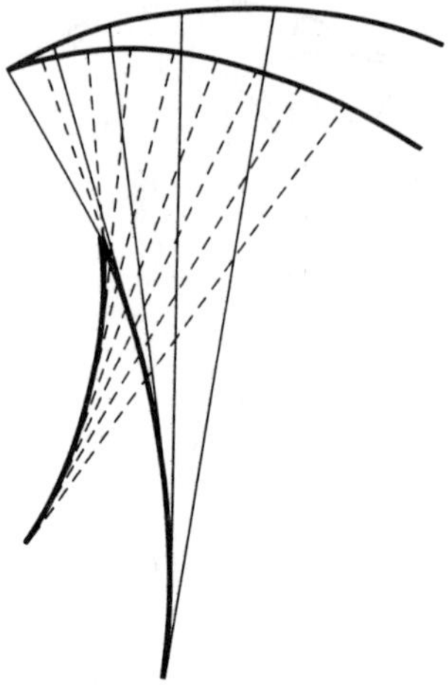

Figure 38.

4. An evolvent is an orthogonal trajectory of the evolute tangents.

5. Every orthogonal trajectory of the tangents to a curve is an evolvent of the curve.

6. An arc length of the evolute is equal to the absolute value of the difference radii of the evolvent curvature at the end-points of the arc.

7. For every inflection point of the evolute, there exists a corresponding cuspidal point of the second kind on the evolvent.

8. For every extremum of the evolvent curvature, there exists a corresponding cuspidal point of the first kind on the evolvent.

9. If the evolute has a cuspidal point of the second kind, then the evolvent also has a cuspidal point of the second kind (Figure 38).

10. If the evolute has a straightness point, then the evolute goes to infinity. If the straightness point is an inflection point of the evolute, then both branches of it go to infinity in different directions of the evolute normal.

11. A radius of the evolute curvature is defined by formula

$$\rho^* = \frac{1}{2}\frac{d(\rho^2)}{ds}$$

where ρ is a radius of curvature of a given curve, and s is an arc length for the corresponding point on the evolvent.

An evolute of the evolute is named *a second evolute*.

A radius of curvature of the second evolute is calculated by formula

$$\rho_2 = \rho\left(\left(\frac{d\rho}{ds}\right)^2 + \rho\frac{d^2\rho}{ds^2}\right)$$

For given curve γ, consider a family of straight lines possessing the following property: every line is deflected from the normal to the curve

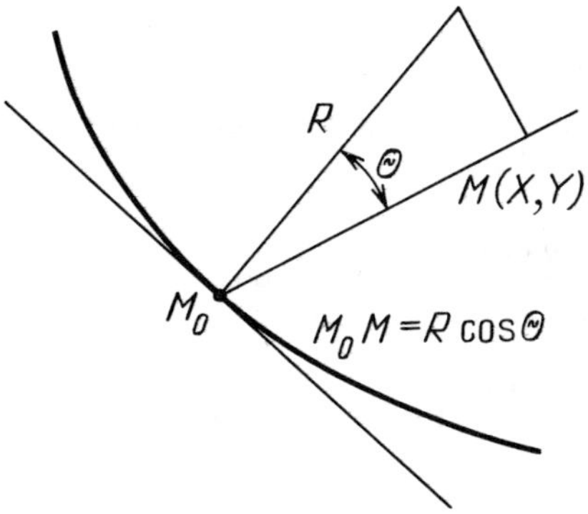

Figure 39.

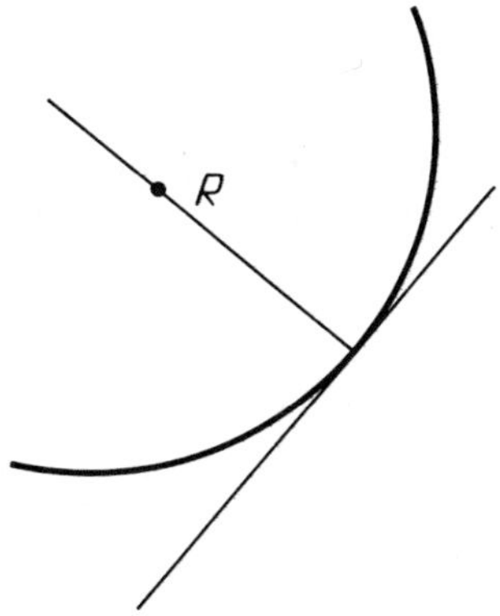

Figure 40.

by given angle Θ. An envelope of the family is called *an evolutoid* of the curve (if Θ is equal to zero, we have an evolute) (Figure 39). Curve γ is an isogonal trajectory of the evolutoid tangents, much like the evolvent is an orthogonal trajectory of the evolute tangents.

Let

$$y = y(x)$$

be an explicit equation of curve γ. Then the evolutoid is defined with the help of relations

$$X = x + \cos^2\Theta \frac{1 + (y'(x))^2}{y''(x)}(\tan\Theta + y'(x))$$

$$Y = y + \cos^2\Theta \frac{1 + (y'(x))^2}{y''(x)}(y'(x)\tan\Theta - 1)$$

A curve which intersects every curve of given family γ_λ

$$\varphi(x, y, \lambda) = 0$$

at given angle Θ is called *an isogonal trajectory*.

The isogonal trajectory is described by a differential equation

$$(\varphi_x \cos\Theta - \varphi_y \sin\Theta)\, dx + (\varphi_x \sin\Theta + \varphi_y \cos\Theta)\, dy = 0$$

(for $\Theta = 0$ we have *an orthogonal trajectory*).

Let γ be a regular curve, M — a point on it and R — its radius of curvature at point M (Figure 40). Let NM be a segment of the curve normal at point M and αR — its length (α is a given number). A set of the end-points of the segments NM is called *an evolute of Braude*.

Let

$$y = y(x)$$

be an explicit equation of curve γ. Then the Braude evolute is described by parametric equations

$$X = \alpha\left(x - \frac{y'(x)(1 + [y'(x)]^2)}{y''(x)}\right) + (1 - \alpha)x$$

$$Y = \alpha\left(y + \frac{y'(x)(1 + [y'(x)]^2)}{y''(x)}\right) + (1 - \alpha)y(x)$$

1.1.17 Characteristic points of regular plane curves

Point M of regular plane curve γ is named *a (simple) vertex* of the curve, if an osculating circle has a tangency of the third order with the curve at this point.

In order for point M_0 of a 4-regular curve γ defined by parametric equations

$$x = x(t), \quad y = y(t)$$

to be a simple vertex, it is necessary and sufficient to have a concurrent compliance with the following three relations:

$$k(t_0) \neq 0, \quad k'(t_0) = 0, \quad k''(t_0) \neq 0$$

(value t_0 of parameter corresponds to point M_0).

If curve γ is a 3-regular curve then the last condition $k''(t_0) \neq 0$ is replaced with the following: derivative $k'(t)$ reverses its sign while moving through point t_0.

A point of a regular curve is said to be its *highest (degenerate) vertex*, if an osculating circle of the curve has a tangency of the higher-than-third order (Figure 41).

A point of a regular curve is said to be its *point of a (simple) inflection* if a tangent to the curve has a second order contact with it and *splits the curve*. That is, if two points of the curve lie in the neighborhood of the point on different sides of it, then these points also lie in the different half-planes defined by the tangent.

A necessary and sufficient condition for point M_0 of a 3-regular curve be a simple inflection point is a concurrent compliance with the following relations:

$$k(t_0) = 0, \quad k'(t_0) \neq 0$$

(value t_0 of parameter corresponds to point M_0).

If curve γ is regular, then the last relation is replaced with the following: curvature $k(t)$ of the curve reverses its sign while moving through point t_0 (Figure 42).

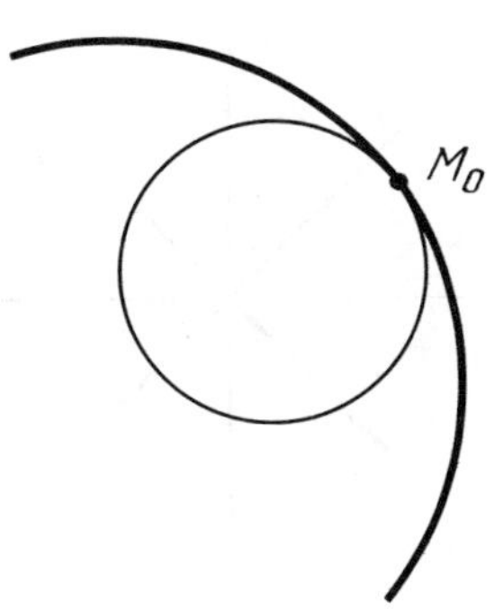

Figure 41.

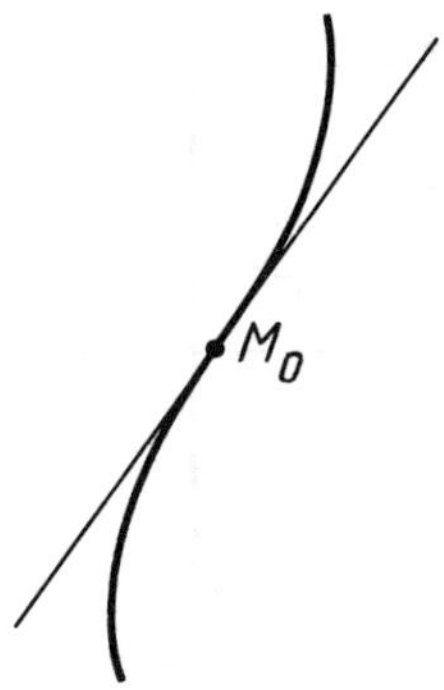

Figure 42.

A point of a regular curve is said to be its *point of a highest (degenerate) inflection*, if the curve tangent has a higher-than-third order contact with it and splits the curve.

Point M_0 of a regular curve is called *a straightness point* of the curve (see 1.1.10) if

$$k(t_0) = 0$$

(value t_0 of parameter corresponds to point M_0).

If a straightness point of a curve is not a point of inflection, then it is also said to be *a parabolic point of straightness*. A curvature of the curve in the neighborhood of the point allows for the following view:

$$k(t) = \frac{1}{(2n)!} k^{(2n)}(t_0)(t - t_0)^{2n} + o(t - t_0)^{2n}$$

where $k^{2n}(t - t_0) \neq 0$.

Curvature $k(t)$ of a curve in the neighborhood of the inflection point has an analogous view

$$k(t) = \frac{1}{(2n-1)!} k^{(2n-1)}(t_0)(t - t_0)^{2n-1} + o(t - t_0)^{2n-1}$$

where $k^{2n-1}(t - t_0) \neq 0$.

1.1.18 Singular points of plane curves

The implicit description. Let a plane curve γ be defined by equation

$$F(x, y) = 0$$

where function $F(x, y)$ has continuous derivatives of every order, up to and including s, that is, $F(x, y)$ is an s-smooth function.

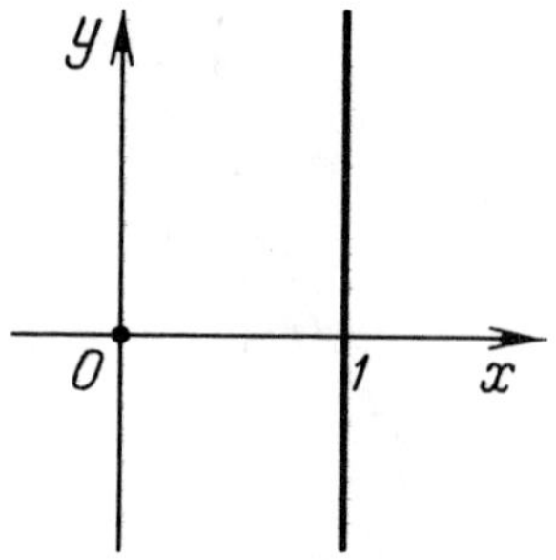

Figure 43.

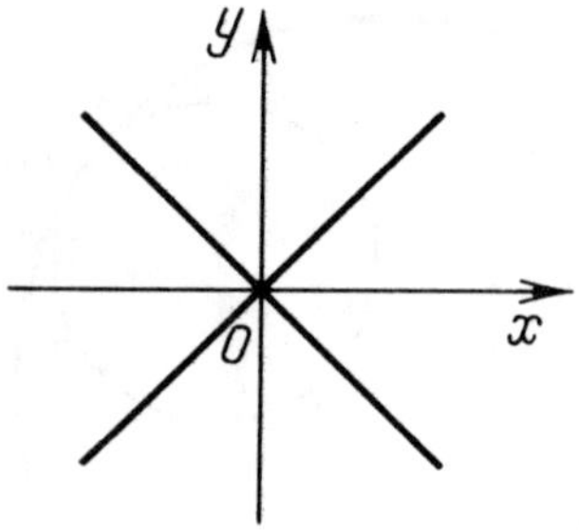

Figure 44.

Point $M_0(x_0, y_0)$ is called *a singular point* of curve γ, if the following three conditions

$$F(x_0, y_0) = 0,$$

$$F_x(x_0, y_0) = 0, \quad F_y(x_0, y_0) = 0$$

are met concurrently.

Let

$$A = F_x x(x_0, y_0), \quad B = F_x y(x_0, y_0), \quad C = F_y y(x_0, y_0),$$

$$\Delta = AC - B^2$$

There exist some principal types of singular points of plane curves.

1. $\Delta > 0$: M_0 is *an isolated point.*

Example.

$$F(x, y) \equiv (x^2 + y^2)(x - 1) = 0$$

(Figure 43). At point M_0 we have

$$F(0,0) = 0, \quad F_x(0,0) = 0, \quad F_y(0,0) = 0,$$

$$A = -2, \quad B = 0, \quad C = -2, \quad \Delta = 4 > 0$$

2. $\Delta < 0$: M_0 is *a double point (node).*

Example.

$$F(x, y) \equiv x^2 - y^2 = 0$$

(Figure 44). At point M_0 we have

$$F(0,0) = 0, \quad F_x(0,0) = 0, \quad F_y(0,0) = 0,$$

$$A = 2, \quad B = 0, \quad C = -2, \quad \Delta = -4 < 0$$

3. $\Delta = 0$: M_0 is *a cuspidal point of the first kind,* or *a cuspidal point of the second kind,* or *a multiple point,* or *a self-intersection point.*

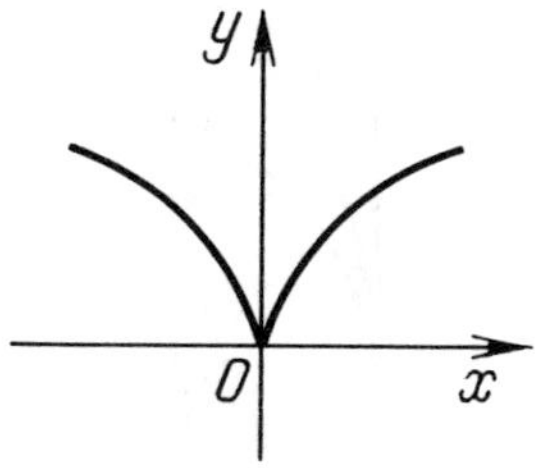

Figure 45.

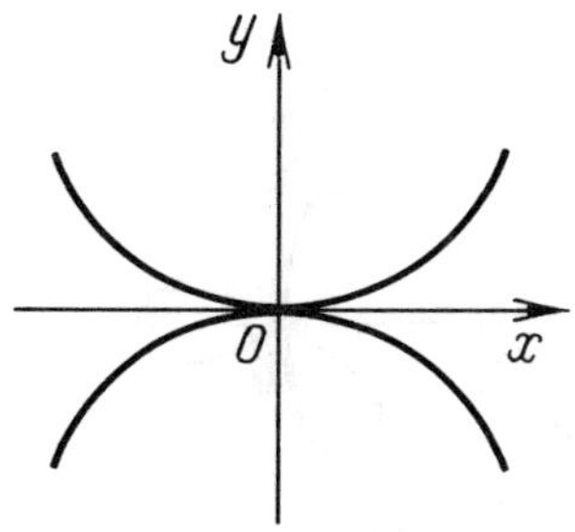

Figure 46.

Examples.

1) $F(x,y) \equiv x^2 - y^3 = 0$ (Figure 45):

$$F(0,0) = 0, \quad F_x(0,0) = 0, \quad F_y(0,0) = 0,$$

$$A = 2, \quad B = 0, \quad C = 0, \quad \Delta = 0$$

Point $M_0(0,0)$ is a cuspidal point of the first kind.

2) $F(x,y) \equiv y^2 - 4x^4 = 0$ (Figure 46):

$$F(0,0) = 0, \quad F_x(0,0) = 0, \quad F_y(0,0) = 0,$$

$$A = 0, \quad B = 0, \quad C = 2, \quad \Delta = 0$$

Point $M_0(0,0)$ is *a point of self-osculation.*

3) $F(x,y) \equiv (x^2+y^2)^2 - x^3 - 3xy^2 = 0$ (Figure 47):

$$F(0,0) = 0, \quad F_x(0,0) = 0, \quad F_y(0,0) = 0,$$

$$A = 0, \quad B = 0, \quad C = 0, \quad \Delta = 0$$

Point $M_0(0,0)$ is a triple point.

4) $F(x,y) \equiv 2x^2 + y^5 - 3x\sqrt{y^5} = 0$ (Figure 48):

$$F(0,0) = 0, \quad F_x(0,0) = 0, \quad F_y(0,0) = 0,$$

$$A = 4, \quad B = 0, \quad C = 2, \quad \Delta = 0$$

Point $M_0(0,0)$ is a cuspidal point of the second kind.

Point M_0 of an implicitly defined curve is the cuspidal point of the first kind if both conditions

$$\Delta\,|_{M_0} = 0, \quad D\,|_{M_0} \neq 0$$

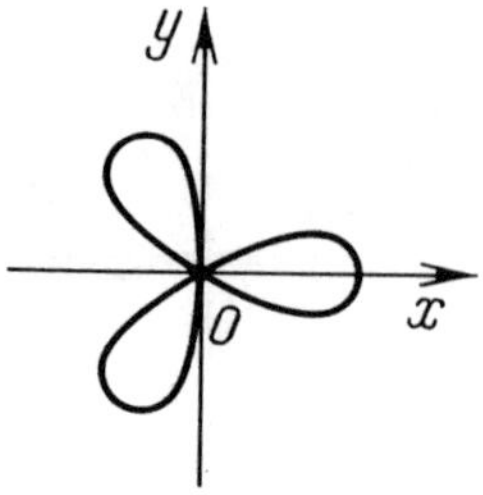

Figure 47.

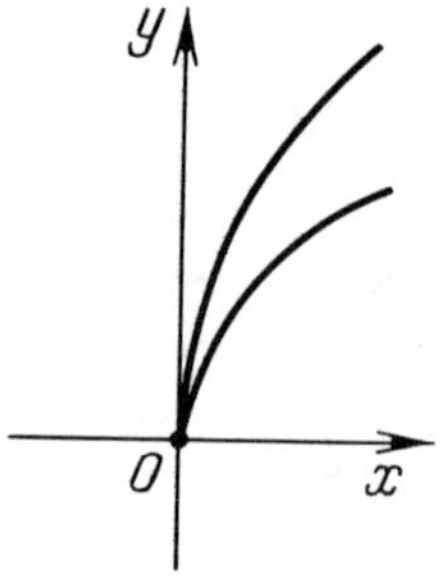

Figure 48.

where

$$D = \begin{vmatrix} F_{xx} & 2F_{xy} & F_{yy} & 0 & 0 \\ 0 & F_{xx} & 2F_{xy} & F_{yy} & 0 \\ 0 & 0 & F_{xx} & 2F_{xy} & F_{yy} \\ F_{xxx} & 3F_{xxy} & 3F_{xyy} & F_{yyy} & 0 \\ 0 & F_{xxx} & 3F_{xxy} & 3F_{xyy} & F_{yyy} \end{vmatrix}$$

are concurrently met.

For the case

$$\Delta\,|_{M_0} = 0, \quad D\,|_{M_0} = 0$$

there exists a great variety of singularities in accordance with the values of the higher order derivatives of function $F(x, y)$ at point $M_0(x_0, y_0)$ and more in-depth research is required.

Parametric description. Let a smooth plane curve γ be defined by equations

$$x = x(t), \quad y = y(t)$$

Point $M_0(x_0, y_0)$ of curve γ corresponding to value t_0 of the parameter, that is, $x_0 = x(t_0), y_0 = y(t_0)$, is called ***a singular point*** of curve γ with respect to the given parametrization if the first derivatives of functions $x(t)$ and $y(t)$ are both zero for $t = t_0$:

$$x'(t_0) = 0, \quad y'(t_0) = 0$$

There exist two general types of singularities of a smoothly parametrized curve.

1. *Singularities of parametrization* (disappear with the appropriate curve reparametrization).

Example. A curve is defined by parametric equations

$$x = t^3, \quad y = at^3, \quad -1 < t < 1$$

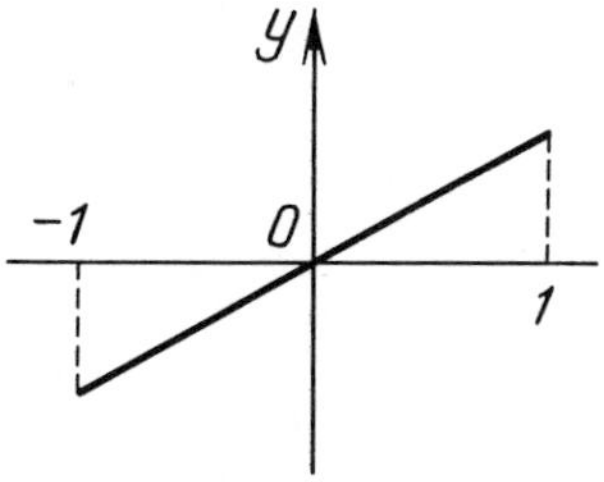

Figure 49.

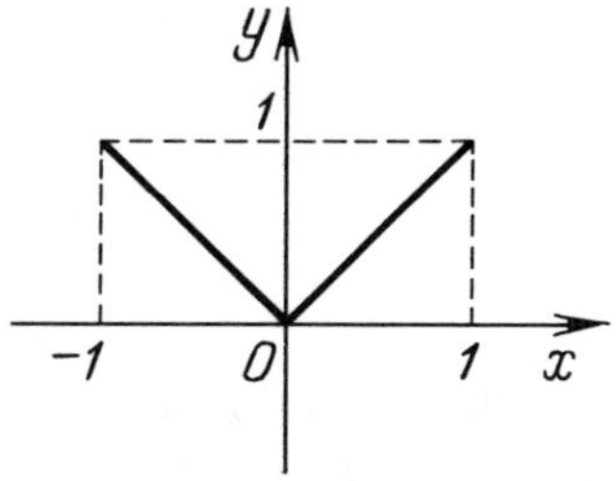

Figure 50.

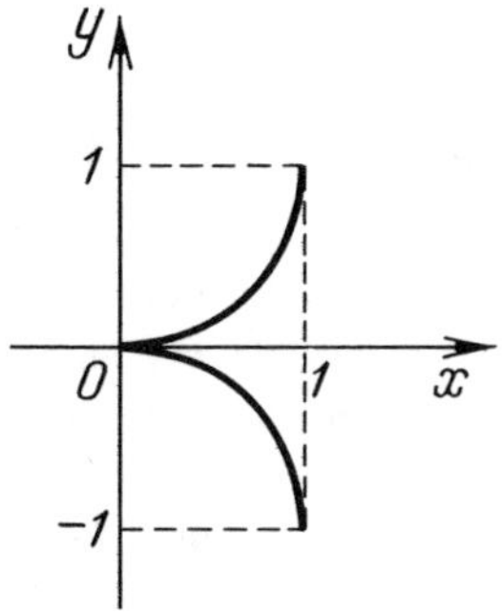

Figure 51.

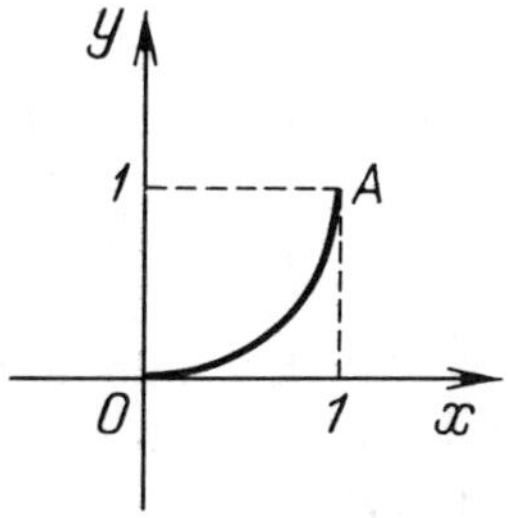

Figure 52.

It is easy to see that $x'(0) = 0$, $y'(0) = 0$. Let $\tau = t^3$. Then we have

$$x = \tau, \quad y = a\tau, \quad -1 < \tau < 1$$

or

$$y = ax$$

(Figure 49).

2. *Geometric singularities* (retained under every admissible reparametrization of a curve: the new functions defining the curve remain smooth).

Examples.

1) $x = t^3, \quad y = |\, t\,|^3, \quad |\, t\,| \leq 1$
(Figure 50).

2) $x = t^2, \quad y = t, \quad |\, t\,| \leq 1$
(Figure 51).

3) $x = t^2, \quad y = t^4, \quad |\, t\,| \leq 1$
(Figure 52).

We have $x'(0) = 0, \quad y'(0) = 0$ in all three cases. In the first case, the singularity is an angular point, in the second — a cuspidal point of the first kind. In the third case, origin $O(0, 0)$ is the only nonmultiple point of the curve (a degenerate cuspidal point of the second kind).

None of the above singularities 1–3 are remediable.

Let analytic curve γ be defined by equations

$$x = at^n,$$
$$y = b_1t^{m_1} + b_2t^{m_2} + \cdots,$$
$$1 \le n \le m_1 < m_2 < \ldots$$

in the neighborhood of origin $O(0,0)$.

For point $O(0,0)$ to be a singular point of the curve it is necessary and sufficient for at least one number m_i not to be divisible by n.

If point $O(0,0)$ is singular and numbers n and m_1 are even, then the point is a cuspidal point of the second kind.

Sufficient criterion for a point on the curve to be singular. Let analytic curve γ be defined by equations

$$x = x(t), \quad y = y(t)$$

where $x(t)$ and $y(t)$ are analytic functions of parameter t in the neighborhood of point M_0. Let the least orders of the function derivatives nonvanished at $t = t_0$ be equal to n and m, $n \neq m$, respectively (for the sake of definition, $n < m$). Then point M_0 is a singular point of the analytic curve if m is not divisible by n. In addition, point M_0 is a cuspidal point of the first kind if n is even and m is odd, and the point is a cuspidal point of the second kind if n and m are both even.

Let γ be a smooth plane curve,

$$\mathbf{r} = \mathbf{r}(t)$$

its parametrization, and $\mathbf{r}'(t_0) = 0$.

Denote the least order derivative different from zero by $\mathbf{r}^{(p)}(t_0)$ and the least order derivative noncollinear to $\mathbf{r}^{(p)}(t_0)$ by $\mathbf{r}^{(q)}(t_0)$.

Then point M_0 corresponding to value t_0 of the parameter may be

1) a regular point (if p is odd, and q is even),

2) an inflection point (if p is odd, and q is odd),

3) a cuspidal point of the first kind (if p is even, and q is odd), or

4) a cuspidal point of the second kind (if both p and q are even).

Note. In cases 1 and 2, singularities may disappear with reparametrization (in some cases, it is possible for a curve to become regular after reparametrization, but often under a lesser smoothness class), and in case 4, a cuspidal point may be degenerate.

In conclusion, it is useful to note that the notions of singular points for the curves defined implicitly and for those defined explicitly are *different*. Let, for example, a curve with self-intersection (multiple) points allow both description ways. In case of the implicit description, the points enter into the set of singular points and may be discovered from the conditions

$$F(x,y) = 0, \quad F_x(x,y) = 0, \quad F_y(x,y) = 0$$

In case of the parametric description, the self-intersection points are ordinary points of the curve.

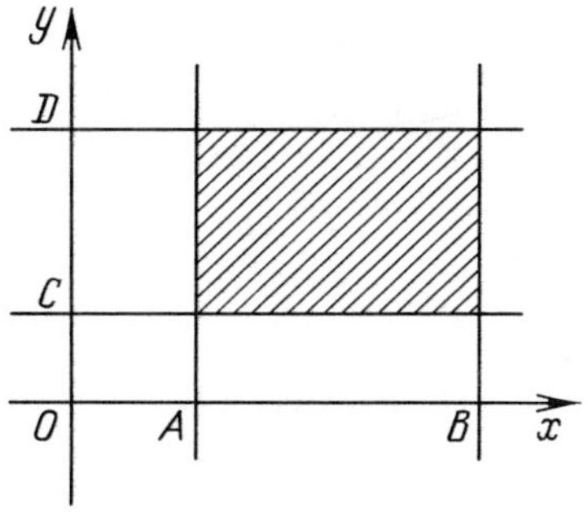

Figure 53.

1.1.19 Envelope of a family of plane curves

Parametric description. Let relations

$$x = x(u, v), \quad y = y(u, v), \quad A < u < B, \quad C < v < D \qquad (*)$$

for every fixed value v of parameter define a (locally) simple curve. Then *a family of the parametrized (locally) simple plane curves* is said to be given.

Let functions $x(u, v)$ and $y(u, v)$ be smooth in the domain of their definition — rectangle $A < u < B$, $C < v < D$ (Figure 53).

A smooth curve

$$X = X(t), \quad Y = Y(t), \quad a < t < b \qquad (\tilde{*})$$

is called *(the portion of) envelope* of the curve family $(*)$ if for every value t of the parameter it has a common tangent with one of the family curves $(*)$, and furthermore, a correspondence between value t and parameter values u and v (for the chosen curve of the family) at the tangency point may be defined by continuous functions

$$u = u(t), \quad v = v(t)$$

(the last two functions describe *the attachment law* for concrete curves of the family to the envelope points), with function $v = v(t)$ differing from the constant at every section of interval (a, b) (Figure 54). Hence, we have

$$X = x\big(u(t), v(t)\big), \quad Y = y\big(u(t), v(t)\big)$$

and for every value t_1 a curve defined by equations

$$x = x\big(u, v(t_1)\big), \quad y = y\big(u, v(t_1)\big)$$

has a common tangent with the curve described by equations $(\tilde{*})$ at point $(X(t_1), Y(t_1))$.

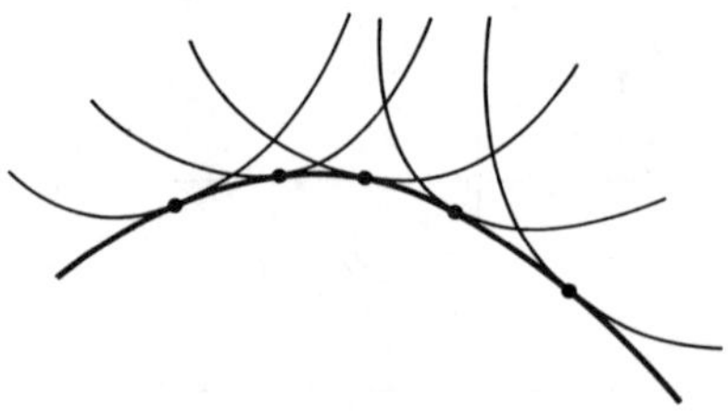

Figure 54.

The necessary criterion for envelopes. Let a family $(*)$ have an envelope $(\tilde{*})$ and point $M_0(x_0, y_0)$ on it correspond to values t_0, $u_0 = u(t_0)$ and $v_0 = v(t_0)$ of the parameters. Then

$$\begin{vmatrix} x_u(u_0, v_0) & x_v(u_0, v_0) \\ y_u(u_0, v_0) & y_v(u_0, v_0) \end{vmatrix} = 0$$

The sufficient criterion for envelopes. Let functions

$$x = x(u, v), \quad y = y(u, v)$$

defining a family of plane curves be 2-smooth in a rectangle $A < u < B$, $C < v < D$, and

$$D = \begin{vmatrix} x_u & x_v \\ y_u & y_v \end{vmatrix} = 0,$$

$$\begin{vmatrix} D_u & D_v \\ x_u & x_v \end{vmatrix}^2 + \begin{vmatrix} D_u & D_v \\ y_u & y_v \end{vmatrix}^2 > 0$$

$$x_u^2 + y_u^2 \neq 0, \quad D_u \neq 0$$

at its point (u_0, v_0) (Figure 55). It then becomes possible to demonstrate a neighborhood of point (u_0, v_0) where:

1) an envelope exists and is defined by equations

$$x = x(u, v), \quad y = y(u, v)$$

with the $D(u, v) = 0$ bond. Here we may use parameter v for parameter t on the envelope; the attachment law $u = u(t)$ will be a smooth function, and value t_0 will correspond to value u_0;

2) a tangent to the envelope at point (u_0, v_0) is parallel to vector

$$\left(\begin{vmatrix} D_u & D_v \\ x_u & x_v \end{vmatrix}, \begin{vmatrix} D_u & D_v \\ y_u & y_v \end{vmatrix} \right)$$

3) every curve of the family is tangent to the envelope at only one characteristic point in the neighborhood (Figure 56).

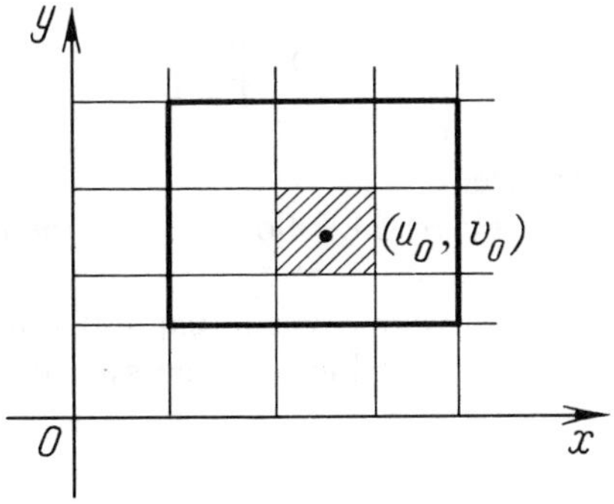

Figure 55.

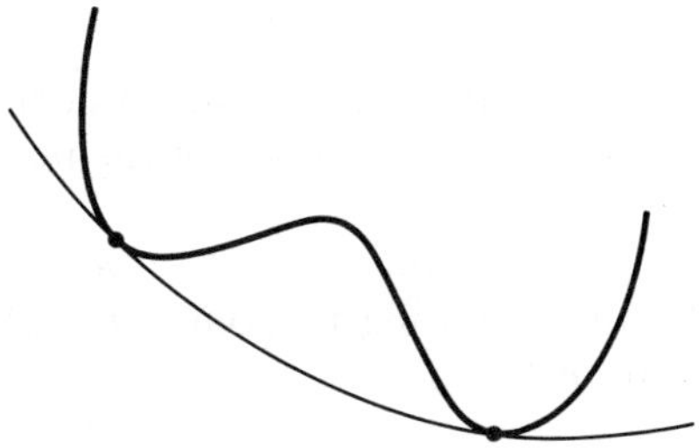

Figure 56.

Every smooth plane curve is an envelope of the one-parameter family of its tangents.

In whole, an envelope may consist of one or more locally simple curves. The curves may be closed, or go to infinity, or approach the boundary of the family definition domain, or break due to the approaching of a singularity where the conditions of the sufficient criterion for the envelope existence cannot be met.

In applications, the singular point of an envelope is often a cusp.

The necessary criterion for cuspidal points on the envelope. If, for a family of plane curves $(*)$, values u_0 and v_0 correspond to cuspidal point $M(x_0, y_0)$ on the envelope, then point (u_0, v_0) lies within one of the following three sets:

1) a set of points (u, v) in the neighborhood of which functions $x = x(u, v)$, $y = y(u, v)$ are nonsmooth;

2) a set of points (u, v) where $D = 0$ and/or $x_u^2 + y_u^2 = 0$, or $D_u = 0$;

3) a set of points (u, v) where

$$D = 0, \quad x_u^2 + y_u^2 \neq 0, \quad D_u \neq 0$$

and

$$\begin{vmatrix} D_u & D_v \\ x_u & x_v \end{vmatrix} = 0, \quad \begin{vmatrix} D_u & D_v \\ y_u & y_v \end{vmatrix} = 0$$

The sufficient criterion for cuspidal points on the envelope. Let functions

$$x = x(u, v), \quad y = y(u, v)$$

defining a family of plane curves $(*)$ be 3-smooth in rectangle $A < y < B$, $C < v < D$ and the following conditions

$$D = 0, \quad x_u^2 + y_u^2 > 0, \quad D_u \neq 0,$$

$$\Delta = \begin{vmatrix} D_u & D_v \\ x_u^2 + y_u^2 & x_u x_v + y_u y_v \end{vmatrix} = 0, \begin{vmatrix} D_u & D_v \\ \Delta_u & \Delta_v \end{vmatrix} \neq 0$$

be met at inner point (u_0, v_0) of the rectangle.

Then a neighborhood of the point exists where the following statements are true:

1. The family of the curves $(*)$ has an envelope with cuspidal point M_0 corresponding to point (u_0, v_0); point M_0 is defined by the system of equations

$$D = 0, \quad \Delta = 0$$

2. On the envelope a parameter t may be chosen so that functions

$$X = X(t), \quad Y = Y(t)$$

defining it, and attachment law

$$u = u(t), \quad v = v(t)$$

are 2-smooth, and, in addition,

$$[X'(t)]^2 + [Y(t)]^2 > 0$$

for $t \neq t_0$ and

$$X'(t_0) = 0, \quad Y'(t_0) = 0$$

where value t_0 corresponds to pair u_0, v_0.

3. Every curve of the family $(*)$ is tangent to the envelope at a (locally) single point.

The implicit description. We say that equality

$$F(x, y, \alpha) = 0 \qquad (*)$$

where

$$A < u < B, \quad C < v < D, \quad E < \alpha < F$$

defines *a family of the implicitly described plane curves*, if function $F(x, y, \alpha)$ is continuous, and for every value α_0 of parameter α, $E < \alpha_0 < F$, a subset of rectangle $A < u < B$, $C < v < D$ points satisfy equality

$$F(x, y, \alpha_0) = 0$$

and provide reason enough to call the subset a curve (Figure 57).

Let function $F(x, y, \alpha)$ be smooth in its domain of definition.

A smooth curve

$$x = x(t), \quad y = y(t), \quad a < t < b \qquad (\tilde{*})$$

is called *a (portion of) envelope* of the curve family $(*)$, if a continuous *attachment law* of the family curves exists to the points of envelope

$$\alpha = \alpha(t), \quad E < \alpha < F$$

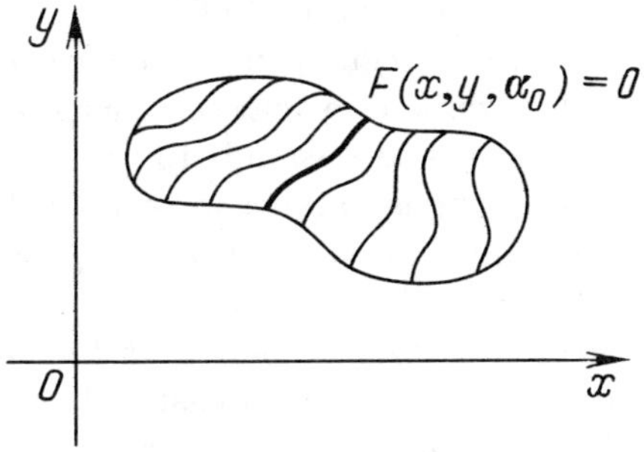

Figure 57.

and the following conditions are met:

1) $F(x(t), y(t), \alpha(t)) = 0$;

2) for every value t_1 of parameter t a curve described by equation

$$F\big(x, y, \alpha(t_1)\big) = 0$$

and a curve defined by the equations $(\tilde{*})$ have a common tangent at point $(x(t_1), y(t_1))$;

3) function $\alpha = \alpha(t)$ is not constant at any section of interval (a, b).

The necessary criterion for envelopes. Let an envelope of the family $(*)$ be described by the equations $(\tilde{*})$, and value t_0 of parameter t correspond to values

$$x_0 = x(t_0), \quad y_0 = y(t_0), \quad \alpha_0 = \alpha(t_0)$$

Then either

$$F(x_0, y_0, \alpha_0) = 0, \quad F_x(x_0, y_0, \alpha_0) = 0, \quad F_y(x_0, y_0, \alpha_0) = 0$$

or

$$F(x_0, y_0, \alpha_0) = 0,$$
$$F_\alpha(x_0, y_0, \alpha_0) = 0$$

The sufficient criterion for the existence of envelopes. Let a family of plane curves

$$F(x, y, \alpha) = 0,$$
$$A < x < B, \quad C < y < D, \quad E < \alpha < F$$

be given, and function $F(x, y, \alpha)$ be 2-smooth. If

$$F = 0, \quad F_\alpha = 0, \quad F_x^2 + F_y^2 > 0, \quad F_{\alpha\alpha} \neq 0,$$

$$\begin{vmatrix} F_x & F_y \\ F_{\alpha x} & F_{\alpha y} \end{vmatrix} \neq 0$$

at point (x_0, y_0, α_0), then a neighborhood of the point exists wherein:

1) the envelope is a simple regular curve described by the system of equations $F = 0, F_\alpha = 0$; parameter α may serve as a regular parameter on the envelope, and point (u_0, v_0) corresponds to value α_0 of the parameter;

2) a tangent of the envelope is parallel to vector $(F_y, -F_x)$;

3) every curve of the family contacts the envelope at a (locally) unique *characteristic* point.

The necessary criterion for cuspidal points on the envelope. If a family of plane curves $(*)$ has an envelope with a cuspidal point $M_0(x_0, y_0)$, then the corresponding value α_0 of parameter α is such that point (x_0, y_0, α_0) enters into one of the following three sets:

1) a set of points (x, y, α), the coordinates of which are bound by the condition that $F(x, y, \alpha) = 0$, but in the vicinity of which function $F(x, y, \alpha)$ is not 2-smooth;

2) a set of points (x, y, α) wherein

$$F = 0, \quad F_\alpha = 0, \quad F_{\alpha\alpha} = 0$$

and either

$$F_x^2 + F_y^2 > 0$$

or

$$\begin{vmatrix} F_x & F_y \\ F_{\alpha x} & F_{\alpha y} \end{vmatrix} = 0;$$

3) a set of points (x, y, α) wherein

$$F = 0, \quad F_\alpha = 0, \quad F_{\alpha\alpha} = 0,$$

$$F_x^2 + F_y^2 > 0,$$

$$\begin{vmatrix} F_x & F_y \\ F_{\alpha x} & F_{\alpha y} \end{vmatrix} \neq 0$$

The sufficient criterion for cuspidal points on the envelope. Let $F(x, y, \alpha)$ be a 3-smooth function in domain

$$A < x < B, \quad C < y < D, \quad E < \alpha < F$$

and conditions

$$F = 0, \quad F_\alpha = 0, \quad F_{\alpha\alpha} = 0,$$

$$F_x^2 + F_y^2 > 0, \quad F_{\alpha\alpha\alpha} \neq 0,$$

$$\begin{vmatrix} F_x & F_y \\ F_{\alpha x} & F_{\alpha y} \end{vmatrix} \neq 0,$$

be met in the neighborhood of its point (x_0, y_0, α_0). Then a neighborhood of the point exists wherein:

1) the family of plane curves (*) has an envelope with cuspidal point $M_0(x_0, y_0)$ defined by the system of equations

$$F = 0, \quad F_\alpha = 0, \quad F_{\alpha\alpha} = 0;$$

2) parameter t on the envelope may be chosen so that the functions

$$x = x(t), \quad x = x(t)$$

defining it and the attachment law are 2-smooth,

$$[x'(t)]^2 + [y(t)]^2 > 0$$

for $t \neq t_0$, and

$$x'(t_0) = 0, \quad y'(t_0) = 0$$

where value t_0 corresponds to the triple x_0, y_0, α_0 (for example, $t \equiv \alpha$);

3) every curve of family (*) and the envelope have a common tangent at a (locally) unique point.

1.2 Some general classes of plane curves

1.2.1 Closed plane curves

Curve γ defined by formulae

$$x = x(t), \quad y = y(t), \quad a \leq t \leq b, \quad a < b$$

where functions $x(t), y(t)$ are continuous on segment $[a, b]$ is called *closed* if

$$x(a) = x(b), \quad y(a) = y(b)$$

A. On every smooth closed curve γ there exist two pairs of points possessing the following properties:

1) every pair divides curve γ into two arcs of the same length;

2) straight lines defined by the pairs are perpendicular.

B. If every normal of a regular closed curve divides it into two arcs of the same length, then the curve is a circle.

C. On every regular closed curve γ there exist a pair of points with the following two properties:

1) the points divide the curve into two arcs of the same length;

2) tangents to curve γ at the points are parallel.

D. If for every point of a regular closed curve there exist a point on the curve so that the pair possesses property C then the curve has a center of symmetry.

E. Every regular closed convex curve has at least three pairs of points possessing property C.

F. For every regular closed convex curve, there exist at least three pairs of points with the following properties:

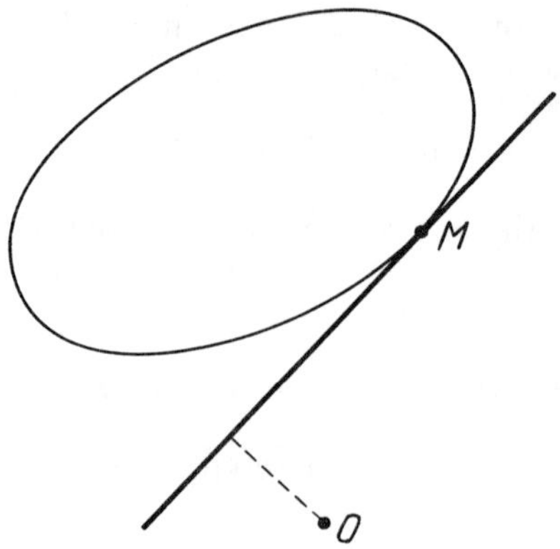

Figure 58.

1) every pair divides the curve into two arcs of equal length;

2) the curvature of the curve at the points of the pair is the same.

G (*The Jordan theorem*). Every simple (see 1.1.1) closed curve divides a plane into two domains — bounded and unbounded — and the curve is a boundary to each of them.

H. For a curve the curvature $k(s)$ of which is a continuous periodic function with period S to be closed, it is necessary and sufficient for the following two conditions to be met:

$$\int_0^S \cos\left(\int_0^s k(\sigma)d\sigma + \alpha_0\right)ds = 0,$$

$$\int_0^S \sin\left(\int_0^s k(\sigma)d\sigma + \alpha_0\right)ds = 0$$

where

$$\alpha_0 = \text{const}$$

I. Let γ be a closed regular curve with curvature $k(s)$. Then

$$\int_\gamma k(s)\,ds = 2\pi$$

(*the Fenchel equality*).

J. Distance $h(M)$ from fixed point O to a tangent of a convex regular closed curve γ at point M is called *the support function*, or *the Minkowski function* of the curve (Figure 58).

Let angle α between a tangent at running point M and the fixed direction be chosen as a parameter on curve γ. Then, radius of curvature $R(\alpha)$

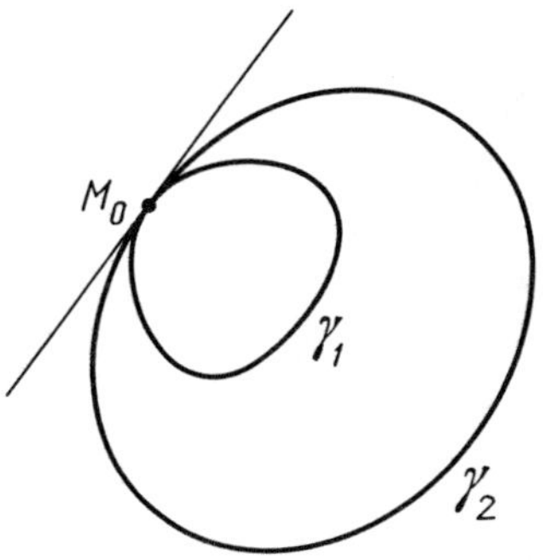

Figure 59.

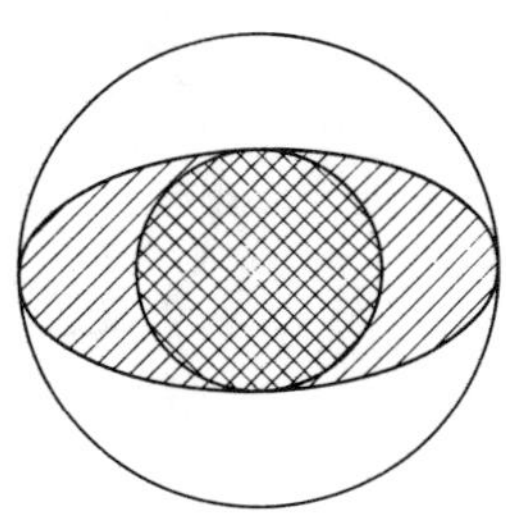

Figure 60.

of the curve is calculated by formula

$$R(\alpha) = h(\alpha) + h''(\alpha)$$

K. An area of domain bounded by regular convex closed curve γ is equal to

$$\frac{1}{2}\int_{\gamma} h(s)ds$$

L. Let two regular closed curves γ_1 and γ_2 have a common tangent at point M_0 and lie on the same side of it. If curvatures k_1 and k_2 of the curves at the points with the respectively parallel tangents follow inequality

$$k_1 \geq k_2$$

then curve γ_1 lies in the domain bounded by curve γ_2 (Figure 59).

M (*Isoperimetric inequalities*). Let a closed rectifiable curve γ with length S bound domain D with area F. Then,

$$S^2 - 4\pi F \geq 0$$

and the equality holds if, and only if, the curve is a circle.

Let r be a radius of *the incircle*, that is, the radius of the largest circle contained within domain $\overline{D}$ (a closure of domain D), and R be a radius of *the circumcircle*, that is, the radius of the smallest circle which contains domain D (Figure 60).

The next inequalities are true:

$$S^2 - 4\pi F \geq \pi^2(R - r)^2$$

(*the Bonnesen inequality*),

$$S^2 - 4\pi F \geq F^2\left(\frac{1}{r} - \frac{1}{R}\right)^2,$$

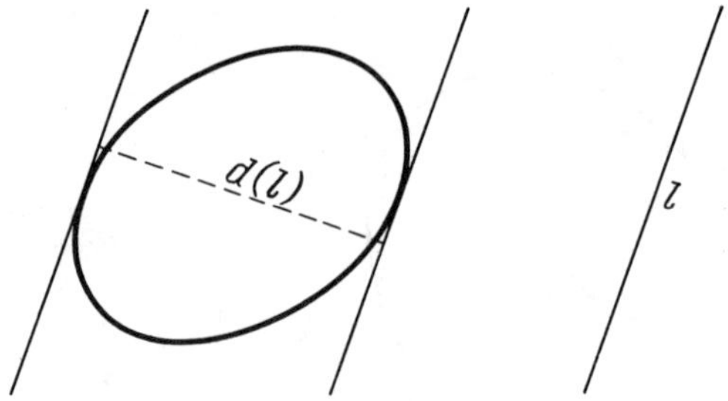

Figure 61.

$$S^2 - 4\pi F \geq S^2 \left(\frac{R-r}{R+r}\right)^2,$$

$$\frac{S - \sqrt{S^2 - 4\pi F}}{2\pi} \leq r \leq R \leq \frac{S + \sqrt{S^2 - 4\pi F}}{2\pi}$$

The equalities hold if, and only if, the curve is a circle.

N. Among the closed curves of the same length, a circle bounds the domain of the largest area.

1.2.2 Ovals

A simple plane closed regular curve the curvature of which is positive at everyone of its point is called *an oval.*

A point of an oval is called *the vertex of an oval* if the oval curvature at the point reaches the local extremum (the local maximum or the local minimum).

A. Every oval has at least four vertices.

B. An oval with exactly four vertices intersects a circle at least at four points.

C. If an oval intersects a circle at $2n$ points, then it has at least $2n$ vertices.

D. For every oval, a unit vector $\mathbf{t}$ of a tangent and its second derivative

$$\frac{d^2\mathbf{t}}{ds^2}$$

are parallel at least at four points.

The distance between two tangents of an oval parallel to direction l is said to be *the width of an oval in direction l* (Figure 61).

E (*The Barbiere theorem*). The length of an oval of constant width d is equal to πd.

F. The sum of curvature radii at the opposite points of an oval of constant width is constant and independent of the a choice of points.

1.2.3 Caustics

Let a beam of rays, either parallel, or issuing from one point, be incident on regular curve γ. An envelope of a family of rays reflected from the curve is called *the caustic* of a curve.

Let curve γ be defined by explicit equation

$$y = y(x)$$

and rays of a given beam have the direction of (unit) vector $\mathbf{e} = (l, m)$. Then the equations of caustic have the following form:

$$X = x - \frac{(m - ly'(x))^2}{2y''(x)}y'(x) - \frac{(m - ly'(x))(l + my'(x))}{2y''(x)},$$

$$Y = y(x) + \frac{(m - ly'(x))^2}{2y''(x)} - \frac{(m - ly'(x))(l + my'(x))}{2y''(x)}y'(x)$$

If $\mathbf{r} = \mathbf{r}(s)$ is a natural parametrization of curve γ, then the equation of caustic has the following form:

$$\mathbf{R} = \mathbf{R}(s) + \frac{(\mathbf{e}, \mathbf{n}(s))}{2k(s)}((\mathbf{e}, \mathbf{n}(s))\mathbf{n}(s) - (\mathbf{e}, \mathbf{t}(s))\mathbf{t}(s))$$

If a beam of rays issuing from origin O is incident on a plane curve with equation $\mathbf{r} = \mathbf{r}(s)$, then the equation of caustic has the following form:

$$\mathbf{R} = \mathbf{r}(s) + \frac{(\mathbf{r}(s), \mathbf{n}(s))}{2k(s)\mathbf{r}^2(s) + (\mathbf{r}(s), \mathbf{n}(s))}((\mathbf{r}(s), \mathbf{n}(s))\mathbf{n}(s) - (\mathbf{r}(s), \mathbf{t}(s))\mathbf{t}(s))$$

An envelope of a family of reflected rays is called *a catacaustic*.

An envelope of a family of refracted rays is called *a diacaustic*.

1.2.4 Conchoidal curves

Let γ be an arbitrary curve, A — a given point exterior to the curve and l — a positive number. Draw ray AM intersecting curve γ at point M and plot a line segment of length L in both directions from the point (Figure 62).

A conchoidal curve is a set of the segment end-points generated by *base* γ of the conchoidal curve and *pole* A.

If the segments are plotted from point M on the straight lines passing through the point at constant angle α to the ray AM then the set of the segment end-points is called *an oblique conchoidal curve* (Figure 63). For $\alpha = \pi/2$ we have *an orthoconchoidal curve*.

Let $x = x(t)$, $y = y(t)$ be the parametric equations of curve γ. Then the parametric equations of a conchoidal curve with base γ and pole $A(a, b)$ have the following form:

$$X = x(t) \pm \frac{x(t) - a}{\sqrt{(x(t) - a)^2 + (y(t) - b)^2}}l,$$

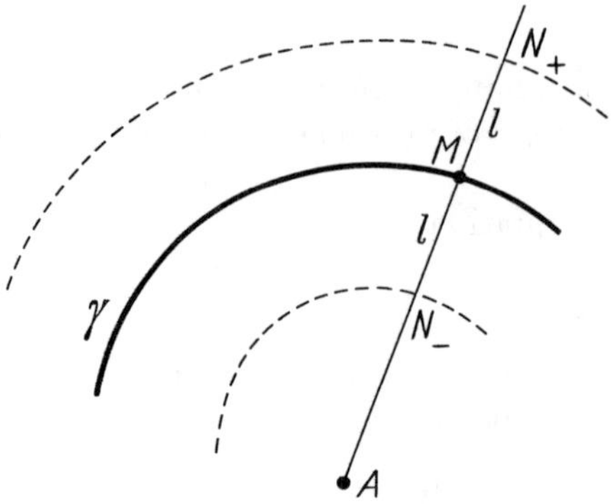

Figure 62.

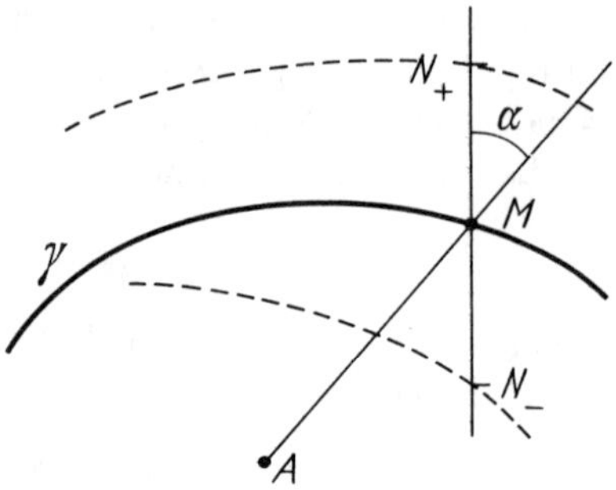

Figure 63.

$$Y = y(t) \pm \frac{y(t) - b}{\sqrt{(x(t) - a)^2 + (y(t) - b)^2}} l$$

1.2.5 Sliding curves

Let γ_1 and γ_2 be regular curves and Δ — a uniform triangle A_1A_2A, the vertices A_1 and A_2 of which slide along curves γ_1 and γ_2, respectively (Figure 64).

The trajectory of third vertex A is called *a sliding curve*, and given curves γ_1 and γ_2 are called *directors* of the sliding curve.

An important degenerate case: triangle Δ degenerates into segment A_1A_2, and A is its inner point (Figure 65).

1.2.6 Parallel curves

Let γ be a regular curve, $\mathbf{r} = \mathbf{r}(s)$ — its natural parametrization, and ε — a fixed positive number.

The curve defined by equation

$$\mathbf{R} = \mathbf{r}(s) \pm \varepsilon \mathbf{n}(s)$$

is said to be *a parallel curve* (*parallel, equdistant curve*) of curve γ positioned at distance ε from the curve. It is obvious that for every ε there exist two different parallels (Figure 66).

If curve γ is defined by parametric equations $x = x(t)$, $y = y(t)$, then its parallel is described by the following equations:

$$X = x(t) \mp \frac{\varepsilon y'(t)}{\sqrt{[x'(t)]^2 + [y'(t)]^2}}, \quad Y = y(t) \pm \frac{\varepsilon x'(t)}{\sqrt{[x'(t)]^2 + [y'(t)]^2}}$$

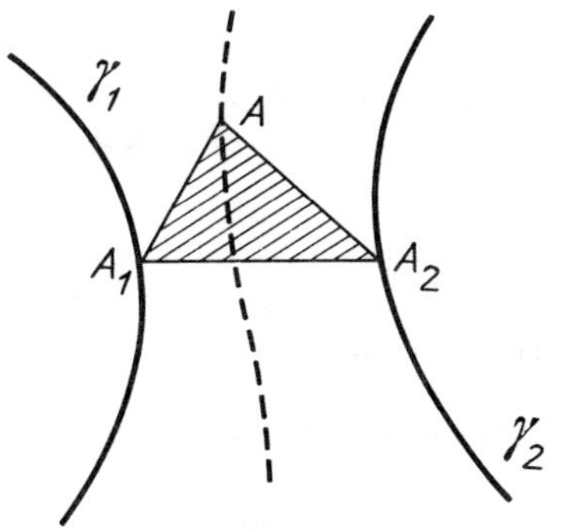

Figure 64.

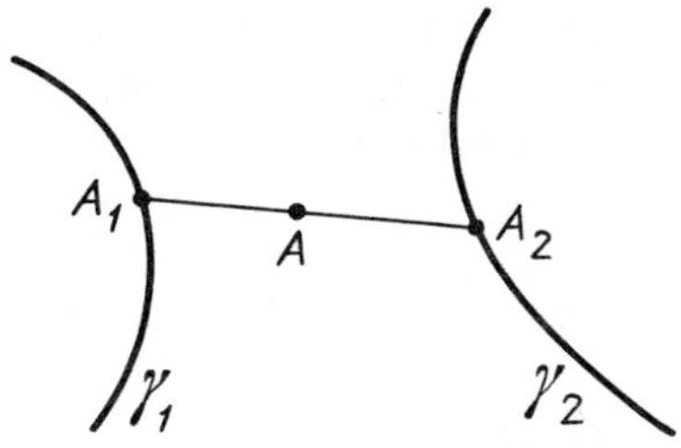

Figure 65.

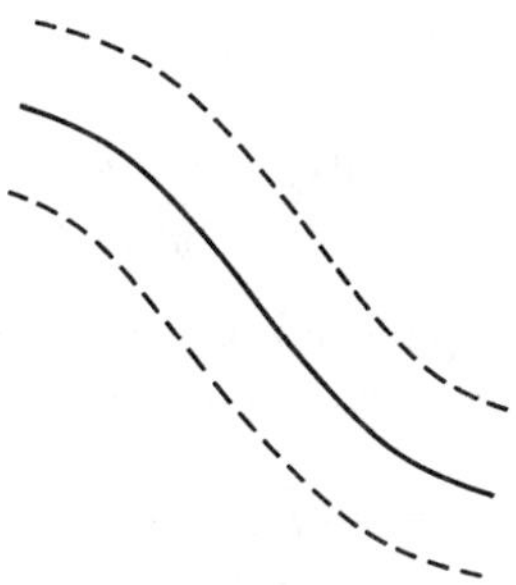

Figure 66.

A parallel is a regular curve for every value of the natural parameter if $|\,\varepsilon k\,| < 1$.

Let parallel closed convex curves γ_1 and γ_2 be positioned apart at distance l. Then the difference of their lengths is equal to $2\pi l$ and the difference of the domain areas bounded by the curves is equal to

$$lS_1 + \pi l^2$$

where S_1 is the length of inner curve γ_1.

Parallel curves have parallel tangents and a common normal at the corresponding points. Every parallel of curve γ is an envelope of a circle family (centers of the circles lie on the curve and radii are constant). The singular points appear on the parallel, if $|\,\varepsilon k\,| = 1$. A tangent to the parallel of a given curve at a cusp (cuspidal point of the first kind) is perpendicular to the curve.

1.2.7 Pedal curves

Let γ be a regular curve, and A — a fixed point. A set of the perpendicular feet dropped from point A to the tangents of the curve is called *a pedal curve* (*pedal, pedal locus, podaire*) (Figure 67).

Let $\mathbf{r} = \mathbf{r}(s)$ be a natural parametrization of curve γ. Then the equation of its pedal has the form

$$\mathbf{R} = (\mathbf{r}(s), \mathbf{n})\mathbf{n}$$

where $\mathbf{n}(s)$ is a unit vector of the normal (here point A coincides with origin O).

If $\mathbf{r}(s) \neq \mathbf{0}$ then a pedal of a regular curve is regular everywhere except for the points corresponding to the inflection points.

If curve γ is described by parametric equations $x = x(t),\ y = y(t)$ then the equations of its pedal have the following form ($A \equiv O$):

$$X = -\frac{x'(t)y(t) - x(t)y'(t)}{[x'(t)]^2 + [y'(t)]^2} y'(t),$$

$$Y = \frac{x'(t)y(t) - x(t)y'(t)}{[x'(t)]^2 + [y'(t)]^2} x'(t)$$

Let $F(x, y) = 0$ be an implicit equation of curve γ. Then the equations of its pedal with respect to point $A(a, b)$ may be written as follows:

$$X = \frac{xF_x^2 + (y - b)F_xF_y + aF_y^2}{F_x^2 + F_y^2},$$

$$Y = \frac{bF_x^2 + (x - a)F_xF_y + yF_y^2}{F_x^2 + F_y^2}$$

The curve for which a given curve is a pedal is said to be *a negative pedal* (*antipedal*).

The antipedal of curve γ with respect to point A is an envelope of the straight line family every line of which is perpendicular to the line connecting the point with the points of the curve.

A set of points symmetric to point A with respect to the tangents of curve γ is called *an orthotomic* (*podoid*) of the curve with respect to the point (Figure 68).

If $x = x(t),\quad y = y(t)$ are the parametric equations of curve γ and $A \equiv O$, then the orthotomic equations may be written as follows:

$$X = -2\frac{x'(t)y(t) - x(t)y'(t)}{[x'(t)]^2 + [y'(t)]^2} y'(t),$$

$$Y = 2\frac{x'(t)y(t) - x(t)y'(t)}{[x'(t)]^2 + [y'(t)]^2} x'(t)$$

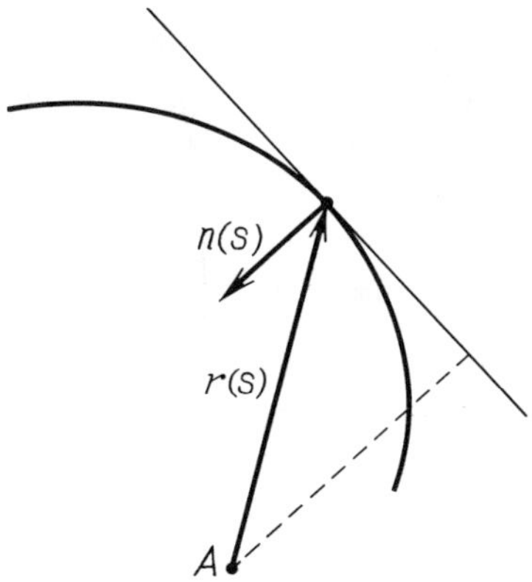

Figure 67.

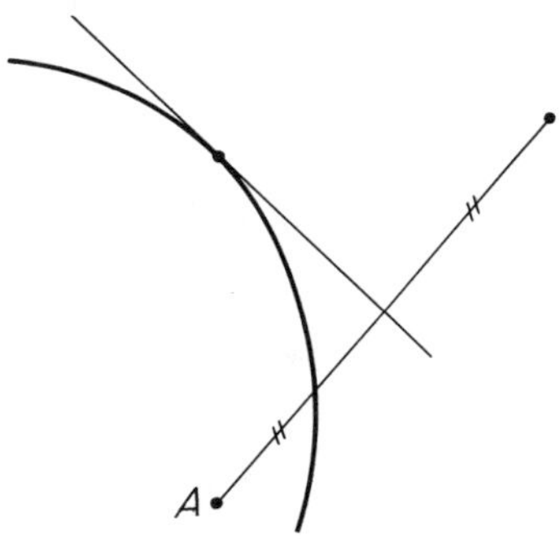

Figure 68.

If $F(x, y) = 0$ is the implicit equation of curve γ, then its orthotomic equations have the following form:

$$X = 2\frac{xF_x + yF_y}{F_x^2 + F_y^2}F_x, \quad Y = 2\frac{xF_x + yF_y}{F_x^2 + F_y^2}F_y$$

An orthotomic of curve γ may be obtained from its pedal by a similarity transformation with point O as the center and the similarity coefficient equal to two.

A curve for which a given curve is an orthotomic is said to be *an antiorthotomic* (*antipodoid*).

1.2.8 Isooptic curves

Let γ be a regular curve, and $\mathcal{A}$ — a given angle of constant value Θ. Let the angle move on the plane so that its sides are tangent to the given curve at all times. The trajectory of the angle running vertex is called *an isooptic curve* (Figure 69). For $\Theta = \pi/2$ we name the trajectory *an orthooptic curve*.

Let curve γ be given as an envelope of the straight line family

$$-x \sin\alpha + y\cos\alpha - f(\alpha) = 0$$

Then the parametric equations of the isooptic curve have the form

$$x = \frac{1}{\sin\Theta}\left(\cos(\alpha + \Theta)f(\alpha) - \cos\alpha f(\alpha + \Theta)\right),$$

$$y = \frac{1}{\sin\Theta}\left(\sin(\alpha + \Theta)f(\alpha) - \sin\alpha f(\alpha + \Theta)\right)$$

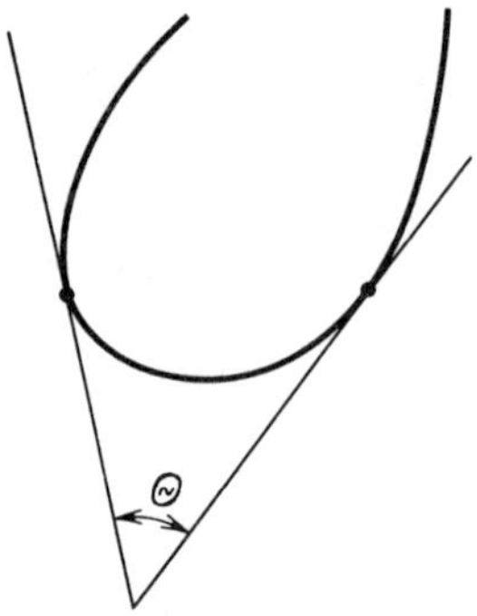

Figure 69.

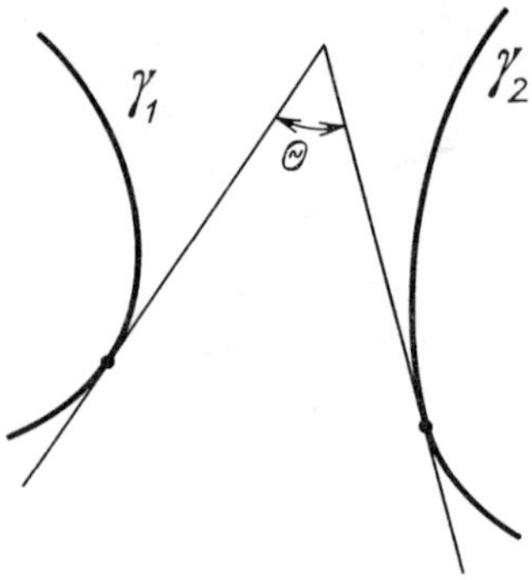

Figure 70.

In particular, an orthooptic curve is described by the equations

$$x = -\sin\alpha f(\alpha) - \cos\alpha f(\alpha + \phi/2),$$

$$y = \cos\alpha f(\alpha) - \sin\alpha f(\alpha + \phi/2)$$

Let γ_1 and γ_2 be regular curves, and $\mathcal{A}$ — a given angle of constant value Θ. Let the angle move on the plane so that one of its sides is tangent to given curve γ_1 and the other to curve γ_2 at all times. The trajectory of the angle running vertex is called *a podaire of the curve γ_1 with respect to curve γ_2* (or *a podaire of curve γ_2 with respect to curve γ_1*) (Figure 70).

1.2.9 Radial curves

Let γ be a regular curve, M — an arbitrary point of the curve, and N — its center of curvature at point M. From a fixed point A, plot line segment AP codirected to segment MN and equal to it in length (Figure 71). A set of end-points AP is called *a radial curve* generated by curve γ.

A curve for which a given curve is radial is said to be *an antiradial curve* with respect to it.

Let $\mathbf{r} = \mathbf{r}(t)$ be the natural parametrization of curve γ. Then a vector equation of a radial curve generated by curve γ has the following view:

$$\mathbf{R} = \frac{\mathbf{r}''(s)}{k^2(s)}$$

If curve γ is given by parametric equations $x = x(t), \quad y = y(t)$, then the equations of the radial curve may be written as follows:

$$X = -\frac{y'(t)}{\sqrt{[x'(t)]^2 + [y'(t)]^2}} \frac{1}{|\,k(t)\,|},$$

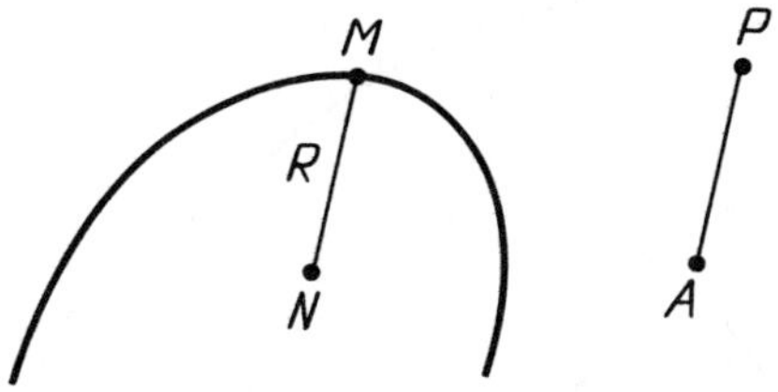

Figure 71.

$$Y = \frac{x'(t)}{\sqrt{[x'(t)]^2 + [y'(t)]^2}} \frac{1}{\mid k(t) \mid}$$

(here, for simplicity point A coincides with origin O).

1.2.10 Cycloidal curves

Let γ be an arbitrary regular curve, and C — a circle the radius of which is subject to condition $r \mid k \mid \leq 1$ where k is the curvature of the curve. Let the circle roll on the curve without slipping, and point A be mounted rigidly to the circle. A trajectory of running point A is called *a cycloidal curve* (Figure 72).

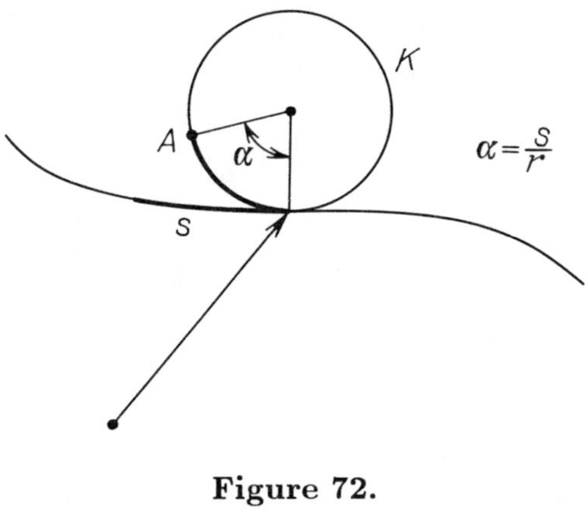

Figure 72.

1.2.11 Cissoidal curves

Let curves γ_1 and γ_2 be given by polar equations of the form $\rho_1 = f_1(\theta)$ and $\rho_2 = f_2(\theta)$. Curve γ described by equation

$$\rho \equiv f(\theta) = f_1(\theta) - f_2(\theta)$$

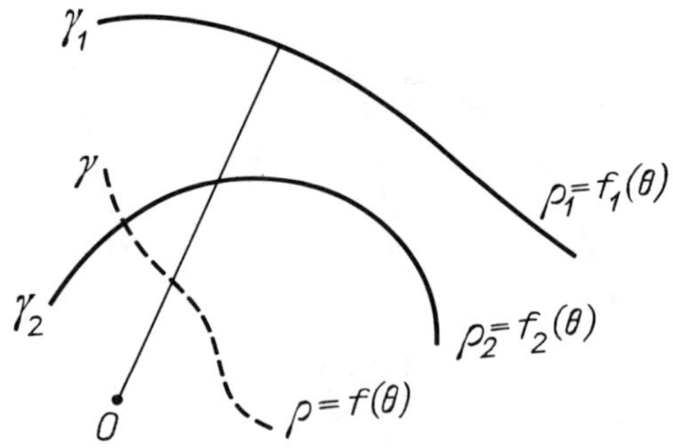

Figure 73.

is called *a cissoidal curve* (*cissoid*) generated by curves γ_1 and γ_2. The curves are called *generating curves* (Figure 73).

A polar subnormal of a cissoidal curve is equal to the difference of the corresponding polar subnormals of the generating curves.

Chapter 2

Atlas of Plane Curves

2.1 Structure of the Atlas

The basis for plane curve classification combines two complementary approaches.

The coordinate approach. Let a Cartesian orthogonal coordinate system be given on a plane, and plane curve γ be defined by equation

$$F(x, y) = 0$$

If function $F(x, y)$ is a polynomial or the function has been reduced to the polynomial by arithmetic means, then curve γ is called *an algebraic curve,* and otherwise — *a transcendent.* Algebraic curves are divided into classes according to the orders of the defining polynomials (recall that the order of algebraic curves does not change under simple coordinate transformations described in Appendix C): *algebraic curves of the second order, algebraic curves of the third order, algebraic curves of the fourth order,* etc.

The geometric approach. Every curve possesses many various characteristic geometric properties such as: boundedness or unboundedness, connectedness or unconnectedness, availability of singular points, their number and types, availability of asymptotes, symmetry, etc. Curves may be divided into classes so that two curves from one class have analogous geometric properties and graphs.

The sampling of the curves has been performed in accodance with the following rules. Every curve included in the Atlas:

1) possesses important and interesting properties which have drawn attention of scientists at different times,

2) may be described with the help of elementary functions, and

3) has a name.

Some of the curves have several names. This fact is explained by a variety reasons. For example, a curve may

1) get its name from the scientist who investigated its properties (sometimes two or three of them),

2) be similar in appearance to a well-known object, and

3) have some important properties.

A complete index of the names of the plane curves in the Atlas is located at the end of the book.

In this chapter an attempt is made to combine both the coordinate and the geometric approaches in an alphabetic order.

It is important to keep in mind that for the sake of brevity it is generally assumed that a curve in the Atlas is, in fact, an infinity of curves described with the help of one equation (two equations in the cases of a parametric description), containing one or more numerical parameters. For example, equation

$$\frac{x^2}{a^2} + \frac{y^2}{b^2} = 1$$

describes a whole class of ellipses: choosing concrete values of parameters a and b determines the specific ellipse.

The majority of the curves in the Atlas may be described in two or three ways. The availability of several ways for description of a curve is very convenient. But it is useful to remember that in some instances one way of description is not equivalent to another. For example, replacement

$$x = \rho \cos \varphi, \quad y = \rho \sin \varphi$$

transforms the equation of scarabaeus

$$(x^2 + y^2)(x^2 + y^2 + cx)^2 = a^2(x^2 - y^2)^2 \tag{A}$$

into an equivalent equation

$$\rho^2 = (a \cos^2 \varphi - c \cos \varphi)^2$$

It is common practice to write the latter in the following simplified, but not equivalent, form

$$\rho = a \cos^2 \varphi - c \cos \varphi \tag{B}$$

Equations (A) and (B) become equivalent when we allow variable ρ to be of any sign.

The Atlas is an integration of plane curve charts. To every chart there corresponds a definite class of plane curves.

CURVE CHARTS. The description of every class follows a common scheme. It includes *the name (names) of the class* (for example, cochleoid, bifolium, bullet-nose curve, spiril curve, etc.), *the equation (equations) defining the class* together with a domain of admissible values of numerical parameters, *additional equations*, and *basic geometric properties* of the curves in the class. By the basic geometric properties we mean such characteristics as *boundedness* or *unboundedness*, *closedness*, *availability of asymptotes* (vertical, horizontal, inclined) and *symmetry* (axial, central), *connectedness* or its absence, *availability of singular points* and *types of singularities*, including *a number of self-intersection points (multiple points)*.

In some cases, characteristic points (vertices, inflection points, etc.) are described. In addition, important *special cases*, *generalizations*, and *areas* of domain naturally connected with the class, as well as the *arc-lengths* of the curves are presented.

Typical graphs of curves from the class and exact drawings form the main part of each chart. The drawings show at a relative scale the behavior of the curves in accordance with the changing of parameter values describing the curve class. In addition, the chart contains equations and a list of admissible values of numerical parameters.

SEARCHING FOR A CURVE IN THE ATLAS.

Searching directions depend on the information available on the curve.

A. You know the name.

It is best to turn to *English-German-French indices* of curve names. With the help of the indices, you find all of the detailed information on the curve contained in the Atlas, including its most frequently used name.

B. You know the equation (equations).

By the equation (equations) you should determine whether the curve is transcendent or algebraic (in the latter case, it is useful to find out the order of the curve). Next, with the help of *the index of curve equations* you obtain the name of the curve.

C. You know the picture.

By refering to *the picture index*, you can find the names of curves with analogous graphs.

D. You know certain geometric characteristics.

You should look at *the indices of curves by basic geometric characteristics*. Through comparing and correlating curves which have been arranged in an alphabetic order, you can find the names of curves with all of the given characteristics.

However, the above suggestions are not, in any way, meant to exclude direct scanning of the whole Atlas of plane curves.

2.2 Atlas of curves

2.2.1 A

Acceleration curve (acceleration)

Equations:

$$(x-y)(x^2+y^2)^3 = 3a^2l^2x^3,$$

$$\rho^4 = \frac{3a^2l^2\cos^3\varphi}{\cos\varphi - \sin\varphi}, \quad a>0, \quad l>0$$

Basic geometric properties:

Unboundedness: present.

Asymptotes: straight line $x = y$ is an inclined asymptote.

Axial symmetry: none.

Central symmetry: point $(0,0)$ is a center of symmetry.

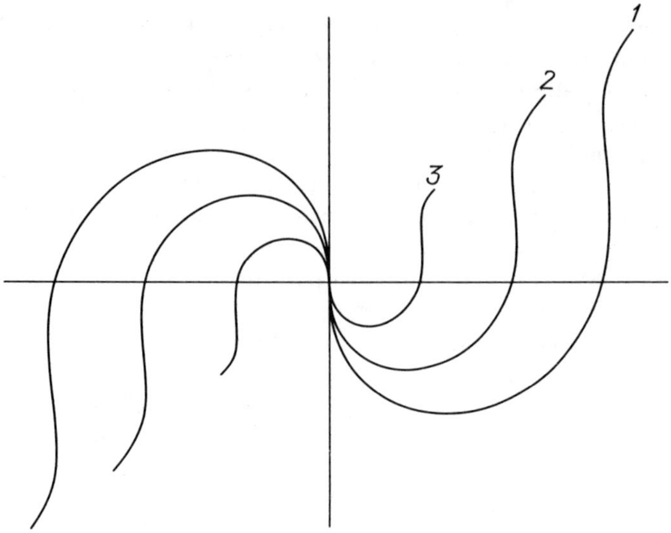

Figure 1. Acceleration curve

N	1	2	3
a	0.50	1.00	1.50
l	0.50	1.00	1.50

Connectedness: present.
Singular points: none.
Self-intersection points: none.
(Figure 1).
ACCELERATION CURVE *is an unbounded connected central-symmetric curve free of singular points with a single inclined asymptote.*

Agnesiana
Equation:

$$y(b^2 + x^2) = ab^2, \quad a > 0, \quad b > 0$$

Additional equation:

$$xy^2 = a^2(b - x)$$

Basic geometric properties:
Unboundedness: present.
Asymptotes: straight line $y = 0$ is a horizontal asymptote.
Axial symmetry: straight line $x = 0$ is an axis of symmetry.
Central symmetry: none.
Connectedness: present.
Singular points: none.
Self-intersection points: none.

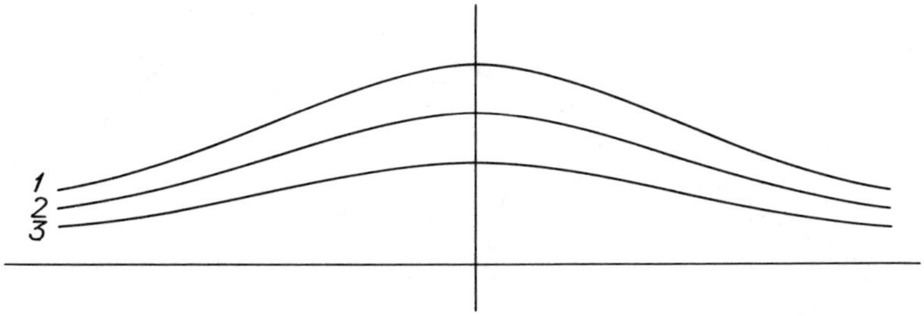

Figure 2. Agnesiana

N	1	2	3
a	1.00	0.75	0.50
b	1.50	1.50	1.50

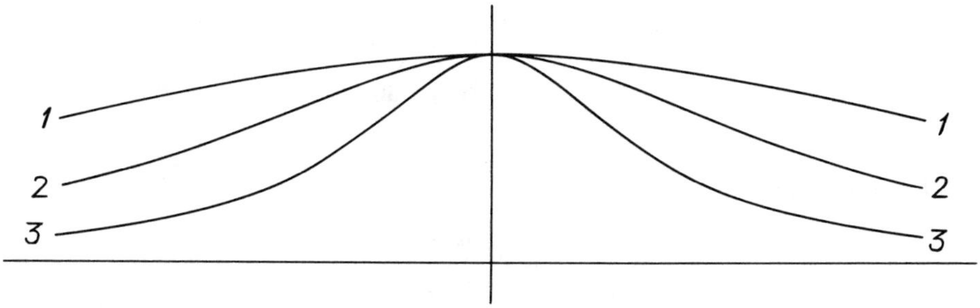

Figure 3. Agnesiana

N	1	2	3
a	1.00	1.00	1.00
b	3.00	1.50	0.75

Characterized points: point $(0, a)$ is located at maximal distance from x-axis; points $(\pm b/\sqrt{3}, 3a/4)$ are inflection points.
Special cases: versiera ($a = b$), pseudoversiera ($a = 2b$).
Area of domain enclosed by the curve and its asymptote is equal to πab. (Figures 2–3).
AGNESIANA *is an unbounded connected curve free of singular points with a single horizontal asymptote and a single symmetry axis.*

Antiversiera
Equations:

$$x^4 - 2rx^3 + 4r^2y^2 = 0,$$

$$y = \pm\frac{1}{2r}\sqrt{(2r - x)x^3}, \quad r > 0$$

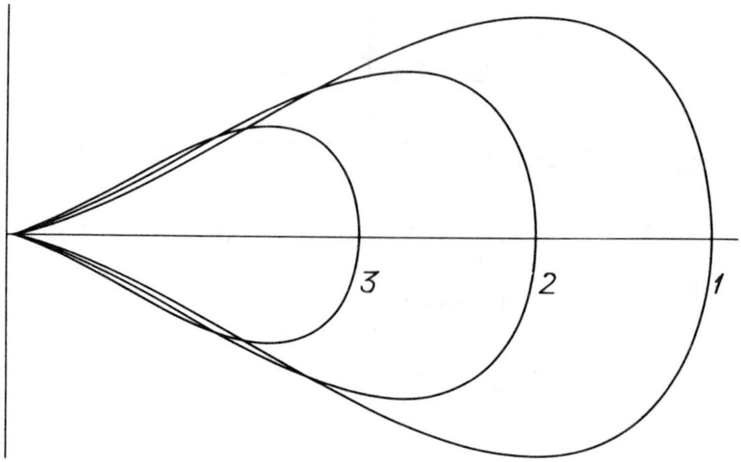

Figure 4. Antiversiera

N	1	2	3
r	1.00	0.75	0.50

Additional equation:

$$x^4 - 2ax^3 + 4\frac{a^4}{b^2}y^2 = 0$$

Basic geometric properties:
Boundedness: present.
Closedness: present.
Axial symmetry: straight line $y = 0$ is an axis of symmetry.
Central symmetry: none.
Connectedness: present.
Singular points: point $(0, 0)$ is a cuspidal point of the first kind, straight line $y = 0$ is a tangent at the point.
Self-intersection points: none.
(Figure 4).

ANTIVERSIERA *is a bounded connected closed curve with a single singular point (a cusp point of the first kind) and a single symmetry axis.*

Arachnida (A) (araneidan (A))
Equation:

$$\rho = 2a\frac{\sin n\varphi}{\sin(n-1)\varphi},$$

$$a > 0, \quad n \in \mathbf{N}, \quad n > 2$$

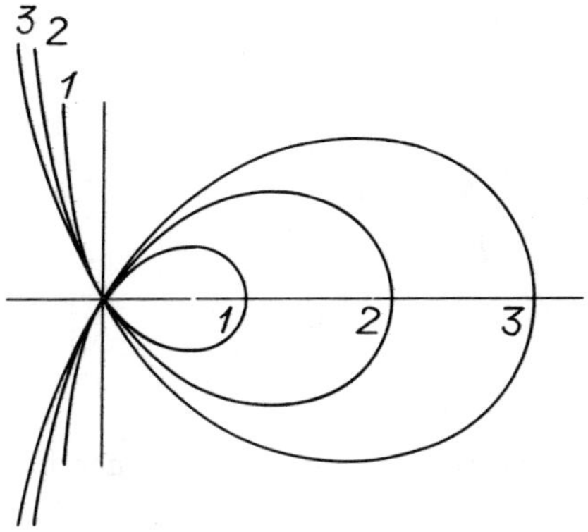

Figure 5. Arachnida (A)

N	1	2	3
a	1.00	2.00	3.00
n	3	3	3

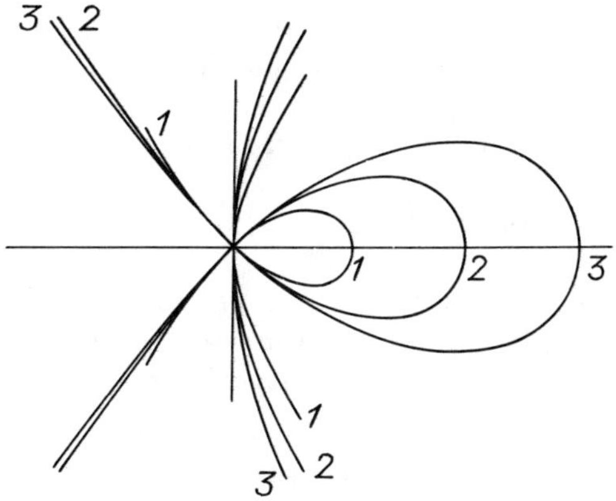

Figure 6. Arachnida (A)

N	1	2	3
a	1.00	2.00	3.00
n	4	4	4

Basic geometric properties:
Unboundedness: present.
Asymptotes: the curve has inclined asymptotes the number and equations of which are defined by values of parameters n and a; for example, for $n = 4$ and $a = 1/2$ the curve has two asymptotes

$$\sqrt{3}x \pm \left(y - \frac{1}{2\sqrt{3}}\right) = 0$$

Axial symmetry: straight line $y = 0$ is an axis of symmetry.
Central symmetry: none.
Connectedness: present.
Singular points: point $(0, 0)$ is a multiple point.
Self-intersection points: one, $(0, 0)$.
Characterized points: point

$$\left(\frac{2an}{n-1}, 0\right)$$

is a point of the loop located at maximal distance from origin $(0, 0)$. (Figures 5–6).

ARACHNIDA (A) *is an unbounded connected curve with a single singular point (multiple point), a single axis of symmetry and inclined asymptotes (their number and equations are defined by values of parameters).*

Arachnida (B) (araneidan (B))
Equation:

$$\rho = 2a\frac{\sin n\varphi}{\sin(n+1)\varphi},$$

$$a > 0,\ n \in \mathbf{N},\ n > 1$$

Basic geometric properties:
Unboundedness: present.
Asymptotes: the curve has inclined asymptotes the number and equations of which are defined by values of parameters n and a; for example, for $n = 3$ and $a = 1/2$ the curve has two asymptotes

$$y = \pm\left(x - \frac{1}{4}\right)$$

Axial symmetry: straight line $y = 0$ is an axis of symmetry.
Central symmetry: none.
Connectedness: none.
Singular points: none (for $n = 2$); point $(0, 0)$ is a node (for $n = 3$), straight lines $y = \pm\sqrt{3}x$ are tangents at the node.
Self-intersection points: none (for $n = 2$); one, $(0, 0)$ (for $n \geq 3$).
Characterized points: point

$$\left(\frac{2an}{n+1}, 0\right)$$

is a point of right branch located at minimal distance from origin $(0, 0)$. (Figures 7–8).

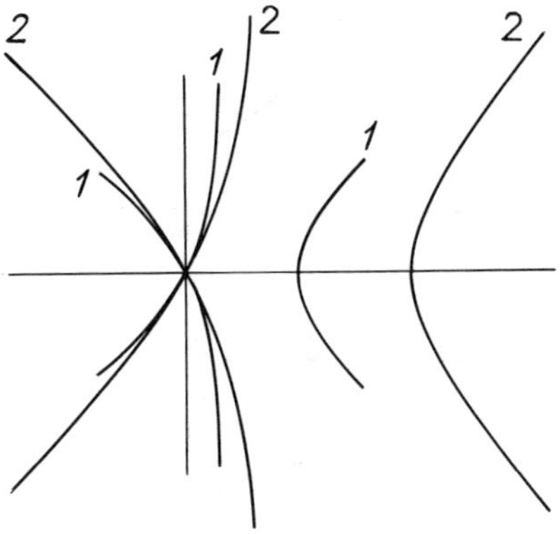

Figure 7. Arachnida (B)

N	1	2
a	1.00	2.00
n	3	3

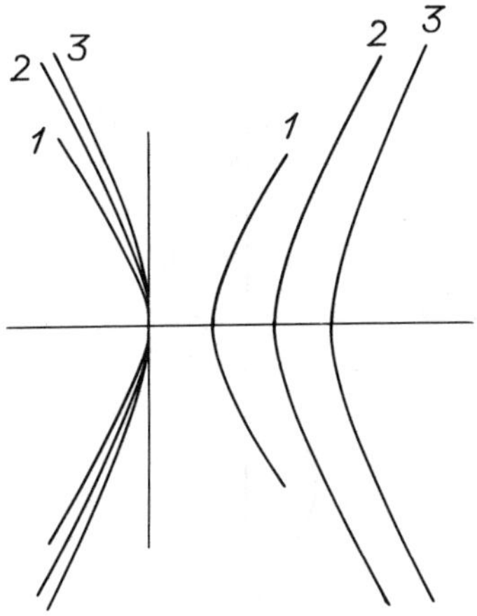

Figure 8. Arachnida (B)

N	1	2	3
a	1.00	2.00	3.00
n	2	2	2

ARACHNIDA (B) *is an unbounded unconnected (consisting of two unbounded branches) curve with a single singular point (multiple point (for $n \geq 3$)), a single axis of symmetry and inclined asymptotes (their number and equations are defined by values of parameters).*

Arc cosinusoid
Equations:

$$y = \operatorname{arccos}(ax) = \pm \arccos(ax) + 2\pi n, \quad a > 0, \quad n \in \mathbf{Z}$$

Basic geometric properties:
Unboundedness: present.
Asymptotes: none.
Axial symmetry: straight lines $y = k\pi, k \in \mathbf{Z}$, are axes of symmetry.
Central symmetry: points $(0, (2k+1)\pi/2), k \in \mathbf{Z}$, are centers of symmetry.
Connectedness: present.
Singular points: none.
Self-intersection points: none.
Characterized points: points $((-1)^k/a, k\pi)), k \in \mathbf{Z}$, are located at maximal distance from y-axis; points $(0, (2k+1)\pi/2), k \in \mathbf{Z}$, are inflection points. (Figure 9).
ARC COSINUSOID *is an unbounded connected central-symmetric curve free of singular points with an infinitely large number of symmetry axes.*

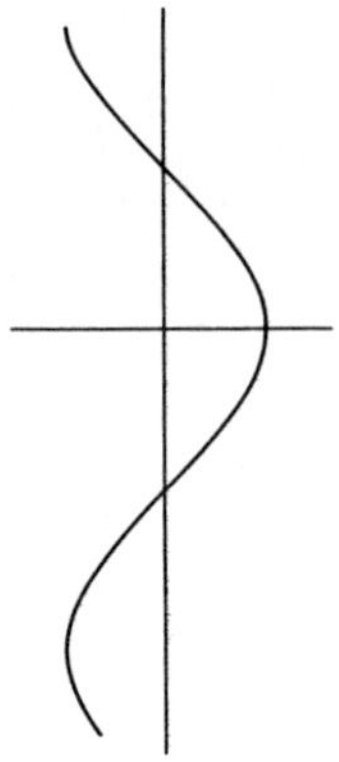

Figure 9. Arc cosinusoid

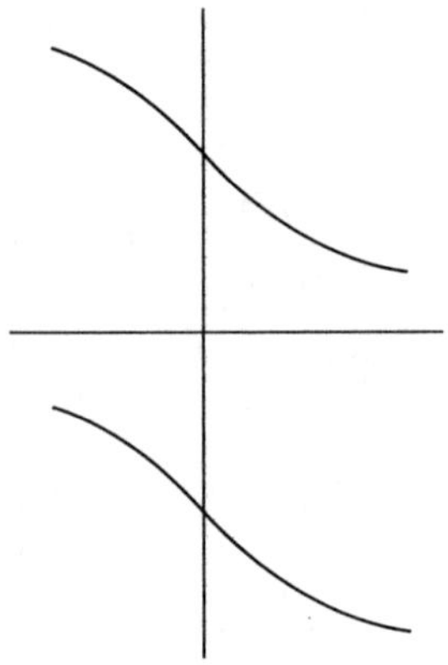

Figure 10. Arc cotangentoid

Arc cotangentoid

Equation:

$$y = \operatorname{arccot}(ax) = \operatorname{arccot}(ax) + \pi n,$$

$$a > 0, \quad n \in \mathbf{Z}$$

Basic geometric properties:
Unboundedness: present.
Asymptotes: straight lines $y = k\pi, k \in \mathbf{Z}$, are horizontal asymptotes.
Axial symmetry: none.
Central symmetry: $(0, k\pi/2)$, $k \in \mathbf{Z}$, are centers of symmetry.
Connectedness: none.
Singular points: none.
Self-intersection points: none.
Characterized points: points $(0, \pi/2 + k\pi)$, $k \in \mathbf{Z}$, are inflection points.
(Figure 10).

ARC COTANGENTOID *is an unbounded unconnected (consisting of an infinitely large number of unbounded branches) central-symmetric curve free of singular points with an infinity of horizontal asymptotes and symmetry centers.*

Arc sinusoid

Equations:

$$y = \arcsin(ax) = (-1)^n \arcsin(ax) + \pi n,$$

$$a > 0, \quad n \in \mathbf{Z}$$

Basic geometric properties:
Unboundedness: present.
Asymptotes: none.

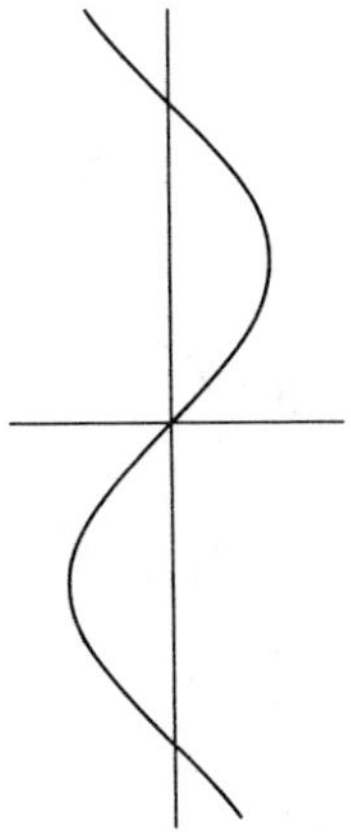

Figure 11. Arc sinusoid

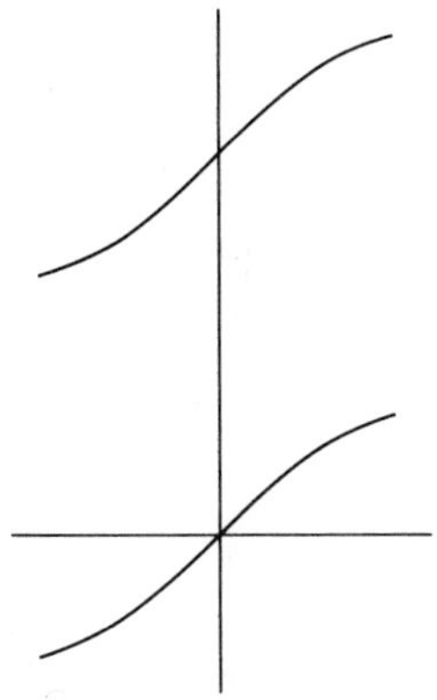

Figure 12. Arc tangentoid

Axial symmetry: straight lines $y = \pi/2 + k\pi$, $k \in \mathbf{Z}$, are axes of symmetry.
Central symmetry: points $(0, k\pi)$, $k \in \mathbf{Z}$, are centers of symmetry.
Connectedness: present.
Singular points: none.
Self-intersection points: none.
Characterized points: points $((-1)^k/b, \pi/2 + k\pi)$, $k \in \mathbf{Z}$, are located at maximal distance from y-axis; points $(0, k\pi)$, $k \in \mathbf{Z}$, are inflection points.
(Figure 11).
ARC SINUSOID *is an unbounded connected central-symmetric curve free of singular points with an infinitely large number of symmetry axes.*

Arc tangentoid
Equations:

$$y = \arctan(ax) = \arctan(ax) + \pi n,$$

$$a > 0, \quad n \in \mathbf{Z}$$

Basic geometric properties:
Unboundedness: present.
Asymptotes: straight lines $y = \pi/2 + k\pi$, $k \in \mathbf{Z}$, are horizontal asymptotes.
Axial symmetry: none.
Central symmetry: points $(0, k\pi/2)$, $k \in \mathbf{Z}$, are centers of symmetry.
Connectedness: none.
Singular points: none.
Self-intersection points: none.
Characterized points: points $(0, k\pi), k \in \mathbf{Z}$, are inflection points.
(Figure 12).
ARC TANGENTOID *is an unbounded unconnected (consisting of an infinitely large number of unbounded branches) central-symmetric curve free of sin-*

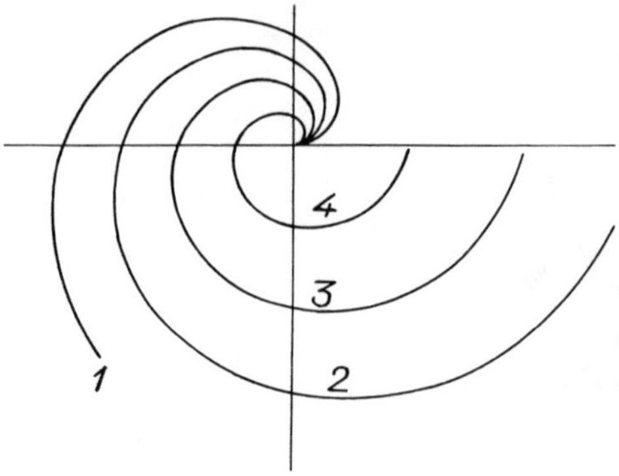

Figure 13. Archimedean spiral

N	1	2	3	4
a	1.00	0.75	0.50	0.25

gular points with an infinity of horizontal asymptotes and symmetry centers.

Archimedean spiral (Archimedean screw, Archimedes' spiral, spiral of Archimedes)
Equations:

$$\rho = a\varphi, \quad \sqrt{x^2 + y^2} = a \arctan \frac{y}{x}, \quad a > 0$$

Basic geometric properties:
Unboundedness: present.
Asymptotes: none.
Axial symmetry: none.
Central symmetry: none.
Connectedness: present.
Singular points: point $(0, 0)$ is a cessation point.
Self-intersection points: none.
(Figure 13).
ARCHIMEDEAN SPIRAL *is an unbounded connected curve with a single singular point (cessation point).*

Area-hyperbolic cosinusoid
Equation:

$$y = \operatorname{arcosh}(ax) = \frac{1}{2} \ln(ax \pm \sqrt{a^2x^2 - 1}), \quad a > 0$$

Basic geometric properties:
Unboundedness: present.

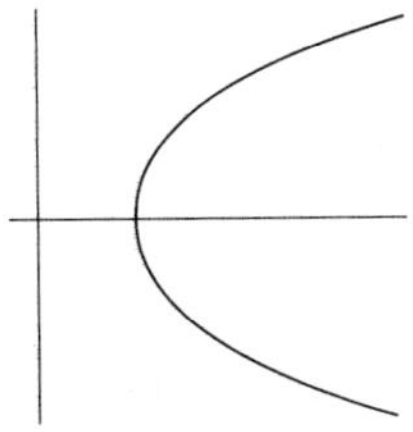

Figure 14.
Area-hyperbolic cosinusoid

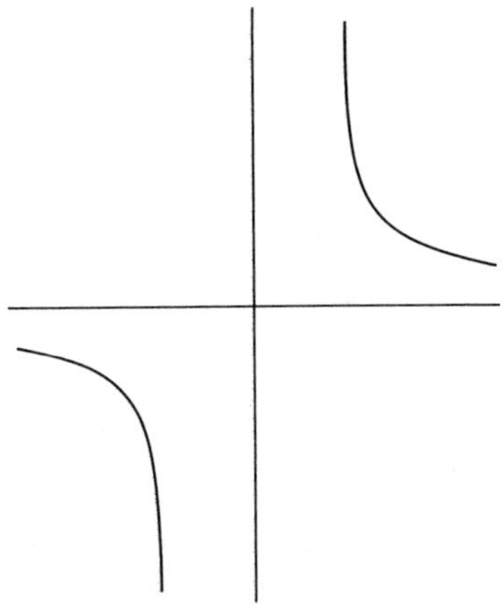

Figure 15.
Area-hyperbolic cotangentoid

Asymptotes: none.
Axial symmetry: straight line $y = 0$ is a symmetry axis.
Central symmetry: none.
Connectedness: present.
Singular points: none.
Self-intersection points: none.
Characterized points: point $(1/a, 0)$ is a vertex.
(Figure 14).
AREA-HYPERBOLIC COSINUSOID *is an unbounded connected curve free of singular points with a single symmetry axis.*

Area-hyperbolic cotangentoid
Equation:

$$y = \operatorname{arcoth}(ax) = \frac{1}{2} \ln \frac{ax+1}{ax-1}, \quad a > 0$$

Basic geometric properties:
Unboundedness: present.
Asymptotes: straight lines $ax \pm 1 = 0$ are vertical asymptotes, straight line $y = 0$ is a horizontal asymptote.
Axial symmetry: none.
Central symmetry: point $(0, 0)$ is a center of symmetry.
Connectedness: none.
Singular points: none.
Self-intersection points: none.
(Figure 15).

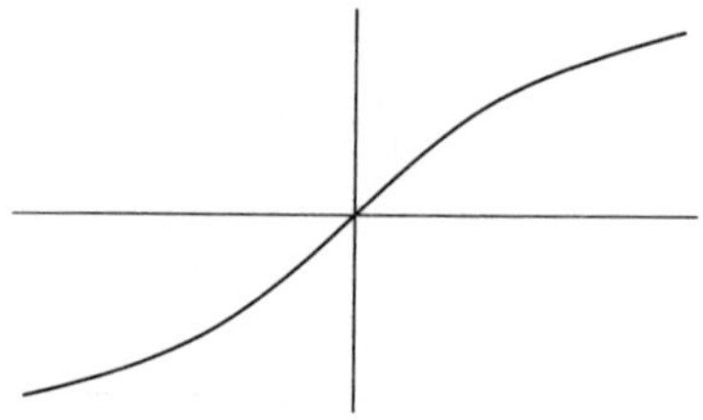

Figure 16.
Area-hyperbolic sinusoid

AREA-HYPERBOLIC COTANGENTOID *is an unbounded unconnected (consisting of two unbounded branches) central-symmetric curve free of singular points with two vertical asymptotes and a single horizontal asymptote.*

Area-hyperbolic sinusoid
Equation:

$$y = \operatorname{arsinh}(ax) = \ln(ax + \sqrt{a^2x^2 + 1}), \quad a > 0$$

Basic geometric properties:
Unboundedness: present.
Asymptotes: none.
Axial symmetry: none.
Central symmetry: point $(0, 0)$ is a center of symmetry.
Connectedness: present.
Singular points: none.
Self-intersection points: none.
Characterized points: point $(0, 0)$ is an inflection point.
(Figure 16).
AREA-HYPERBOLIC SINUSOID *is an unbounded connected central-symmetric curve free of singular points.*

Area-hyperbolic tangentoid
Equation:

$$y = \operatorname{artanh}(ax) = \frac{1}{2}\ln\frac{1 + ax}{1 - ax}, \quad a > 0$$

Basic geometric properties:
Unboundedness: present.
Asymptotes: straight lines $ax \pm 1 = 0$ are vertical asymptotes.
Axial symmetry: none.
Central symmetry: point $(0, 0)$ is a center of symmetry.
Connectedness: present.

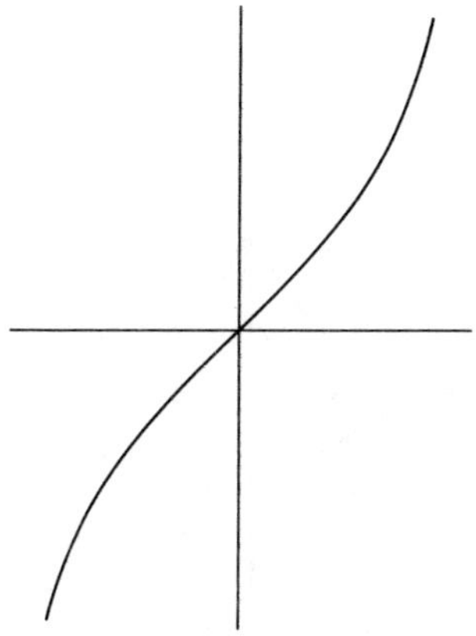

Figure 17.
Area-hyperbolic tangentoid

Singular points: none.
Self-intersection points: none.
Characterized points: point $(0, 0)$ is an inflection point.
(Figure 17).
AREA-HYPERBOLIC TANGENTOID *is an unbounded connected central-symmetric curve free of singular points with two vertical asymptotes.*

Astroid (asteroid, four-cusped hypocycloid, tetracuspid)
Equations:

$$| x |^{2/3} + | y |^{2/3} = l^{2/3},$$

$$x = l \cos^3 t, \quad y = l \sin^3 t, \quad l > 0$$

Additional equation:

$$(x^2 + y^2 - t^2)^3 + 27x^2y^2 = 0$$

Basic geometric properties:
Boundedness: present.
Closedness: present.
Axial symmetry: straight lines $x = 0$, $y = 0$, $x - y = 0$, $x + y = 0$ are axes of symmetry.
Central symmetry: point $(0, 0)$ is a center of symmetry.
Connectedness: present.
Singular points: points $(\pm l, 0)$ are cuspidal points of the first kind, straight line $y = 0$ is a tangent at the cusp points; points $(0, \pm l)$ are cuspidal points of the first kind, straight line $x = 0$ is a tangent at the cusp points.

Figure 18. Astroid

N	1	2	3	4
l	1.00	0.75	0.50	0.25

Self-intersection points: none.
Characterized points: points

$$(\pm l/(2\sqrt{2}), l/(2\sqrt{2})), \quad (\pm l/(2\sqrt{2}), -l/(2\sqrt{2}))$$

are located at minimal distance from origin $(0,0)$.
Area of domain bounded by the curve: $3\pi l^2/8$.
Length of the curve: $6l$.
(Figure 18).
ASTROID *is a bounded connected closed central-symmetric curve with four singular points (cuspidal points of the first kind) and four symmetry axes.*

Atriphtaloid (A) (atriphtothlassic curve (A))
Equations:

$$x^4(x^2+y^2) = (x^2-k^3)^2, \quad \rho^2(\rho-h) + \frac{k^3}{\cos^2\varphi} = 0,$$

$$y = \pm\sqrt{\frac{(hx^2-k^3)}{x^4} - x^2}, \quad h>0, \quad k>0$$

Basic geometric properties:
Unboundedness: present.

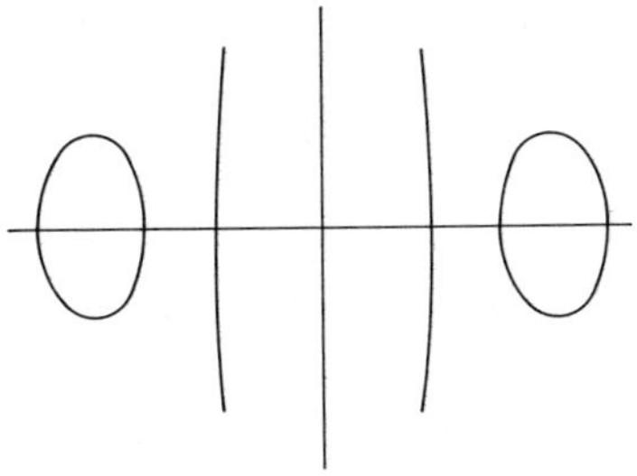

Figure 19. Atriphtaloid (A)
$k = 1.00 \quad h = 2.00$

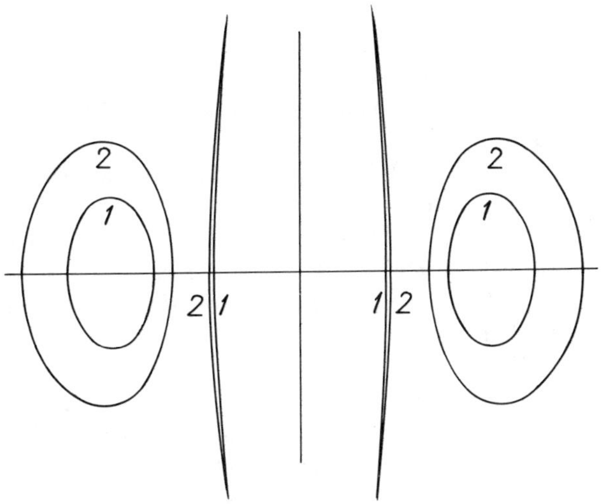

Figure 20. Atriphtaloid (A)

N	1	2
k	1.00	1.00
h	2.00	2.20

Asymptotes: straight line $x = 0$ is a vertical asymptote.
Axial symmetry: straight lines $x = 0, y = 0$ are symmetry axes.
Central symmetry: point $(0, 0)$ is a center of symmetry.
Connectedness: none.
Singular points: points $(\pm 2h/3, 0)$ are isolated points (for $4h^3 = 27k^3$ only).
Self-intersection points: none.
Special cases: for $4h^3 - 27k^3 > 0$ the curve consists of two unbounded branches and two ovals; for $4h^3 - k^3 = 0$ the curve consists of two unbounded branches and two isolated points; for $4h^3 = 27k^3$ the curve consists of two unbounded branches.
(Figures 19–20).

ATRIPHTALOID (A) *is an unbounded unconnected (consisting of two unbounded branches and according to the parameter values two ovals or two isolated points) central-symmetric curve with a single vertical asymptote, two axes of symmetry and two isolated points (the last for* $4h^3 = 27k^3$ *only).*

Atriphtaloid (B) (atriphtothlassic curve (B))
Equations:

$$(x^2 + y^2)^2 x^2 = (hx^2 - k^3)^2,$$

$$y = \pm\sqrt{\frac{hx^2 - k^3}{x} - x^2}$$

Basic geometric properties:
Unboundedness: present.
Asymptotes: straight line $x = 0$ is a vertical asymptote.
Axial symmetry: straight lines $x = 0, y = 0$ are symmetry axes.
Central symmetry: point $(0, 0)$ is a center of symmetry.
Connectedness: none.
Singular points: points $(\pm 2h/3, 0)$ are isolated points (for $4h^3 = 27k^3$ only).
Self-intersection points: none.
Special cases: for $4h^3 - 27k^3 > 0$ the curve consists of two unbounded branches and two ovals; for $4h^3 - k^3 = 0$ the curve consists of two unbounded branches and two isolated points; for $4h^3 = 27k^3$ the curve consists of two unbounded branches.
(Figures 21–22).
ATRIPHTALOID (B) *is an unbounded unconnected (consisting of two unbounded branches and according to the parameter values two ovals or two isolated points) central-symmetric curve with a single vertical asymptote, two axes of symmetry and two isolated points (the last for* $4h^3 = 27k^3$ *only).*

2.2.2 B

Bellshaped curve (error curve)
Equation:

$$y = e^{-x^4}$$

Basic geometric properties:
Unboundedness: present.
Asymptotes: straight line $y = 0$ is a horizontal asymptote.
Axial symmetry: straight line $x = 0$ is an axis of symmetry.
Central symmetry: none.
Connectedness: present.
Singular points: none.

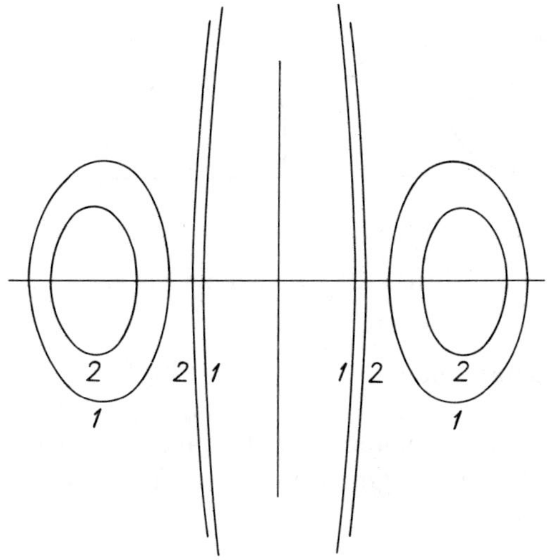

Figure 21. Atriphtaloid (B)

N	1	2
k	0.90	2.00
h	2.00	2.00

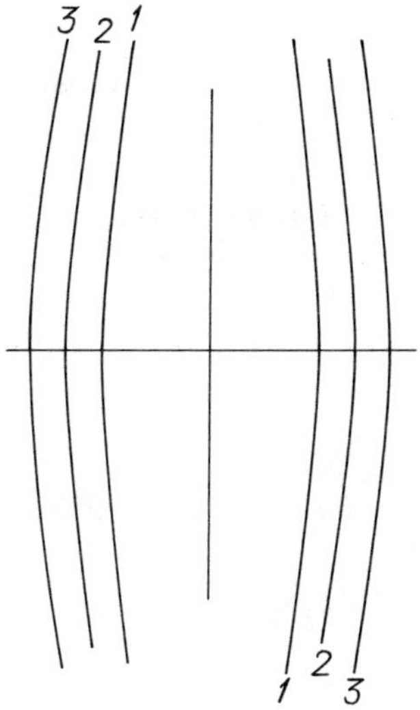

Figure 22. Atriphtaloid (B)

N	1	2	3
k	0.80	1.00	1.20
h	1.00	1.00	1.00

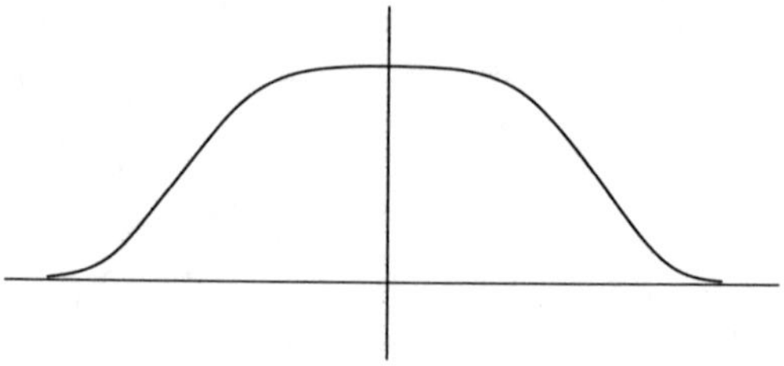

Figure 23. Bellshaped curve

Self-intersection points: none.
(Figure 23).
BELLSHAPED CURVE *is an unbounded connected curve free of singular points with a single axis of symmetry and a single horizontal asymptote.*

Besace
Equations:

$$x^4 - 4bx^2y + 4(a^2+b^2)y^2 - 2(a^2-b^2-r^2)x^2$$

$$-4b(a^2+b^2-r^2)^2 = 0,$$

$$x = \pm\sqrt{a^2+b^2+r^2+2r(a\cos t + b\sin t)},$$

$$y = b + r\sin t,$$

$$a > 0, \quad b > 0, \quad r > 0$$

Additional equation:

$$(x^2 - by)^2 + a^2(y^2 - x^2) = 0$$

Basic geometric properties:
Boundedness: present.
Closedness: present.
Axial symmetry: straight line $x = 0$ is a symmetry axis.
Central symmetry: none.
Connectedness: present (for $a^2+b^2=r^2$), none (for $a^2+b^2 \neq r^2$).
Singular points: point $(0,0)$ is a node (for $a^2+b^2=r^2$ only), straight lines $ax \pm ry = 0$ are tangents at the point.

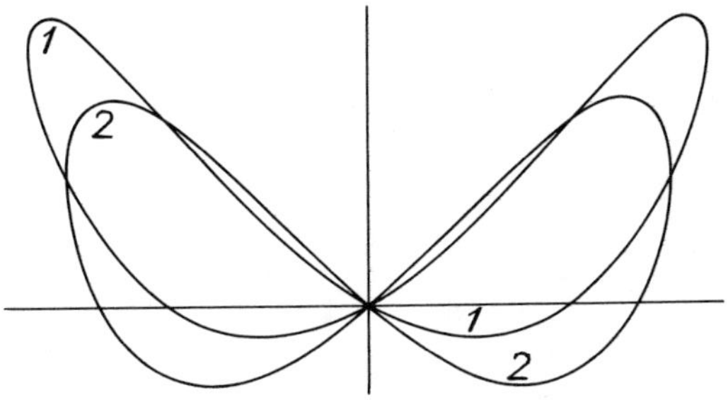

Figure 24. Besace

N	1	2
a	3.00	2.00
b	4.00	1.00
r	5.000	2.236

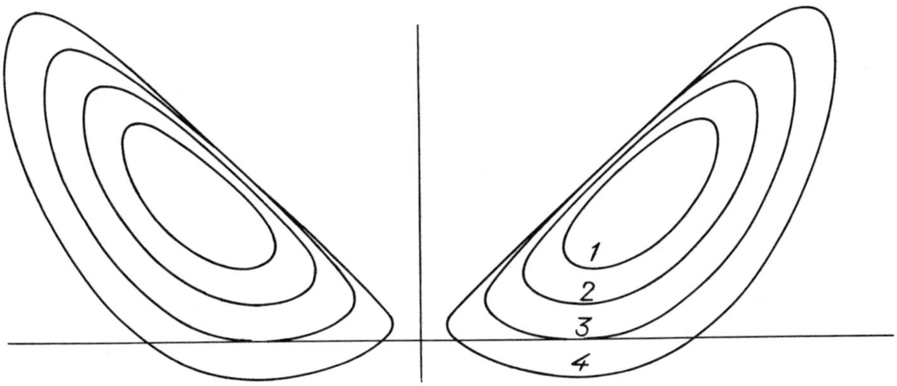

Figure 25. Besace

N	1	2	3	4
r	0.50	0.75	1.00	1.25
a	1.00	1.00	1.00	1.00
b	1.00	1.00	1.00	1.00

Self-intersection points: one, $(0,0)$ (for $a^2+b^2=r^2$), none (for $a^2+b^2 \neq r^2$). (Figures 24–27).

BESACE is a bounded unconnected (consisting of two closed curves, $a^2+b^2 \neq r^2$) curve with a single symmetry axis and (for $a^2+b^2=r^2$ only) a single singular point (node).

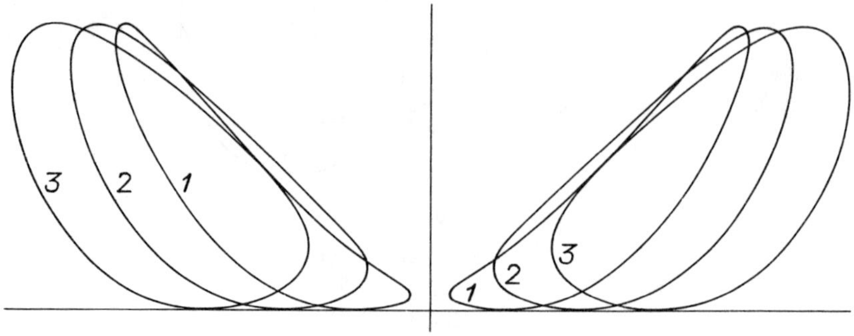

Figure 26. Besace

N	1	2	3
a	0.50	1.00	1.50
b	1.00	1.00	1.00
r	1.00	1.00	1.00

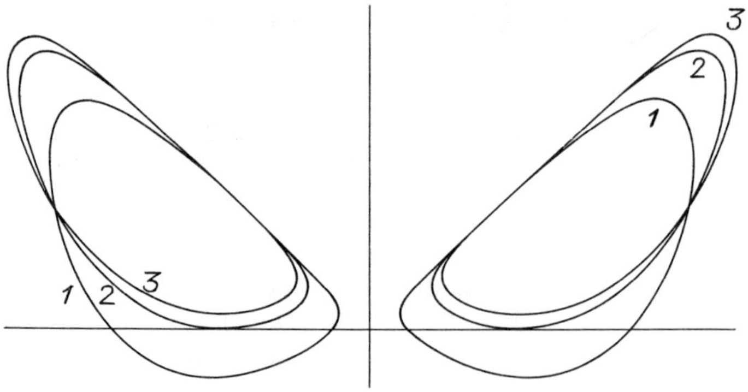

Figure 27. Besace

N	1	2	3
b	0.65	1.00	1.10
a	1.00	1.00	1.00
r	1.00	1.00	1.00

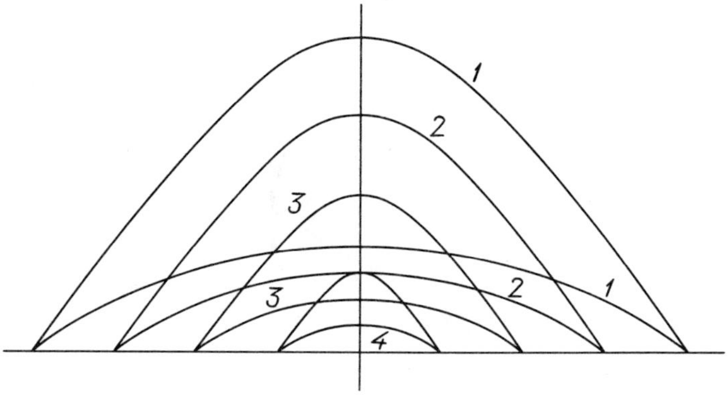

Figure 28. Bicorn curve

N	1	2	3	4
a	1.00	0.75	0.50	0.25

Bicorn curve (bicorn, cocked hat)
Equations:

$$(x^2 + 2ay - a^2)^2 = y^2(a^2 - x^2),$$

$$x = a\cos t, \quad y = \frac{a\sin^2 t}{2 + \sin t}, \quad a > 0$$

Basic geometric properties:
Boundedness: present.
Closedness: present.
Axial symmetry: straight line $x = 0$ is an axis of symmetry.
Central symmetry: none.
Connectedness: present.
Singular points: point $(a, 0)$ is a cuspidal point of the first kind, straight line $x + y - a = 0$ is a tangent at the cusp point; point $(-a, 0)$ is a cuspidal point of the first kind, straight line $x - y + a = 0$ is a tangent at the cusp point.
Self-intersection points: none.
Characterized points: points $(0, a/3)$ and $(0, a)$ are vertices.
(Figure 28).
BICORN CURVE *is a bounded connected closed curve with two singular points (cuspidal points of the first kind) and a single axis of symmery.*

Bifolium (double folium, right bifolium, right double folium)
Equations:

$$(x^2 + y^2)^2 + 4rxy^2 = 0, \quad \rho = -4r\cos\varphi\sin^2\varphi, \quad r > 0$$

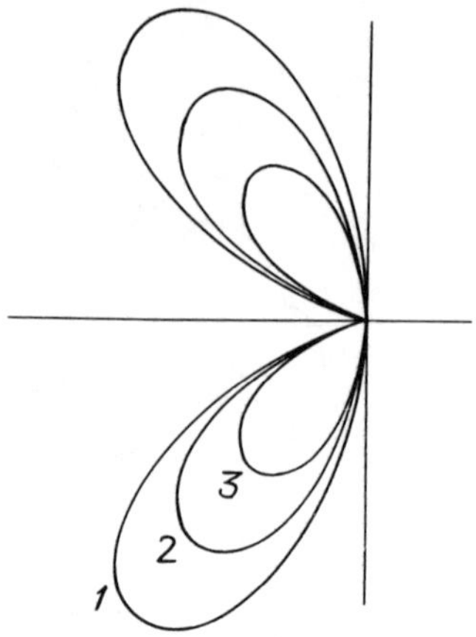

Figure 29. Bifolium

N	1	2	3
r	1.00	0.75	0.50

Additional equations:

$$(x^2 + y^2)^2 = bx^2y,$$

$$\rho = b\cos^2\varphi \sin\varphi$$

Basic geometric properties:
Boundedness: present.
Closedness: present.
Axial symmetry: straight line $y = 0$ is an axis of symmetry.
Central symmetry: none.
Connectedness: present.
Singular points: point $(0, 0)$ is a degenerate triple point, straight lines $x = 0$, $y = 0$ are tangents at the triple point.
Self-intersection points: one, $(0, 0)$.
Area of each of two petals: $\pi r^2/4$.
(Figure 29).
BIFOLIUM *is a bounded connected closed curve with a single singular point (degenerate triple point) and a single axis of symmetry.*

Biquadratic cubic parabola
Equations:

$$(x + y)^4 = ax^3,$$

$$x = \frac{a}{(1+t)^4}, \quad y = \frac{at}{(1+t)^4}, \quad a > 0$$

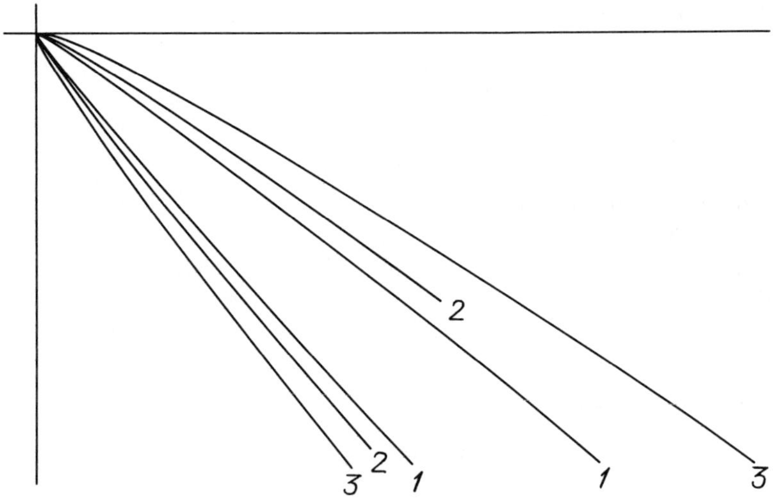

Figure 30.
Biquadratic cubic parabola

N	1	2	3
a	0.25	1.00	4.00

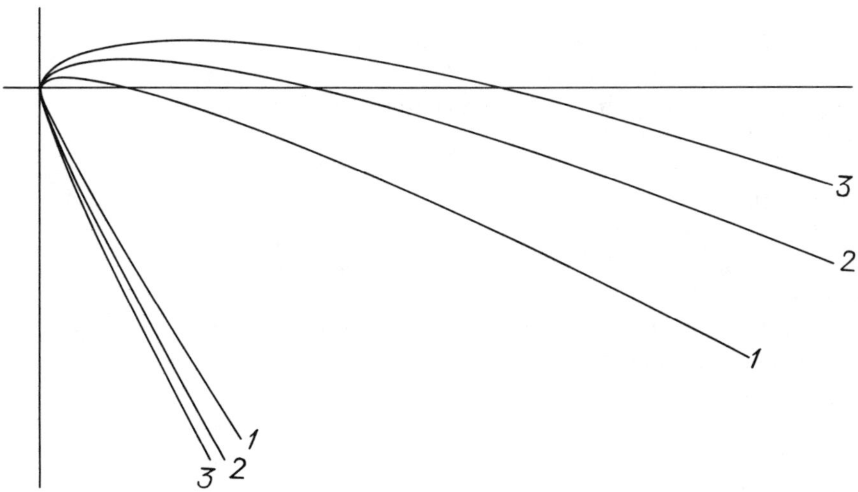

Figure 31.
Biquadratic cubic parabola

N	1	2	3
a	1.00	3.00	5.00

Basic geometric properties:
Unboundedness: present.
Asymptotes: none.
Axial symmetry: none.
Central symmetry: none.
Connectedness: present.
Singular points: none.
Self-intersection points: none.
(Figures 30–31).
BIQUADRATIC CUBIC PARABOLA *is an unbounded connected curve free of singular points.*

Bullet-nose curve
Equations:

$$a^2y^2 - b^2x^2 = x^2y^2,$$

$$x = a\cos t, \quad y = b\cot t,$$

$$a > 0, \quad b > 0$$

Additional equation:

$$\frac{a^2}{x^2} - \frac{b^2}{y^2} = 1$$

Basic geometric properties:
Unboundedness: present.
Asymptotes: straight lines $x \pm a = 0$ are vertical asymptotes.
Axial symmetry: straight lines $x = 0$, $y = 0$ are axes of symmetry.
Central symmetry: point $(0,0)$ is a center of symmetry.
Connectedness: present.
Singular points: point $(0,0)$ is a node, straight lines $bx \pm ay = 0$ are tangents at the node.
Self-intersection points: one, $(0,0)$.
Special cases: equilateral bullet-nose curve ($a = b$).
(Figures 32–34).
BULLET-NOSE CURVE *is an unbounded connected central-symmetric curve with a single singular point (node), two vertical asymptotes and two axes of symmetry.*

2.2.3 C

Cappa (cappa curve, kappa curve)
Equations:

$$y^2(x^2 + y^2) = a^2x^2, 0$$

$$\rho = a\cot\varphi, \quad a > 0$$

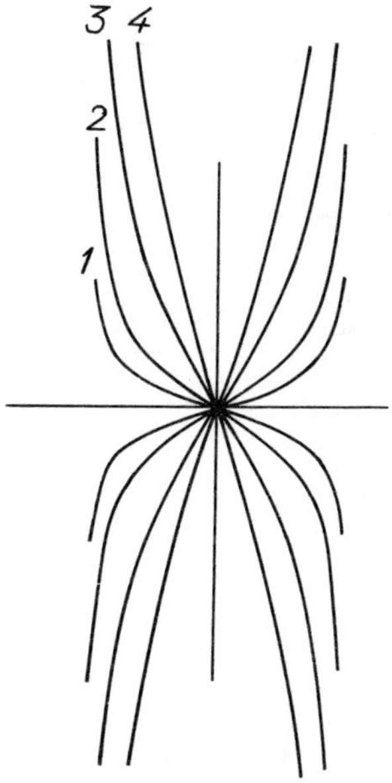

Figure 32. Bullet-nose curve

N	1	2	3	4
b	0.50	1.00	1.50	2.50
a	1.00	1.00	1.00	1.00

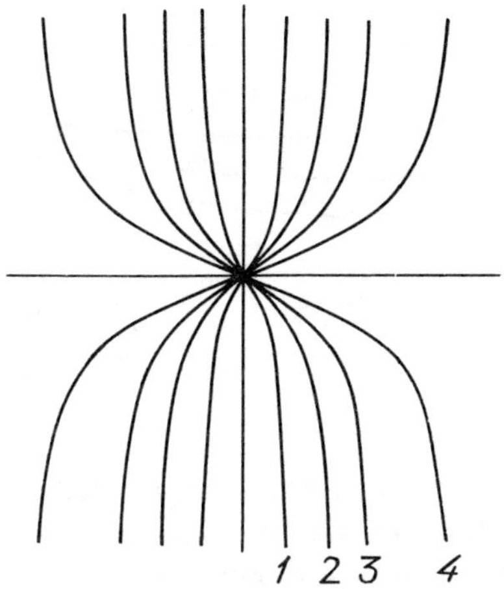

Figure 33. Bullet-nose curve

N	1	2	3	4
a	0.50	1.00	1.50	2.50
b	1.00	1.00	1.00	1.00

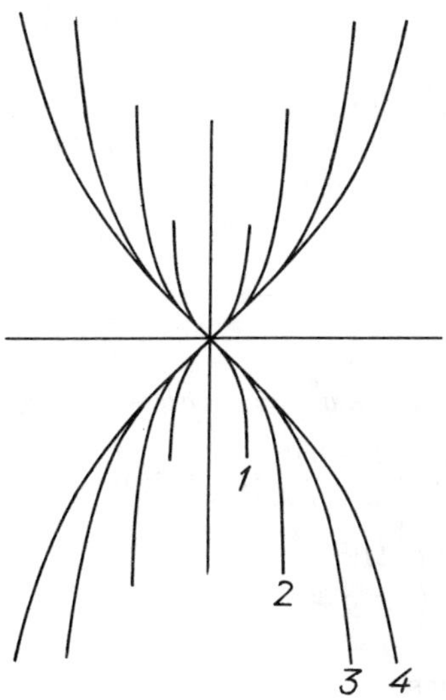

Figure 34. Bullet-nose curve

N	1	2	3	4
a	0.50	1.00	2.00	3.00
b	0.50	1.00	2.00	3.00

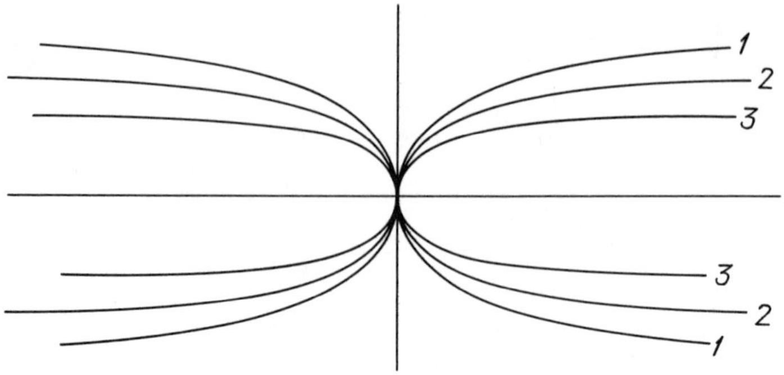

Figure 35. Cappa

N	1	2	3
a	1.00	0.75	0.50

Basic geometric properties:
Unboundedness: present.
Asymptotes: straight lines $y \pm a = 0$ are horizontal asymptotes.
Axial symmetry: straight lines $x = 0$, $y = 0$ are axes of symmetry.
Central symmetry: point $(0, 0)$ is a center of symmetry.
Connectedness: present.
Singular points: point $(0, 0)$ is a degenerate node, straight line $x = 0$ is a tangent at the node.
Self-intersection points: one, $(0, 0)$.
Area of domain enclosed by the curve and one of its asymptotes: πa^2.
(Figure 35).
CAPPA *is an unbounded connected central-symmetric curve with a single singular point (degenerate node), two horizontal asymptotes and two axes of symmetry.*

Capricornoid
Equations:

$$g^2x^2(x^2 + y^2) = c^2(gy - x^2 - y^2)^2,$$

$$\rho = \frac{gc \sin\varphi}{c + g\cos\varphi}, \quad g > 0, \quad c > 0$$

Basic geometric properties:
Unboundedness: present (for $g \geq c$).
Boundedness: present (for $g < c$).
Asymptotes: straight lines

$$y = \pm\frac{\sqrt{g^2 - c^2}}{c}x + g, \quad g > c$$

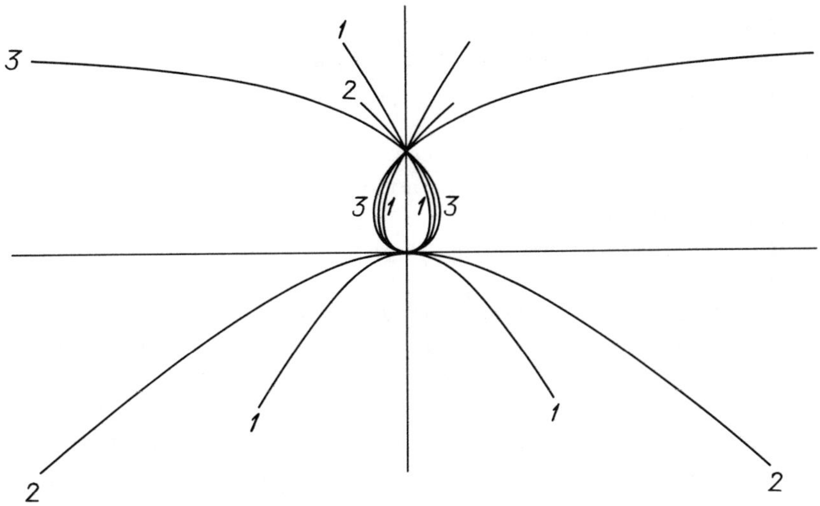

Figure 36. Capricornoid

N	1	2	3
c	0.50	0.75	1.00
g	1.00	1.00	1.00

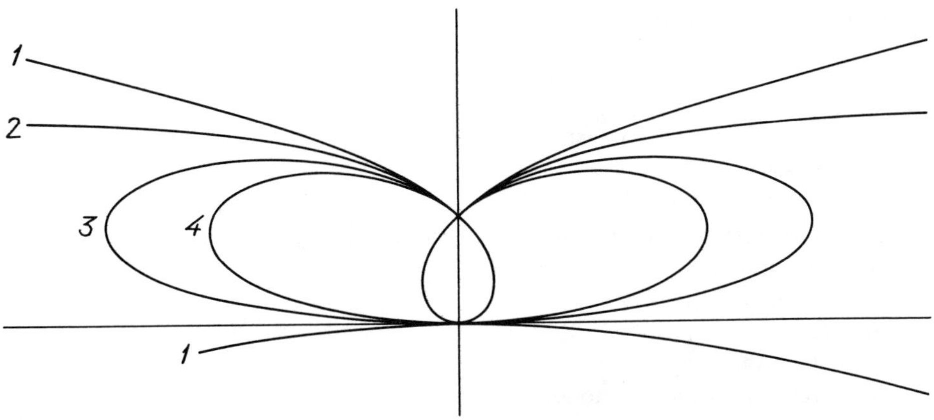

Figure 37. Capricornoid

N	1	2	3	4
c	0.95	1.00	1.05	1.10
g	1.00	1.00	1.00	1.00

are inclined asymptotes; straight line $y = 2g, (g = c)$ is a horizontal asymptote.
Closedness: present (for $g < c$).
Axial symmetry: straight line $x = 0$ is an axis of symmetry.
Central symmetry: none.
Connectedness: present.
Singular points: point $(0, 0)$ is a degenerate node, straight line $y = 0$ is a tangent at the node; point $(0, g)$ is a node, straight lines $\pm gx + cy - cg = 0$ are tangents at the node.
Self-intersection points: two, $(0, 0), (0, g)$.
(Figures 36–37).
CAPRICORNOID ($g \geq c$) *is an unbounded connected curve with two singular points (node, degenerate node), two inclined asymptotes (for $g = c$ one of them is horizontal) and a single axis of symmetry.*
CAPRICORNOID ($g < c$) *is a bounded connected closed curve with two singular points (node, degenerate node) and a single axis of symmetry.*

Cardioid (cardioid curve, cardioid line, heart contour, heart line, heart shape)
Equations:

$$(x^2 + y^2 + 2rx)^2 = 4r^2(x^2 + y^2),$$

$$\rho = 2r(1 - \cos\varphi),$$

$$r > 0$$

Additional equations:

$$x = 2r\cos t - r\cos 2t, \quad y = 2r\sin t - r\sin 2t$$

Basic geometric properties:
Boundedness: present.
Closedness: present.
Axial symmetry: straight line $y = 0$ is an axis of symmetry.
Central symmetry: none.
Connectedness: present.
Singular points: point $(0, 0)$ is a cuspidal point, straight line $y = 0$ is a tangent at the cusp point.
Self-intersection points: none.
Characterized points: points $(0, 0)$ and $(-4r, 0)$ are intersection points of the curve and x-axis.
Area of domain enclosed by the curve: $6\pi r^2$.
(Figure 38).
CARDIOID *is a bounded connected closed curve with a single singular point (cuspidal point of the first kind) and a single axis of symmetry.*

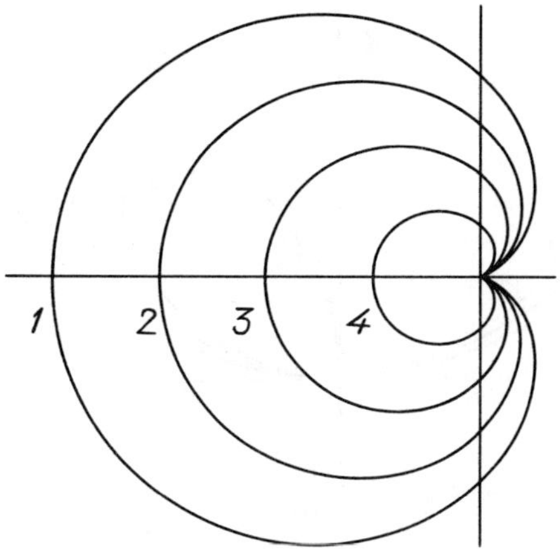

Figure 38. Cardioid

N	1	2	3	4
r	1.00	0.75	0.50	0.25

Cartesian folium (Descartes' folium, folium of Descartes)
Equations:

$$x^3 + y^3 = 3axy,$$

$$\rho = \frac{3a \sin\varphi \cos\varphi}{\sin^3\varphi + \cos^3\varphi},$$

$$x = \frac{3at}{1+t^3}, \quad y = \frac{3at^2}{1+t^3}, \quad a > 0$$

Additional equations:

$$x(x^2 + 3y^2) = l(x^2 - y^2), \quad l = \frac{3a}{\sqrt{2}}$$

(straight line $x = 0$ is an axis of symmetry, straight line $3x + l = 0$ is an asymptote),

$$x = \frac{l \sin 3t}{3 \sin t}, \quad y = \frac{l \sin 3t}{3\sqrt{3} \cos t}$$

Basic geometric properties:
Unboundedness: present.
Asymptotes: straight line $x + y + a = 0$ is an inclined asymptote.
Axial symmetry: straight line $x - y = 0$ is an axis of symmetry.
Central symmetry: none.
Connectedness: present.
Singular points: point $(0,0)$ is a node, straight lines $x = 0$, $y = 0$ are tangents at the node.
Self-intersection points: one, $(0,0)$.

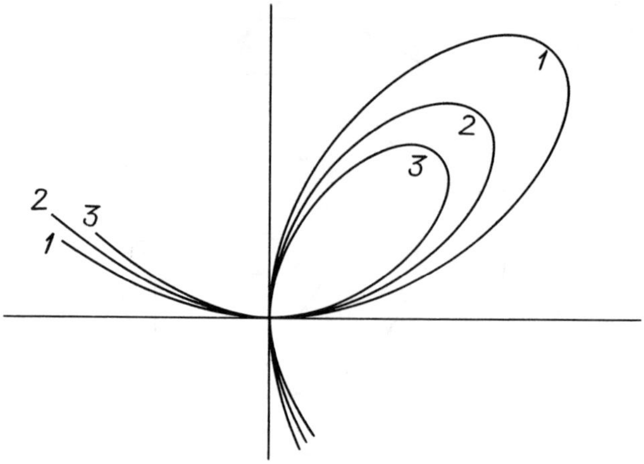

Figure 39. Cartesian folium

N	1	2	3
a	0.(3)	0.25	0.20

Characterized points: point $(a\sqrt[3]{2}, a\sqrt[3]{4})$ is located on a curve's loop at maximal distance from x-axis; point $(a\sqrt[3]{4}, a\sqrt[3]{2})$ is located on the curve's loop at maximal distance from y-axis; point $(3a/2, 3a/2)$ is located on the curve's loop at maximal distance from origin $(0,0)$ (vertex).
Areas: area of the curve's loop is equal to $3a^2/2$, area of domain enclosed by the curve and the inclined asymptote is equal to $3a^2/2$.
(Figure 39).
CARTESIAN FOLIUM *is an unbounded connected curve with a single singular point (node), a single inclined asymptote and a single axis of symmetry.*

Cartesian oval
Equations:

$$m\sqrt{x^2+y^2} + n\sqrt{(x-d)^2+y^2} = c,$$
$$(n^2-m^2)\rho^2 + 2\rho(mc - n^2 d\cos\varphi)n^2d^2 - c^2 = 0,$$
$$m>0, \quad n>0, \quad c>0, \quad d>0$$

Additional equation:

$$(x^2+y^2-2rx)^2 - l^2(x^2+y^2) - k = 0$$

Basic geometric properties:
Boundedness: present.
Closedness: present.
Axial symmetry: straight line $y=0$ is an axis of symmetry.
Central symmetry: none.
Connectedness: present (for $nd=c$, $n \geq m$), none (for $nd \neq c$).

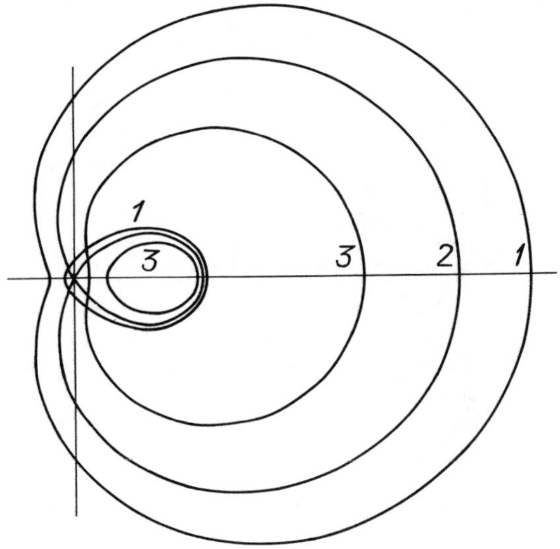

Figure 40. Cartesian oval

N	1	2	3
n	1.80	2.00	2.50
c	2.00	2.00	2.00
d	1.00	1.00	1.00
m	1.00	1.00	1.00

Singular points: point $(0,0)$ is a node (for $nd = c$, $n > m$), straight lines

$$y = \pm x\sqrt{4\left(\frac{r}{l}\right)^2 - 1}$$

are tangents at the node.
Self-intersection points: none (for $nd \neq c$).
Special cases: Pascal's limaçon ($nd = c$).
(Figures 40–42).
CARTESIAN OVAL *is a bounded unconnected (consisting of two connected closed branches,* $nd \neq c$*) curve free of singular points with a single axis of symmetry.*

Cassinian oval (Cassinian ellipse, oval of Cassini)
Equations:

$$(x^2 + y^2)^2 - 2c^2(x^2 - y^2) = a^4 - c^4,$$

$$\rho = c\sqrt{\cos 2\varphi \pm \sqrt{\cos^2 2\varphi + \frac{a^4}{c^4} - 1}},$$

$$y = \pm\sqrt{\sqrt{4c^2x^2 + a^4} - x^2 - c^2}, \quad a > 0, \quad c > 0$$

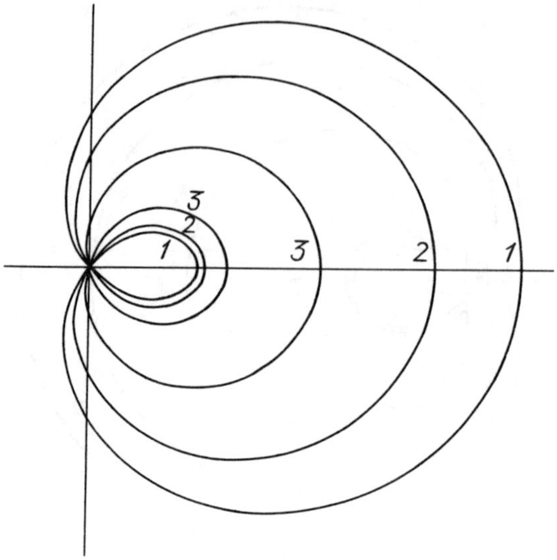

Figure 41. Cartesian oval

N	1	2	3
m	0.50	1.00	1.20
c	2.00	2.00	2.00
d	2.00	2.00	2.00
n	2.00	2.00	2.00

Basic geometric properties:

Boundedness: present.
Closedness: present.
Axial symmetry: straight lines $x = 0, y = 0$ are axes of symmetry.
Central symmetry: point $(0, 0)$ is a center of symmetry.
Connectedness: present (for $a \geq c$), none (for $a < c$).
Singular points: point $(0, 0)$ is a node, straight lines $x \pm y = 0$ are tangents at the node.
Self-intersection points: none (for $a \neq c$).
Characterized points: points

$$\left(\pm \frac{\sqrt{4c^4 - a^4}}{2c}, \pm \frac{a^2}{2c} \right)$$

are located at maximal distance from x-axis (for $a < c\sqrt{2}$), points

$$(0, \pm\sqrt{a^2 - c^2}), \quad (\pm\sqrt{a^2 + c^2}, 0)$$

are vertices.
Special cases: lemniscate of Bernoulli ($a = c$).
(Figures 43–44).

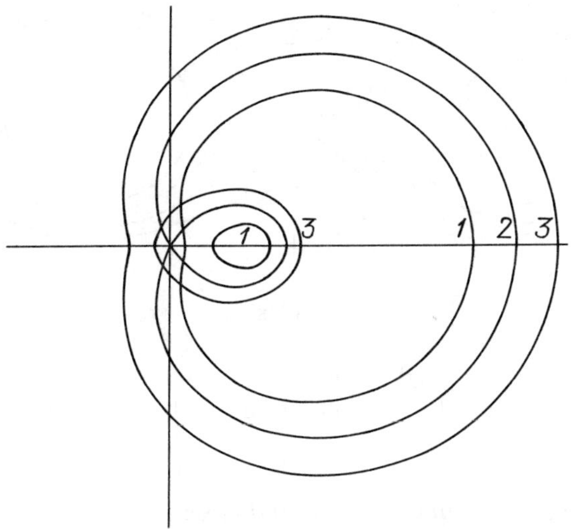

Figure 42. Cartesian oval

N	1	2	3
c	1.50	2.00	2.50
d	1.00	1.00	1.00
m	1.00	1.00	1.00
n	2.00	2.00	2.00

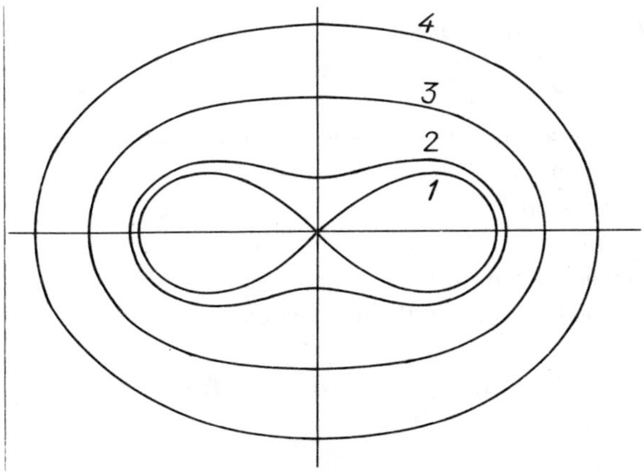

Figure 43. Cassinian oval

N	1	2	3	4
a	1.00	1.10	1.50	2.00
c	1.00	1.00	1.00	1.00

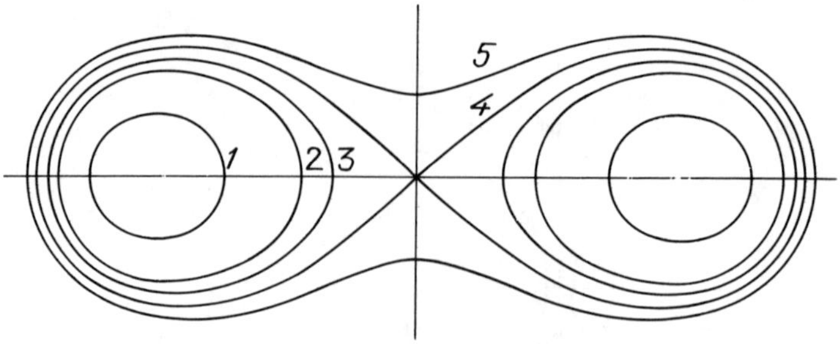

Figure 44. Cassinian oval

N	1	2	3	4	5
a	0.70	0.90	0.95	1.00	1.05
c	1.00	1.00	1.00	1.00	1.00

CASSINIAN OVAL *is a bounded connected closed (for $a > c$) or unconnected (consisting of two ovals (for $a < c$)) convex (for $a \geq c\sqrt{2}$) central-symmetric curve free of singular points with two axes of symmetry.*

Catenary curve (catenary, chainette, funicular curve, funicular line)
Equation:

$$y = a \cosh \frac{x}{a}, \quad a > 0$$

Basic geometric properties:
Unboundedness: present.
Asymptotes: none.
Axial symmetry: straight line $x = 0$ is an axis of symmetry.
Central symmetry: none.
Connectedness: present.
Singular points: none.
Self-intersection points: none.
Characterized points: point $(a, 0)$ is a vertex.
(Figure 45).
CATENARY *is an unbounded connected curve free of singular points with a single axis of symmetry .*

Catenary curve of equal resistance
Equation:

$$e^{y/a} \left| \cos \frac{x}{a} \right| = 1, \quad a > 0$$

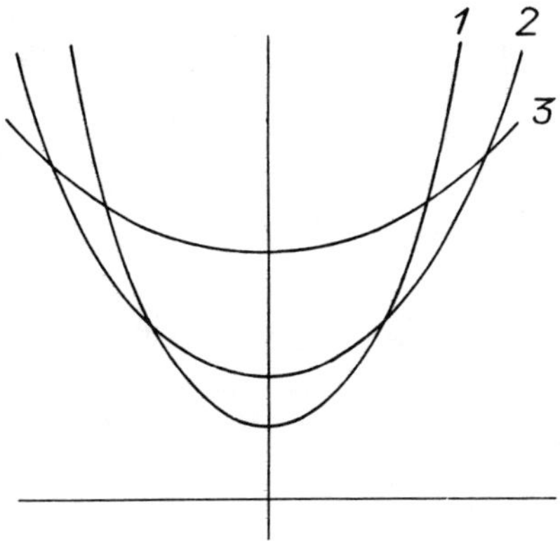

Figure 45. Catenary curve

N	1	2	3
a	0.60	1.00	2.00

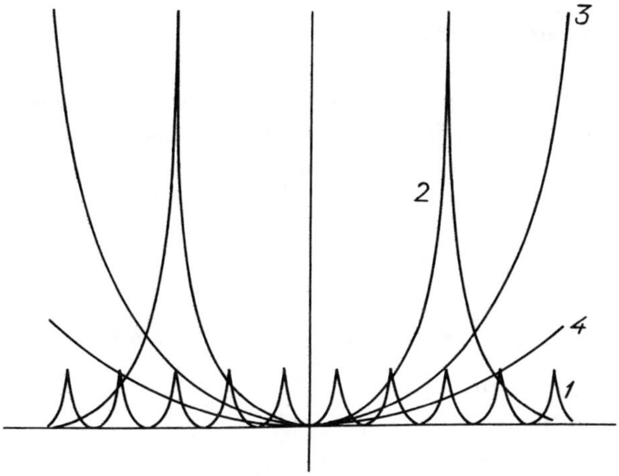

Figure 46.
Catenary curve of equal resistance

N	1	2	3	4
a	0.10	0.50	1.00	2.00

Basic geometric properties:
Unboundedness: present.
Asymptotes: straight lines

$$x = (2k+1)\frac{\pi a}{2}, \quad k \in \mathbf{Z}$$

are vertical asymptotes.

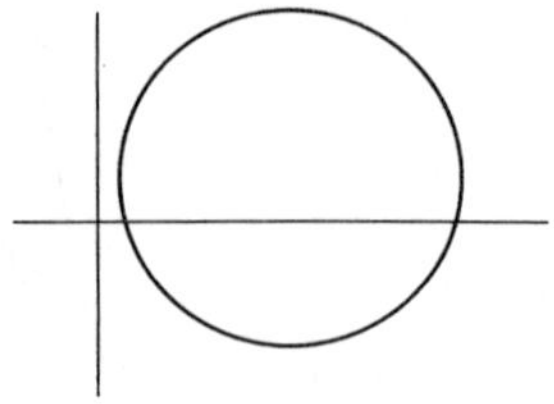

Figure 47. Circle

Axial symmetry: straight lines

$$x = l\frac{\pi a}{2}, \quad l \in \mathbf{Z}$$

are axes of symmetry.
Central symmetry: none.
Connectedness: none.
Singular points: none.
Self-intersection points: none.
Characterized points: points $(m\pi a, 0)$, $m \in \mathbf{Z}$, are vertices.
(Figure 46).
CATENARY CURVE OF EQUAL RESISTANCE *is an unbounded unconnected (consisting of an infinitely large number of unbounded branches) curve free of singular points with an infinitely large number of the symmetry axes.*

Circle
Equation:

$$(x-a)^2 + (y-b)^2 = R^2, \quad R > 0$$

Basic geometric properties:
Boundedness: present.
Closedness: present.
Axial symmetry: every straight line described by equation $A(x-a)+B(y-b)=0$ is an axis of symmetry.
Central symmetry: point (a, b) is a center of symmetry.
Connectedness: present.
Singular points: none.
Self-intersection points: none.
(Figure 47).
CIRCLE *is a bounded connected closed central-symmetric curve free of singular points with an infinitely large number of symmetry axes.*

Circular tractrix
Equations:

$$\rho = 2a\cos t, \quad \varphi = t - \tan t,$$

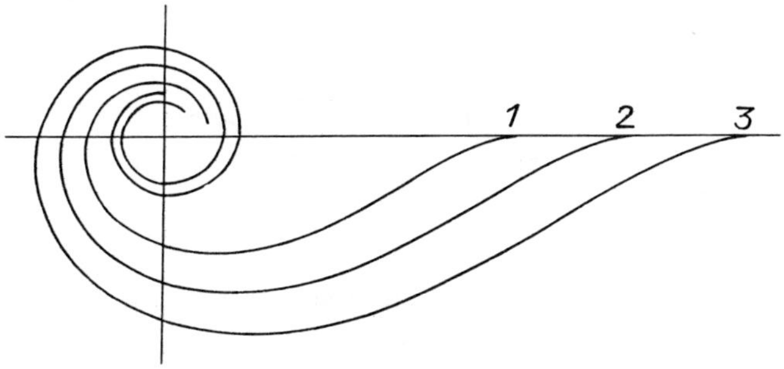

Figure 48. Circular tractrix

N	1	2	3
a	0.75	1.00	1.25

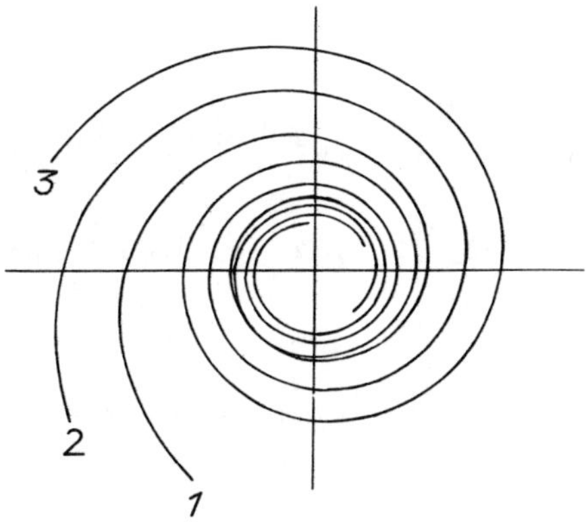

Figure 49. Circular tractrix

N	1	2	3
a	0.75	1.00	1.25

$$\varphi = \arctan\frac{\sqrt{4a^2-\rho^2}}{\rho} - \frac{\sqrt{4a^2-\rho^2}}{\rho}, \quad a > 0$$

Additional equation:

$$\varphi = -\arcsin\frac{\rho}{2a} + \frac{\sqrt{4a^2-\rho^2}}{\rho}$$

Basic geometric properties:

Boundedness: present.
Closedness: none.
Axial symmetry: straight line $y = 0$ is axis of symmetry.
Central symmetry: none.
Connectedness: present.

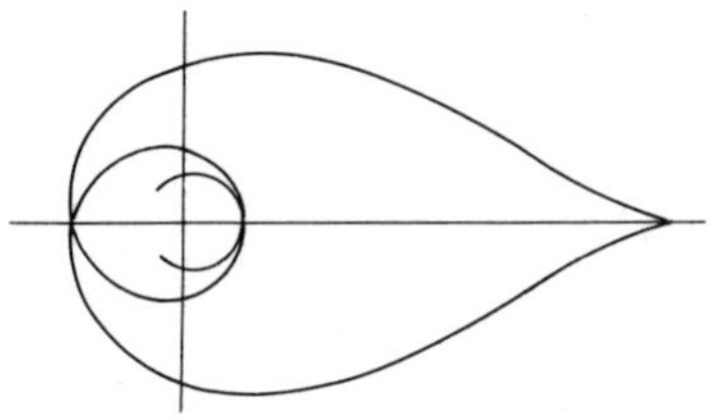

Figure 50. Circular tractrix

Singular points: point $(2a, 0)$ is a cuspidal point of the first kind, straight line $y = 0$ is a tangent at the point; point $(0, 0)$ is an asymptotic point.
Self-intersection points: none.
Generalizations:

$$\rho = \sqrt{a^2 + l^2 + 2al\frac{1-t^2}{1+t^2}},$$

$$\varphi = \begin{cases} \arctan \dfrac{2lt}{(a+l)+(a-l)t^2} + \frac{2l}{n} \arctan \sqrt{\dfrac{l-a}{l+a}}t, & a < l \\ \arctan \dfrac{2lt}{(a+l)+(a-l)t^2} - \frac{l}{n} \ln \dfrac{a+l+nt}{a+l-nt}, & a > l \end{cases}$$

(Figures 48–50).

CIRCULAR TRACTRIX *is a bounded nonclosed connected curve with two singular points (cuspidal point of the first kind, asymptotic point) and a single axis of symmetry.*

Cissoid (cissoid of Diocles, Diokles' cissoid)
Equations:

$$x(x^2 + y^2) = 2ay^2,$$

$$\rho = 2a\frac{\sin^2 \varphi}{\cos \varphi},$$

$$x = \frac{2a}{1+t^2}, \quad y = \frac{2a}{t(1+t^2)},$$

$$a > 0$$

Additional equation:

$$y^2 = \frac{x^3}{2a - x}$$

Basic geometric properties:
Unboundedness: present.
Asymptotes: straight line $x = 2a$ is a vertical asymptote.

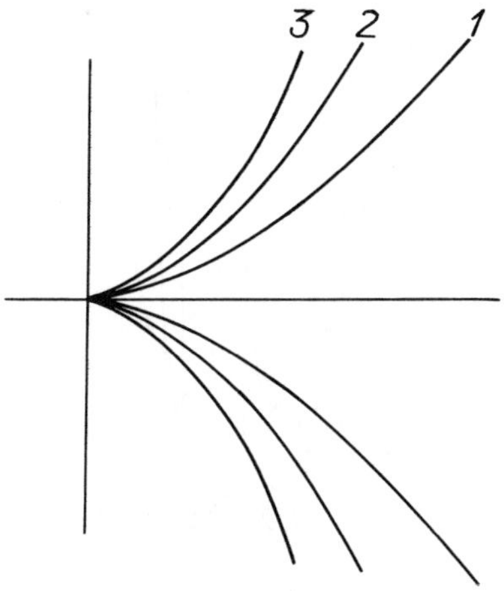

Figure 51. Cissoid

N	1	2	3
a	1.00	0.50	0.(3)

Axial symmetry: straight line $y = 0$ is an axis of symmetry.
Central symmetry: none.
Connectedness: present.
Singular points: point $(0, 0)$ is a cuspidal point of the first kind, straight line $y = 0$ is a tangent at the cusp point.
Self-intersection points: none.
Area of domain enclosed between the curve and the asymptote: $3\pi a^2$.
(Figure 51).
CISSOID *is an unbounded connected curve with a single singular point (cuspidal point of the first kind), a single vertical asymptote and a single axis of symmetry.*

Cissoid of the fourth order
Equations:

$$y^2(k^2 - (a + x)^2) = (x^2 + k^2 - a^2)^2,$$

$$y = \pm \frac{x^2 + k^2 - a^2}{\sqrt{k^2 - (a + x)^2}},$$

$$a > 0, \quad k > 0, \quad a \neq k$$

Basic geometric properties:
Unboundedness: present.
Asymptotes: straight lines $x + a \pm k = 0$ are vertical asymptotes.
Axial symmetry: straight line $y = 0$ is an axis of symmetry.
Central symmetry: none.
Connectedness: none.

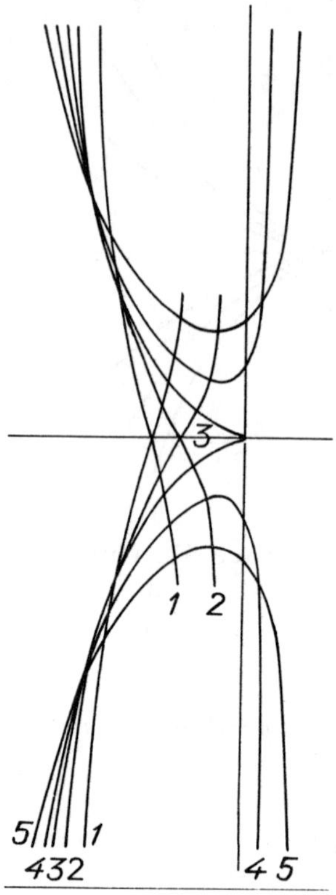

Figure 52. Cissoid of the fourth order

N	1	2	3	4	5
k	0.50	0.80	1.00	1.50	2.00
a	1.00	1.00	1.00	1.00	1.00

Singular points: point $(-\sqrt{a^2-k^2}, 0)$ is a node (for $a > k$), straight lines

$$y = \pm(x + \sqrt{a^2-k^2})\sqrt{\frac{2(a^2-k^2)}{a-\sqrt{a^2-k^2}}}$$

are tangents at the node; point $(\sqrt{a^2-k^2}, 0)$ is an isolated point (for $a > k$).
Self-intersection points: one, $(-\sqrt{a^2-k^2}, 0)$ (for $a > k$), none (for $a < k$).
Characterized points: $(0, \pm\sqrt{k^2-a^2})$ are points of intersection between the curve and y-axis (for $k > a$ only).
(Figure 52).
CISSOID OF THE FOURTH ORDER *is an unbounded unconnected (consisting of two branches) curve with a single axis of symmetry possessing either two singular points (node, isolated point) (for $a > k$) or free of singularities (for $a < k$).*

Cissoids of second-order curves
Equations:

$$(x-k)(y^2-qx^2)+2px^2=0,$$

$$\rho=\frac{k}{\cos\varphi}-\frac{2p\cos\varphi}{\sin^2\varphi-q\cos^2\varphi},$$

$$x=k+\frac{2p}{q-t^2},\quad y=kt+\frac{2pt}{q-t^2},$$

$$p>0,\quad k\neq 0$$

generating curves:

$$y^2=2px+qx^2$$

(ellipse ($q<0$), parabola (for $q=0$), hyperbola (for $q>0$)) and straight line $x=k$.

Basic geometric properties:

$q<0$
Unboundedness: present.
Asymptotes: straight line $x=k$ is a vertical asymptote.
Axial symmetry: straight line $y=0$ is a symmetry axis.
Central symmetry: none.
Connectedness: present (for $0<k\le -2p/q$).
Singular points: point $(0,0)$ is a node (for $q+2p/k>0$), straight lines

$$y=\pm x\sqrt{q+\frac{2p}{k}}$$

are tangents at the node, or a cuspidal point of the first kind (for $qk+2p=0$), straight line $y=0$ is a tangent at the cusp point, or an isolated point (for $k<0$ or $q+2p/k<0$).
Self-intersection points: one, $(0,0)$ (for $q+2p/k>0$).

$q=0$
Unboundedness: present.
Asymptotes: straight line $x=k$ is a vertical asymptote.
Axial symmetry: straight line $y=0$ is a symmetry axis.
Central symmetry: none.
Connectedness: present (for $k>0$), none (for $k<0$).
Singular points: point $(0,0)$ is a node (for $k>0$), straight lines

$$y=\pm x\sqrt{\frac{2p}{k}}$$

are tangents at the node, or an isolated point (for $k<0$).
Self-intersection points: one, $(0,0)$ (for $k>0$).

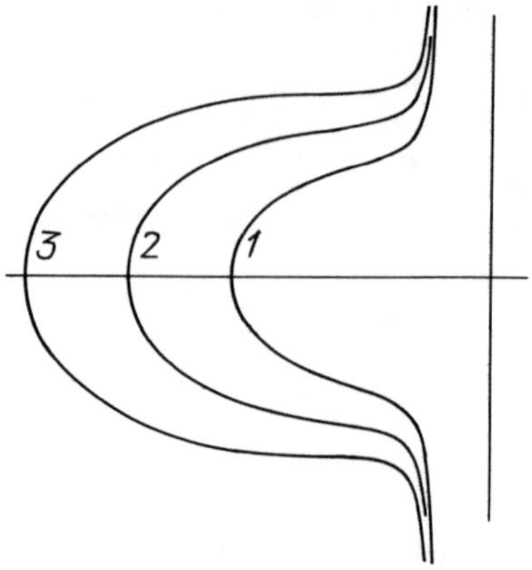

Figure 53.
Cissoids of second-order curves

N	1	2	3
p	1.00	1.50	2.00
k	−1.00	−1.00	−1.00
q	0.50	0.50	0.50

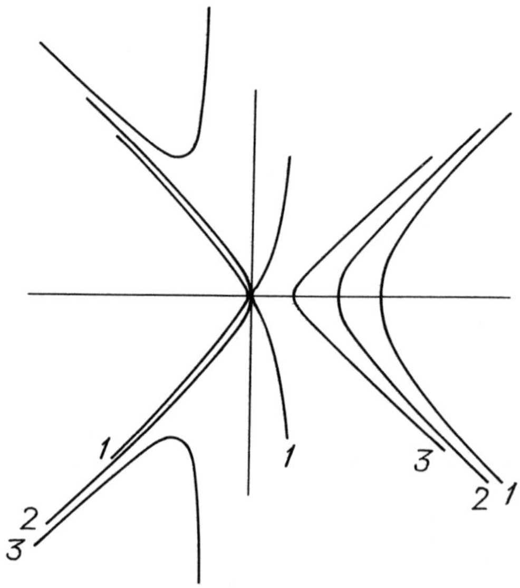

Figure 54.
Cissoids of second-order curves

N	1	2	3
p	1.00	1.00	1.00
k	−1.00	0.00	1.00
q	1.00	1.00	1.00

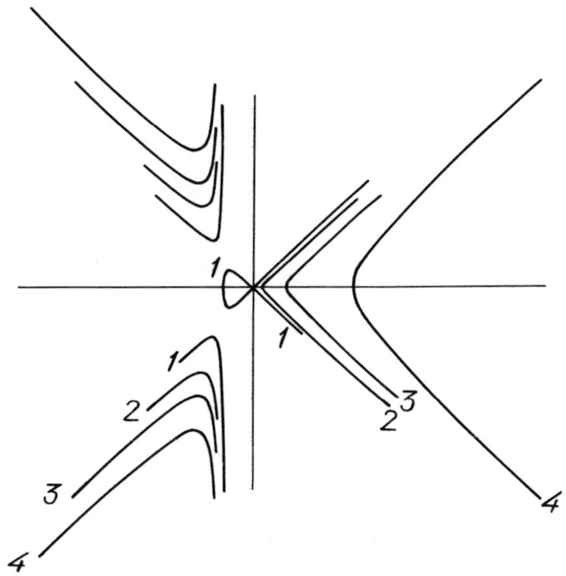

Figure 55.

Cissoids of second-order curves

N	1	2	3	4
p	0.05	0.50	1.00	2.00
k	−1.00	−1.00	−1.00	−1.00
q	1.00	1.00	1.00	1.00

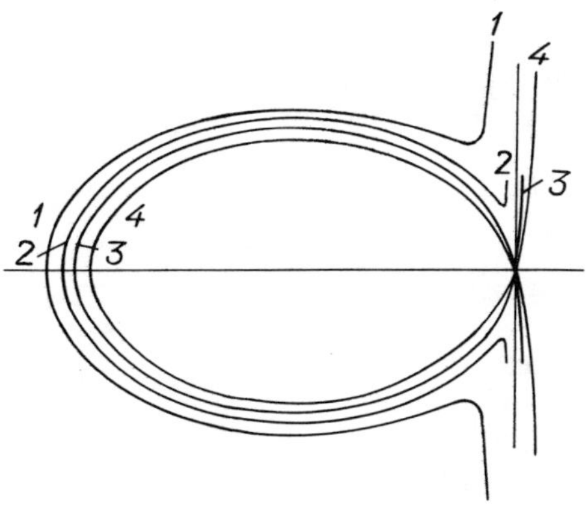

Figure 56.

Cissoids of second-order curves

N	1	2	3	4
p	1.00	1.00	1.00	1.00
k	−0.20	−0.05	0.05	0.20
q	−0.50	−0.50	−0.50	−0.50

$q > 0$
Unboundedness: present.
Asymptotes: straight line $x = k$ is a vertical asymptote, straight lines

$$y = \pm\sqrt{q}\left(x + \frac{p}{\sqrt{q}}\right)$$

are inclined asymptotes.
Axial symmetry: straight line $y = 0$ is a symmetry axis.
Central symmetry: none.
Connectedness: none.
Singular points:

$k < 0$
point $(0, 0)$ is a node (for $q + 2p/k > 0$), straight lines

$$y = \pm x\sqrt{q + \frac{2p}{k}}$$

are tangents at the node, or a cuspidal point of the first kind (for $qk + 2p = 0$), straight line $y = 0$ is a tangent at the cusp point, or an isolated point (for $q + 2p/k < 0$).

$k > 0$
point $(0, 0)$ is a node, straight lines

$$y = \pm x\sqrt{q + \frac{2p}{k}}$$

are tangents at the node.
Self-intersection points: one, $(0, 0)$ (for $q + 2p/k > 0$).
Special cases:
Cartesian folium

$$\left(p = -\frac{\sqrt{2}}{3}a, \quad q = -\frac{1}{3}, \quad k = \frac{\sqrt{2}}{2}a\right)$$

trisectrix of Longchamps

$$\left(p = \frac{2}{9}a, \quad q = \frac{1}{3}, \quad k = -\frac{1}{3}a\right)$$

mixed cubic

$$\left(p = -a, \quad q = 0, \quad k = \frac{1}{2}a\right)$$

(Figures 53–58).

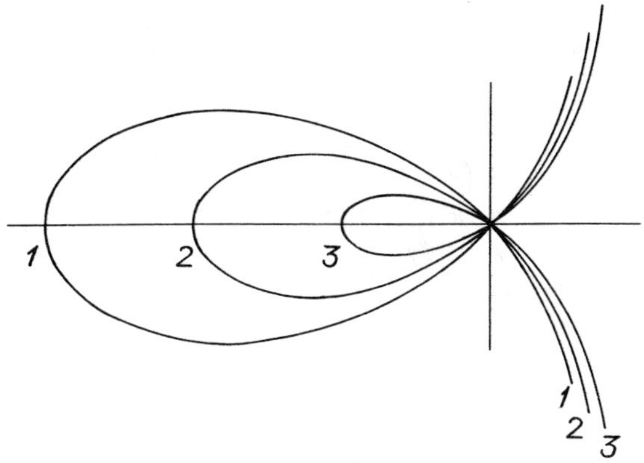

Figure 57.
Cissoids of second-order curves

N	1	2	3
p	1.00	0.75	0.50
k	1.00	1.00	1.00
q	−0.50	−0.50	−0.50

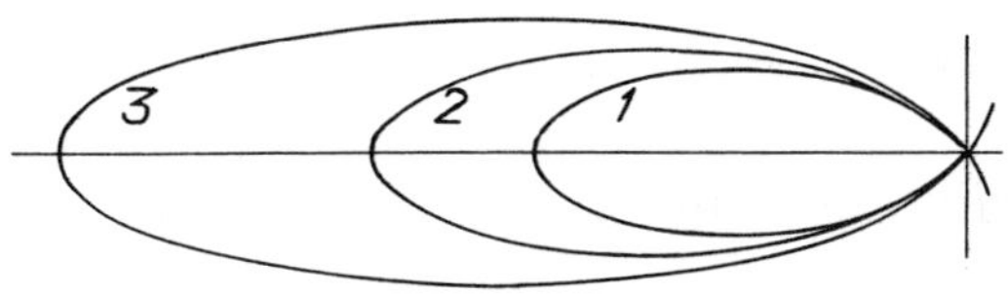

Figure 58.
Cissoids of second-order curves

N	1	2	3
p	1.00	1.00	1.00
k	1.00	1.00	1.00
q	−0.20	−0.15	−0.10

CISSOID OF ELLIPSE $(q < 0)$ *is an unbounded connected or unconnected (consisting of an unbounded branch and an isolated point (for* $0 < k \leq -2p/q$*)) curve with a single singular point (type of the singular point depends on relationship between parameters* p, q *and* k*), a single vertical asymptote and a single symmetry axis.*

CISSOID OF PARABOLA $(q = 0)$ *is an unbounded connected (for* $k > 0$*) or unconnected (consisting of two unbounded branches and an isolated point (for* $k < 0$*)) curve with a single singular point (type of the singular point depends on sign of parameter* k*), a single vertical asymptote and a single symmetry axis.*

CISSOID OF HYPERBOLA $(q > 0)$ *is an unbounded unconnected (consisting of two or three unbounded branches) curve with a single singular point (type of the singular point depends on relationship between parameters*

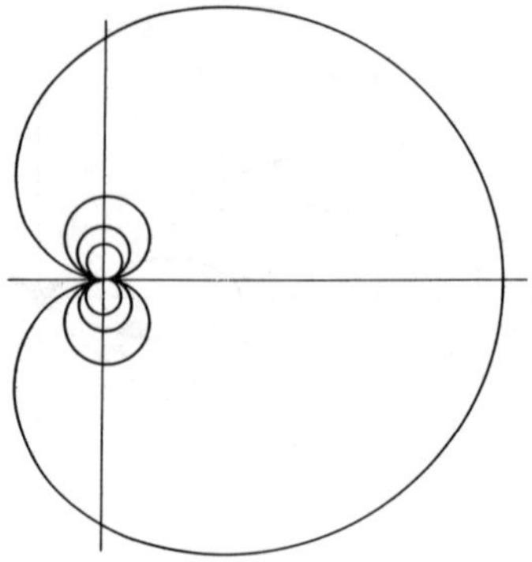

Figure 59. Cochleoid

p, q and k), a single symmetry axis, a single vertical asymptote and two inclined asymptotes.

Cochleoid
Equations:

$$(x^2 + y^2) \arctan \frac{y}{x} = ay, \quad \rho = a\frac{\sin\varphi}{\varphi}, \quad a > 0$$

Basic geometric properties:
Boundedness: present.
Closedness: present.
Axial symmetry: straight line $y = 0$ is an axis of symmetry.
Central symmetry: none.
Connectedness: present.
Singular points: point $(0,0)$ is a tacnode, straight line $y = 0$ is a tangent at the osculation point.
Self-intersection points: one, $(0,0)$.
(Figure 59).
COCHLEOID *is a bounded connected closed curve with a single singular point (tacnode) and a single axis of symmerty.*

Conchal
Equations:

$$(x + a)^2((x - a)^2 + y^2) = a^2k^2,$$

$$y = \pm\sqrt{\frac{a^2k^2}{(x + a)^2} - (x - a)^2}, \quad a > 0, \quad k > 0$$

Basic geometric properties:
Unboundedness: present.

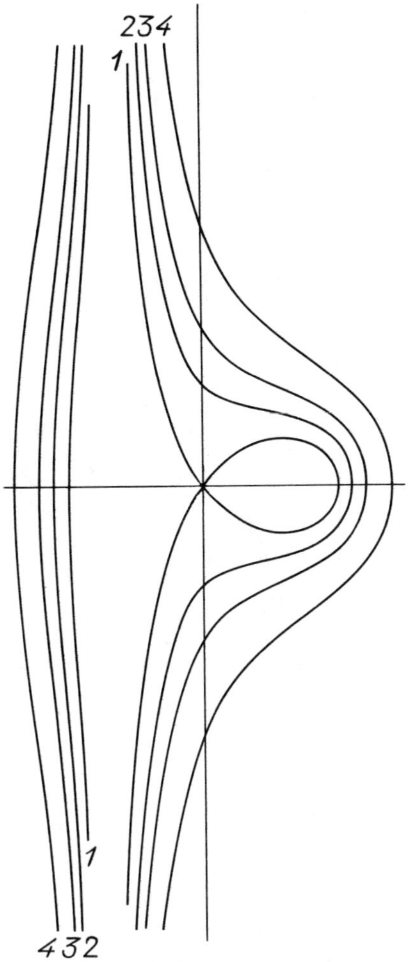

Figure 60. Conchal

N	1	2	3	4
k	1.00	1.50	2.00	3.00
a	1.00	1.00	1.00	1.00

Asymptotes: straight line $x + a = 0$ is a vertical asymptote.
Axial symmetry: straight line $y = 0$ is an axis of symmetry.
Central symmetry: none.
Connectedness: none.
Singular points: point $(0, 0)$ is a node (for $k = a$ only), straight lines $y = \pm x\sqrt{2}$ are tangents at the node.
Self-intersection points: none (for $k \neq a$).
Characterized points: points

$$(\pm\sqrt{a(a \pm k)}, 0)$$

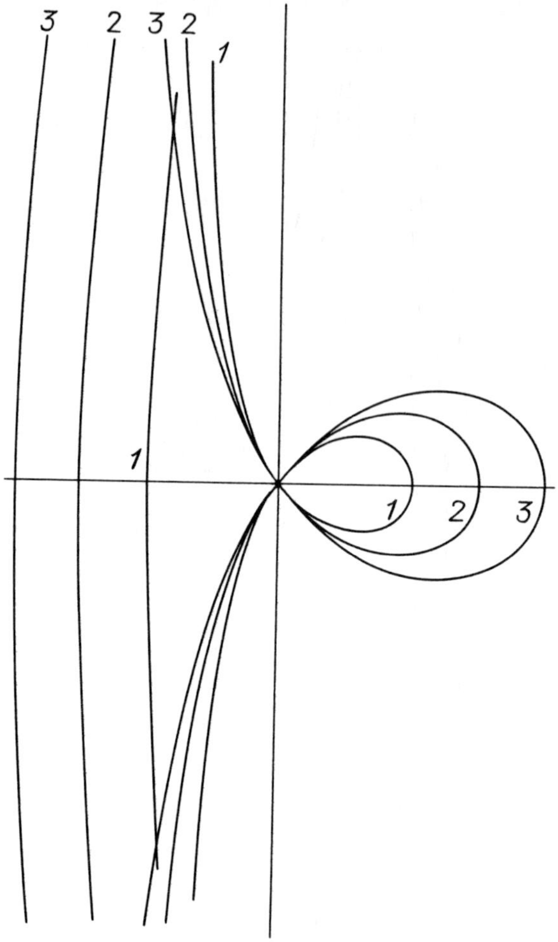

Figure 61. Conchal

N	1	2	3
a	1.00	1.50	2.00
k	1.00	1.50	2.00

are the intersection points of the curve and x-axis ($k < a$); points

$$(\pm a\sqrt{2}, 0)$$

are the intersection points of the curve and x-axis ($k = a$); points

$$(\pm\sqrt{a(a+k)}, 0)$$

are intersection points of the curve and x-axis ($k > a$).
(Figures 60–62).

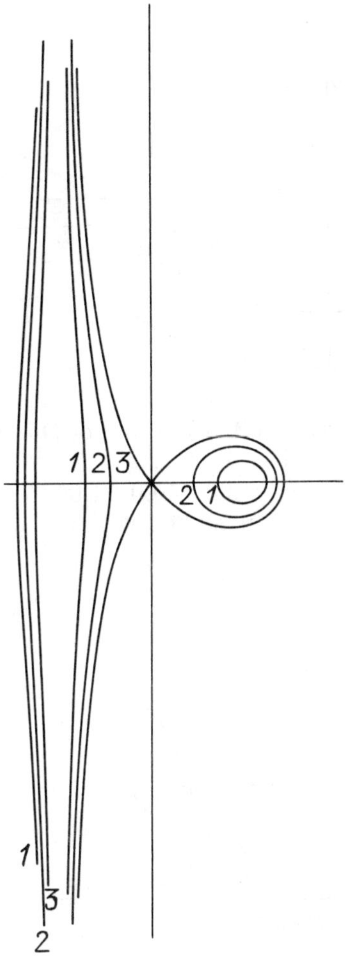

Figure 62. Conchal

N	1	2	3
a	1.00	1.00	1.00
k	0.50	0.80	1.00

CONCHAL *is an unbounded unconnected (consisting of two unbounded branches and an oval ($k < a$) or of two unbounded branches only ($k \geq a$)) curve with a single singular point (node, for $k = a$ only) and a single axis of symmetry.*

Conchoid of Dürer

Equations:

$$(xy + b^2 - y^2)^2 = (x + y - a)^2(b^2 - y^2),$$

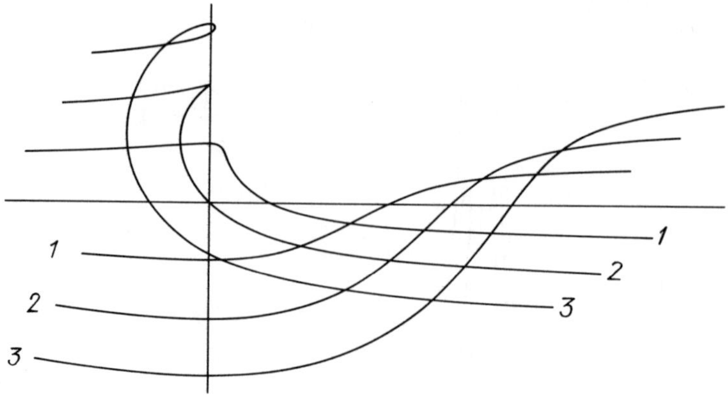

Figure 63. Conchoid of Dürer

N	1	2	3
a	1.00	1.00	1.00
b	0.50	1.00	1.50

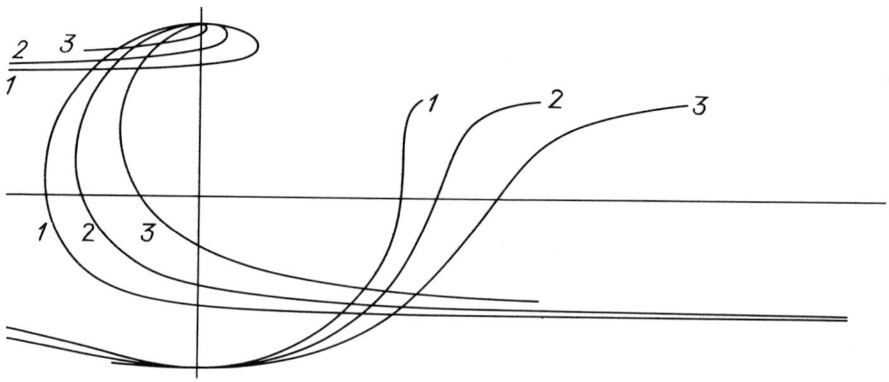

Figure 64. Conchoid of Dürer

N	1	2	3
a	0.20	0.50	1.00
b	1.50	1.50	1.50

$$x = \frac{a \cos t}{\cos t - \sin t} + b \cos t, \quad y = b \sin t, \quad a > 0, \quad b > 0$$

Basic geometric properties:
Unboundedness: present.
Asymptotes: straight lines $\sqrt{2y \pm b = 0}$ are horizontal asymptotes.
Axial symmetry: none.
Central symmetry: none.
Connectedness: present (for $b \geq a$), none (for $b < a$).

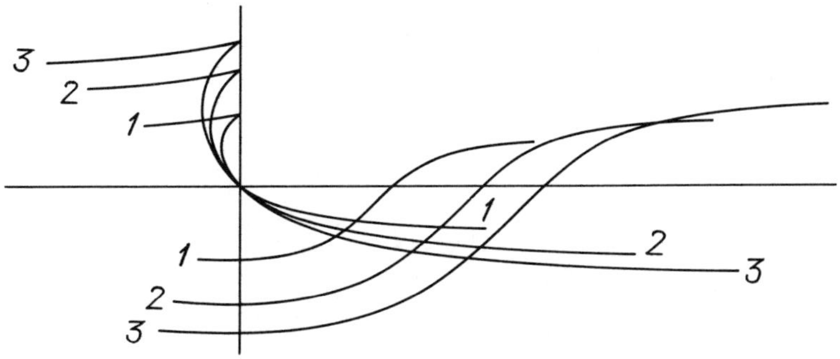

Figure 65. Conchoid of Dürer

N	1	2	3
a	0.50	0.80	1.00
b	0.50	0.80	1.00

Singular points: point $(x(t_1), y(t_1))$ where

$$\sin t_1 = \frac{a - \sqrt{a^2 + 8b^2}}{4b}, \quad \cos t_1 = \sqrt{1 - \sin^2 t_1}$$

is a node; point $(x(t_2), y(t_2))$ where

$$\sin t_2 = \frac{a + \sqrt{a^2 + 8b^2}}{4b}, \quad \cos t_2 = -\sqrt{1 - \sin^2 t_2}$$

is a node (for $(b > a)$; point $(0, a)$ is a cuspidal point of the first kind (for $a = b$), straight line $x - 2y - a = 0$ is a tangent at the cusp point; for $a > b$ the curve has an isolated point.
Self-intersection points: one (for $a \geq b$), two (for $a < b$).
Characterized points: point $(a \pm b, 0)$ is an intersection point of the curve and x-axis; points

$$\left(0, \frac{a \pm \sqrt{2b^2 - a^2}}{2}\right)$$

are the intersection points of the curve and y-axis.
(Figures 63–65).
CONCHOID OF DÜRER *is an unbounded connected (for $b \geq a$) or unconnected (for $b < a$) curve with two singular points (one of them is a node, type of the other depends on relationship between the parameter values) and two horizontal asymptotes.*

Conchoid of higher-order circle

Equation:

$$\rho = a\frac{\sin \frac{m+1}{m}\varphi}{\sin \frac{\varphi}{m}}, \quad a > 0, \quad m \in \mathbf{N}, \quad m \geq 2$$

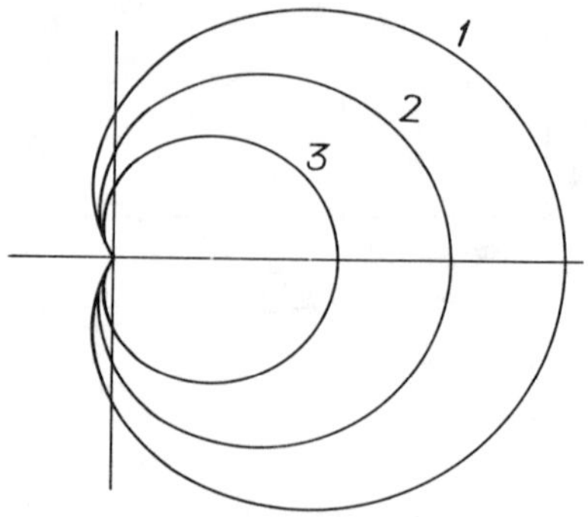

Figure 66.
Conchoid of higher-order circle

N	1	2	3
a	1.00	0.75	0.50
n	2	2	2

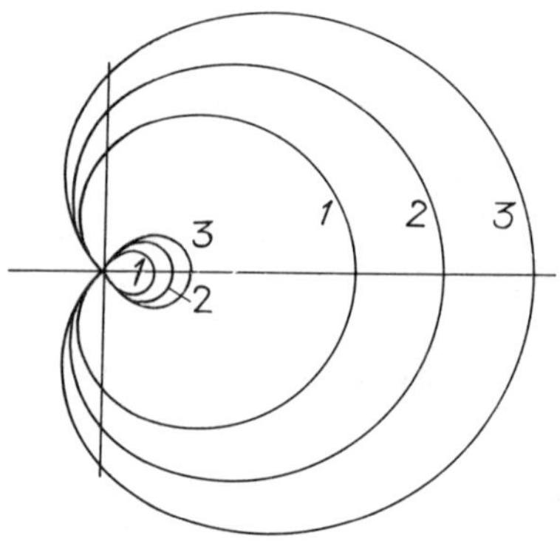

Figure 67.
Conchoid of higher-order circle

N	1	2	3
a	0.75	1.00	1.25
n	4	4	4

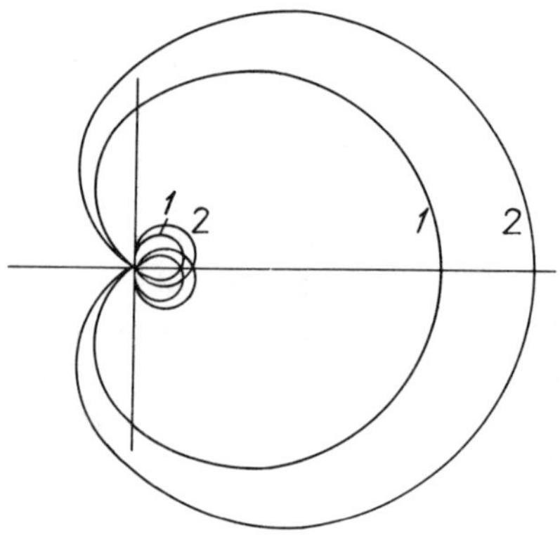

Figure 68.
Conchoid of higher-order circle

N	1	2
a	1.00	1.30
n	6	6

Basic geometric properties:
Boundedness: present.
Closedness: present.
Axial symmetry: straight line $y = 0$ is an axis of symmetry.
Central symmetry: none.
Connectedness: present.
Singular points: point $(0, 0)$ is an angular point ($m = 2$), straight lines $\sqrt{3}x \pm y = 0$ are tangents at the angular point, or a multiple point ($m \geq 3$); point $(\rho(2\pi), 0)$ is a multiple point ($m \neq 2, 4$).

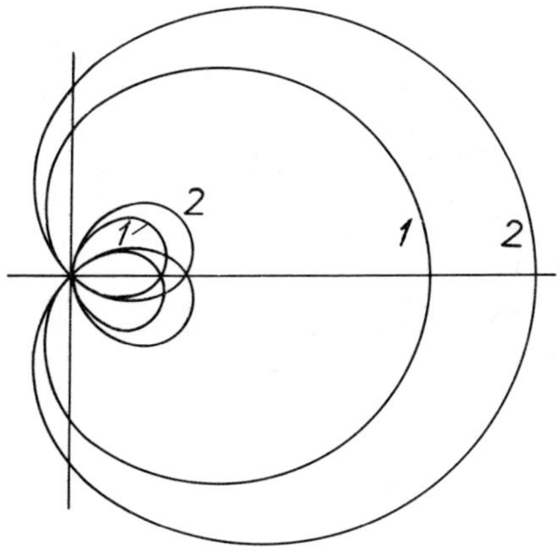

Figure 69.
Conchoid of higher-order circle

N	1	2
a	1.00	1.30
n	3	3

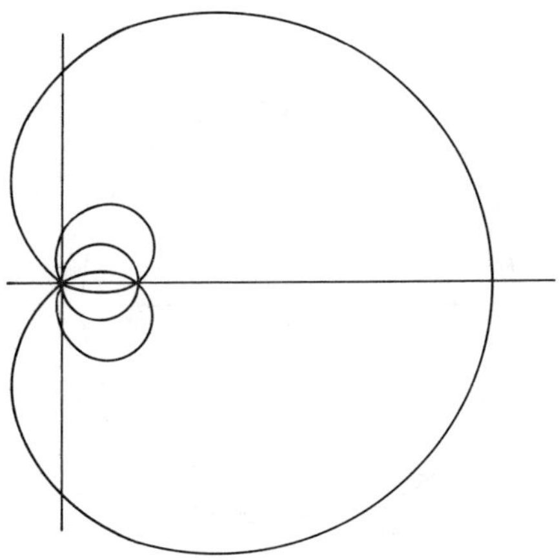

Figure 70.
Conchoid of higher-order circle
$a = 1, \quad n = 5$

Self-intersection points: two, $(0,0)$, $(\rho(2\pi), 0)$ $(m \neq 2, 4)$; one, $(0,0)$, $(m = 4)$; none $(m = 2)$.
(Figures 66–70).
CONCHOID OF HIGHER-ORDER CIRCLE *is a bounded connected closed curve with a single axis of symmetry; number and type of its singular points depends on values of parameter* m.

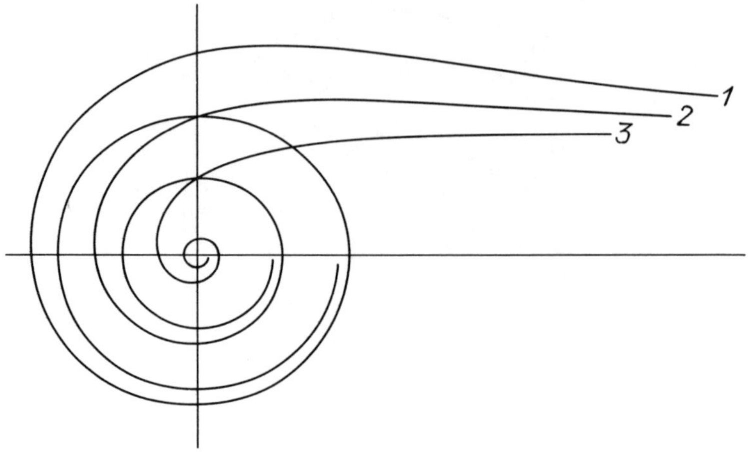

Figure 71.
Conchoid of hyperbolic spiral

N	1	2	3
a	1.00	1.00	1.00
l	1.00	0.50	0.00

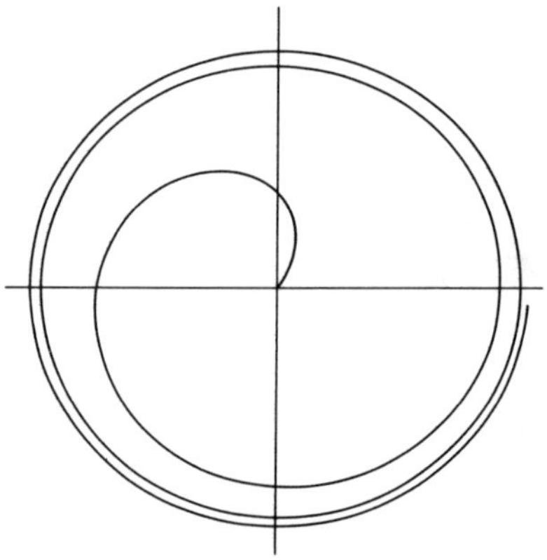

Figure 72.
Conchoid of hyperbolic spiral
$a = -1.00, \quad l = 1.00$

Conchoid of hyperbolic spiral
Equation:

$$\rho = \frac{a}{\varphi} + l, \quad a \neq 0, \quad l > 0$$

Basic geometric properties:
Unboundedness: present (for $a > 0$).
Asymptotes: straight line $y = a$ is a horizontal asymptote (for $a > 0$).

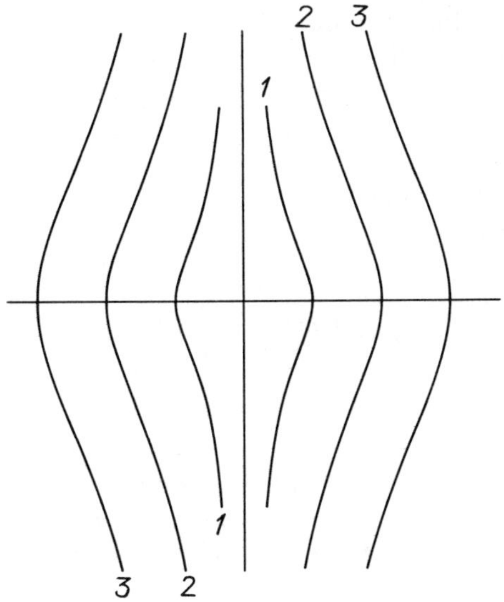

Figure 73. Conchoid of Külp

N	1	2	3
a	1.00	2.00	3.00

Boundedness: present (for $a < 0$).
Closedness: none.
Axial symmetry: none.
Central symmetry: none.
Connectedness: present.
Singular points: point $(0, 0)$ is a cessation point.
Self-intersection points: none.
Special cases: hyperbolic spiral ($l = 0$).
(Figure 71–72).
CONCHOID OF HYPERBOLIC SPIRAL ($a > 0$) *is an unbounded connected curve free of singular points with a single horizontal asymptote.*
CONCHOID OF HYPERBOLIC SPIRAL ($a < 0$) *is a bounded connected non-closed curve with a single singular point (point of cessation).*

Conchoid of Külp
Equations:

$$x^2y^2 = a^2(a^2 - x^2),$$

$$x = a\cos t, \quad y = \tan t,$$

$$a > 0$$

Basic geometric properties:
Unboundedness: present.

Asymptotes: straight line $x = 0$ is a horizontal asymptote.
Axial symmetry: straight lines $x = 0, y = 0$ are axes of symmetry.
Central symmetry: point $(0,0)$ is a center of symmetry.
Connectedness: none.
Singular points: none.
Self-intersection points: none.
Characterized points: points

$$\left(\pm a\sqrt{\frac{2}{3}}, \pm\frac{a}{\sqrt{2}}\right)$$

are inflection points.
(Figure 73).
CONCHOID OF KÜLP *is an unbounded unconnected (consisting of two unbounded branches) central-symmetric curve free of singular points with a single vertical asymptote and two axes of symmetry.*

Conchoid of Nicomedes (cochloid, conchoid of line, Nicomedean conchoid)
Equations:

$$(x^2+y^2)(y-a)^2 - l^2y^2 = 0, \quad \rho = \frac{a}{\sin\varphi} \pm l,$$

$$x = a\cot t \pm l\cos t, \quad y = a \pm l\sin t, \quad a > 0, \quad l > 0$$

Additional equations:

$$(x^2+y^2)(x-a)^2 - l^2x^2 = 0, \quad \rho = \frac{a}{\cos\varphi} \pm l$$

Basic geometric properties:
Unboundedness: present.
Asymptotes: straight line $y = a$ is a horizontal asymptote.
Axial symmetry: straight line $x = 0$ is an axis of symmetry.
Central symmetry: none.
Connectedness: none.
Singular points: point $(0,0)$ is an isolated point (for $l < a$), or a cuspidal point of the first kind (for $l = a$), straight line $x = 0$ is a tangent at the cusp point or a node (for $l > a$), straight lines

$$y = \pm\frac{a}{\sqrt{l^2+a^2}}x$$

are tangents at the node.
Characterized points: points $(0, a \pm l)$ are vertices.
(Figures 74–75).

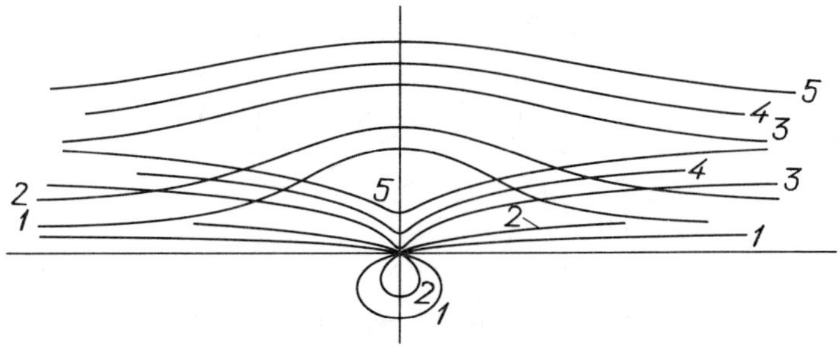

Figure 74. Conchoid of Nicomedes

N	1	2	3	4	5
a	0.50	1.00	2.00	2.50	5.00
l	2.00	2.00	2.00	2.00	2.00

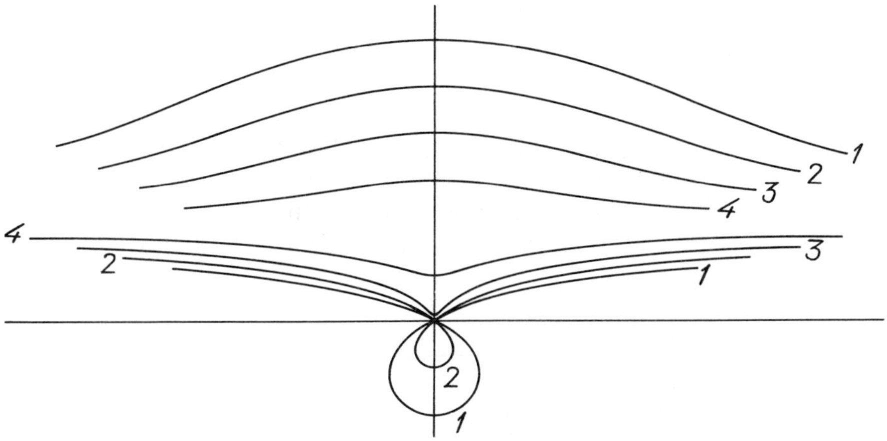

Figure 75. Conchoid of Nicomedes

N	1	2	3	4
a	1.00	1.00	1.00	1.00
l	2.00	1.50	1.00	0.50

CONCHOID OF NICOMEDES *is an unbounded unconnected (consisting of two unbounded branches or of two unbounded branches and an isolated point) curve with a single singular point (type of singularity depends on relationship between parameter values), a single axis of symmetry and a single horizontal asymptote.*

Conchoid of Slüse (+)
Equations:

$$a(x-a)(x^2+y^2)=k^2x^2, \quad a(\rho\cos\varphi-a)=k^2\cos^2\varphi,$$

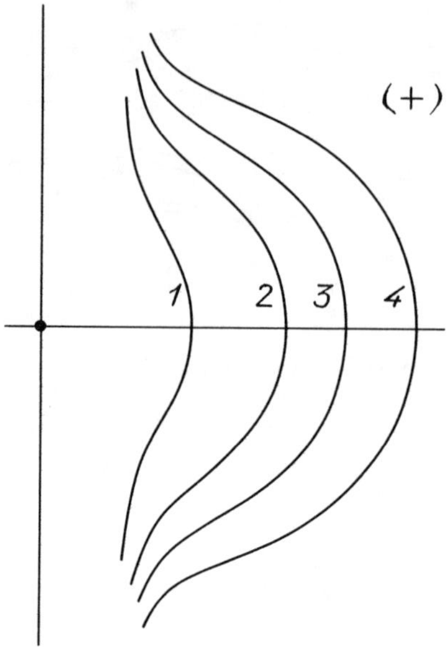

Figure 76. Conchoid of Slüse (+)

N	1	2	3	4
a	1.00	1.00	1.00	1.00
k	1.00	1.50	1.75	2.00

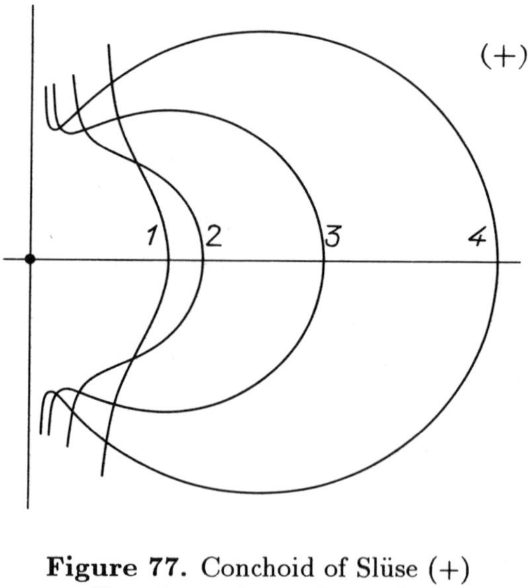

Figure 77. Conchoid of Slüse (+)

N	1	2	3	4
a	1.00	0.50	0.25	0.15
k	1.00	1.00	1.00	1.00

$$x = a + \frac{k^2}{a(1+t^2)}, \quad y = \left(a + \frac{k^2}{a(1+t^2)}\right) t,$$

$$a > 0, \quad k > 0$$

Basic geometric properties:
Unboundedness: present.
Asymptotes: straight line $x = a$ is a vertical asymptote.
Axial symmetry: straight line $y = 0$ is an axis of symmetry.
Central symmetry: none.
Connectedness: none.
Singular points: point $(0, 0)$ is an isolated point.
Self-intersection points: none.
Characterized points: point

$$\left(\frac{a^2 + k^2}{a}, 0\right)$$

is located at maximal distance from y-axis.
(Figures 76–77).
CONCHOID OF SLÜSE (+) *is an unbounded unconnected (consisting of an unbounded branch and an isolated point) curve with a single singular point (isolated point), a single vertical asymptote and a single axis of symmetry.*

Conchoid of Slüse (−)
Equations:

$$a(x-a)(x^2+y^2) = -k^2x^2,$$

$$a(\rho\cos\varphi - a) = -k^2\cos^2\varphi,$$

$$x = a - \frac{k^2}{a(1+t^2)}, \quad y = \left(a - \frac{k^2}{a(1+t^2)}\right) t,$$

$$a > 0, \quad k > 0$$

Basic geometric properties:
Unboundedness: present.

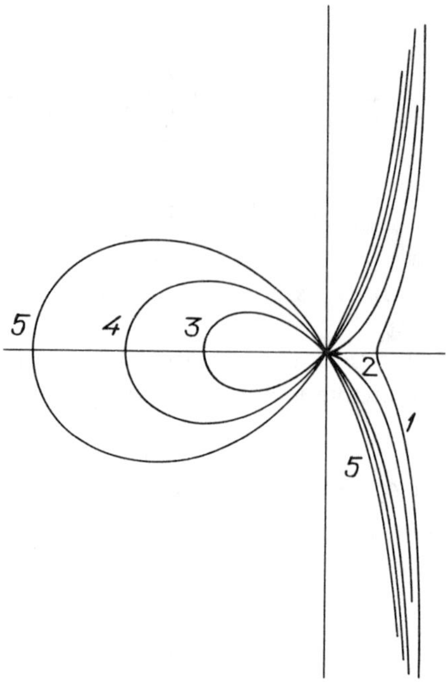

Figure 78. Conchoid of Slüse (−)

N	1	2	3	4	5
a	1.00	1.00	1.00	1.00	1.00
k	0.70	1.00	1.50	1.75	2.00

Asymptotes: straight line $x = a$ is a vertical asymptote.
Axial symmetry: straight line $y = 0$ is an axis of symmetry.
Central symmetry: none.
Connectedness: present (for $k \geq a$), none (for $k < a$).
Singular points: point $(0, 0)$ is an isolated point (for $k < a$), or a cuspidal point of the first kind (for $k = a$), straight line $y = 0$ is a tangent at the cusp point, or a node (for $k > a$), straight lines $kx \pm ay = 0$ are tangents at the node.
Self-intersection points: none (for $k \leq a$); one, $(0, 0)$ (for $k > a$).
Characterized points: point

$$\left(\frac{a^2 - k^2}{a}, 0\right)$$

is located at maximal distance from y-axis.
(Figures 78–79).
CONCHOID OF SLÜSE (−) *is an unbounded connected (for $k \geq a$) or unconnected (for $k < a$) curve with a single singular point (type of singularity*

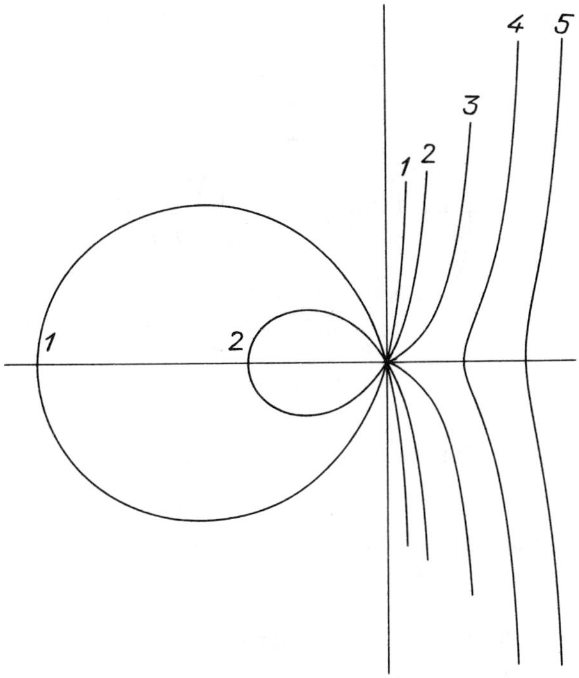

Figure 79. Conchoid of Slüse (–)

N	1	2	3	4	5
a	0.25	0.50	1.00	1.50	2.00
k	1.00	1.00	1.00	1.00	1.00

depends on relationship between parameter values), a single vertical asymptote and a single axis of symmetry.

Convolute of circle

Equations:

$$\varphi = \frac{1}{a}\sqrt{\rho^2 - \frac{a^2}{4}} + \arcsin\frac{a}{2\rho},$$

$$\varphi = \frac{1}{a}\sqrt{\rho^2 - \frac{a^2}{4}} + \arctan\frac{a}{2\sqrt{\rho^2 - \frac{a^2}{4}}}, \quad a > 0$$

Basic geometric properties:

Unboundedness: present.

Asymptotes: none.

Axial symmetry: none.

Central symmetry: none.

Connectedness: present.

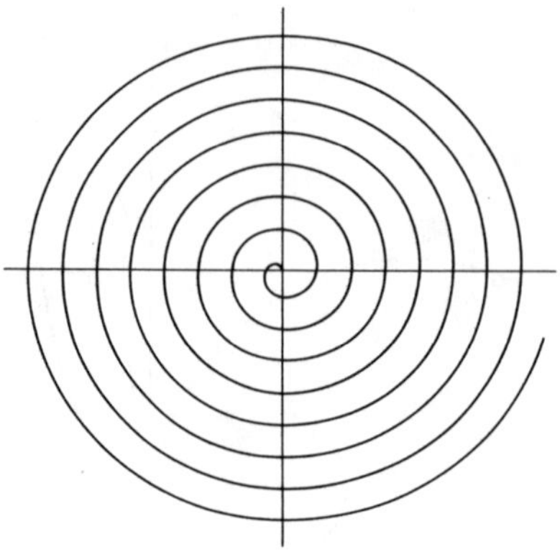

Figure 80. Convolute of circle

Singular points: point $(0, a/2)$ is a cessation point.
Self-intersection points: none.
(Figures 80–83).
CONVOLUTE OF CIRCLE *is an unbounded connected curve with a single singular point (point of cessation).*

Cornoid
Equations:

$$x = r\cos t(1 - 2\sin^2 t), \quad y = r\sin t(1 + 2\cos^2 t), \quad r > 0$$

Basic geometric properties:
Boundedness: present.
Closedness: present.
Axial symmetry: straight lines $x = 0, y = 0$ are axes of symmetry.
Central symmetry: point $(0, 0)$ is a center of symmetry.
Connectedness: present.
Singular points: points $(0, \pm r\sqrt{2})$ are points of self-osculation, straight line $y = \pm r\sqrt{2}$ are tangents at the points.
Self-intersection points: two, $(0, \pm r\sqrt{2})$.
(Figure 84).
CORNOID *is a bounded connected closed central-symmetric curve with two singular points (points of self-osculation) and two axes of symmetry.*

Cosecantoid
Equations:

$$y = \csc(ax) = \frac{1}{\sin(ax)}, \quad a > 0$$

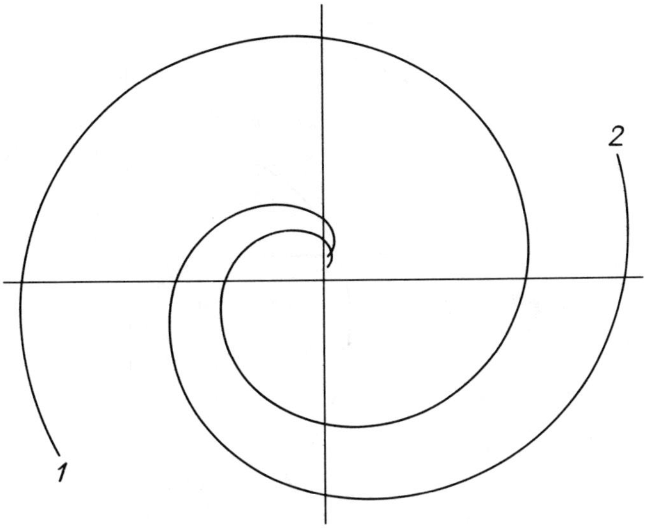

Figure 81. Convolute of circle

N	1	2
a	2.00	3.00

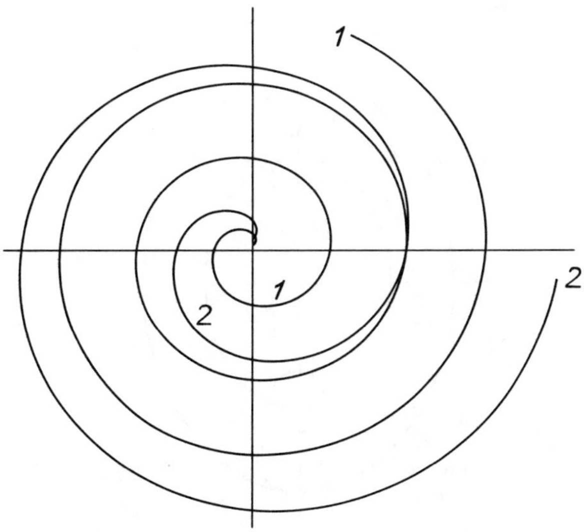

Figure 82. Convolute of circle

N	1	2
a	1.00	2.00

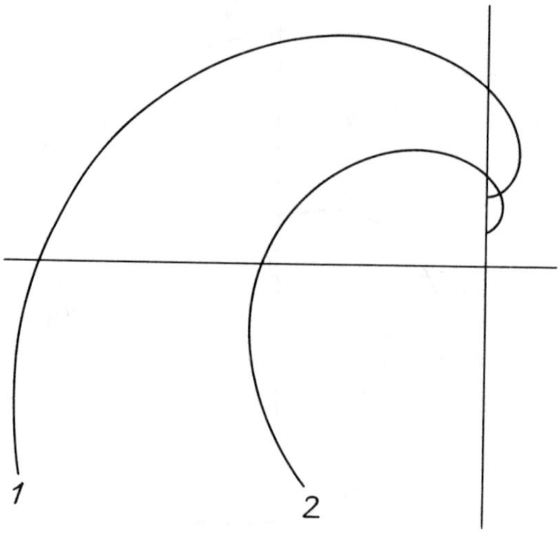

Figure 83. Convolute of circle

N	1	2
a	2.00	1.00

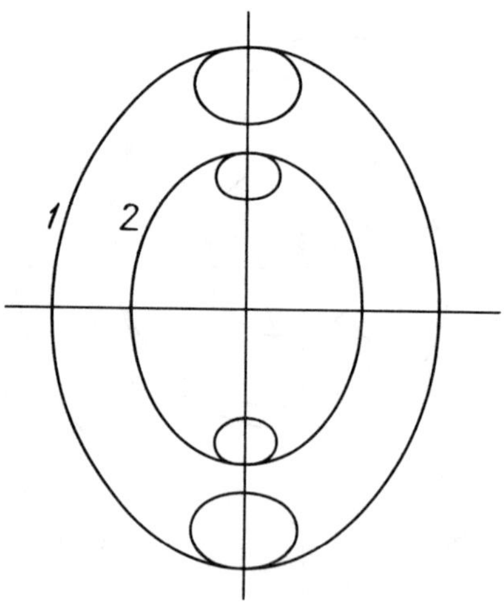

Figure 84. Cornoid

N	1	2
r	1.00	0.50

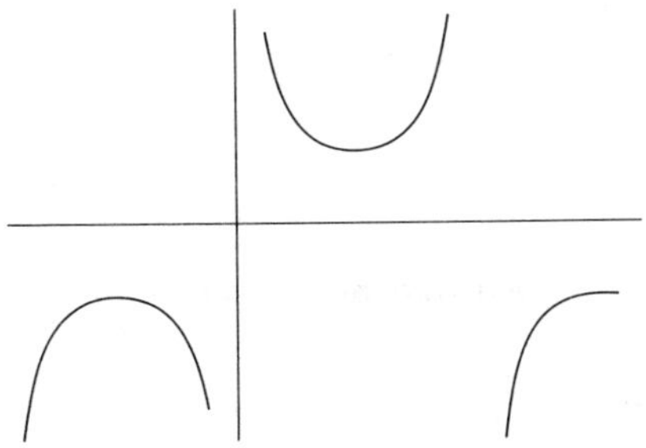

Figure 85. Cosecantoid

Basic geometric properties:
Unboundedness: present.
Asymptotes: straight lines $ax = \pi k,\ k \in \mathbf{Z}$, are vertical asymptotes.
Axial symmetry: straight lines $2ax = (2k+1)\pi,\ k \in \mathbf{Z}$, are axes of symmetry.
Central symmetry: points $(\pi k/a, 0),\ k \in \mathbf{Z}$, are centers of symmetry.
Connectedness: none.
Singular points: none.
Self-intersection points: none.
Characterized points: points

$$\left(\frac{(2k+1)\pi}{2a}, (-1)^k)\right), \quad k \in \mathbf{Z}$$

are located at minimal distance from x-axis (vertices).
(Figure 85).
COSECANTOID *is an unbounded unconnected (consisting of an infinitely large number of unbounded branches) central-symmetric curve free of singular points with an infinitely large number of symmetry axes and vertical asymptotes.*

Cosinusoid
Equation:

$$y = \cos(ax), \quad a > 0$$

Basic geometric properties:
Unboundedness: present.
Asymptotes: none.
Axial symmetry: straight lines $ax = \pi k,\ k \in \mathbf{Z}$, are axes of symmetry.
Central symmetry: points

$$\left(\frac{(2k+1)\pi}{2a}, 0\right), \quad k \in \mathbf{Z}$$

are centers of symmetry.

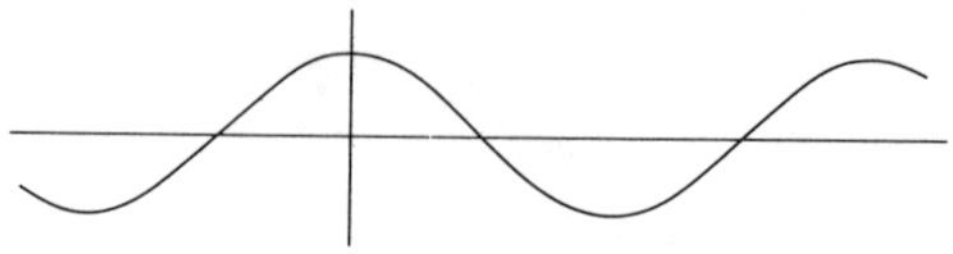

Figure 86. Cosinusoid

Connectedness: present.
Singular points: none.
Self-intersection points: none.
Characterized points: points

$$\left(\frac{\pi k}{a}, (-1)^k\right), \quad k \in \mathbf{Z}$$

are located at maximal distance from x-axis (vertices); points

$$\left(\frac{(2k+1)\pi}{2a}, 0\right), \quad k \in \mathbf{Z}$$

are inflection points.
(Figure 86).
COSINUSOID *is an unbounded connected central-symmetric curve free of singular points with an infinitely large number of axes and centers of symmetry.*

Cotangentoid
Equation:

$$y = \cot(ax), \quad a > 0$$

Basic geometric properties:
Unboundedness: present.
Asymptotes: straight lines $ax = \pi k$, $k \in \mathbf{Z}$, are vertical asymptotes.
Axial symmetry: none.
Central symmetry: points

$$\left(\frac{\pi k}{2a}, 0\right), \quad k \in \mathbf{Z}$$

are centers of symmetry.
Connectedness: present.
Singular points: none.
Self-intersection points: none.
Characterized points: points

$$\left(\frac{(2k+1)\pi}{2a}, 0\right), \quad k \in \mathbf{Z}$$

are inflection points.

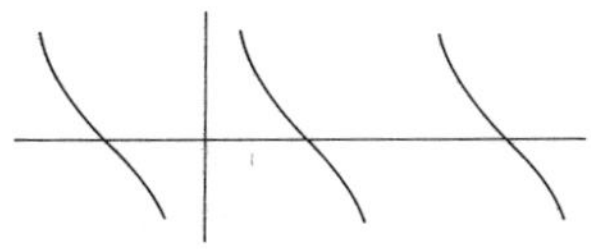

Figure 87. Cotangentoid

(Figure 87).
COTANGENTOID *is an unbounded unconnected (consisting of an infinitely large number of unbounded branches) central-symmetric curve free of singular points with an infinitely large number of vertical asymptotes.*

Côtes' spiral
Equation:

$$\rho = \frac{a}{\cos k\varphi},$$

$$a > 0, \quad k = \frac{p}{q}, \quad p \in \mathbf{N}, \quad q \in \mathbf{N}, \quad \text{G.C.D.}(p, q) = 1$$

Additional equation:

$$\rho = \frac{a}{\sin k\varphi}$$

Basic geometric properties:
Unboundedness: present.
Asymptotes: straight lines

$$x \sin \frac{2l+1}{2p}\pi = y \cos \frac{2l+1}{2p}\pi, \quad l = 0, 1, \ldots, p-1$$

are asymptotes.
Axial symmetry:

m is a positive integer:
straight lines

$$x \sin \frac{2\pi k}{m} = y \cos \frac{2\pi k}{m}, \quad k = 0, 1, \ldots, m-1$$

are axes of symmetry (for odd m); straight lines

$$x \sin \frac{\pi k}{m} = y \cos \frac{\pi k}{m}, \quad k = 0, 1, \ldots, 2m-1$$

are axes of symmetry (for even m);

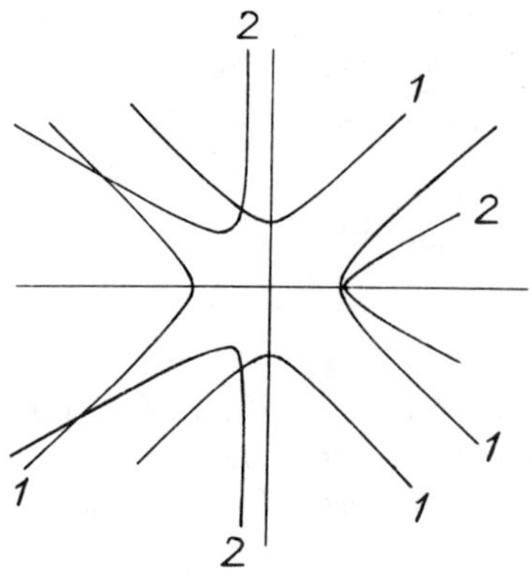

Figure 88. Côtes' spiral

N	1	2
k	2	3
a	1.00	1.00

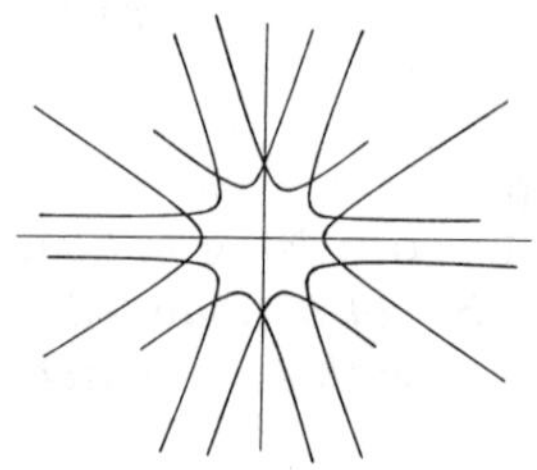

Figure 89. Côtes' spiral

$k = 5/2$
$a = 1.00$

m is rational:
straight lines

$$x \sin \frac{2\pi l}{p} = y \cos \frac{2\pi l}{p}, \quad l = 0, 1, \ldots, p-1$$

are axes of symmetry (for odd pq); straight lines

$$x \sin \frac{\pi l}{p} = y \cos \frac{\pi l}{p}, \quad l = 0, 1, \ldots, 2p-1$$

are axes of symmetry (for even pq).
Central symmetry: point $(0,0)$ is a center of symmetry (for even pq only).
Connectedness: none (for $q = 1, p \neq 1$), present (for $q > 1$).
Singular points: none (for $q = 1$), nodes (for $q > 1$).
Self-intersection points: finite (for $q > 1$), none (for $q = 1$).
Special cases: trisectrix of Maclaurin ($k = 1/3$), trisectrix of Delanges ($k = 1/2$), cruciform curve ($k = 2$), trisectrix of Longchamps ($k = 3$). (Figures 88–89).
CÔTES' SPIRAL *is an unbounded connected ($q > 1$) or unconnected ($q = 1$) curve with a finite number of singular points (nodes, $q > 1$), a center of symmetry (for even p only), p axes of symmetry and p asymptotes.*

Cranioid

Equation:

$$\rho = a \cos \varphi + b\sqrt{1 - m^2 \sin^2 \varphi} + \sqrt{1 - k^2 \sin^2 \varphi},$$

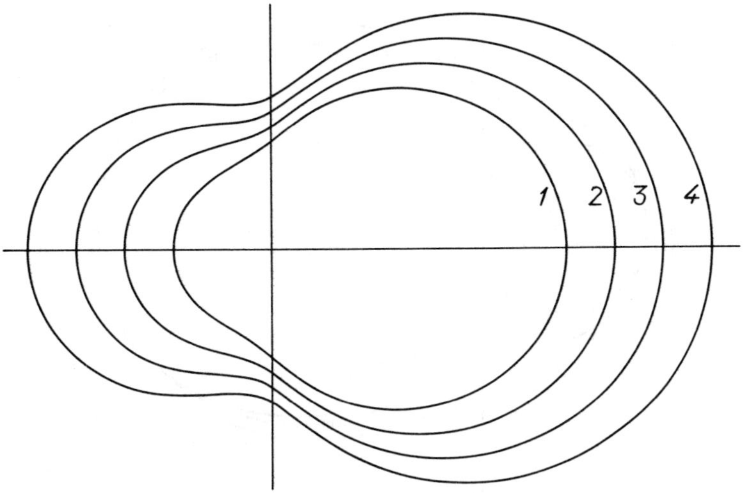

Figure 90. Cranioid

N	1	2	3	4
c	1.00	1.50	2.00	2.50
a	1.00	1.00	1.00	1.00
b	1.00	1.00	1.00	1.00
k	0.95	0.95	0.95	0.95
m	0.50	0.50	0.50	0.50

$$a > 0, \quad b > 0, \quad c > 0, \quad a < b + c,$$
$$0 < m < 1, \quad 0 < k < 1, \quad m \neq k$$

Additional equation:

$$(gy - x^2 - y^2)^2 = g^2x^2\left(\frac{x^2+y^2}{c^2} - \frac{x^2+y^2+d^2-l^2}{cd}\right.$$
$$\left. + \frac{x^2+y^2+d^2-l^2}{4d^2}(x^2+y^2-2gy+g^2)\right)$$

Basic geometric properties:
Boundedness: present.
Closedness: present.
Axial symmetry: straight line $y = 0$ is an axis of symmetry.
Central symmetry: none.
Connectedness: present.
Singular points: none.
Self-intersection points: none.
(Figures 90–92).
CRANIOID *is a bounded connected closed curve free of singular points with a single symmetry axis.*

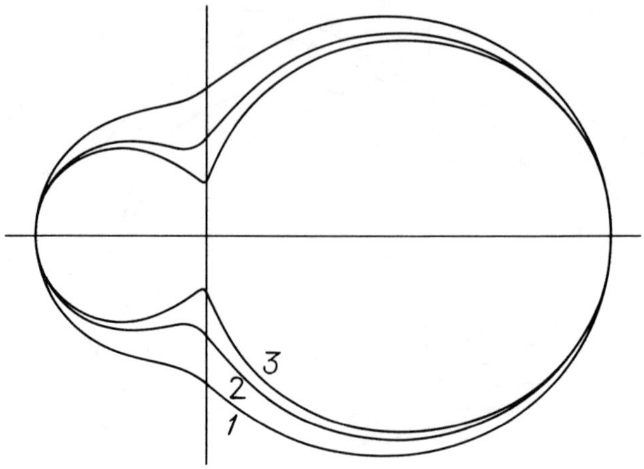

Figure 91. Cranioid

N	1	2	3
m	0.5000	0.9000	0.9999
a	1.00	1.00	1.00
b	1.00	1.00	1.00
k	0.95	0.95	0.95
c	1.50	1.50	1.50

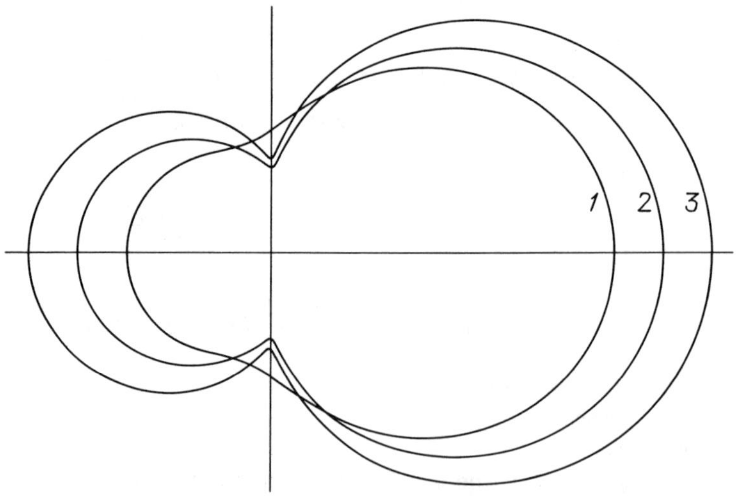

Figure 92. Cranioid

N	1	2	3
m	0.50	0.50	0.25
a	1.00	1.00	1.00
b	1.00	1.00	1.00
k	0.950	0.999	0.999
c	1.50	2.00	2.50

Cruciform curve (cross curve)
Equations:

$$\frac{a^2}{x^2} + \frac{b^2}{y^2} = 1,$$

$$x = \frac{a}{\cos t}, \quad y = \frac{b}{\sin t},$$

$$a > 0, \quad b > 0$$

Additional equation:

$$b^2x^2 + a^2y^2 = x^2y^2$$

Basic geometric properties:
Unboundedness: present.
Asymptotes: straight lines $x \pm a = 0$ are vertical asymptotes; straight lines $y \pm b = 0$ are horizontal asymptotes.
Axial symmetry: straight lines $x = 0$ and $y = 0$ are axes of symmetry.
Central symmetry: point $(0, 0)$ is a center of symmetry.
Connectedness: none.
Singular points: none.
Self-intersection points: none.
Special cases: circular cruciform curve ($a = b$).
(Figures 93-94).

CRUCIFORM CURVE *is an unbounded unconnected (consisting of four unbounded branches) central-symmetric curve free of singular points with two horizontal asymptotes, two vertical asymptotes and two axes of symmetry.*

Cubic of Tschirnhausen (trisectrix of Catalan)
Equations:

$$2(2p + x)^3 = 27p(x^2 + y^2),$$

$$x = p\frac{\cos t - \cos 2t}{(1 - \cos t)^2}, \quad y = p\frac{\sin t - \sin 2t}{(1 - \cos t)^2},$$

$$p > 0$$

Additional equations:

$$x = \frac{p}{2}\frac{\sin\frac{3t}{2}}{\sin^3\frac{t}{2}}, \quad y = \frac{p}{2}\frac{\cos\frac{3t}{2}}{\sin^3\frac{t}{2}},$$

$$27py^2 = (x - 4p)^2(2x + p),$$

$$27py^2 = (x - 2p)^2(4x + p)$$

Basic geometric properties:
Unboundedness: present.
Asymptotes: none.

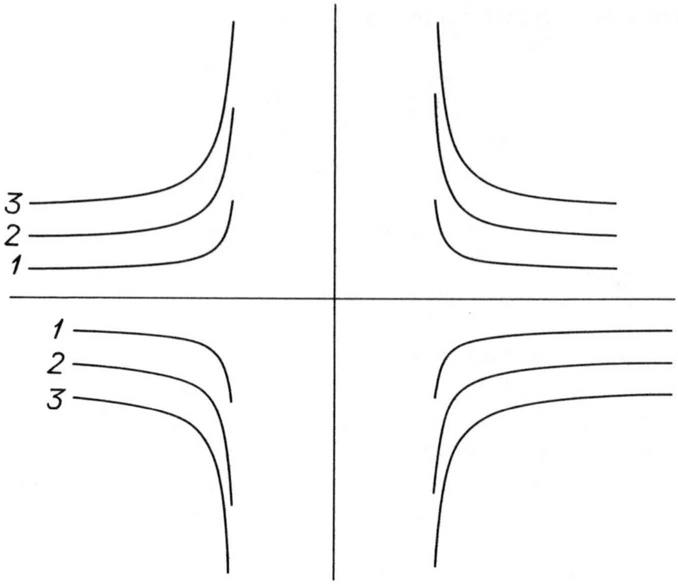

Figure 93. Cruciform curve

N	1	2	3
b	0.50	1.00	1.50
a	1.50	1.50	1.50

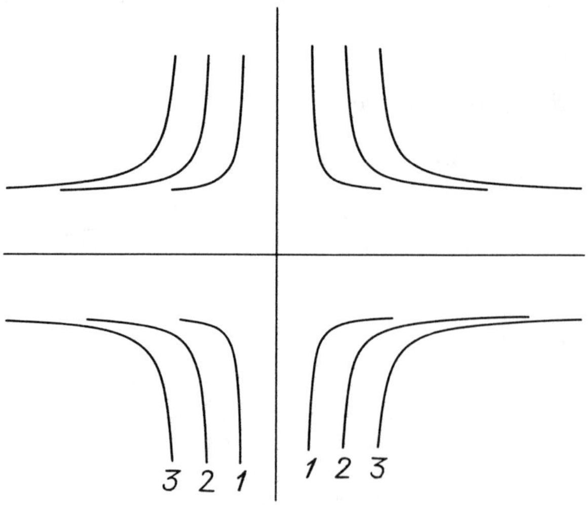

Figure 94. Cruciform curve

N	1	2	3
a	0.50	1.00	1.50
b	1.00	1.00	1.00

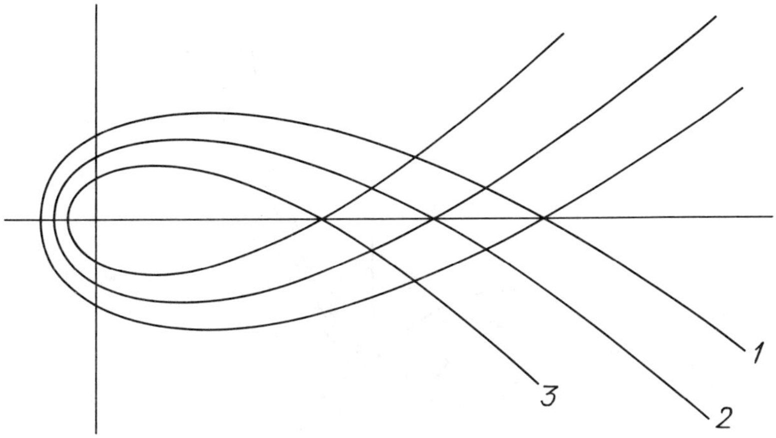

Figure 95. Cubic of Tschirnhausen

N	1	2	3
p	1.00	0.75	0.50

Axial symmetry: straight line $y = 0$ is a symmetry axis.
Central symmetry: none.
Connectedness: present.
Singular points: point $(4p, 0)$ is a node, straight lines

$$y = \pm\frac{1}{\sqrt{3}}(x - 4p)$$

are tangents at the node.
Self-intersection points: one, $(0, 0)$.
Characterized points: point $(-p/2, 0)$ is a vertex; points $(p, \pm p)$ of the loop are located at maximal distance from x-axis.
(Figure 95).
CUBIC OF TSCHIRNHAUSEN *is an unbounded connected curve with a single singular point (node) and a single symmetry axis.*

Curtate and prolate cissoids
Equation:

$$y^2 = \frac{2r - a + x}{a - x}x^2,$$
$$a > 0, \quad r > 0$$

Additional equation:

$$y = \pm x\sqrt{\frac{2r - a + x}{a - x}}$$

Basic geometric properties:
Unboundedness: present.

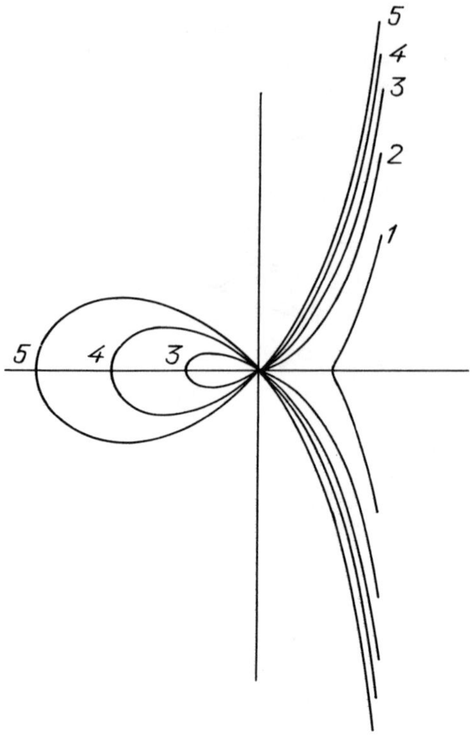

Figure 96.

Curtate and prolate cissoids

N	1	2	3	4	5
r	0.25	0.50	0.75	1.00	1.25
a	1.00	1.00	1.00	1.00	1.00

Asymptotes: straight line $x - a = 0$ is a vertical asymptote.
Axial symmetry: straight line $y = 0$ is an axis of symmetry.
Central symmetry: none.
Connectedness: present (for $a \leq 2r$), none (for $a > 2r$).
Singular points: point $(0, 0)$ is a node (for $a < 2r$), straight lines

$$y = \pm x\sqrt{\frac{2r - a}{a}}$$

are tangents at the node, or a cuspidal point of the first kind (for $a = 2r$), straight line $y = 0$ is a tangent at the cusp point, or an isolated point (for $a > 2r$).
Self-intersection points: none (for $a \geq 2r$), present (for $a < 2r$).
Special cases: strophoid ($a = r$), prolate cissoid ($a < 2r$), (right) cissoid ($a = 2r$), curtate cissoid ($a > 2r$).
(Figures 96–97).

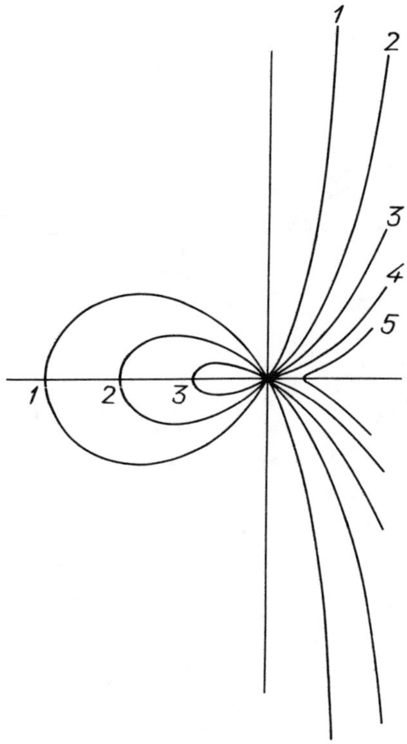

Figure 97.
Curtate and prolate cissoids

N	1	2	3	4	5
a	0.50	1.00	1.50	2.00	2.25
r	1.00	1.00	1.00	1.00	1.00

CURTATE CISSOID ($a < 2r$) *is an unbounded connected curve with a single singular point (node), a single axis of symmetry and a single vertical asymptote.*
PROLATE CISSOID ($a > 2r$) *is an unbounded unconnected (consisting of an unbounded branch and an isolated point) curve with a single singular point (isolated point), a single axis of symmetry and a single vertical asymptote.*

Curtate and prolate cycloids
Equations:

$$x = rt - d\sin t, \quad y = r - d\cos t,$$

$$x = r\arccos\frac{r-y}{d} - \sqrt{(d+r-y)(d-r+y)},$$

$$r > 0, \quad d > 0, \quad r \neq d$$

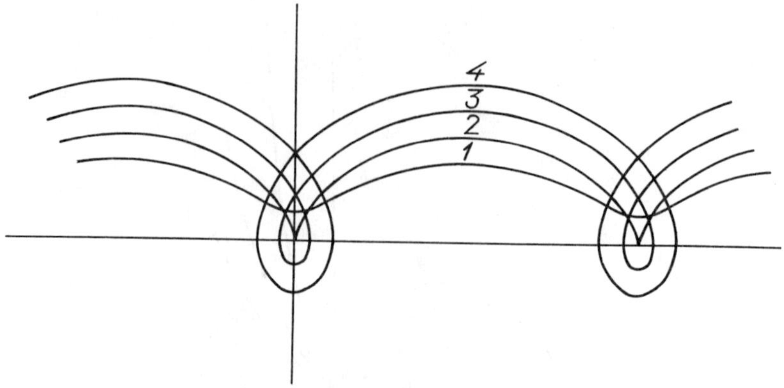

Figure 98.
Curtate and prolate cycloids

N	1	2	3	4
d	0.50	1.00	1.50	2.00
r	1.00	1.00	1.00	1.00

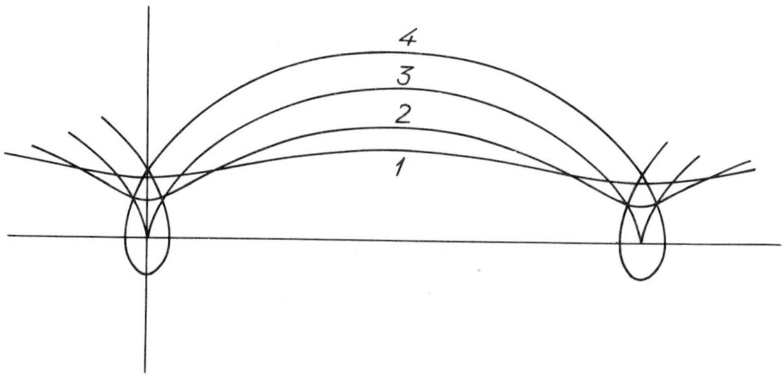

Figure 99.
Curtate and prolate cycloids

N	1	2	3	4
d	0.30	0.50	1.00	1.50
r	1.00	1.00	1.00	1.00

Additional equations:

$$x = d\sin\psi - p\psi, \quad y = d\cos\psi - q\psi$$

Basic geometric properties:
Unboundedness: present.
Asymptotes: none.

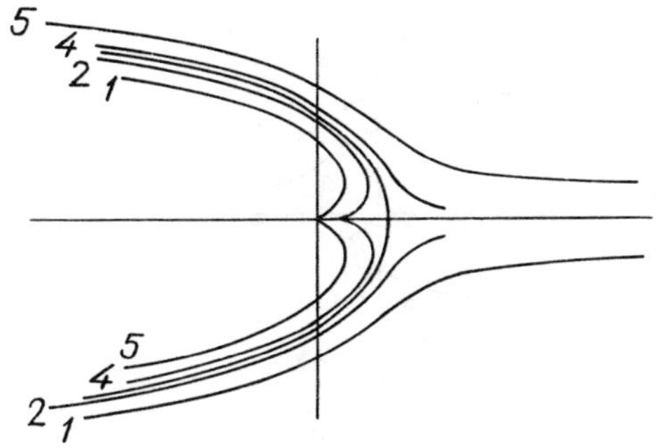

Figure 100.

Curtate and prolate quadratrices

N	1	2	3	4	5
l	0.40	0.10	0.00	−0.10	−0.40
a	1.00	1.00	1.00	1.00	1.00

Axial symmetry: straight lines $x = r\pi k,\ k \in \mathbf{Z}$, are axes of symmetry.
Central symmetry: none.
Connectedness: present.
Singular points: points $(2\pi lr, y_0)$, $l \in \mathbf{Z}$, are nodes (for $d > r$), straight lines

$$\frac{x - 2\pi lr}{y_0} = \pm\frac{y - y_0}{d \sin t_0}$$

are tangents at the nodes; here $y_0 = r - d\cos t_0$, where t_0 is the least positive solution of an equation

$$\frac{\sin t_0}{t_0} = \frac{r}{d}$$

Self-intersection points: infinite, (for $d > r$), none (for $d < r$).
Special cases: prolate cycloid ($d > r$), curtate cycloid ($d < r$).
(Figures 98–99).
CURTATE CYCLOID $d > r$ *is an unbounded connected curve with an infinitely large number of singular points (nodes) and axes of symmetry.*
PROLATE CYCLOID $d < r$ *is an unbounded connected curve free of singular points with an infinitely large number of the symmetry axes.*

Curtate and prolate quadratrices
Equations:

$$\rho = a\frac{\varphi \pm \alpha}{\sin\varphi},$$

$$a > 0, \quad 0 < \alpha < \pi$$

Basic geometric properties:
Unboundedness: present.

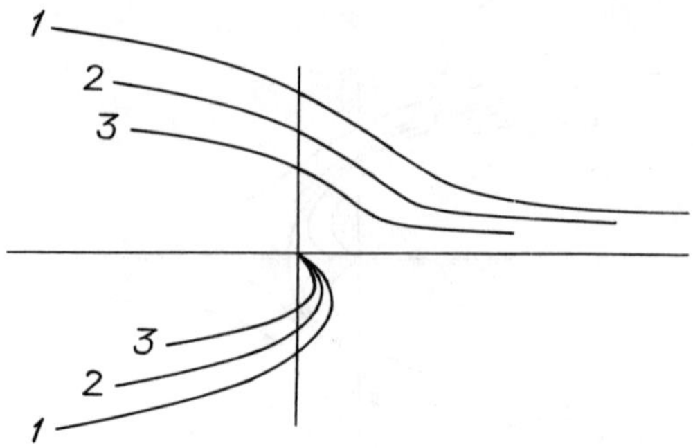

Figure 101.
Curtate and prolate quadratrices

N	1	2	3
a	1.30	1.00	0.70
l	0.40	0.40	0.40

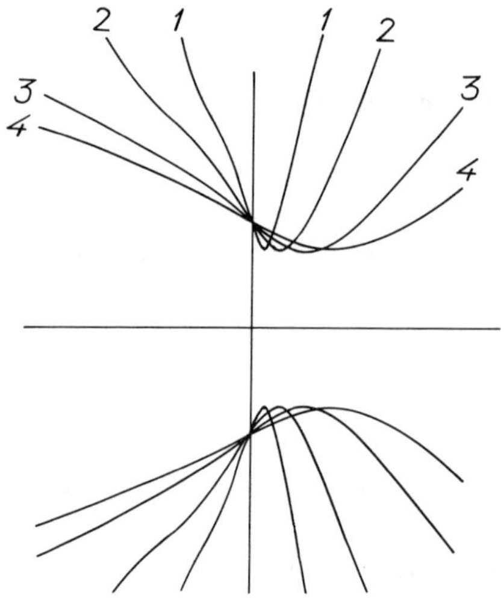

Figure 102. Curve of Baum

N	1	2	3	4
a	0.25	0.50	1.00	1.50
b	1.00	1.00	1.00	1.00
c	1.00	1.00	1.00	1.00

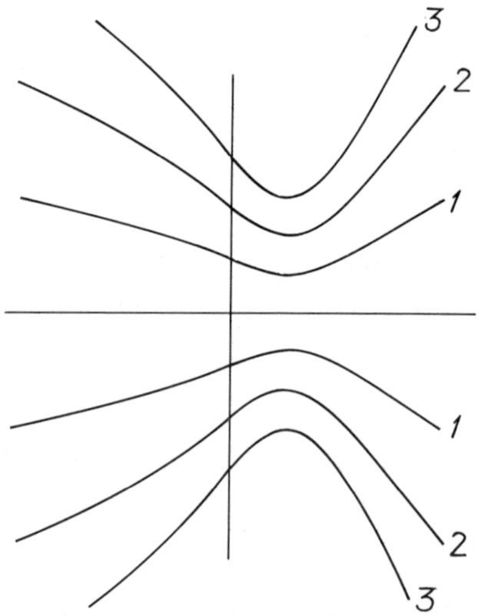

Figure 103. Curve of Baum

N	1	2	3
b	0.50	1.00	1.50
a	1.00	1.00	1.00
c	1.00	1.00	1.00

Asymptotes: straight lines $y = \pm\alpha a$ are horizontal asymptotes.
Axial symmetry: none.
Central symmetry: none.
Connectedness: none.

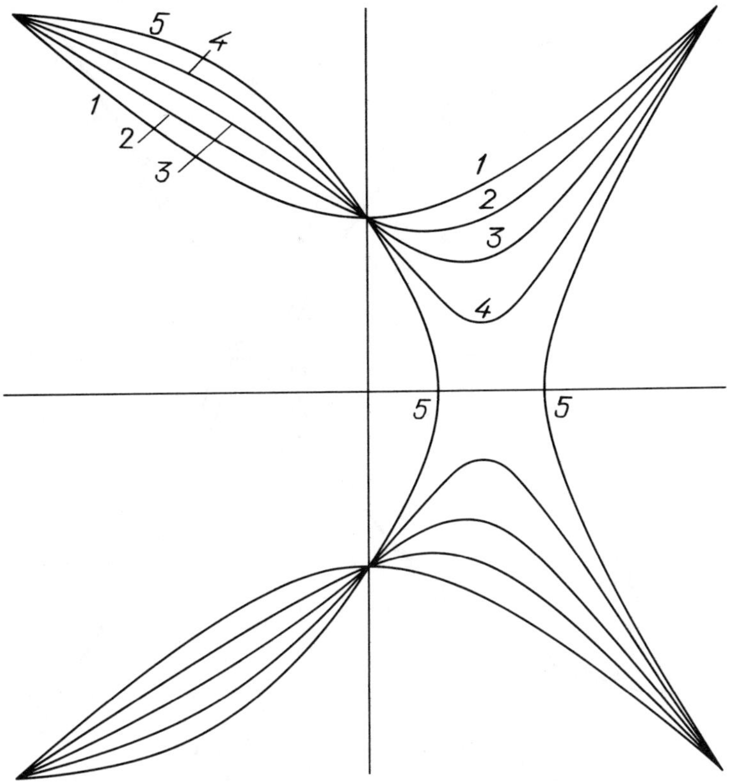

Figure 104. Curve of Baum

N	1	2	3	4	5
c	0.00	0.50	1.00	1.50	2.00
a	1.00	1.00	1.00	1.00	1.00
b	1.00	1.00	1.00	1.00	1.00

Singular points: point $(0,0)$ is a cessation point.
Self-intersection points: none.
(Figures 100–101).
CURTATE (−) AND PROLATE (+) QUADRATRICES *is an unbounded unconnected (consisting of two unbounded branches) curve with a single singular point (cessation point).*

Curve of Baum
Equation:

$$\frac{y^2}{b^2} = \frac{x^2}{a^2} - c\sin\frac{\pi x}{2a} + 1, \quad a > 0, \quad b > 0, \quad c > 0$$

Basic geometric properties:
Unboundedness: present.

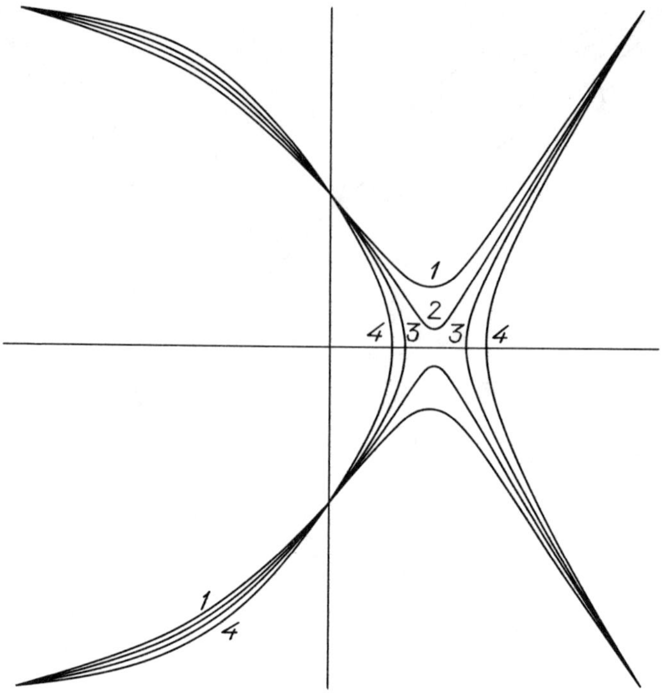

Figure 105. Curve of Baum

N	1	2	3	4
c	1.50	1.66	1.80	2.00
a	1.00	1.00	1.00	1.00
b	1.00	1.00	1.00	1.00

Asymptotes: none.
Axial symmetry: straight line $y = 0$ is an axis of symmetry.
Central symmetry: none.
Connectedness: present (for $c = 1.6685...$).
Singular points: none (for $c \neq 1.6685...$).
Self-intersection points: none (for $c \neq 1.6685...$).
(Figures 102–105).
CURVE OF BAUM *is an unbounded unconnected (consisting of two unbounded branches for* $c \neq 1.6685...$*) curve with a single axis of symmetry.*

Curve of Caraccioli
Equation:

$$\rho \frac{\cos \varphi}{a} = 1 - \left(\frac{2\varphi}{\pi} \right)^{m/n},$$

$$a > 0, \quad 0 \leq \varphi < \frac{3\pi}{2}, \quad m \in \mathbf{N}, \quad n \in \mathbf{N}, \quad \text{G.C.D.}(m, n) = 1$$

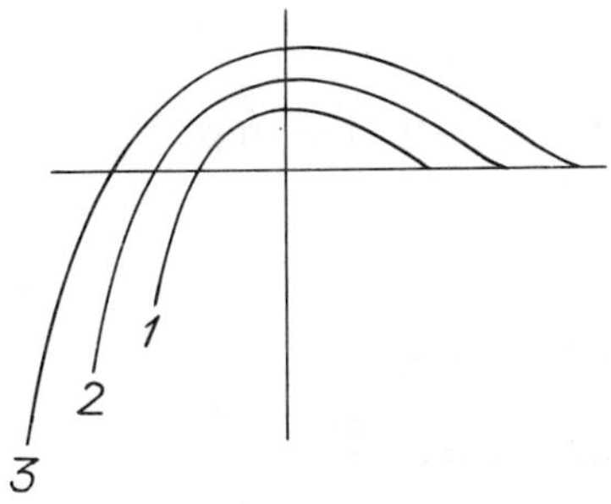

Figure 106. Curve of Caraccioli

N	1	2	3
a	0.50	0.75	1.00
m	2	2	2
n	3	3	3

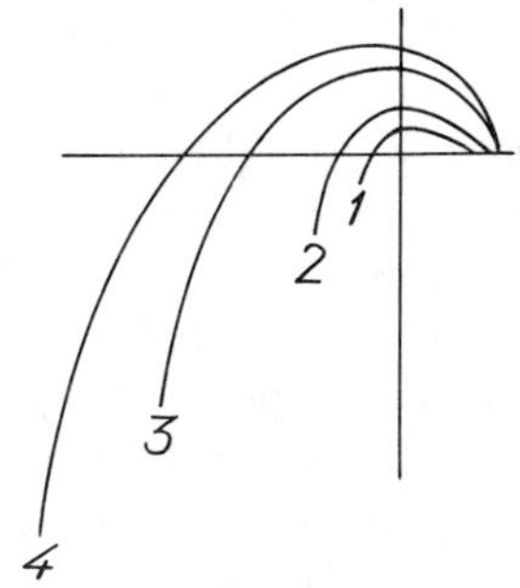

Figure 107. Curve of Caraccioli

N	1	2	3	4
a	1.00	1.00	1.00	1.00
m	1	2	4	5
n	3	3	3	3

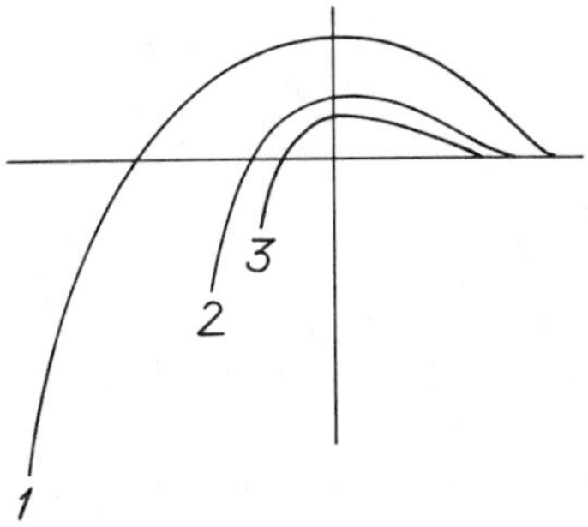

Figure 108. Curve of Caraccioli

N	1	2	3
a	1.00	1.00	1.00
m	1	1	1
n	1	2	3

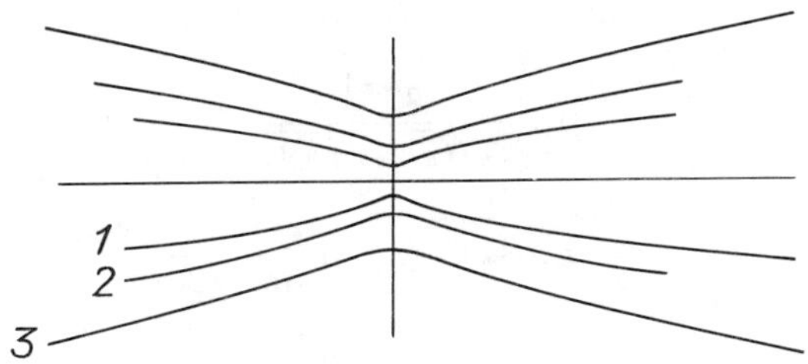

Figure 109. Curve of Clairaut

N	1	2	3
a	0.50	1.00	2.00

Basic geometric properties:
Unboundedness: present.
Asymptotes: straight line $x = a(1 - 3^{m/n})$ is a vertical asymptote.
Axial symmetry: none.
Central symmetry: none.
Connectedness: present.
Singular points: point $(a, 0)$ is a point of cessation.
Self-intersection points: none.
Characterized points: point $(a, 0)$ is a point of cessation.
Special cases: Dinostratus' quadratrix ($m = n = 1$).
(Figures 106–108).
CURVE OF CARACCIOLI *is an unbounded connected curve with a single singular point (cessation point) and a single vertical asymptote.*

Curve of Clairaut (curve of Eudoxus, kampyle of Eudoxus)
Equations:

$$y^4 = a^2(x^2 + y^2),$$
$$\rho = \frac{a}{\sin^2 \varphi}, \quad a > 0$$

Basic geometric properties:
Unboundedness: present.
Asymptotes: none.
Axial symmetry: straight lines $x = 0$, $y = 0$ are axes of symmetry.
Central symmetry: point $(0, 0)$ is a center of symmetry.
Connectedness: none.
Singular points: point $(0, 0)$ is an isolated point.
Self-intersection points: none.
Generalizations:

$$y^m = a^{m-1}\sqrt{x^2 + y^2}, \quad a^m = y^{m-1}\sqrt{x^2 + y^2}$$

Analogues:

$$y^{n+1} = a\sqrt{(a^2 - x^2)^n},$$
$$a^{n+1} = y\sqrt{(a^2 - x^2)^n},$$
$$y = \frac{a^{n+1}}{\sqrt{(a^2 - x^2)^n}},$$
$$x = a\cos t, \quad y = a\frac{1}{\sin n/2t}$$

(Figure 109).
CURVE OF CLAIRAUT *is an unbounded unconnected (consisting of two unbounded branches and isolated point) central-symmetric curve with a single singular point (isolated point) and two axes of symmetry.*

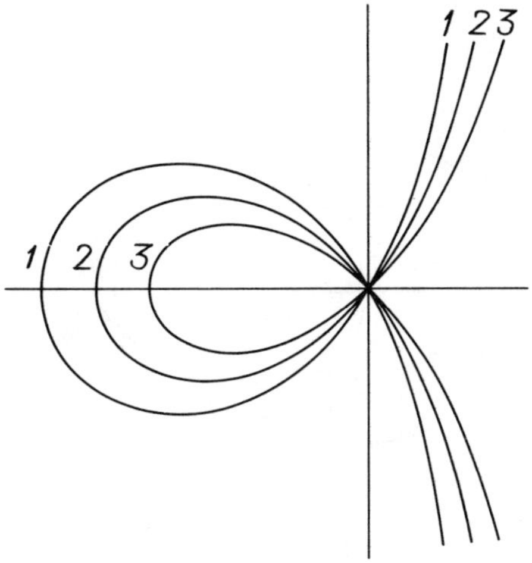

Figure 110. Curve of Cramer

N	1	2	3
l	1.00	0.50	0.00
r	2.00	2.00	2.00

Curve of Cramer
Equations:

$$x(x^2+y^2) = -(r+l)x^2 + (r-l)y^2,$$

$$\rho = -\frac{lr\cos 2\varphi}{\cos\varphi}, \quad r > l > 0$$

Additional equation:

$$x(x^2+y^2) = (r+l)x^2 - (r-l)y^2$$

Basic geometric properties:
Unboundedness: present.
Asymptotes: straight line $x = r - l$ is a vertical asymptote.
Axial symmetry: straight line $y = 0$ is a symmetry axis.
Central symmetry: none.
Connectedness: present.
Singular points: point $(0,0)$ is a node, straight lines

$$y = \pm x\sqrt{\frac{r+l}{r-l}}$$

are tangents at the node.
Self-intersection points: one, $(0,0)$.
Characterized points: point $(-r-l,0)$ is a loop point located at maximal distance from point $(0,0)$.

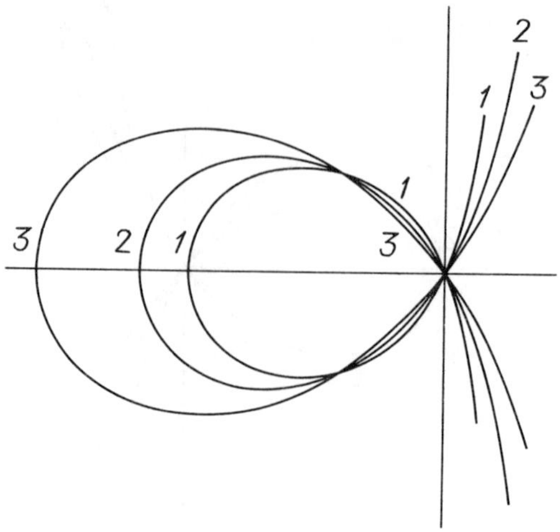

Figure 111. Curve of Cramer

N	1	2	3
r	1.50	2.00	3.00
l	1.00	1.00	1.00

Special cases: trisectrix of Maclaurin ($r = 2l$).
(Figures 110–111).
CURVE OF CRAMER *is an unbounded connected curve with a single singular point (node), a single vertical asymptote and a single symmetry axis.*

Curve of damped oscillations
Equation:

$$y = Ae^{-kx} \sin(\omega x + a),$$

$$A > 0, \quad k > 0, \quad \omega > 0$$

Basic geometric properties:
Unboundedness: present.
Asymptotes: straight line $y = 0$ is a horizontal asymptote.
Axial symmetry: none.
Central symmetry: none.
Connectedness: none.
Singular points: none.
Self-intersection points: none.
(Figures 112–114).
CURVE OF DAMPED OSCILLATIONS *is an unbounded connected curve free of singular points with a single horizontal asymptote.*

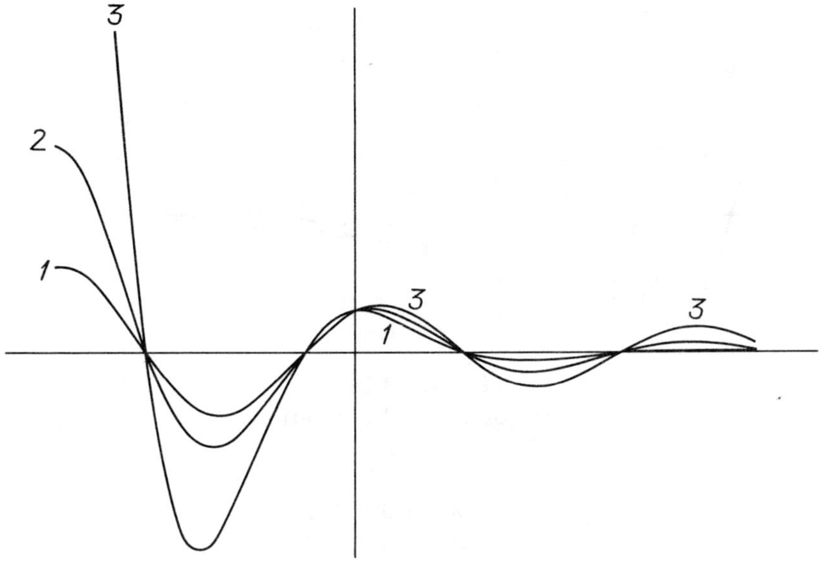

Figure 112.
Curve of damped oscillations

N	1	2	3
k	0.10	0.25	0.50
a	1.00	1.00	1.00
A	1.00	1.00	1.00
ω	1.00	1.00	1.00

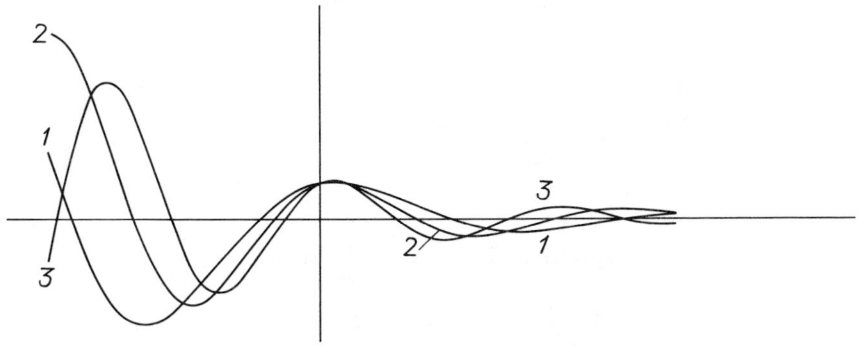

Figure 113.
Curve of damped oscillations

N	1	2	3
k	0.25	0.25	0.25
a	1.00	1.00	1.00
A	1.00	1.00	1.00
ω	0.75	1.00	1.25

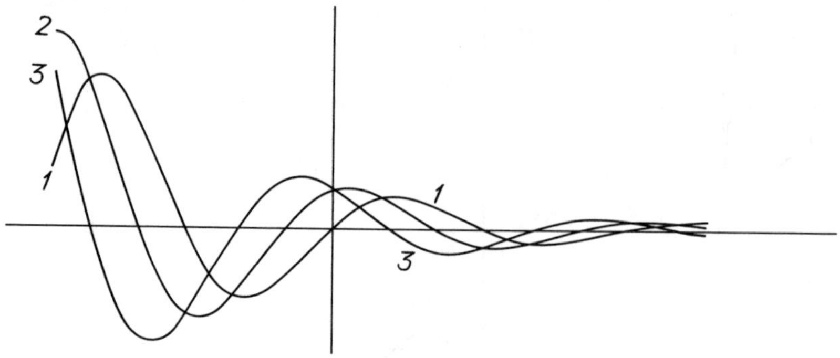

Figure 114.
Curve of damped oscillations

N	1	2	3
k	0.25	0.25	0.25
a	0.00	1.00	2.00
A	1.00	1.00	1.00
ω	1.00	1.00	1.00

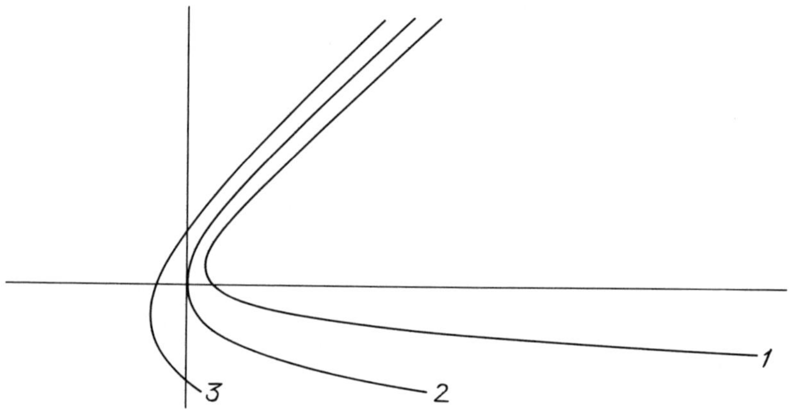

Figure 115. Curve of Debeaune

N	1	2	3
b	0.50	1.00	1.50
a	1.00	1.00	1.00
c	1.00	1.00	1.00

Curve of Debeaune
Equation:

$$x = \frac{y}{a} - \frac{b}{a} + ce^{-y/b},$$

$$a > 0, \quad b > 0, \quad c > 0$$

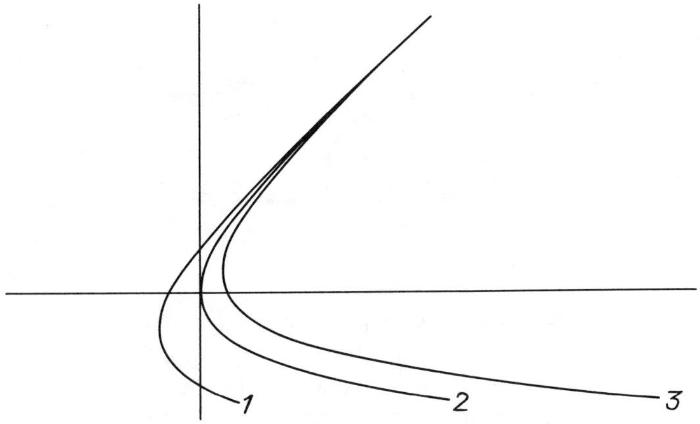

Figure 116. Curve of Debeaune

N	1	2	3
c	0.50	1.00	1.50
a	1.00	1.00	1.00
b	1.00	1.00	1.00

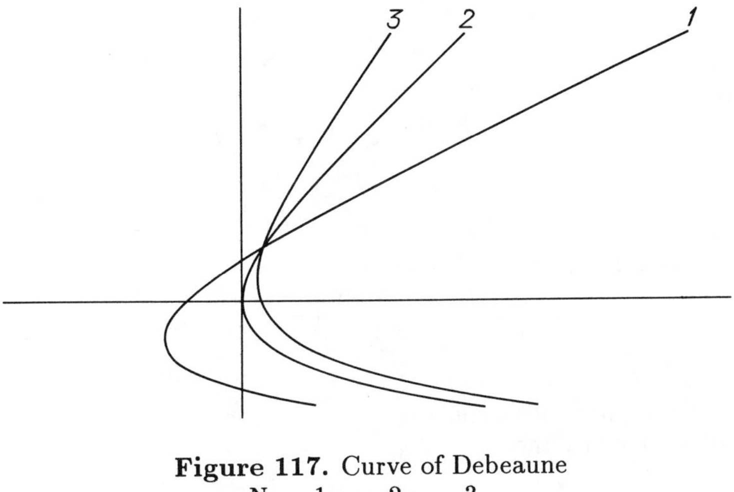

Figure 117. Curve of Debeaune

N	1	2	3
a	0.50	1.00	1.50
b	1.00	1.00	1.00
c	1.00	1.00	1.00

Basic geometric properties:

Unboundedness: present.

Asymptotes: straight line $ax - y + b = 0$ is an inclined asymptote.

Axial symmetry: none.

Central symmetry: none.

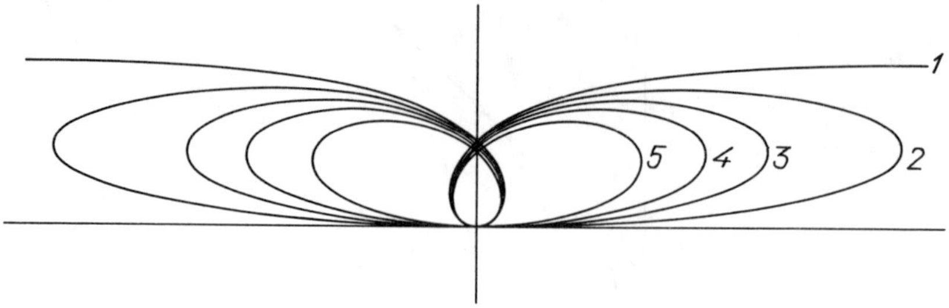

Figure 118. Curve of Gutschoven

N	1	2	3	4	5
l	0.50	0.49	0.48	0.47	0.45
a	0.50	0.50	0.50	0.50	0.50

Connectedness: present.
Singular points: none.
Self-intersection points: none.
(Figures 115–117).
CURVE OF DEBEAUNE *is an unbounded connected curve free of singular points with a single inclined asymptote.*

Curve of Gutschoven
Equations:

$$l^2x^2(x^2+y^2) = a^2(x^2+y^2-ly)^2,$$

$$\rho = \frac{a\sin\varphi}{a+\cos\varphi}, \quad a>0, \quad l>0$$

Additional equation:

$$\rho = \frac{al\sin\varphi}{a-l\cos\varphi}$$

Basic geometric properties:
Unboundedness: present (for $a > l$).
Asymptotes: straight lines

$$y = \pm x\sqrt{\frac{l^2}{a^2}-1}+l, \quad l>a$$

are inclined asymptotes: straight line $y = l$ is a horizontal asymptote (for $l = a$).
Boundedness: present (for $a > l$).
Closedness: present (for $a > l$).
Axial symmetry: straight line $x = 0$ is an axis of symmetry.
Central symmetry: none.

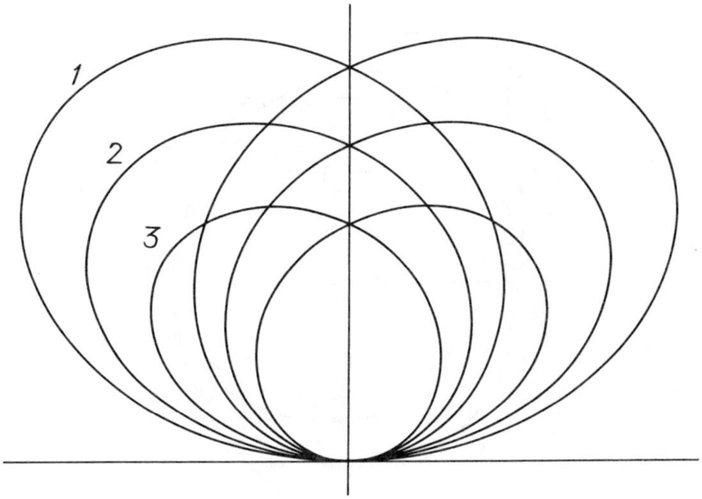

Figure 119. Curve of Gutschoven

N	1	2	3
a	1.00	0.80	0.60
l	0.50	0.40	0.30

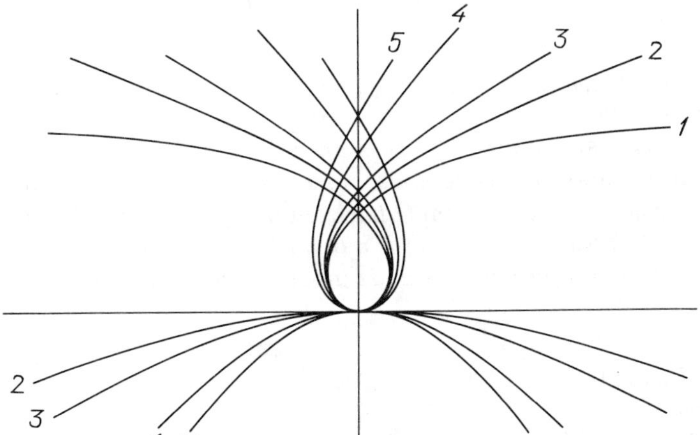

Figure 120. Curve of Gutschoven

N	1	2	3	4	5
l	0.50	0.55	0.60	0.80	1.00
a	0.50	0.50	0.50	0.50	0.50

Connectedness: present.
Singular points: point $(0,0)$ is a degenerate node, straight line $y = 0$ is a tangent at the node; point $(0,l)$ is a node, straight lines $\pm lx + ay + al = 0$ are tangents at the node.

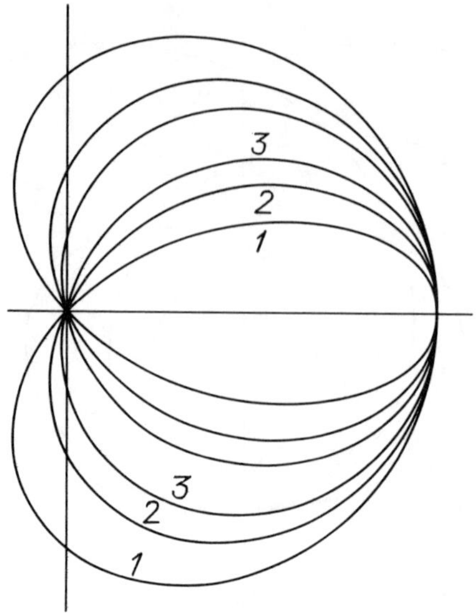

Figure 121. Curve of Jeřabek

N	1	2	3
r	1.50	2.50	5.00
a	1.00	1.00	1.00

Self-intersection points: two, $(0,0)$, $(0,l)$.
(Figures 118–120).
CURVE OF GUTSCHOVEN ($l \geq a$) *is an unbounded connected curve with two singular points (multiple node, node), a single axis of symmetry and two inclined asymptotes (a single horizontal asymptote for* $l = a$*).*
CURVE OF GUTSCHOVEN ($l < a$) *is a bounded connected closed curve with two singular points (degenerate node, node) and a single axis of symmetry.*

Curve of Jeřabek
Equations:

$$r^2(x^2+y^2-ax)^2 = a^2(x^2+y^2)(x-a)^2,$$

$$\rho = a\frac{r\cos\varphi - a}{r - a\cos\varphi}, \quad a>0, \quad r>0, \quad a \neq r$$

Basic geometric properties:
Unboundedness: present (for $r < a$).
Asymptotes: straight lines

$$y = \pm\frac{a^2-r^2}{r}(x+a)$$

are inclined asymptote ($r < a$).

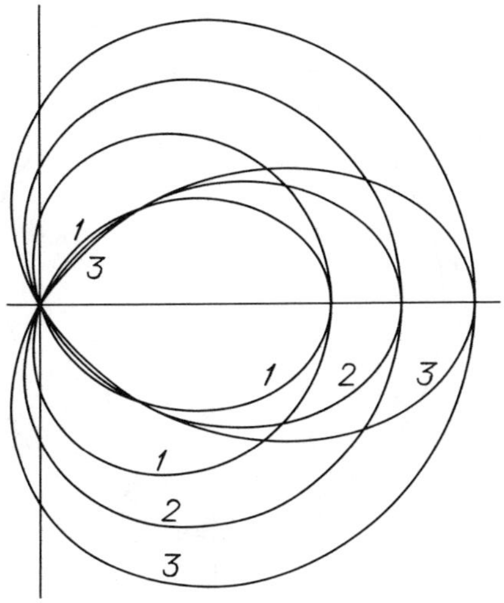

Figure 122. Curve of Jeřabek

N	1	2	3
a	0.80	1.00	1.20
r	2.50	2.50	2.50

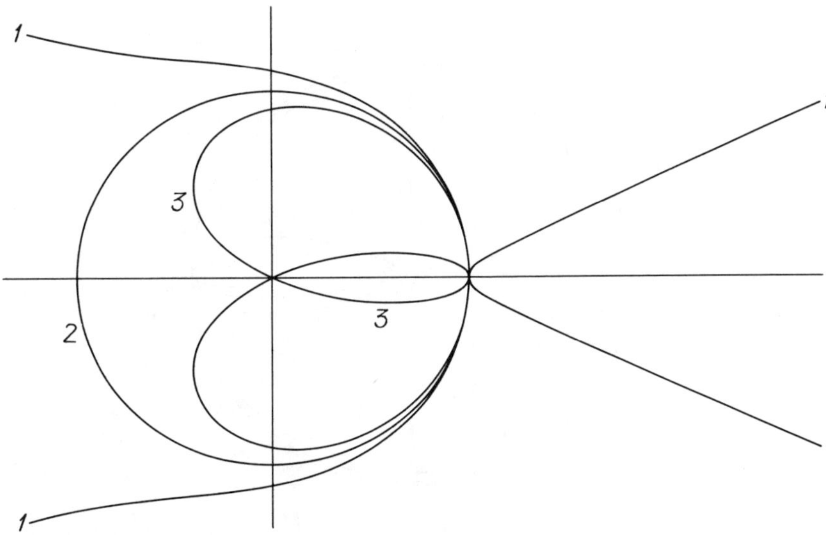

Figure 123. Curve of Jeřabek

N	1	2	3
r	0.90	1.00	1.10
a	1.00	1.00	1.00

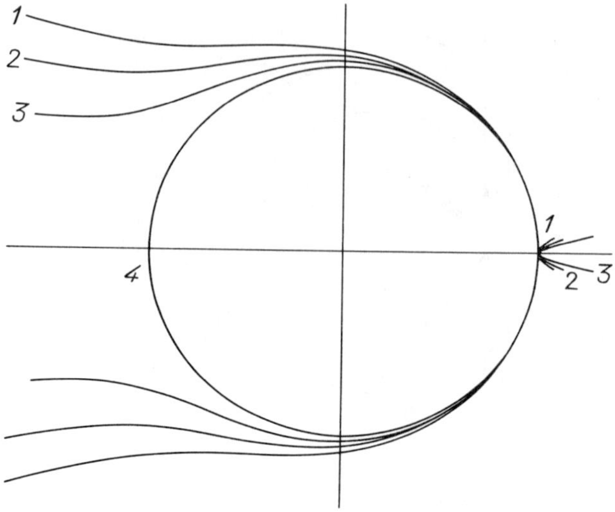

Figure 124. Curve of Jeřabek

N	1	2	3	4
r	0.425	0.950	0.975	1.000
a	1.00	1.00	1.00	1.00

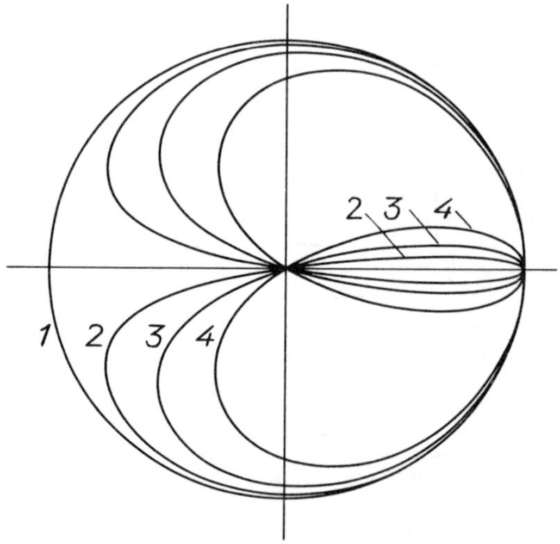

Figure 125. Curve of Jeřabek

N	1	2	3	4
r	1.00	1.01	1.05	1.20
a	1.00	1.00	1.00	1.00

Boundedness: present (for $r > a$).
Closedness: present (for $r > a$).
Axial symmetry: straight line $y = 0$ is an axis of symmetry.
Central symmetry: none.
Connectedness: present (for $r > a$), none (for $r < a$).
Singular points: point $(0,0)$ is an isolated point (for $r < a$), or a node ($r > a$), straight lines

$$y = \pm\frac{\sqrt{r^2 - a^2}}{a}x$$

are tangents at the node; point $(a,0)$ is a degenerate node (tacnode), straight line $x = a$ is a tangent at the node.
Self-intersection points: two, $(0,0)$ and $(a,0)$ (for $r > a$); one, $(a,0)$ (for $r < a$).
(Figures 121–125).
CURVE OF JEŘABEK ($r < a$) *is an unbounded connected curve with two singular points (isolated point, degenerated node), two inclined asymptotes and a single axis of symmetry.*
CURVE OF JEŘABEK ($r > a$) *is a bounded connected closed curve with two singular points (node, degenerated node) and a single axis of symmetry.*

Curve of Kiepert

Equation:

$$\rho^3 = l^3 \cos 3\varphi, \quad l > 0$$

Basic geometric properties:
Boundedness: present.
Closedness: present.
Axial symmetry: straight lines $y = 0$, $y = \pm x\sqrt{3}$ are axes of symmetry.
Central symmetry: none.
Connectedness: present.
Singular points: point $(0,0)$ is a triple point, straight lines $x = 0$, $x \pm y\sqrt{3} = 0$ are tangents at the point.
Self-intersection points: one, $(0,0)$.
(Figure 126).
CURVE OF KIEPERT *is a bounded connected closed curve with a single singular point (triple point) and three axes of symmetry.*

Curve of Lamet (A)

Equation:

$$\left(\frac{x}{a}\right)^m + \left(\frac{y^m}{b}\right) = 1,$$

$$m = \frac{p}{q}, \quad p \in \mathbf{N}, \quad q \in \mathbf{N}, \quad \text{G.C.Q.}(p,q) = 1, \quad a > 0, \quad b > 0$$

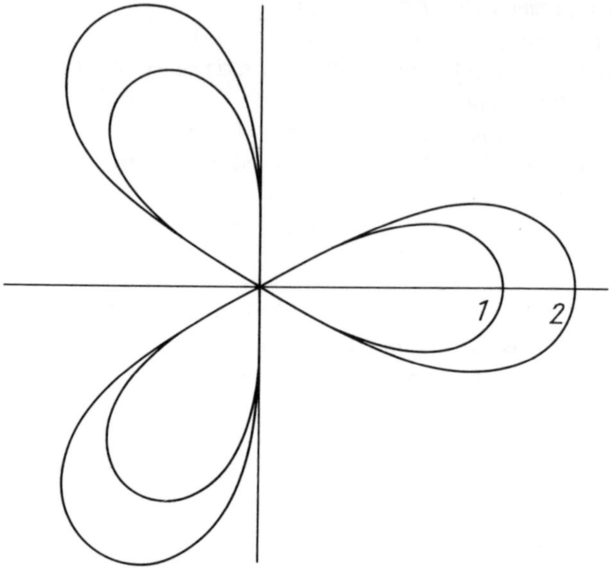

Figure 126. Curve of Kiepert

N	1	2
l	1.00	1.30

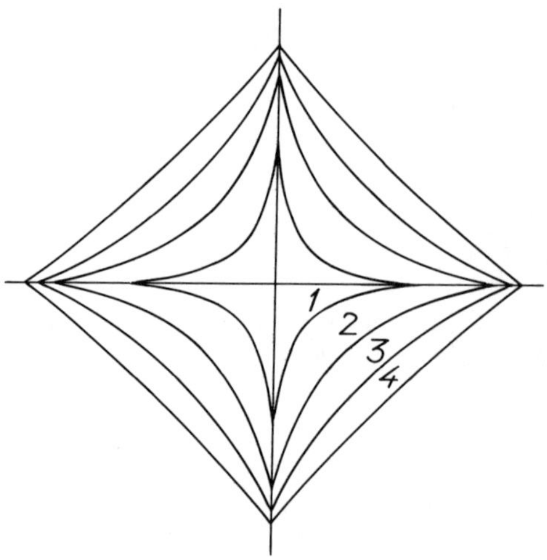

Figure 127. Curve of Lamet (A)

N	1	2	3	4
p	4	6	8	10
q	11	11	11	11
a	1.00	1.00	1.00	1.00
b	1.00	1.00	1.00	1.00

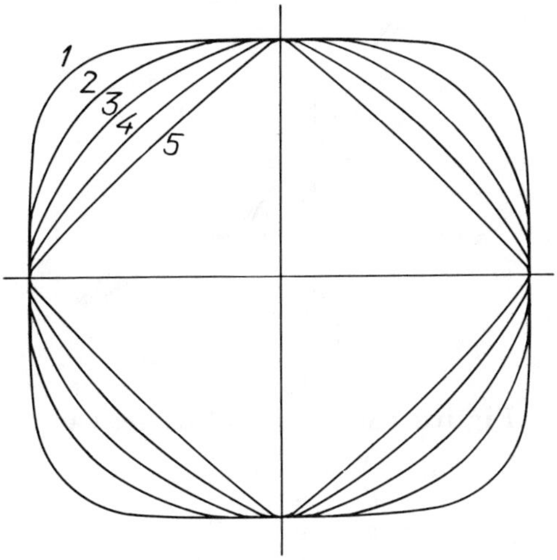

Figure 128. Curve of Lamet (A)

N	1	2	3	4	5
q	3	5	7	9	11
p	12	12	12	12	12
a	1.00	1.00	1.00	1.00	1.00
b	1.00	1.00	1.00	1.00	1.00

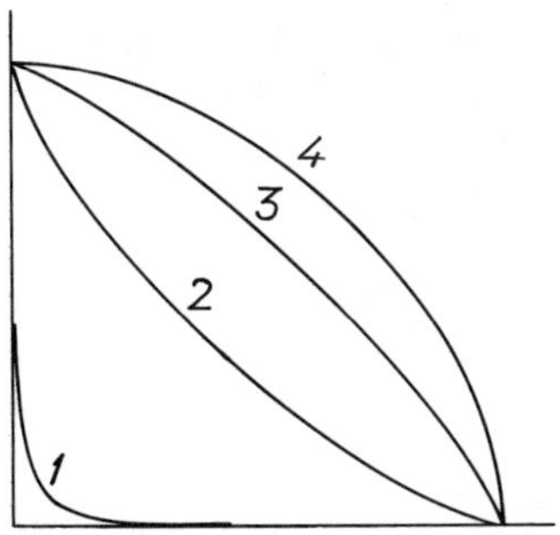

Figure 129. Curve of Lamet (A)

N	1	2	3	4
p	1	3	5	7
q	4	4	4	4
a	1.00	1.00	1.00	1.00
b	1.00	1.00	1.00	1.00

Basic geometric properties:

p is even
Boundedness: present.

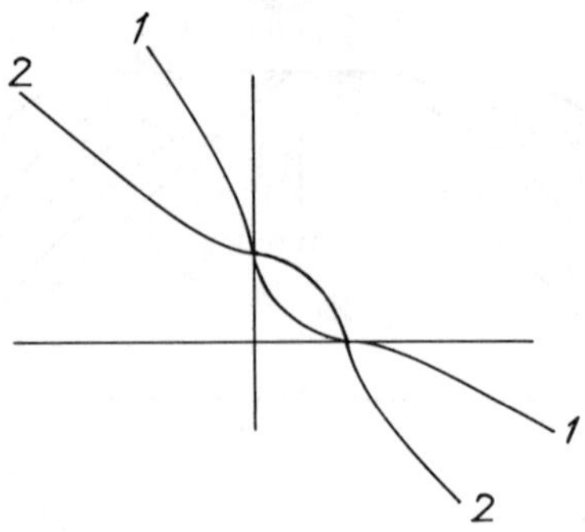

Figure 130. Curve of Lamet (A)

N	1	2
p	3	5
q	5	3
a	1.00	1.00
b	1.00	1.00

Closedness: present.
Axial symmetry: straight lines $x = 0,\ y = 0$ are axes of symmetry.
Central symmetry: point $(0, 0)$ is a center of symmetry.
Connectedness: present.
Singular points: points $(\pm a, 0)$ are cuspidal points of the first kind $(p < q)$, straight line $y = 0$ is a tangent at the cusp points; $(0, \pm b)$ are cuspidal points of the first kind $(p < q)$, straight line $x = 0$ is a tangent at the cusp points.
Self-intersection points: none.
Special cases: ellipse $(p = 2,\ q = 1)$.

q is even
Boundedness: present.
Closedness: none.
Axial symmetry: straight line $x = y$ is an axis of symmetry $(a = b)$.
Central symmetry: none.
Connectedness: present.
Singular points: points $(a, 0), (0, b)$ are points of cessation.
Self-intersection points: none.

pq is odd
Unboundedness: present.
Asymptotes: straight line $bx + ay = 0$ is an inclined asymptote.
Axial symmetry: straight line $x = y$ is an axis of symmetry $(a = b)$.
Central symmetry: none.
Connectedness: present.
Singular points: none.

Self-intersection points: none.
(Figures 127–130).
CURVE OF LAMET (A) *is a bounded connected closed central-symmetric curve with four singular points (cusp points of the first kind* $(p < q)$*) and two axes of symmetry for even* p*, a bounded connected nonclosed curve with two singular points (points of cessation) and a single axis of symmetry (the last for* $a = b$ *only) for even* q*, an unbounded connected curve free of singular points with a single inclined asymptote and a single axis of symmetry (the last for* $a = b$ *only) for odd* pq.

Curve of Lamet (B)
Equation:

$$\left(\frac{x}{a}\right)^m + \left(\frac{y^m}{b}\right) = 1,$$

$$m = -\frac{p}{q}, \quad p \in \mathbf{N}, \quad q \in \mathbf{N}, \quad \text{G.C.Q.}(p,q) = 1, \quad a > 0, \quad b > 0$$

Basic geometric properties:

p is even
Unboundedness: present.
Asymptotes: straight lines $x = \pm a$ are vertical asymptotes; straight lines $y = \pm b$ are horizontal asymptotes.
Axial symmetry: straight lines $x = 0$, $y = 0$ are axes of symmetry.
Central symmetry: point $(0,0)$ is a center of symmetry.
Connectedness: none.
Singular points: none.
Self-intersection points: none.
Special cases: cruciform curve ($p = 2$, $q = 1$).

q is even
Unboundedness: present.
Asymptotes: straight line $x = a$ is a vertical asymptote; straight line $y = b$ is a horizontal asymptote.
Axial symmetry: straight line $x = y$ is an axis of symmetry ($a = b$).
Central symmetry: none.
Connectedness: present.
Singular points: none.
Self-intersection points: none.

pq is odd
Unboundedness: present.
Asymptotes: straight line $x = a$ is a vertical asymptote; straight line $y = b$ is a horizontal asymptote.
Axial symmetry: straight line $x = y$ is an axis of symmetry ($a = b$).

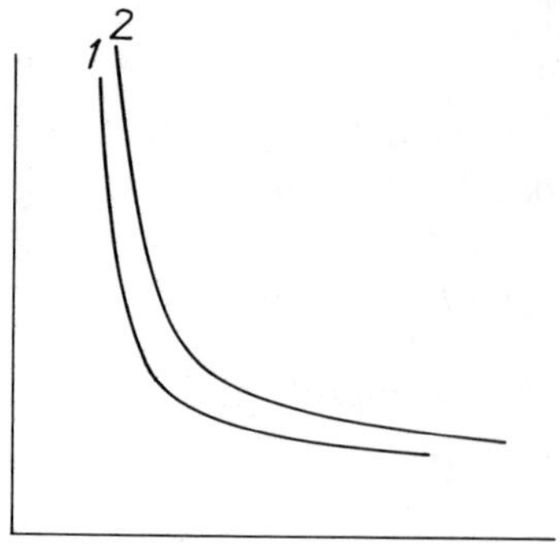

Figure 131.
Curve of Lamet (B)

N	1	2
p	1	5
q	−4	−8
a	1.00	1.00
b	1.00	1.00

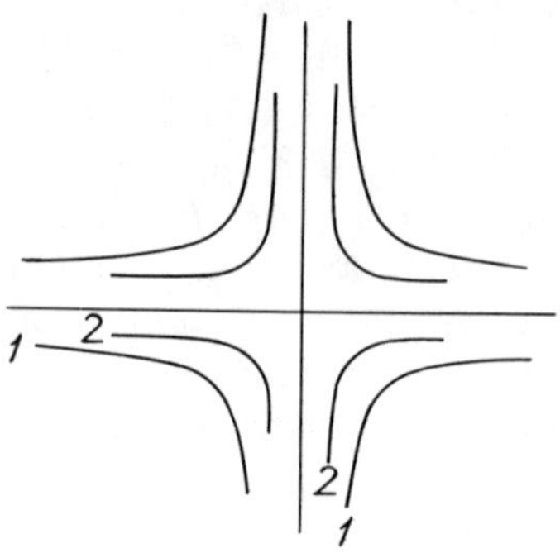

Figure 132.
Curve of Lamet (B)

N	1	2
p	2	4
q	−3	−3
a	1.00	1.00
b	1.00	1.00

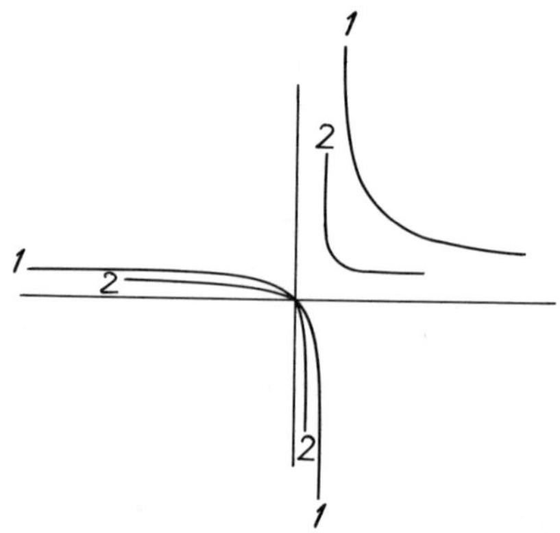

Figure 133. Curve of Lamet (B)

N	1	2
p	3	5
q	−5	−3
a	1.00	1.00
b	1.00	1.00

Central symmetry: none.
Connectedness: none.
Singular points: none.
Self-intersection points: none.
(Figures 131–133).

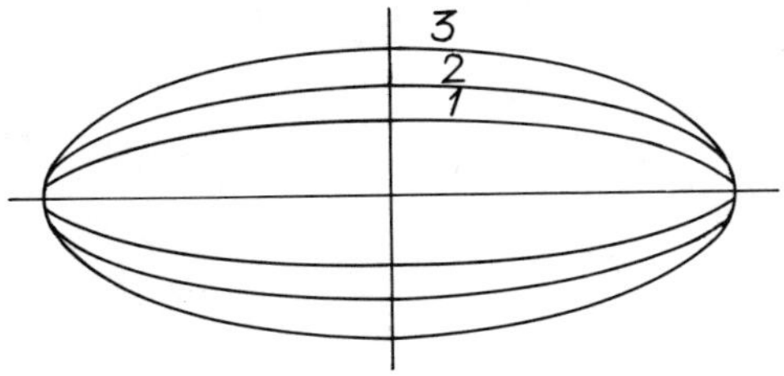

Figure 134. Curve of Legendre

N	1	2	3
b	0.20	0.30	0.40
m	0.40	0.40	0.40
h	1.00	1.00	1.00

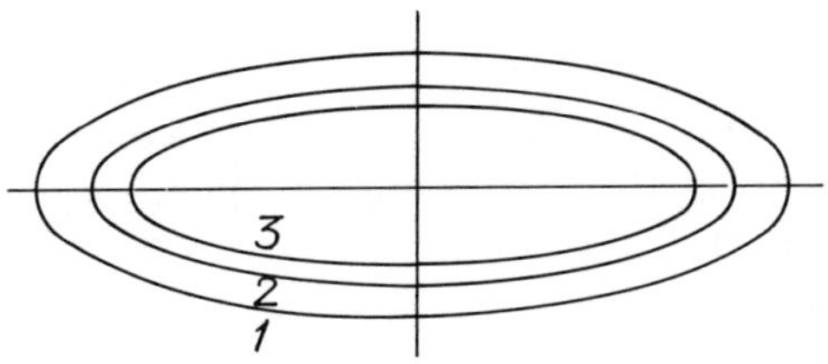

Figure 135. Curve of Legendre

N	1	2	3
m	1.00	0.40	0.01
b	0.30	0.30	0.30
h	1.00	1.00	1.00

CURVE OF LAMET (B) *is an unbounded unconnected (consisting of four unbounded branches) central-symmetric curve free of singular points with horizontal asymptotes, two vertical asymptotes and two axes of symmetry for even p, an unbounded connected curve free of singular points with a single horizontal asymptote, a single vertical asymptote and a single axis of symmetry (the last for $a = b$ only) for even q, an unbounded unconnected (consisting of two unbounded branches) curve free of singular points with a single vertical asymptote, a single horizontal asymptote and a single axis of symmetry (the last for $a = b$ only) for odd pq.*

Curve of Legendre (Legendre's curve of the sixth order, Legendre's sextic)
Equations:

$$x = h \sin t \left(1 + \frac{m}{3} \sin^2 t\right), \quad y = bh \cos \left(1 + m - \frac{m}{3} \cos^2 t\right),$$

$$m = \frac{2k^2}{1-k^2}, \quad h = \frac{1-2k^2}{b^2}, \quad k \neq 0, \quad b \neq 0$$

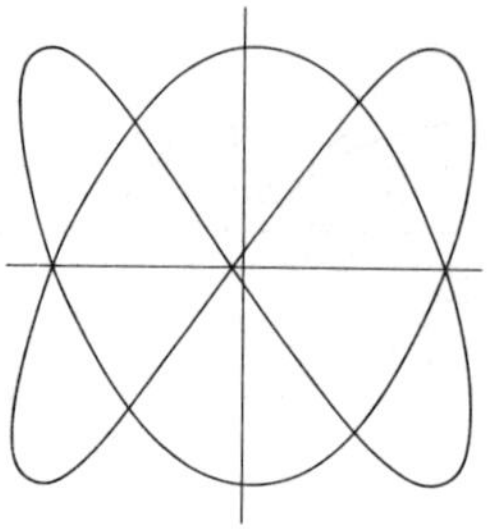

Figure 136. Curve of Lissajous
$a = b = 1.00, \quad m = 2, \quad n = 3,$
$\gamma = 0, \quad \delta = 1.5$

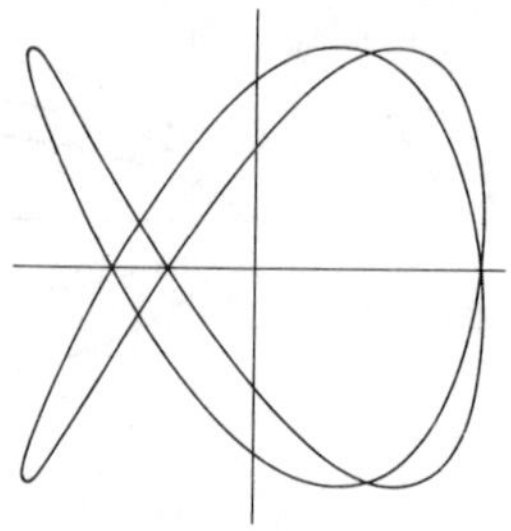

Figure 137. Curve of Lissajous
$a = b = 1.00, \quad m = 2, \quad n = 3,$
$\gamma = 0, \quad \delta = 1.0$

Basic geometric properties:
Boundedness: present.
Closedness: present.
Axial symmetry: straight lines $x = 0$, $y = 0$ are axes of symmetry.
Central symmetry: point $(0, 0)$ is a center of symmetry.
Connectedness: present.
Singular points: none.
Self-intersection points: none.
(Figures 134–135).
CURVE OF LEGENDRE *is a bounded connected closed central-symmetric curve free of singular points with two axes of symmetry.*

Curve of Lissajous (Bowditch curve)
Equations:

$$x = a\sin(mt + \gamma), \quad y = \sin(nt + \delta),$$

$$a > 0, \quad b > 0, \quad m = \frac{p}{q}, \quad n = \frac{r}{s},$$

$$p \in \mathbf{N}, \quad q \in \mathbf{N}, \quad r \in \mathbf{N}, \quad s \in \mathbf{N}$$

$$\gamma = \frac{c}{d}, \quad \delta = \frac{e}{f},$$

$$c \in \mathbf{Z}, \quad d \in \mathbf{N}, \quad e \in \mathbf{Z}, \quad f \in \mathbf{N}$$

Basic geometric properties:
Boundedness: present.
Closedness: present.
Unboundedness: present.
Axial symmetry: availability or absence of the symmetry axes depends on relationship between values of parameters m, n, γ and δ.
Central symmetry: availability or absence of the symmetry centers depends on relationship between values of parameters m, n, γ and δ.

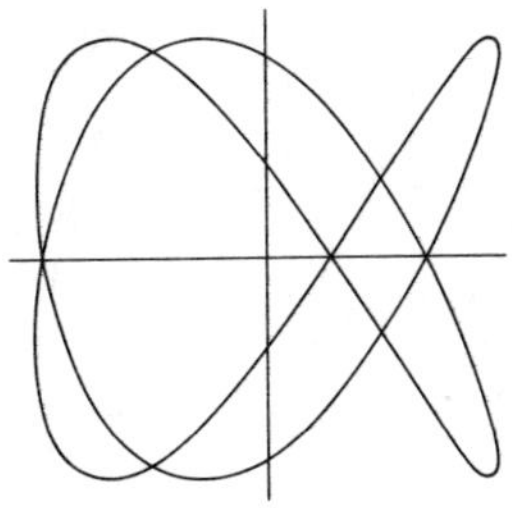

Figure 138. Curve of Lissajous
$a = b = 1.00, \quad m = 2, \quad n = 3,$
$\gamma = 0, \quad \delta = 2.0$

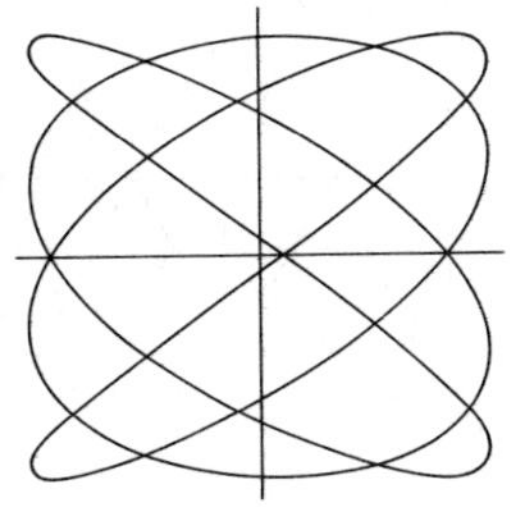

Figure 139. Curve of Lissajous
$a = b = 1.00, \quad m = 4, \quad n = 3,$
$\gamma = 0, \quad \delta = 1.5$

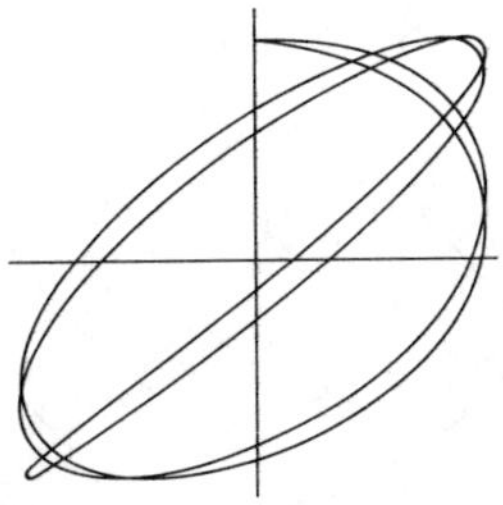

Figure 140. Curve of Lissajous
$a = b = 1.00, \quad m = 7/2, \quad n = 3,$
$\gamma = 0, \quad \delta = 1.5$

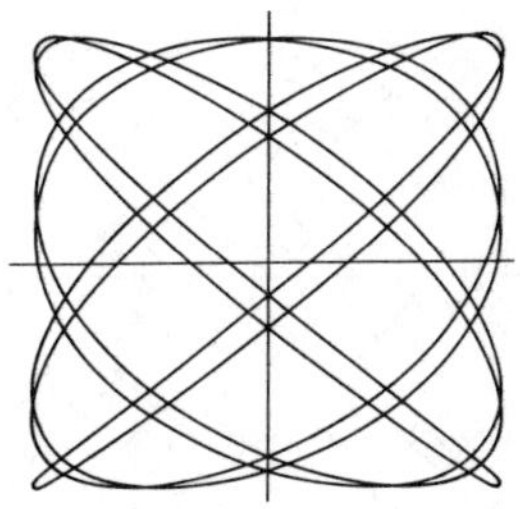

Figure 141. Curve of Lissajous
$a = b = 1.00, \quad m = 7/2, \quad n = 3,$
$\gamma = 0, \quad \delta = 1.5$

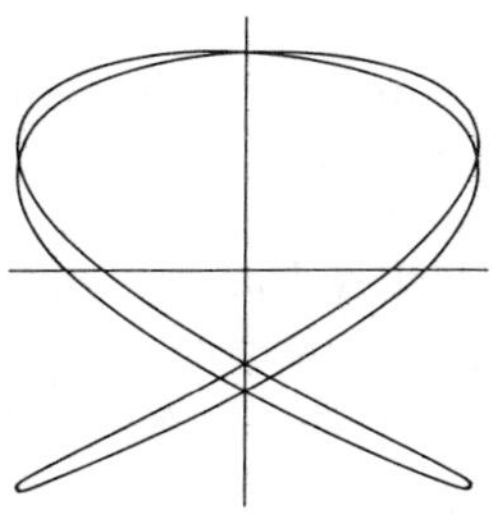

Figure 142. Curve of Lissajous
$a = b = 1.00, \quad m = 9/2, \quad n = 3,$
$\gamma = 0, \quad \delta = 1.5$

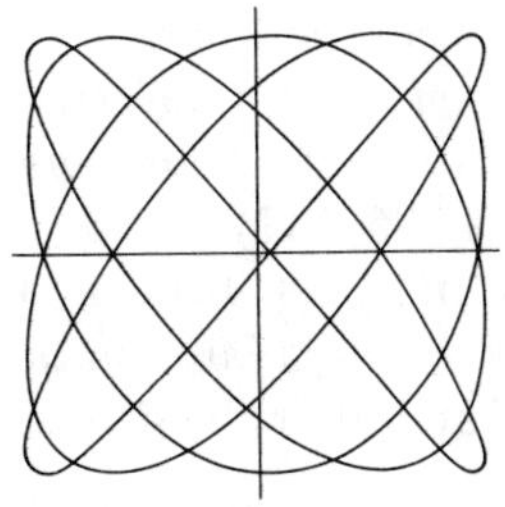

Figure 143. Curve of Lissajous
$a = b = 1.00, \quad m = 2, \quad n = 5/2,$
$\gamma = 0, \quad \delta = 1.5$

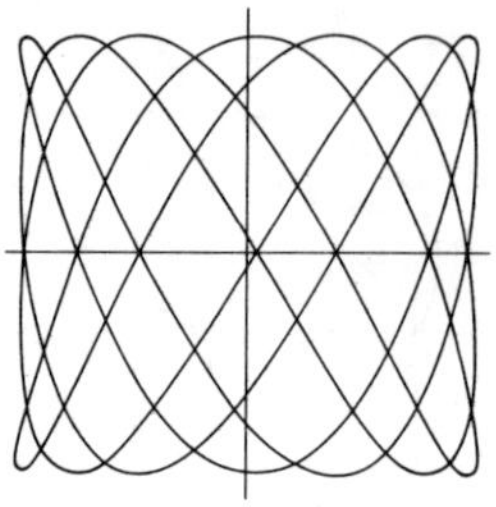

Figure 144.
Curve of Lissajous
$a = b = 1.00, \quad m = 2,$
$n = 7/2, \quad \gamma = 0, \quad \delta = 1.5$

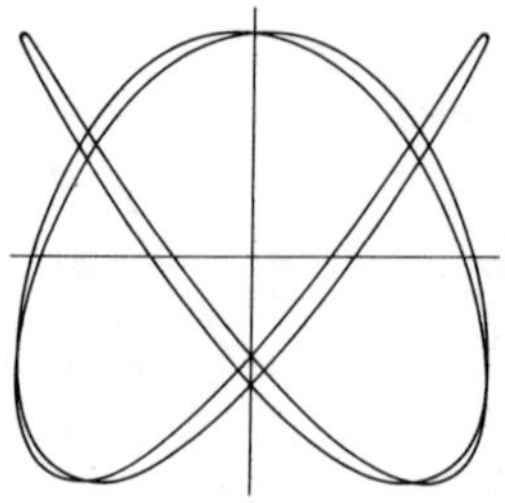

Figure 145.
Curve of Lissajous
$a = b = 1.00, \quad m = 5/2,$
$n = 10/3, \quad \gamma = 0, \quad \delta = 1.5$

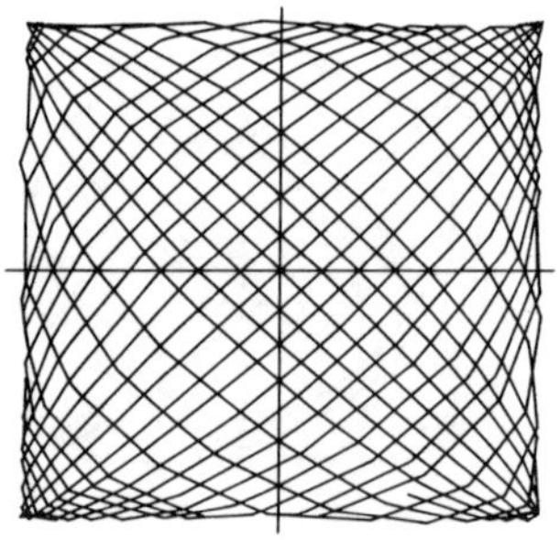

Figure 146.
Curve of Lissajous
$a = b = 1.00, \quad m = 7/2,$
$n = 10/3, \quad \gamma = 0, \quad \delta = 1.5$

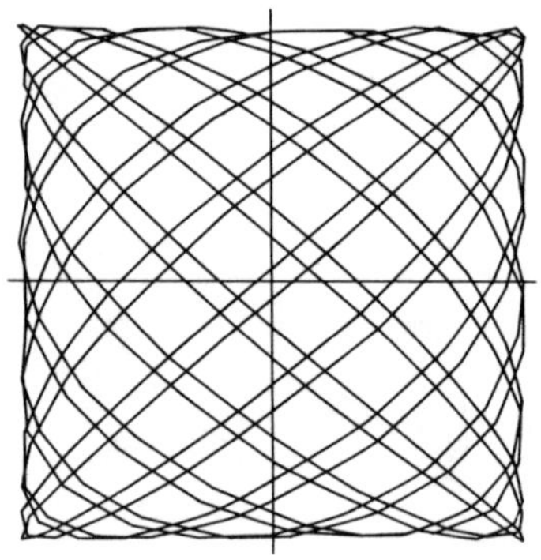

Figure 147.
Curve of Lissajous
$a = b = 1.00, \quad m = 11/10,$
$n = 13/10, \quad \gamma = 0, \quad \delta = 1.5$

Connectedness: present.
Singular points: none.
Number of self-intersection points is finite and depends on relationship between values of parameters m, n, γ and δ.
(Figures 136–147).
CURVE OF LISSAJOUS *is a bounded connected closed curve with a finite number of singular points (nodes); availability or absence of center and axes of symmetry depends on the relationship between parameter values.*

Curve of Mascheroni
Equations:

$$(x^2 + y^2)x^2 = (lx + ay)^2,$$

$$\rho = a \tan \varphi + l, \quad a > 0, \quad l > 0$$

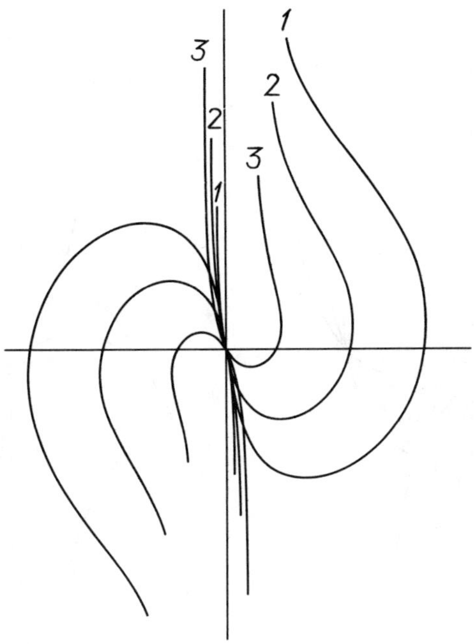

Figure 148. Curve of Mascheroni

N	1	2	3
l	8.00	5.00	2.00
a	1.00	1.00	1.00

Basic geometric properties:
Unboundedness: present.
Asymptotes: straight line $x = 0$ is a vertical asymptote.
Axial symmetry: none.
Central symmetry: point $(0, 0)$ is a center of symmetry.
Connectedness: present.
Singular points: point $(0, 0)$ is a degenerate node, straight line $lx + ay = 0$ is a tangent at the node.
Self-intersection points: one, $(0, 0)$.
(Figures 148–150).
CURVE OF MASCHERONI *is an unbounded connected central-symmetric curve with a single singular point (degenerate node) and a single vertical asymptote.*

Curve of Ozanam
Equation:

$$x = 2a \sin^2 \frac{y}{2a}, \quad a > 0$$

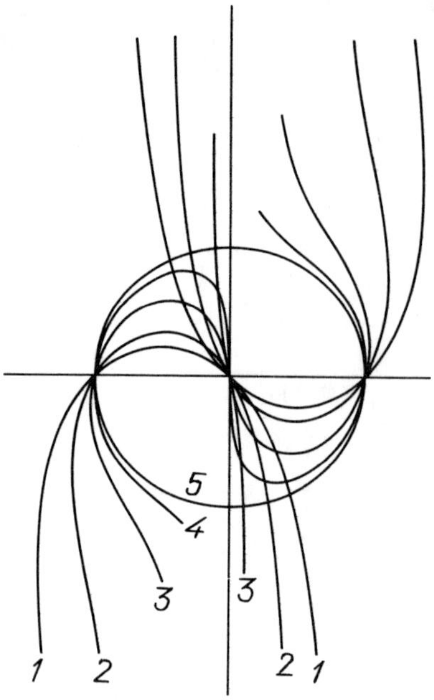

Figure 149.
Curve of Mascheroni

N	1	2	3	4	5
a	5.00	3.00	1.00	0.25	0.00
l	5.00	5.00	5.00	5.00	5.00

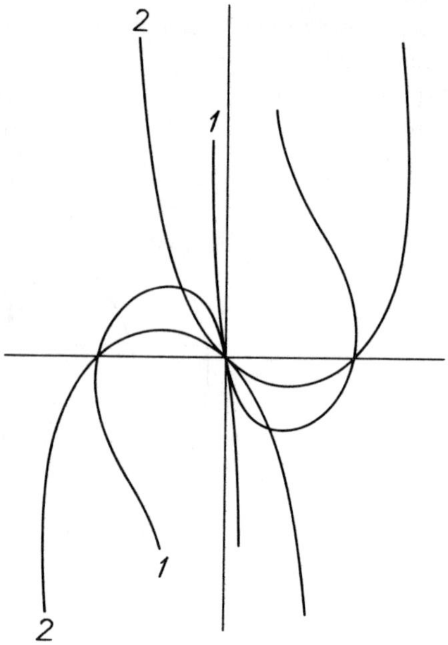

Figure 150.
Curve of Mascheroni

N	1	2
a	1.00	5.00
l	5.00	5.00

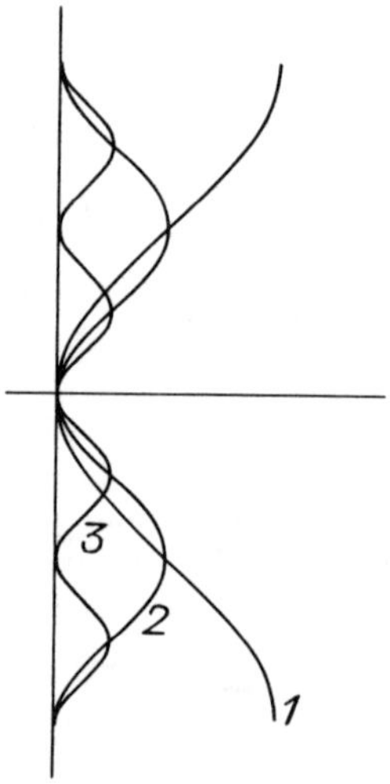

Figure 151. Curve of Ozanam

N	1	2	3
a	2.00	1.00	0.50

Basic geometric properties:
Unboundedness: present.
Asymptotes: none.
Axial symmetry: straight lines $y = k\pi a,\ k \in \mathbf{Z}$, are axes of symmetry.
Central symmetry: none.
Connectedness: present.
Singular points: none.
Self-intersection points: none.
(Figure 151).
CURVE OF OZANAM *is an unbounded connected curve free of singular points with an infinitely large number of axes of symmetry.*

Curve of Rolle (Rolle's curve of the third order)
Equations:

$$xy^2 = a(x+y)^2,$$

$$\rho = a\frac{\cos\varphi + \sin\varphi}{\cos\varphi\sin^2\varphi}, \quad a > 0$$

Additional equation:

$$xy^2 = a(y - mx)^2$$

Basic geometric properties:
Unboundedness: present (for $a > l$).
Asymptotes: straight line $x = a$ is a vertical asymptote.
Axial symmetry: none.
Central symmetry: none.
Connectedness: none.
Singular points: point $(0,0)$ is a cuspidal point, straight line $x + y = 0$ is a tangent at the cusp point.
Self-intersection points: none.
Characterized points: point $(a, -a/2)$ is an intersection point of the curve and the asymptote; point $(9a, 9a/2)$ is an inflection point; point $(4a, 4a)$ is a point of positive minimum.
(Figure 152).
CURVE OF ROLLE *is an unbounded unconnected (consisting of two unbounded branches) curve with a single singular point (cuspidal point of the first kind) and a single vertical asymptote.*

Curve of Slüse (lemniscate of Slüse)
Equations:

$$x^4 - r^2x^2 + a^2y^2 = 0,$$

$$y = \pm\frac{1}{a}\sqrt{(r^2 - x^2)x^2}, \quad a > 0, \quad r > 0$$

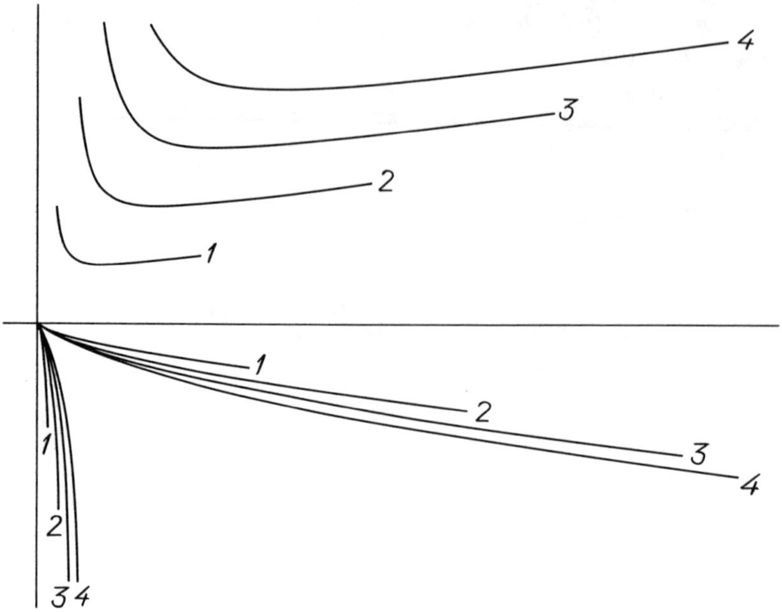

Figure 152. Curve of Rolle

N	1	2	3	4
a	0.50	1.00	1.50	2.00

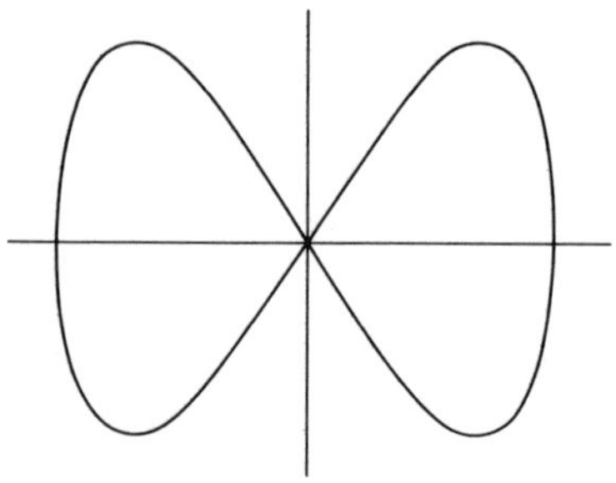

Figure 153. Curve of Slüse

Basic geometric properties:
Boundedness: present.
Closedness: present.
Axial symmetry: straight lines $x = 0$, $y = 0$ are axes of symmetry.
Central symmetry: point $(0, 0)$ is a center of symmetry.
Connectedness: present.
Singular points: point $(0, 0)$ is a node, straight lines $rx \pm ay = 0$ are tangents at the node.
Self-intersection points: one, $(0, 0)$.
(Figure 153).

CURVE OF SLÜSE *is a bounded connected closed central-symmetric curve with a single singular point (node) and two axes of symmetry.*

Curve of Watt (lemniscatoid, lemniscoid)
Equations:

$$(x^2+y^2)(x^2+y^2-a^2-b^2+c^2)^2+4a^2y^2(x^2+y^2-b^2)=0,$$

$$\rho^2=b^2-(a\sin\varphi-\sqrt{c^2-a^2\cos^2\varphi})^2,$$

$$a>0,\quad b>0,\quad c>0,$$

$$d=a^2+b^2-c^2$$

Basic geometric properties:
Boundedness: present.
Closedness: present.
Axial symmetry: straight lines $x=0,\ y=0$ are axes of symmetry.
Central symmetry: point $(0,0)$ is a center of symmetry.
Connectedness: present (for $4a^2b^2>d$), none (for $4a^2b^2<d$).
Singular points: point $(0,0)$ is a node (for $|a-b|<c<a+b$), straight lines

$$y=\pm\frac{d}{\sqrt{4a^2b^2-d^2}}x$$

are tangents at the node, or a degenerate node (for $|a-b|=c$ or $a+b=c$), straight line $x=0$ is a tangent at the degenerate node, or an isolated point (for $|a-b|>c$ or $a+b<c$).
In case $d>0$
point $(\sqrt{d},0)$ is a node (for $a<c$), straight lines

$$(x-\sqrt{d})d^2\pm ay\sqrt{c^2-a^2}=0$$

are tangents at the node, or a degenerate node (for $a=c$), straight line $x=b$ is a tangent at the degenerate node, or an isolated point for $a>c$); point $(-\sqrt{d,0})$ is a node (for $a<c$), straight lines

$$(x+\sqrt{d})d^2\pm ay\sqrt{c^2-a^2}=0$$

are tangents at the node, or degenerate node (for $a=c$), straight line $x=-b$ is a tangent at the degenerate node, or an isolated point (for $a>c$).
Self-intersection points: three, $(0,0),(\pm d,0)$ (for $|a-b|<c<a+b,\ a\leq c$); one, $(0,0)$ (for $|a-b|<c<a+b, a>c$); none (for $4a^2b^2<d^2$).
(Figures 154–159).
CURVE OF WATT *is a bounded closed central-symmetric curve with two axes of symmetry; its connectedness, number and type of singular points are defined by relationship between parameter values.*

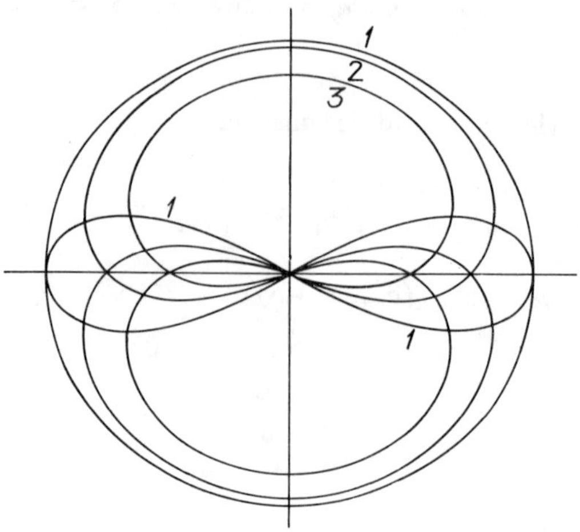

Figure 154. Curve of Watt

N	1	2	3
a	0.50	0.75	1.00
b	1.00	1.00	1.00
c	1.00	1.00	1.00

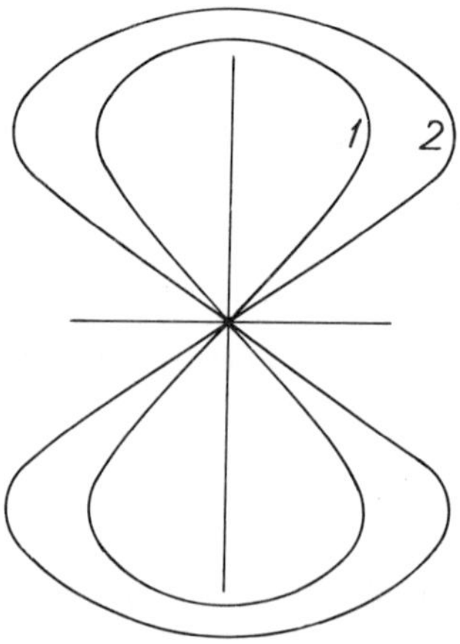

Figure 155. Curve of Watt

N	1	2
a	1.25	1.50
b	1.00	1.00
c	1.00	1.00

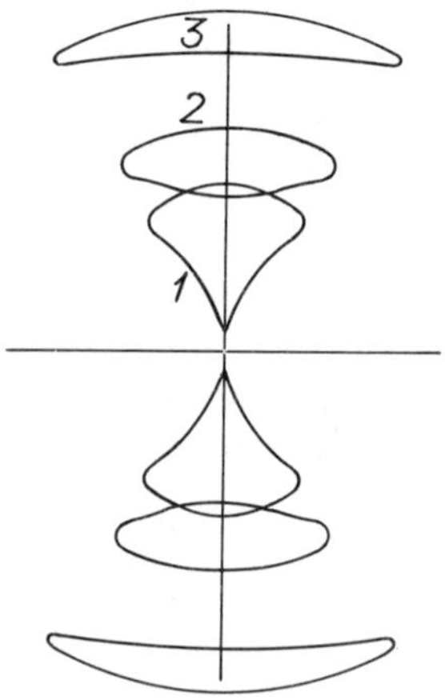

Figure 156. Curve of Watt

N	1	2	3
b	1.50	2.00	3.00
a	1.00	1.00	1.00
c	0.50	0.50	0.50

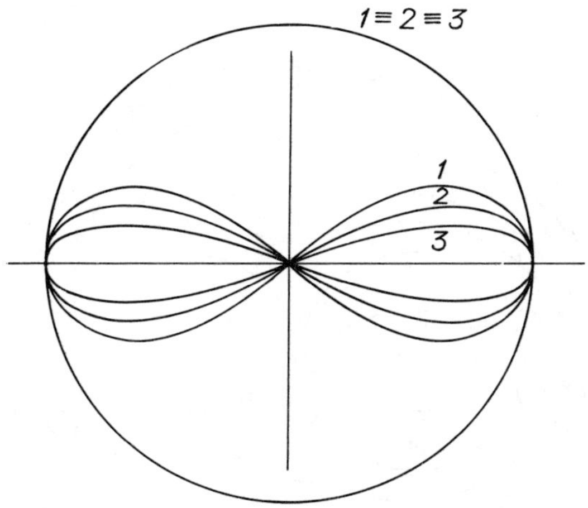

Figure 157. Curve of Watt

N	1	2	3
a	0.75	1.00	1.50
c	0.75	1.00	1.50
b	1.00	1.00	1.00

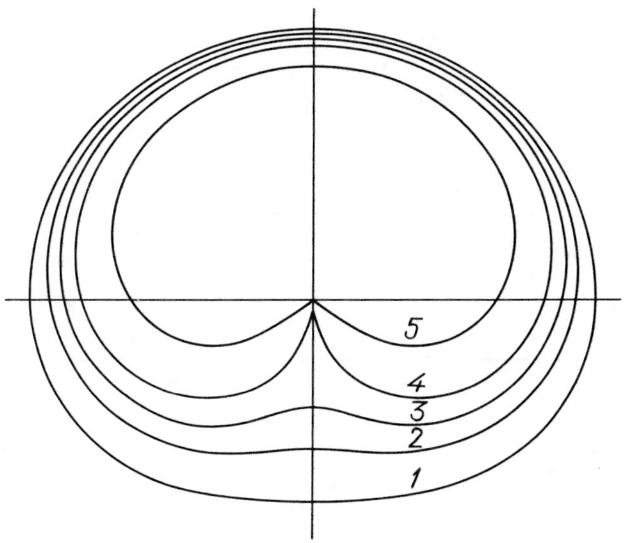

Figure 158. Curve of Watt

N	1	2	3	4	5
c	1.00	1.50	1.75	2.00	2.50
a	1.00	1.00	1.00	1.00	1.00
b	3.00	3.00	3.00	3.00	3.00

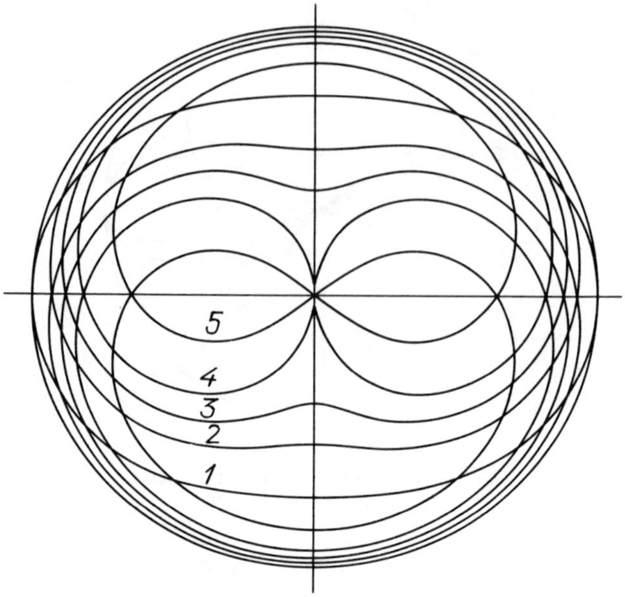

Figure 159. Curve of Watt

N	1	2	3	4	5
c	1.00	1.50	1.75	2.00	2.50
a	1.00	1.00	1.00	1.00	1.00
b	3.00	3.00	3.00	3.00	3.00

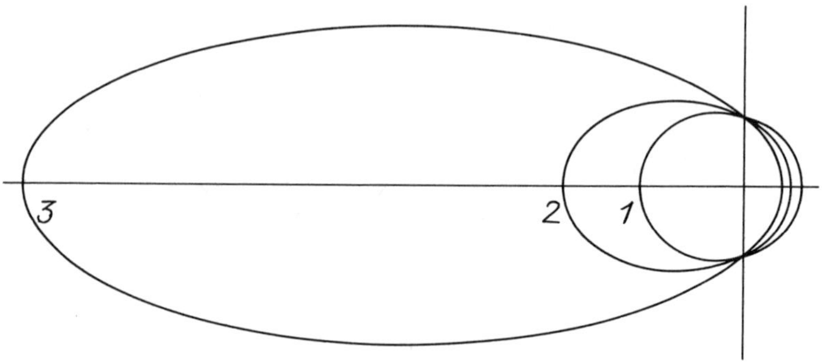

Figure 160. Curve with convexities

N	1	2	3
e	0.30	0.60	0.90
m	1	1	1
p	1.00	1.00	1.00

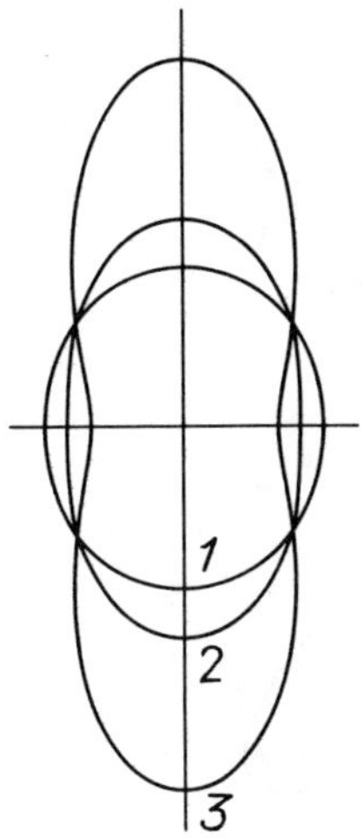

Figure 161.
Curve with convexities

N	1	2	3
e	0.10	0.30	0.60
m	2	2	2
p	1.00	1.00	1.00

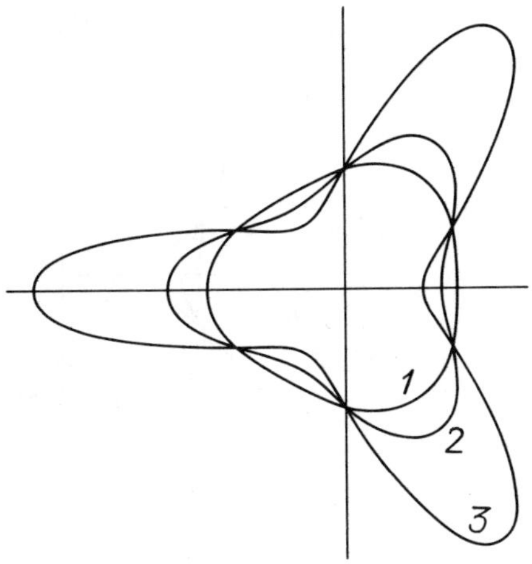

Figure 162.
Curve with convexities

N	1	2	3
e	0.10	0.30	0.60
m	3	3	3
p	1.00	1.00	1.00

Curve with convexities
Equation:

$$\rho = \frac{p}{1 + \varepsilon \cos m\varphi},$$

$$p > 0, \quad 0 < \varepsilon < 1, \quad m \in \mathbf{N}, \quad m > 1$$

Basic geometric properties:
Boundedness: present.
Closedness: present.
Axial symmetry: straight lines

$$x \sin \frac{\pi}{m} k - y \cos \frac{\pi}{m} k = 0,$$

$$k = 0, 1, \ldots, m - 1,$$

are axes of symmetry.
Central symmetry: point $(0, 0)$ is a center of symmetry (for even m).
Connectedness: present.
Singular points: none.
Self-intersection points: none.
(Figures 160–166).
CURVE WITH CONVEXITIES *is a bounded connected closed central-symmetric (for even m only) curve free of singular points with m axes of symmetry.*

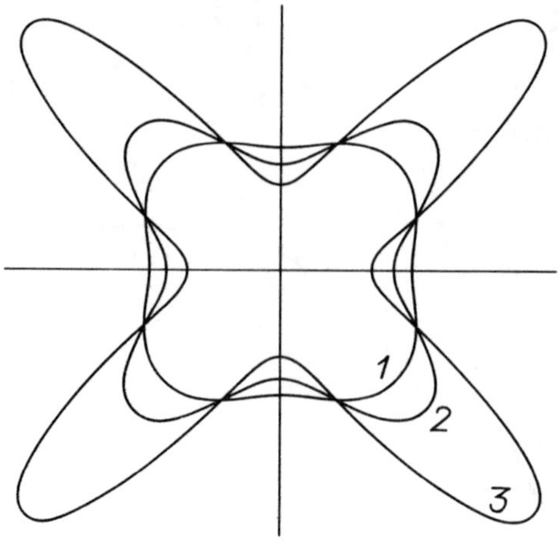

Figure 163.
Curve with convexities

N	1	2	3
e	0.10	0.30	0.60
m	4	4	4
p	1.00	1.00	1.00

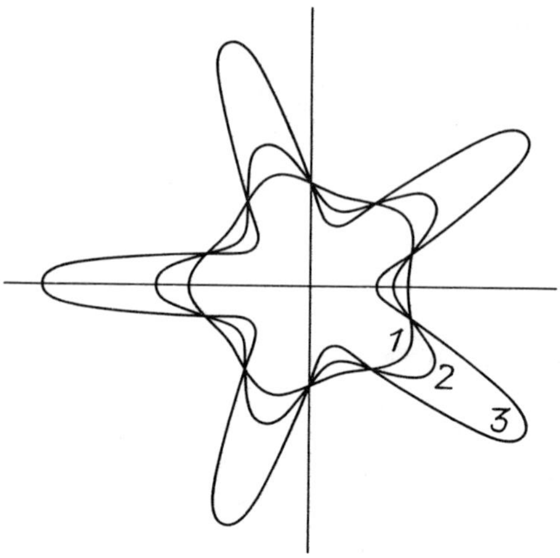

Figure 164.
Curve with convexities

N	1	2	3
e	0.10	0.30	0.60
m	5	5	5
p	1.00	1.00	1.00

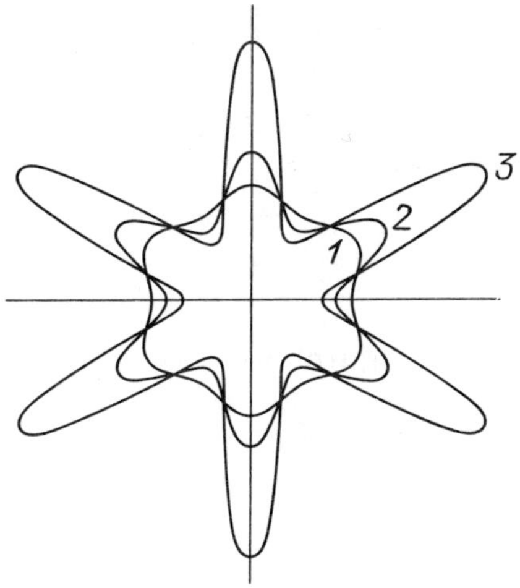

Figure 165. Curve with convexities

N	1	2	3
e	0.10	0.30	0.60
m	6	6	6
p	1.00	1.00	1.00

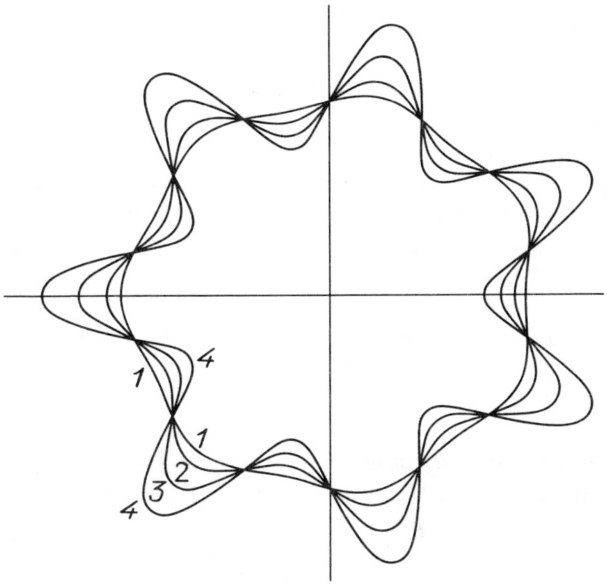

Figure 166. Curve with convexities

N	1	2	3	4
e	0.04	0.10	0.20	0.30
m	7	7	7	7
p	1.00	1.00	1.00	1.00

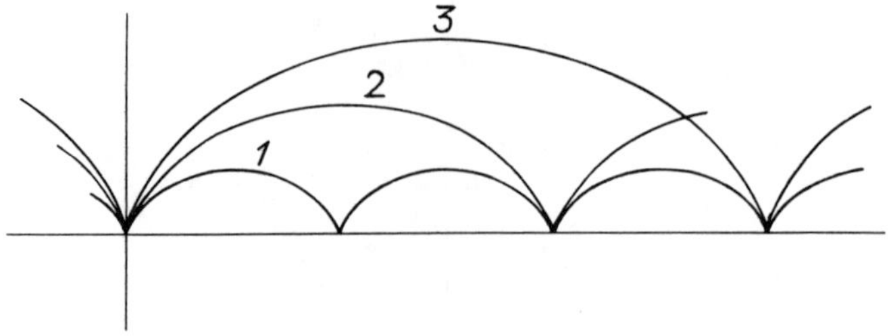

Figure 167. Cycloid

N	1	2	3
r	0.50	1.00	1.50

Cycloid (common cycloid)
Equations:

$$x = rt - r\sin t, \quad y = r - r\cos t, \quad r > 0$$

Additional equations:

$$x = r \arccos \frac{r - y}{y} \sqrt{2ry - y^2}$$

Basic geometric properties:
Unboundedness: present.
Asymptotes: none.
Axial symmetry: straight lines $x = k\pi r$, $k \in \mathbf{Z}$, are axes of symmetry.
Central symmetry: none.
Connectedness: present.
Singular points: points $(2\pi kr, 0)$, $k \in \mathbf{Z}$, are cuspidal points of the first kind, straight lines $x = 2\pi kr$, $k \in \mathbf{Z}$, are tangents at the cusp points.
Self-intersection points: none.
Generalization:

$$x = lr(t - \sin t), \quad y = r(1 - \cos t), \quad l \neq 1$$

is cycloid of Fermat.
(Figure 167).
CYCLOID *is an unbounded connected curve with an infinitely large number of singular points (cuspidal points of the first kind) and an infinitely large number of the symmetry axes.*

2.2.4 D

Double-heart curve
Equations:

$$y^4 + 2x^2y^2 + x^4 - 6axy^2 - ax^3 + a^2x^2 = 0,$$

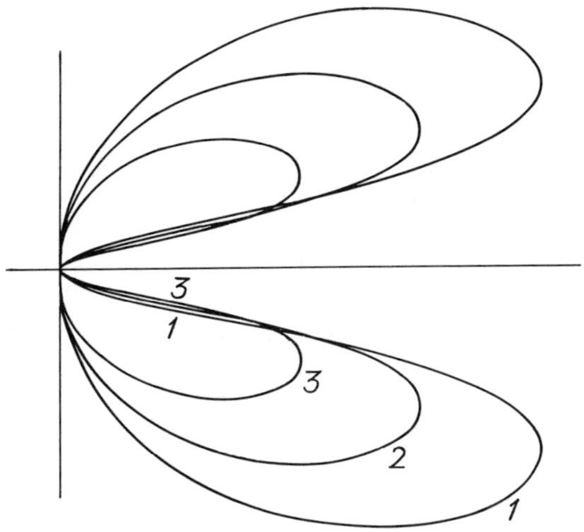

Figure 168. Double-heart curve

N	1	2	3
a	0.80	0.60	0.40

$$y = \pm\frac{1}{2}(\sqrt{x(2a-x)} \pm \sqrt{x(4a-x)}, \quad a > 0$$

Basic geometric properties:
Boundedness: present.
Closedness: present.
Axial symmetry: straight line $y = 0$ is an axis of symmetry.
Central symmetry: none.
Connectedness: none.
Singular points: point $(0,0)$ is a degenerate node (tacnode), straight line $x = 0$ is a tangent at the tacnode.
Self-intersection points: one, $(0,0)$.
(Figure 168).
DOUBLE-HEART CURVE *is a bounded connected closed curve with a single singular point (degenerate node) and a single axis of symmetry.*

Double sinusoid
Equation:

$$y = a\sin(\sin x), \quad x > 0$$

Basic geometric properties:
Unboundedness: present.
Asymptotes: none.

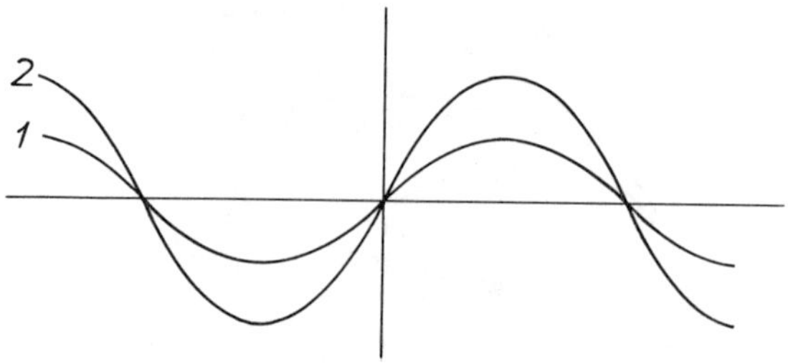

Figure 169. Double sinusoid

N	1	2
a	1.00	2.00

Axial symmetry: straight lines

$$x = \frac{\pi}{2} + \pi k, \quad k \in \mathbf{Z}$$

are axes of symmetry.
Central symmetry: points $(\pi k, 0)$, $k \in \mathbf{Z}$, are centers of symmetry.
Connectedness: present.
Singular points: none.
Self-intersection points: none.
(Figure 169).
DOUBLE SINUSOID *is an unbounded connected central-symmetric curve free of singular points with an infinitely large number of axes and centers of symmetry.*

Duplicatrix cubic (duplicatrix)
Equations:

$$x^3 = l(x^2 + y^2),$$

$$\rho = \frac{l}{\cos^3 \varphi},$$

$$x = l(1 + t^2), \quad y = lt(1 + t^2),$$

$$l > 0$$

Basic geometric properties:
Unboundedness: present.
Asymptotes: none.
Axial symmetry: straight line $y = 0$ is an axis of symmetry.
Central symmetry: none.
Connectedness: none.
Singular points: point $(0, 0)$ is an isolated point.
Self-intersection points: none.

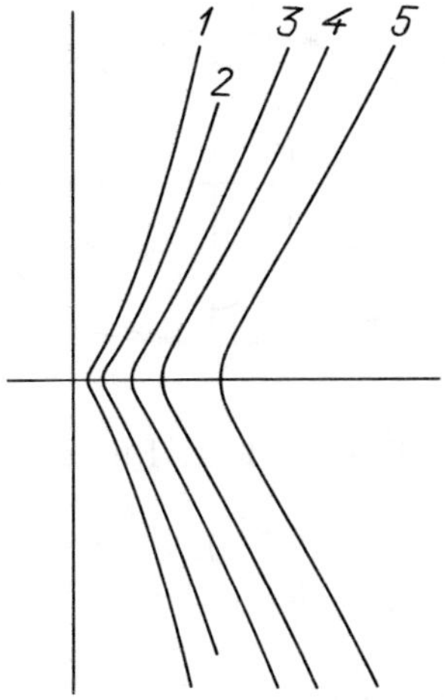

Figure 170. Duplicatrix cubic

N	1	2	3	4	5
l	0.50	1.00	2.00	3.00	5.00

Characterized points: point $(l,0)$ is an intersection point of the curve and x-axis; points

$$\left(\frac{4l}{3}, \pm\frac{4l}{3\sqrt{3}}\right)$$

are inflection points.
(Figure 170).
DUPLICATRIX CUBIC *is an unbounded unconnected (consisting of an unbounded branch and an isolated point) curve with a single singular point (isolated point) and a single axis of symmetry.*

2.2.5 E

Ellipse
Equations:

$$\frac{x^2}{a^2} + \frac{y^2}{b^2} = 1,$$

$$x = a\cos\varphi, \quad y = b\sin\varphi,$$

$$a > 0, \quad b > 0, \quad a \neq b$$

Additional equations:

$$y^2 = 2px + qx^2, \quad q < 0,$$

$$\rho = \frac{p}{1 + \varepsilon\cos\varphi}, \quad 0 < \varepsilon < 1$$

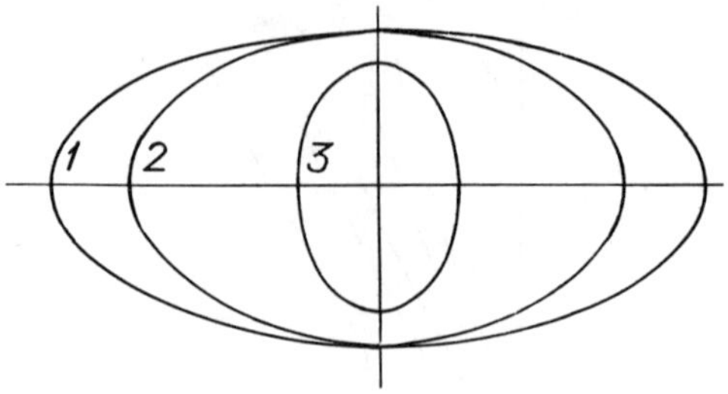

Figure 171. Ellipse

N	1	2	3
a	1.00	0.75	0.25
b	0.50	0.50	0.40

Basic geometric properties:
Boundedness: present.
Closedness: present.
Axial symmetry: straight lines $x = 0$, $y = 0$ are axes of symmetry.
Central symmetry: point $(0, 0)$ is a center of symmetry.
Connectedness: present.
Singular points: none.
Self-intersection points: none.
Characterized points: points $(\pm a, 0)$, $(0, \pm b)$ are vertices.
(Figure 171).
ELLIPSE *is a bounded connected closed central-symmetric curve free of singular points with two axes of symmetry.*

Elliptic lemniscate
Equations:

$$(x^2 + y^2)^2 = a^2x^2 + b^2y^2,$$

$$\rho = \sqrt{a^2 \cos^2 \varphi + b^2 \sin^2 \varphi}, \quad a > 0, \quad b > 0, \quad a \neq b$$

Basic geometric properties:
Boundedness: present.
Closedness: present.
Axial symmetry: straight lines $x = 0$, $y = 0$ are axes of symmetry.
Central symmetry: point $(0, 0)$ is a center of symmetry.
Connectedness: none.
Singular points: point $(0, 0)$ is an isolated point.
Self-intersection points: none.
Area of domain enclosed by the curve: $\pi(a^2 + b^2)/2$.
(Figures 172–173).

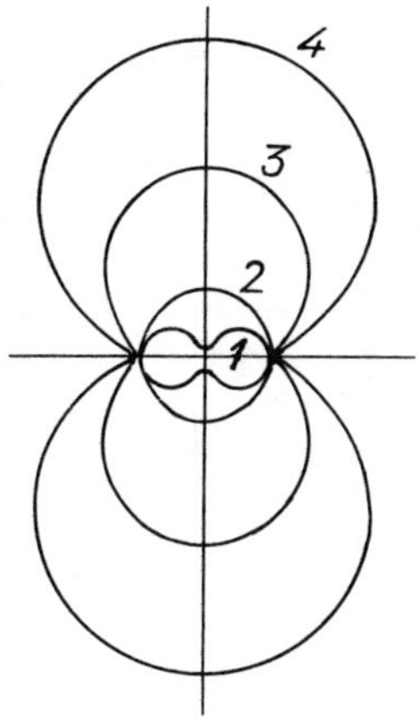

Figure 172. Elliptic lemniscate

N	1	2	3	4
b	0.30	1.00	3.00	5.00
a	1.00	1.00	1.00	1.00

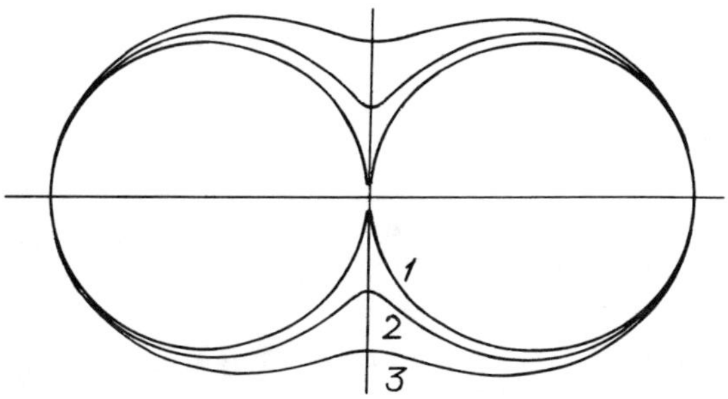

Figure 173. Elliptic lemniscate

N	1	2	3
b	0.10	0.30	0.50
a	1.00	1.00	1.00

ELLIPTIC LEMNISCATE *is a bounded unconnected closed central-symmetric curve with a single singular point (isolated point) and two axes of symmetry.*

Epicycloid

Equations:

$$x = (R + mR)\cos mt + mR\cos(t + mt),$$

$$y = (R + mR)\sin mt - mR\sin(t + mt),$$

$$\rho^2 = (R+)^2 + h^2 - 2(R + r)h\cos\frac{R}{r}\varphi$$

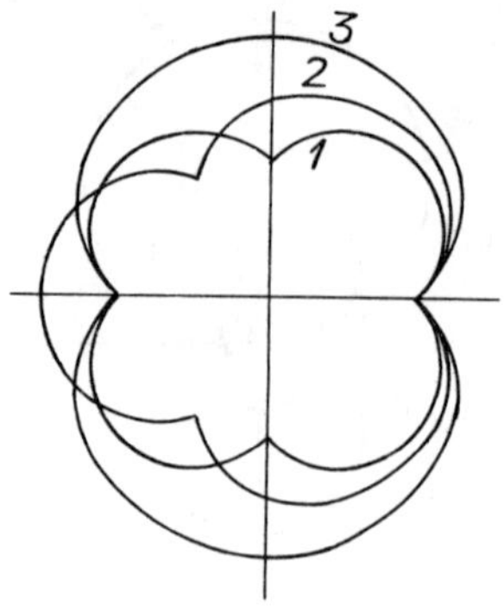

Figure 174. Epicycloid

N	1	2	3
m	1/2	1/3	1/4
R	1.00	1.00	1.00

$$R > 0, \quad m = \frac{p}{q}, \quad p \in \mathbf{N}, \quad q \in \mathbf{N}, \quad \text{G.C.D.}(p, q) = 1$$

Basic geometric properties:

Boundedness: present.
Closedness: present.
Axial symmetry: straight lines

$$x \sin \frac{\pi}{q} k = y \cos \frac{\pi}{q} k, \quad k = 0, 1, \ldots, q - 1$$

are axes of symmetry.
Central symmetry: point $(0, 0)$ is a center of symmetry (q is even).
Connectedness: present.
Singular points: points

$$\left(R \cos \frac{2\pi}{q} k, R \sin \frac{2\pi}{q} k \right), \quad k = 0, 1, \ldots, q - 1$$

are cuspidal points of the first kind, straight lines

$$x \sin \frac{2\pi}{q} k = y \cos \frac{2\pi}{q} k, \quad k = 0, 1, \ldots, q - 1$$

are tangents at the corresponding cusp points; epicycloid may possess $(p - 1)q$ nodes of the form

$$\left(\rho_l \cos \frac{2k + 1}{q} \pi, \rho_l \sin \frac{2k + 1}{q} \pi \right), \quad l = 1, 3, \ldots,$$

$$\left(\rho_n \cos \frac{2k}{q} \pi, \rho_n \sin \frac{2k}{q} \pi \right), \quad n = 2, 4, \ldots, \quad k = 0, 1, \ldots, q - 1$$

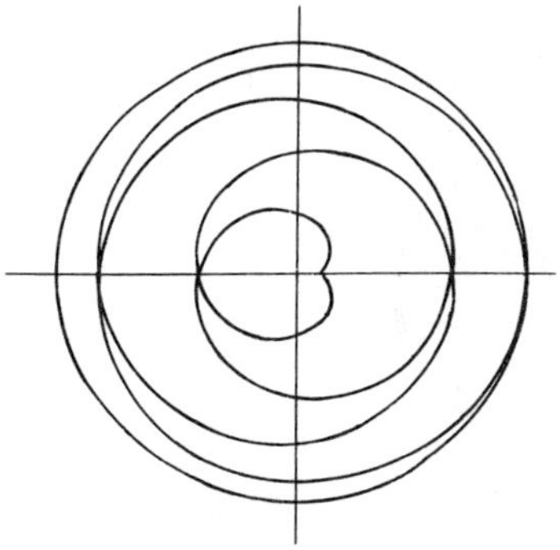

Figure 175. Epicycloid
$m = 5, \quad R = 1.00$

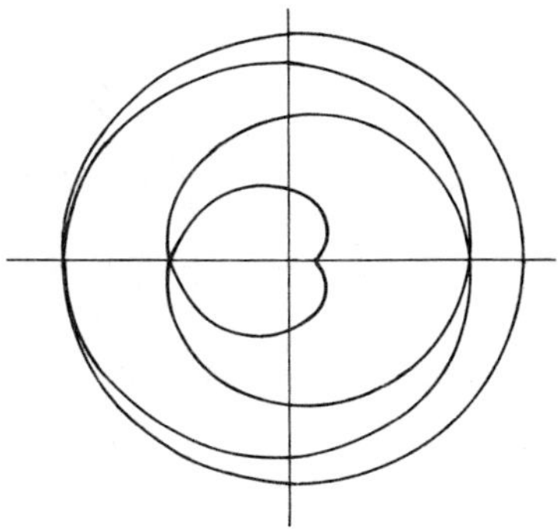

Figure 176. Epicycloid
$m = 4, \quad R = 1.00$

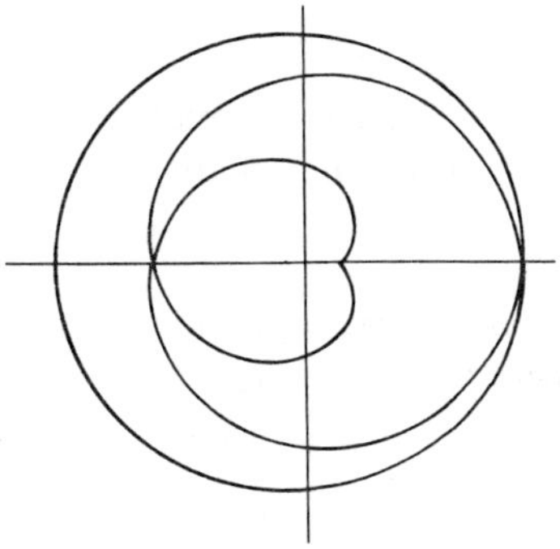

Figure 177. Epicycloid
$m = 3, \quad R = 1.00$

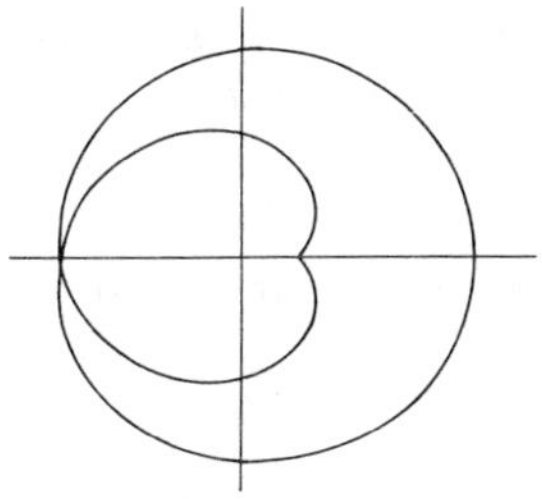

Figure 178. Epicycloid
$m = 2, \quad R = 1.00$

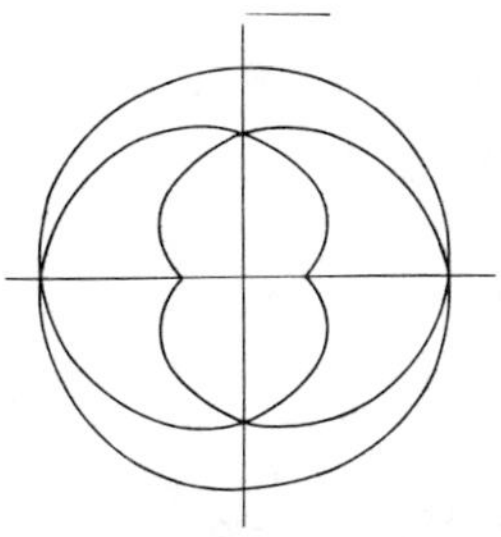

Figure 179. Epicycloid
$m = 3/2, \quad R = 1.00$

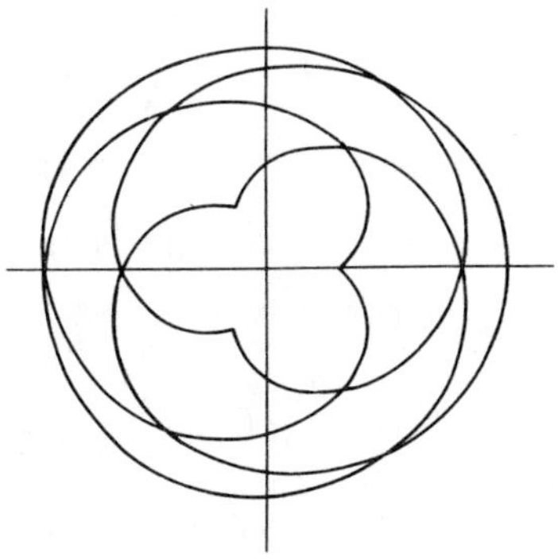

Figure 180. Epicycloid
$m = 4/3, \quad R = 1.00$

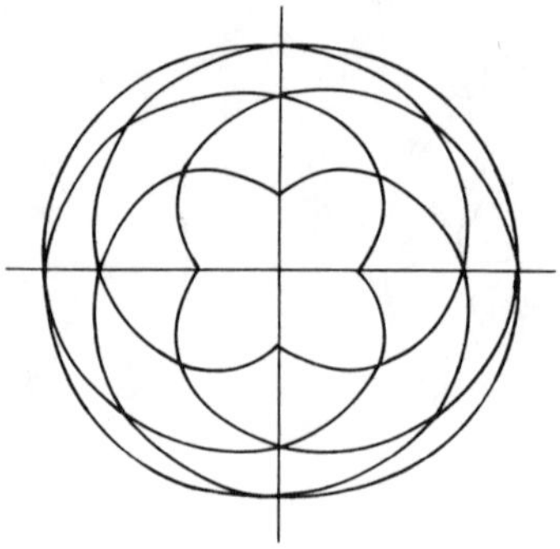

Figure 181. Epicycloid
$m = 5/4, \quad R = 1.00$

Number of self-intersection points is finite and depends on values of parameters R and m.
(Figures 174–181).
EPICYCLOID *is a bounded connected closed central-symmetric (the last for even q only) curve with q cuspidal points of the first kind, $(p-1)q$ nodes and q axes of symmetry.*

In case of irrational m the hypocycloid is a transcendent unclosed curve with an infinitely large number of nodes and cusp points.

Epitrochoid

Equations:

$$x = (R+mR)\cos mt - h\cos(t+mt), \quad y = (R+mR)\sin mt - m\sin(t+mt),$$

$$\rho^2 = (R+mR)^2 + h^2 - 2(R+mR)h\cos t,$$

$$R > 0, \quad m = \frac{p}{q},$$

$$p \in \mathbf{N}, \quad q \in \mathbf{N},$$

$$\text{G.C.D.}(p,q) = 1,$$

$$h > 0, \quad h \neq mR, \quad m \neq -1$$

Additional equations:

$$x = (R+r)\cos\varphi - h\cos\frac{R+r}{r}\varphi, y = (R+r)\sin\varphi - h\sin\frac{R+r}{r}\varphi,$$

$$\rho^2 = (R+r)^2 + h^2 - 2(R+r)h\cos\frac{R}{r}\varphi$$

Basic geometric properties:
Boundedness: present.

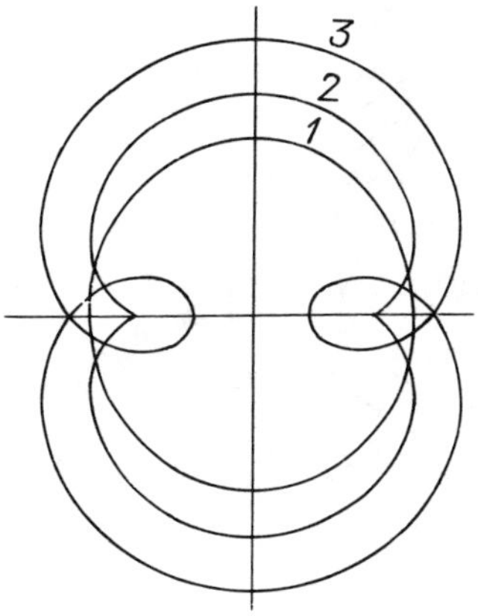

Figure 182. Epitrochoid

N	1	2	3
h	0.10	0.50	1.00
m	1/2	1/2	1/2
R	1.00	1.00	1.00

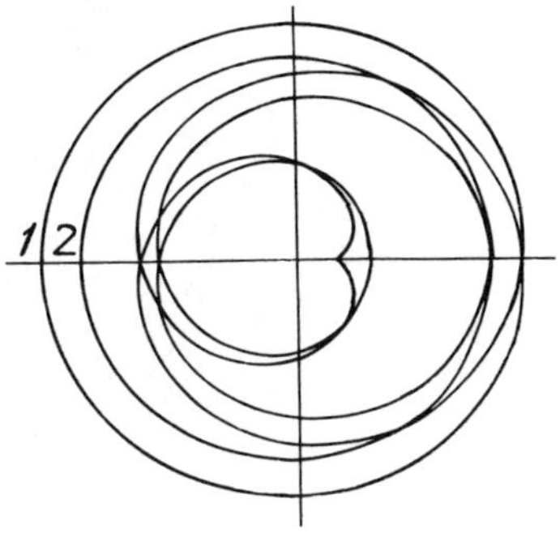

Figure 183. Epitrochoid

N	1	2
h	3.00	2.00
m	3	3
R	1.00	1.00

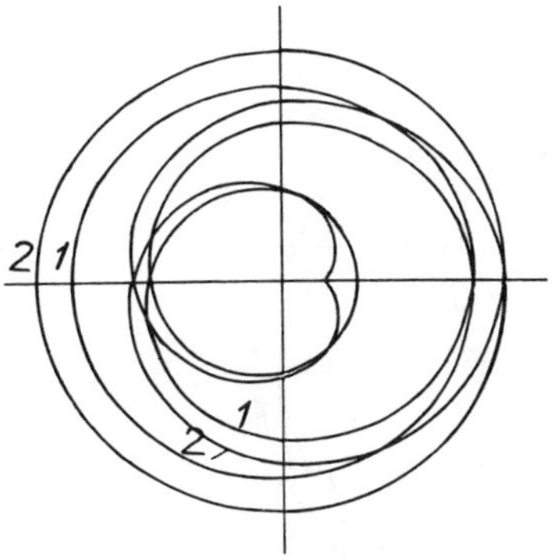

Figure 184. Epitrochoid

N	1	2
h	3.00	5.00
m	3	3
R	1.00	1.00

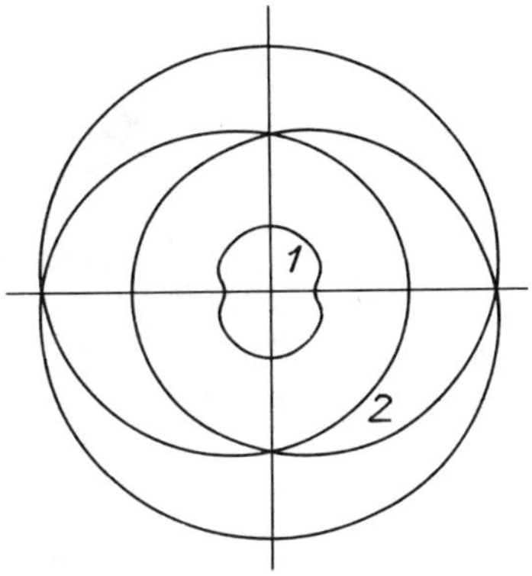

Figure 185. Epitrochoid

N	1	2
h	1.25	1.00
m	5/4	5/4
R	1.00	1.00

Closedness: present.
Axial symmetry: straight lines

$$x \sin \frac{\pi}{q} k = y \cos \frac{\pi}{q} k, \quad k = 0, 1, \ldots, q-1$$

are axes of symmetry.

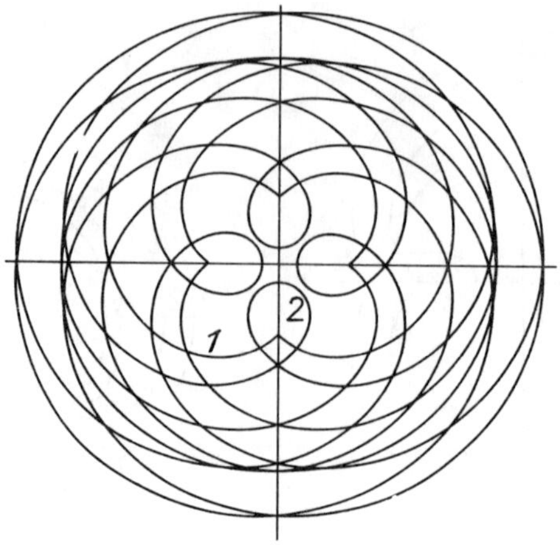

Figure 186. Epitrochoid

N	1	2
h	1.25	2.00
m	5/4	5/4
R	1.00	1.00

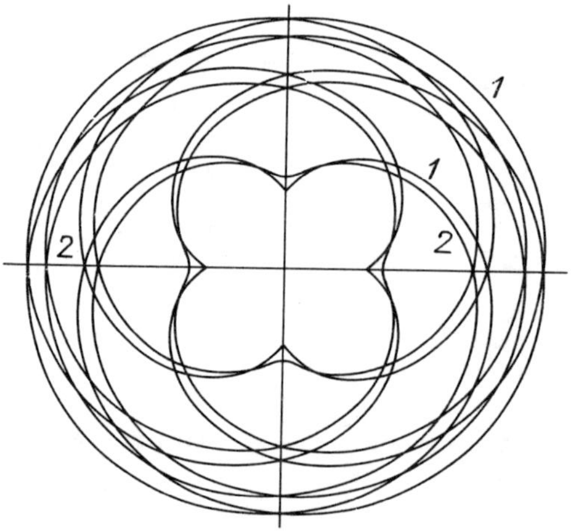

Figure 187. Epitrochoid

N	1	2
h	1.25	5.00
m	1/2	1/2
R	1.00	1.00

Central symmetry: point $(0,0)$ is a center of symmetry (q is even).
Connectedness: present.
Singular points: epitrochoid may have nodes and cuspidal points of the first

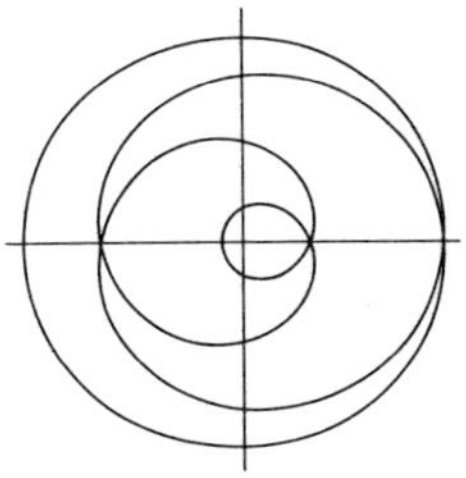

Figure 188. Epitrochoid
$h = 5.00, \quad m = 3, \quad R = 1.00$

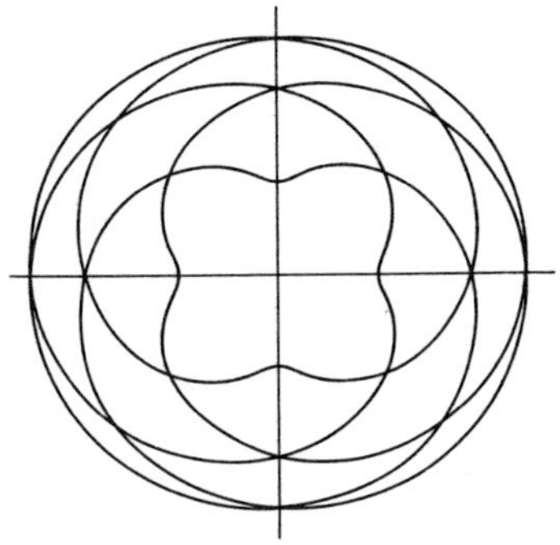

Figure 189. Epitrochoid
$h = 1.00, \quad m = 5/4, \quad R = 1.00$

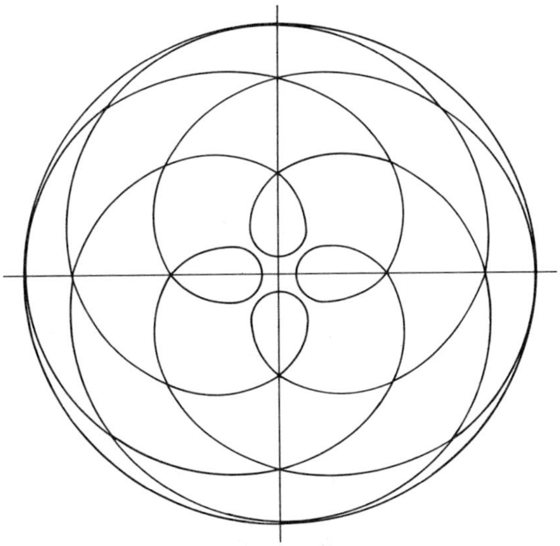

Figure 190. Epitrochoid
$h = 2.00, \quad m = 5/4, \quad R = 1.00$

kind (their number depends on values of parameters R, m and h) located on the symmetry axes.
Number of self-intersection points is finite and depends on values of parameters R, m and h.
Special cases: prolate epitrochoid ($h > Rm$), curtate epitrochoid ($h < Rm$), epicycloid ($h = Rm$).
(Figures 182–193).
EPITROCHOID *is a bounded connected closed central-symmetric (the last for even q only) curve with n_1q nodes, n_2q cuspidal points of the first kind and q axes of symmetry.*

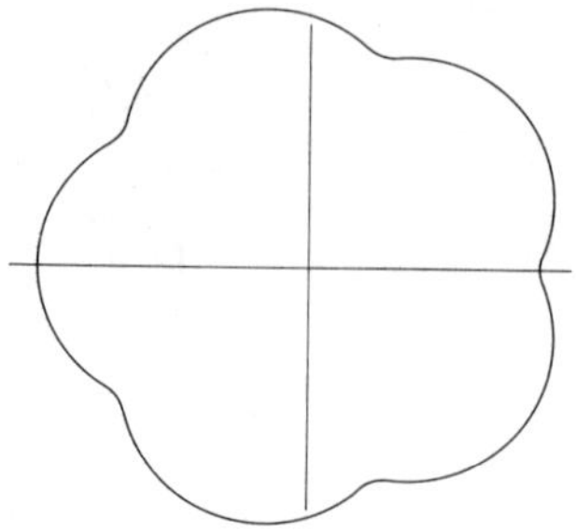

Figure 191. Epitrochoid
$h = 0.10$,
$m = 1/5$,
$R = 1.00$

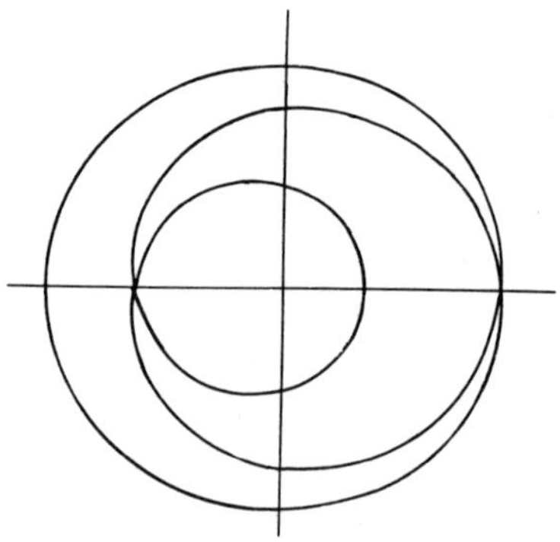

Figure 192. Epitrochoid
$h = 2.00$,
$m = 3$,
$R = 1.00$

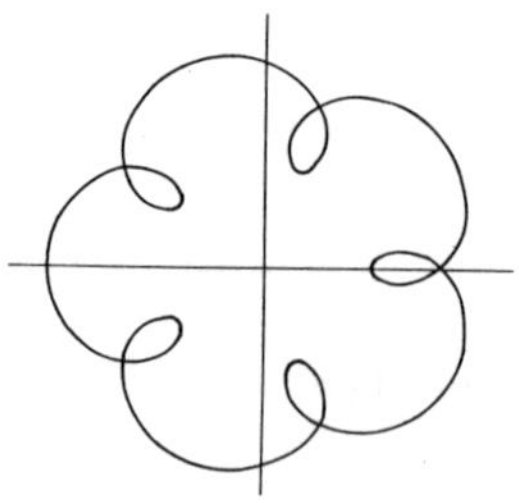

Figure 193. Epitrochoid
$h = 0.40$,
$m = 1/5$,
$R = 1.00$

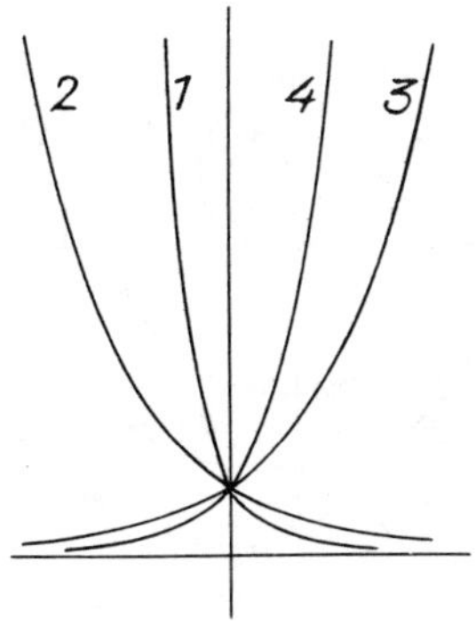

Figure 194. Exponential curve

N	1	2	3	4
a	0.10	0.50	2.00	4.00

In case of irrational m the hypocycloid is a transcendent unclosed curve with an infinitely large number of nodes and cusp points.

Exponential curve (exponent)
Equations:

$$y = a^x, \quad a > 0, \quad a \neq 1$$

Basic geometric properties:
Unboundedness: present.
Asymptotes: straight lines $y = 0$ is a horizontal asymptote.
Axial symmetry: none.
Central symmetry: none.
Connectedness: present.
Singular points: none.
Self-intersection points: none.
(Figure 194).
EXPONENTIAL CURVE *is an unbounded connected curve free of singular points with a single horizontal asymptote.*

2.2.6 F

Fermat's spiral (parabolic spiral)
Equation:

$$\rho = a\sqrt{\varphi}, \quad a > 0$$

Basic geometric properties:
Unboundedness: present.
Asymptotes: none.

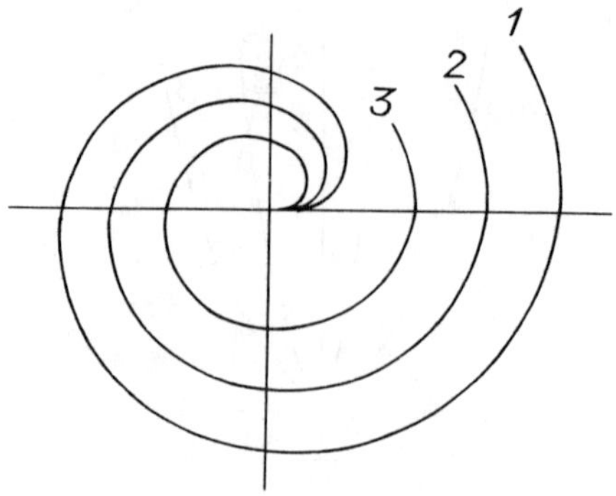

Figure 195. Fermat's spiral

N	1	2	3
a	1.00	0.75	0.50

Axial symmetry: none.
Central symmetry: none.
Connectedness: present.
Singular points: point $(0,0)$ is a cessation point.
Self-intersection points: none.
(Figure 195).
FERMAT'S SPIRAL *is an unbounded connected curve with a single singular point (point of cessation).*

First pearl
Equation:

$$x^2(a-x)=b^2y, \quad a>0, \quad b>0$$

Basic geometric properties:
Unboundedness: present.
Asymptotes: none.
Axial symmetry: none.
Central symmetry: none.
Connectedness: present.
Singular points: none.
Self-intersection points: none.
(Figures 196–197).
FIRST PEARL *is an unbounded connected curve free of singular points.*

Folium (simple folium)
Equations:

$$(x^2+y^2)^2=4rx^3, \quad \rho=4r\cos^3\varphi, \quad r>0$$

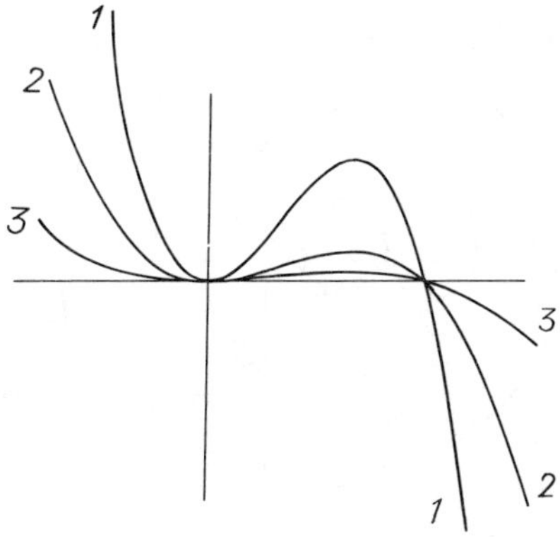

Figure 196. First pearl

N	1	2	3
b	0.50	1.00	2.00
a	1.00	1.00	1.00

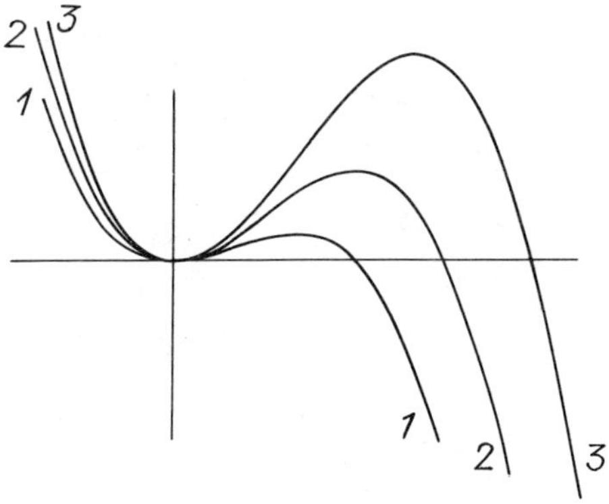

Figure 197. First pearl

N	1	2	3
a	0.50	1.00	2.00
b	1.00	1.00	1.00

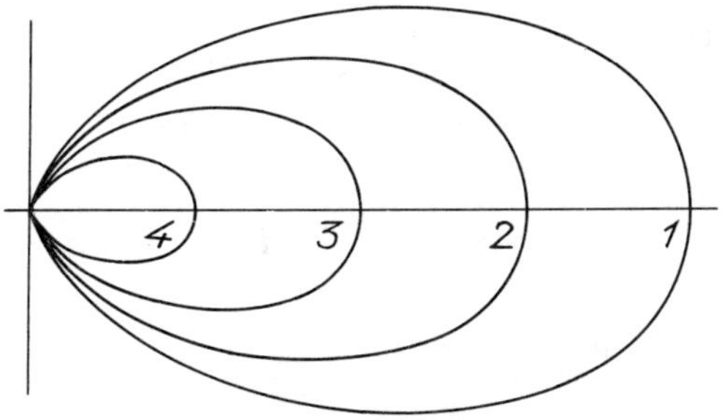

Figure 198. Folium

N	1	2	3	4
r	1.00	0.75	0.50	0.25

Basic geometric properties:
Boundedness: present.
Closedness: present.
Axial symmetry: straight line $y = 0$ is an axis of symmetry.
Central symmetry: none.
Connectedness: present.
Singular points: none.
Self-intersection points: none.
Area of domain enclosed by the curve: $5\pi r^2/2$.
(Figure 198).
FOLIUM *is a bounded connected closed convex curve free of singular points with a single axis of symmetry.*

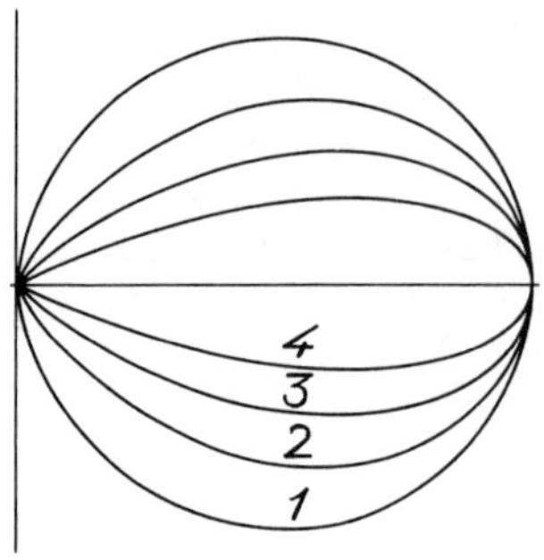

Figure 199. Galileo's spiral

N	1	2	3	4
b	0.50	1.00	2.00	5.00
a	1.00	1.00	1.00	1.00

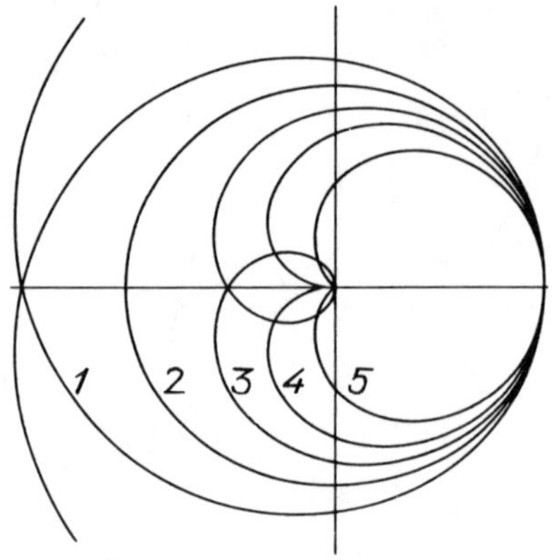

Figure 200. Galileo's spiral

N	1	2	3	4	5
b	−0.05	0.00	0.05	0.10	0.20
a	1.00	1.00	1.00	1.00	1.00

2.2.7 G

Galileo's spiral
Equation:

$$\rho = a - b\varphi^2, \quad a \neq 0, \quad b \neq 0$$

Additional equation:

$$\rho = a\varphi^2 - l$$

Basic geometric properties:
Unboundedness: present (for $b < 0$).
Asymptotes: none.
Boundedness: present (for $a > 0,\ b > 0$).
Closedness: present (for $a > 0,\ b > 0$).
Axial symmetry: straight line $y = 0$ is a symmetry axis.
Central symmetry: none.
Connectedness: present.
Singular points: point $(0,0)$ is an angular point, straight lines

$$y = \pm \tan \frac{a}{b} x$$

are tangents at the point (for $a = 0,\ b > 0$ point $(0,0)$ is a cuspidal point of the first kind); points

$$((-1)^k(a - b(k\pi)^2), 0),$$

$$k \in \mathbf{Z}, \quad 0 <| k | \frac{1}{\pi}\sqrt{\frac{a}{b}}$$

are nodes.

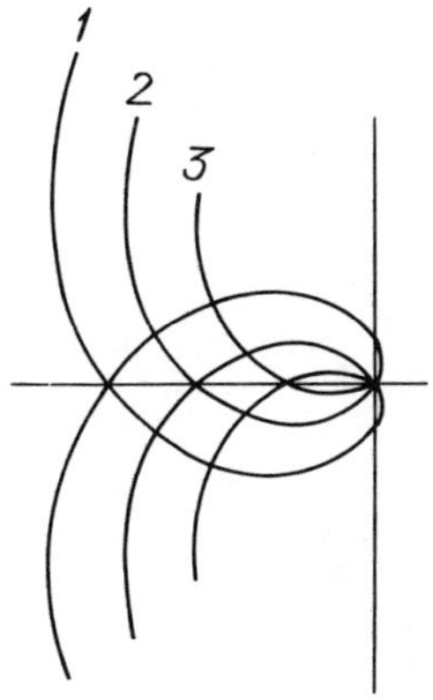

Figure 201. Galileo's spiral

N	1	2	3
a	−1.00	−4.00	−7.00
b	−1.00	−1.00	−1.00

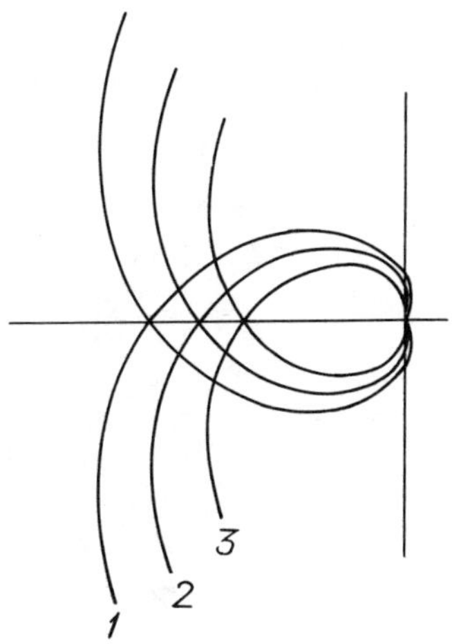

Figure 202. Galileo's spiral

N	1	2	3
b	−1.00	−0.80	−0.20
a	−1.00	−1.00	−1.00

Self-intersection points: none (for $\sqrt{a} \le \pi\sqrt{b}$);

$$m = \left[\frac{1}{\pi}\sqrt{\frac{a}{b}}\right]$$

of the form $((-1)^k(a - b(k\pi)^2), 0), l = 1, 2, \ldots, m$ (for $\sqrt{a} > \pi\sqrt{b}$) (here $[x]$ is an integer part (entier) of number x); infinite (for $b < 0$).
(Figures 199–202).

GALILEO'S SPIRAL *is a connected curve with an angular point and a single axis of symmetry; number of nodes depends on sign of parameter b (for $b < 0$ the curve is unbounded and has an infinitely large number of nodes, for $b > 0$ the curve is bounded and has a finite number of nodes).*

Generalized hyperbola (A)

Equation:

$$\left(\frac{x}{a}\right)^m - \left(\frac{y}{b}\right)^m = 1,$$

$$m = \frac{p}{q}, \quad p \in \mathbf{N}, \quad q \in \mathbf{N}, \quad \text{G.C.D.}(p, q) = 1,$$

$$a > 0, \quad b > 0$$

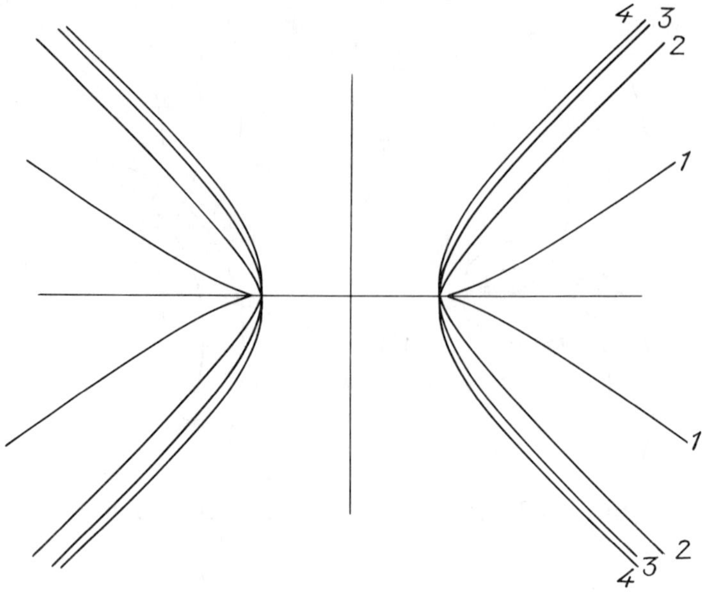

Figure 203. Generalized hyperbola (A)

N	1	2	3	4
p	2	4	6	8
q	3	3	3	3
a	1.00	1.00	1.00	1.00
b	1.00	1.00	1.00	1.00

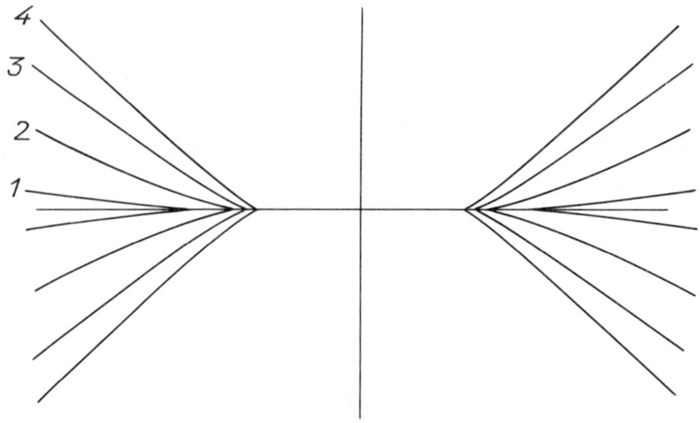

Figure 204. Generalized hyperbola (A)

N	1	2	3	4
p	4	6	8	10
q	11	11	11	11
a	1.00	1.00	1.00	1.00
b	1.00	1.00	1.00	1.00

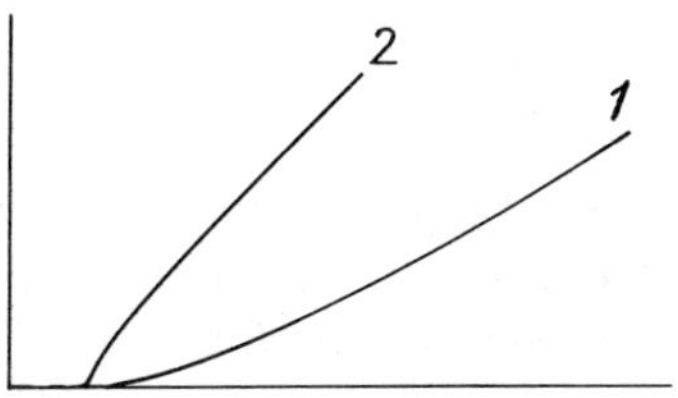

Figure 205.
Generalized hyperbola (A)

N	1	2
p	1	3
q	2	2
a	1.00	1.00
b	1.00	1.00

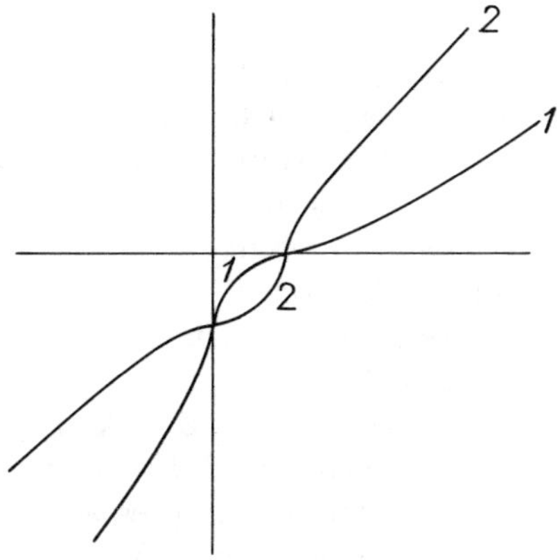

Figure 206.
Generalized hyperbola (A)

N	1	2
p	3	5
q	5	3
a	1.00	1.00
b	1.00	1.00

Basic geometric properties:

p is even
Unboundedness: present.
Asymptotes: straight lines $bx \pm ay = 0$ are inclined asymptotes.
Axial symmetry: straight lines $x = 0$, $y = 0$ are axes of symmetry.
Central symmetry: point $(0, 0)$ is a center of symmetry.
Connectedness: none.
Singular points: points $(\pm a, 0)$ are cuspidal points of the first kind (for $p < q$), straight line $y = 0$ is a tangent at the cusp points.
Self-intersection points: none.
Special cases: hyperbola ($p = 2$, $q = 1$).

q is even
Unboundedness: present.
Asymptotes: straight line $bx - ay = 0$ is an inclined asymptote.
Axial symmetry: none.
Central symmetry: none.
Connectedness: present.
Singular points: point $(a, 0)$ is a cessation point.
Self-intersection points: none.

pq is odd
Unboundedness: present.
Asymptotes: straight line $bx - ay = 0$ is an inclined asymptote.
Axial symmetry: straight line $x + y = 0$ is an axis of symmetry.
Central symmetry: none.
Connectedness: present.
Singular points: none.
Self-intersection points: none.
(Figures 203–206).
GENERALIZED HYPERBOLA (A) *is an unbounded unconnected (consisting of two unbounded branches) central-symmetric curve free of singular points with two inclined asymptotes and two axes of symmetry (for even p), an unbounded connected curve with a single singular point (point of cessation) and a single inclined asymptote (for even q), an unbounded connected curve free of singular points with a single inclined asymptote and a single axis of symmetry (the last for $a = b$ only) (for odd pq).*

Generalized hyperbola (B):
Equation:

$$\left(\frac{x}{a}\right)^m - \left(\frac{y}{b}\right)^m = 1,$$

$$m = -\frac{p}{q}, \quad p \in \mathbf{N}, \quad q \in \mathbf{N}, \quad \text{G.C.D.}(p, q) = 1, \quad a > 0, \quad b > 0$$

Basic geometric properties:

p is even
Unboundedness: present.
Asymptotes: straight lines $x = \pm a$ are vertical asymptotes.
Axial symmetry: straight lines $x = 0,\ y = 0$ are axes of symmetry.
Central symmetry: point $(0, 0)$ is a center of symmetry.
Connectedness: none.
Singular points: none.
Self-intersection points: none.
Special cases: bullet-nose curve ($p = 2,\ q = 1$).

q is even
Unboundedness: present.
Asymptotes: straight line $x = a$ is a vertical asymptote.
Axial symmetry: none.
Central symmetry: none.
Connectedness: present.
Singular points: point $(a, 0)$ is a cessation point.
Self-intersection points: none.

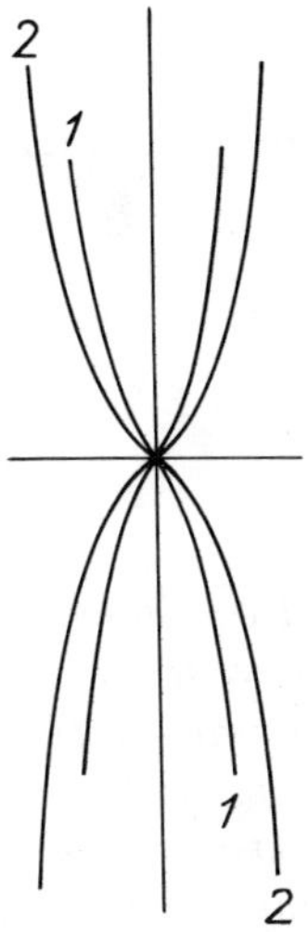

Figure 207.
Generalized hyperbola (B)

N	1	2
p	2	4
q	−3	−3
a	1.00	1.00
b	1.00	1.00

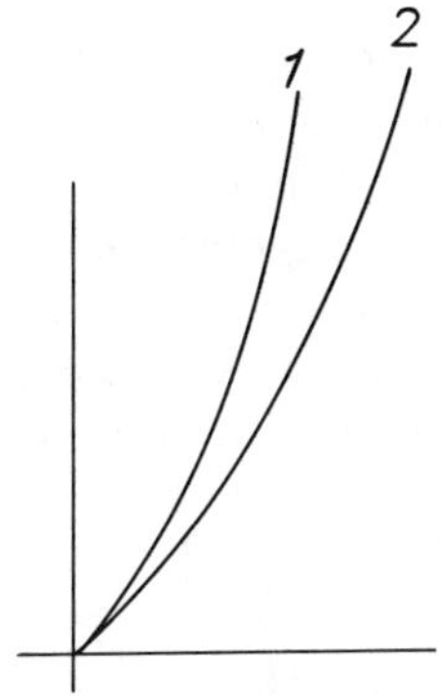

Figure 208.
Generalized hyperbola (B)

N	1	2
p	−3	−5
q	4	4
a	1.00	1.00
b	1.00	1.00

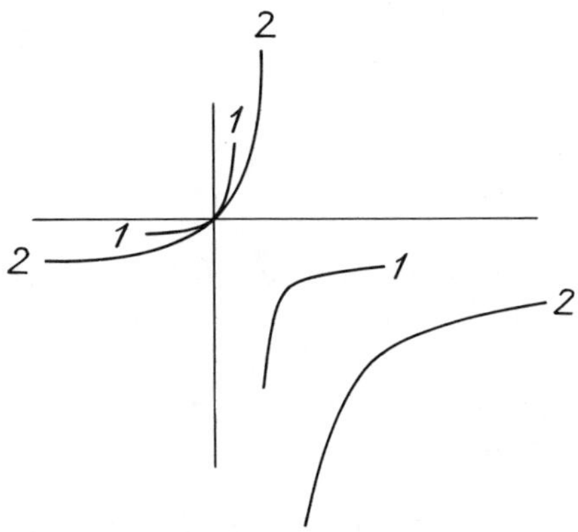

Figure 209.
Generalized hyperbola (B)

N	1	2
p	−3	−5
q	5	3
a	1.00	1.00
b	1.00	1.00

pq is odd
Unboundedness: present.
Asymptotes: straight line $x = a$ is a vertical asymptote, straight line $y = -b$ is a horizontal asymptote.
Axial symmetry: none (for $a \neq b$).
Central symmetry: none (for $a \neq b$).
Connectedness: none.
Singular points: none.
Self-intersection points: none.
(Figures 207–209).
GENERALIZED HYPERBOLA (B) *is an unbounded unconnected (consisting of four unbounded branches) central-symmetric curve free of singular points with two vertical asymptotes and two axes of symmetry (for even p), an unbounded connected curve free of singular points with a single vertical asymptote (for even q), an unbounded unconnected (consisting of three unbounded branches) curve free of singular points with a single vertical asymptote and a single horizontal asymptote (for odd pq).*

Generalized kampyle (A)
Equations:

$$\frac{x^2}{a^2} + \frac{y^2}{b^2} = \frac{x^4}{a^4},$$

$$y = \pm b\sqrt{\frac{x^4}{a^4} - \frac{x^2}{a^2}}, \quad a > 0, \quad b > 0$$

Basic geometric properties:
Unboundedness: present.
Asymptotes: none.
Axial symmetry: straight lines $x = 0$, $y = 0$ are axes of symmetry.
Central symmetry: point $(0,0)$ is a center of symmetry.
Connectedness: none.
Singular points: point $(0,0)$ is an isolated point.
Self-intersection points: none.
(Figures 210–211).
GENERALIZED KAMPYLE (A) *is an unbounded unconnected (consisting of two unbounded branches and an isolated point) central-symmetric curve with a single singular point (isolated point) and two axes of symmetry.*

Generalized kampyle (B)
Equations:

$$b^2x^2y^2 = (x^2 - a^2)(x^2 - a^2 + b^2)^2,$$

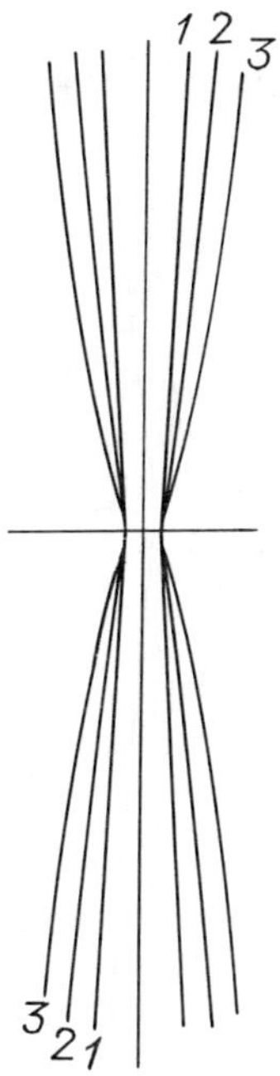

Figure 210.
Generalized kampyle (A)

N	1	2	3
b	1.00	3.00	5.00
a	1.00	1.00	1.00

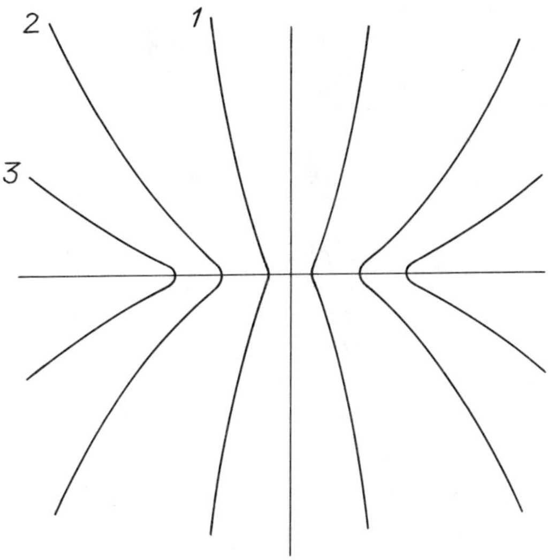

Figure 211.
Generalized kampyle (A)

N	1	2	3
a	1.00	3.00	5.00
b	1.00	1.00	1.00

$$y = \pm\frac{1}{bx}\sqrt{x^2 - a^2}(x^2 - a^2 + b^2),$$

$$a > 0, \quad b > 0$$

Basic geometric properties:
Unboundedness: present.
Asymptotes: none.
Axial symmetry: straight lines $x = 0$, $y = 0$ are axes of symmetry.
Central symmetry: point $(0, 0)$ is a center of symmetry.
Connectedness: none.
Singular points: points $(\pm\sqrt{a^2 - b^2}, 0)$ are isolated points (for $a \geq b$).
Self-intersection points: none.
(Figures 212–213).
GENERALIZED KAMPYLE (B) *is an unbounded unconnected (consisting of two unbounded branches and two isolated points (the last for $a \geq b$ only)) central-symmetric curve with a single (for $a = b$) or two (for $a > b$) singular points (isolated points) and two axes of symmetry.*

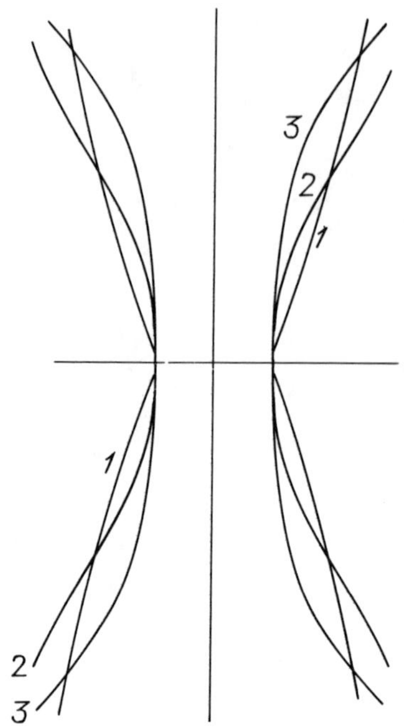

Figure 212.
Generalized kampyle (B)

N	1	2	3
b	1.00	3.00	5.00
a	1.00	1.00	1.00

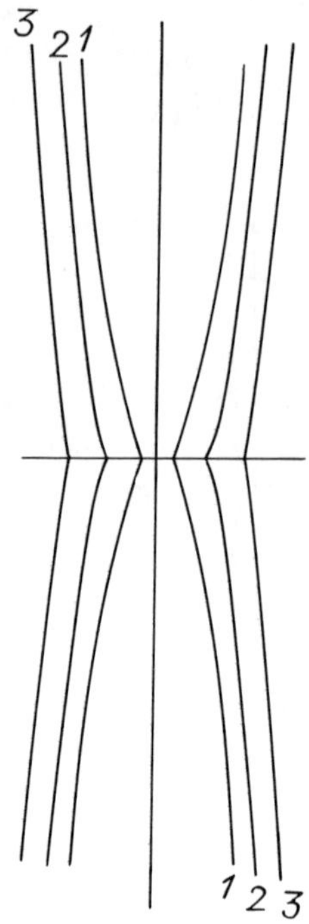

Figure 213.
Generalized kampyle (B)

N	1	2	3
a	1.00	3.00	5.00
b	1.00	1.00	1.00

Geometric petal (A) (botanic curve (A))
Equations:

$$\rho = a + b\cos 2n\varphi,$$

$$a > 0, \quad b \neq 0, \quad n \in \mathbf{N}$$

Basic geometric properties:
Boundedness: present.
Closedness: present.
Axial symmetry: straight lines

$$x \sin\frac{k\pi}{n} = y\cos\frac{k\pi}{n},$$

$$k = 0, 1, \ldots, n-1$$

are axes of symmetry.
Central symmetry: point $(0,0)$ is a center of symmetry.
Connectedness: present.

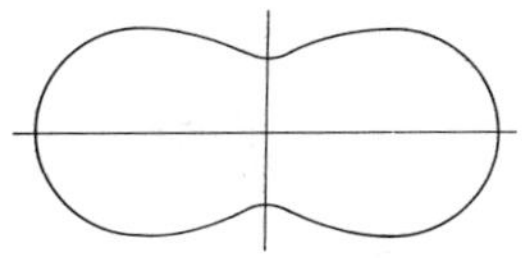

Figure 214. Geometric petal (A)
$a = 2.00, \quad b = 1.00, \quad n = 1$

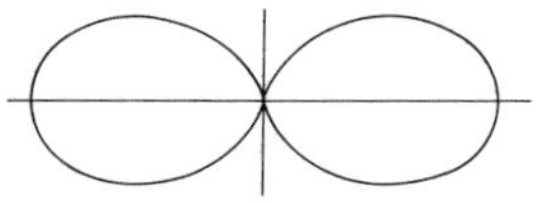

Figure 215. Geometric petal (A)
$a = 2.00, \quad b = 2.00, \quad n = 1$

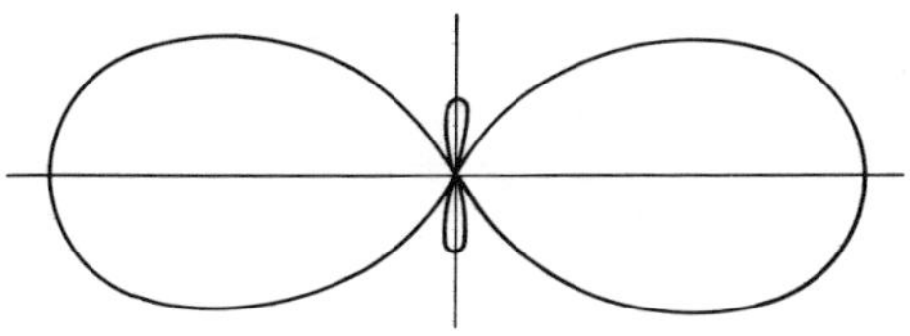

Figure 216. Geometric petal (A)
$a = 2.00, \quad b = 3.00, \quad n = 1$

Figure 217. Geometric petal (A)
$a = 2.00, \quad b = 1.00, \quad n = 2$

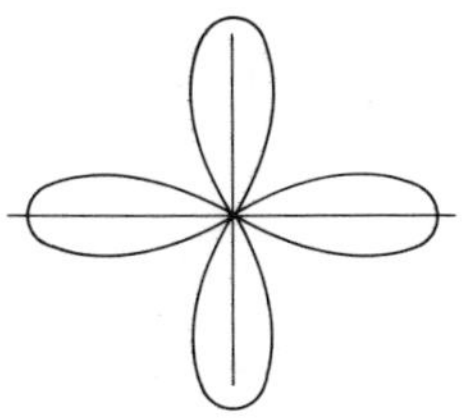

Figure 218. Geometric petal (A)
$a = 2.00, \quad b = 2.00, \quad n = 2$

Figure 219. Geometric petal (A)
$a = 2.00, \quad b = 5.00, \quad n = 2$

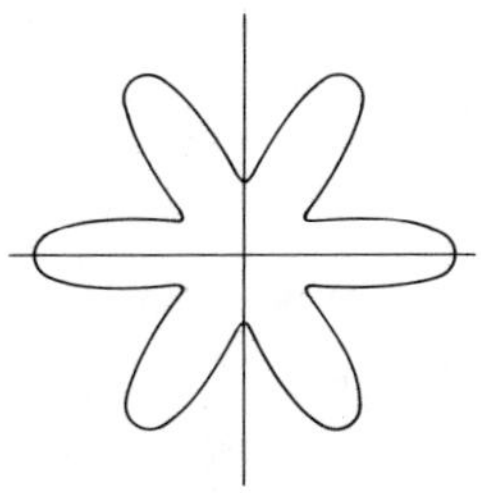

Figure 220. Geometric petal (A)
$a = 2.00, \quad b = 1.00, \quad n = 3$

Singular points: none (for $a >| b |$); point $(0,0)$ is a multiple point (for $a \leq| b |$).
Self-intersection points: none (for $a >| b |$); one, $(0,0)$ (for $a \leq| b |$).
(Figures 214–222).

Figure 221. Geometric petal (A)
$a = 2.00, \quad b = 2.00, \quad n = 3$

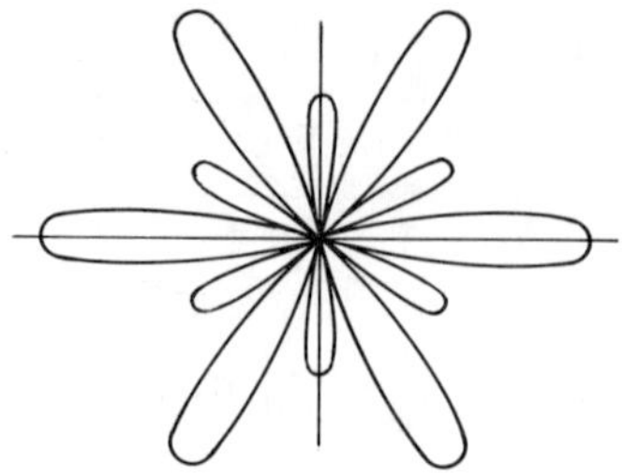

Figure 222. Geometric petal (A)
$a = 2.00, \quad b = 7.00, \quad n = 3$

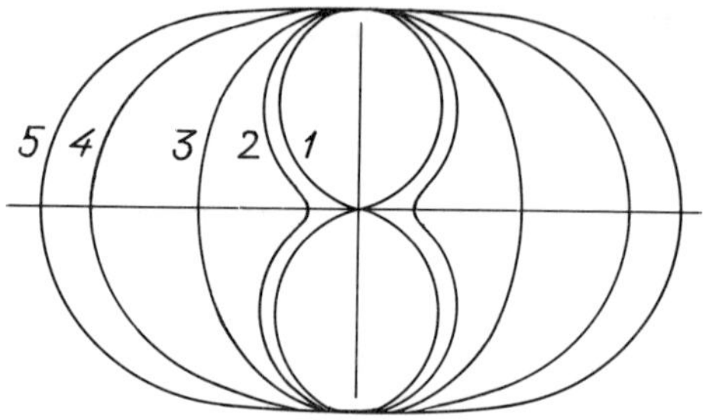

Figure 223. Geometric petal (B)

N	1	2	3	4	5
a	2.00	2.00	2.00	2.00	2.00
b	−2.00	−1.50	−0.50	0.50	1.00
n	1	1	1	1	1

GEOMETRIC PETAL (A) *is a bounded connected closed central-symmetric curve with a single singular point (multiple point for* $|\, b \,| \geq a$ *only) and no less than two axes of symmetry (number of the symmetry axes depends on values of parameter* n*).*

Geometric petal (B) (botanic curve (B))
Equations:

$$\rho = a + b\cos^{2n}\varphi,$$

$$a > 0, \quad b \neq 0, \quad n \in \mathbf{N}$$

Basic geometric properties:
Boundedness: present.
Closedness: present.
Axial symmetry: straight lines $x = 0, \ y = 0$ are axes of symmetry.
Central symmetry: point $(0, 0)$ is a center of symmetry.
Connectedness: present.

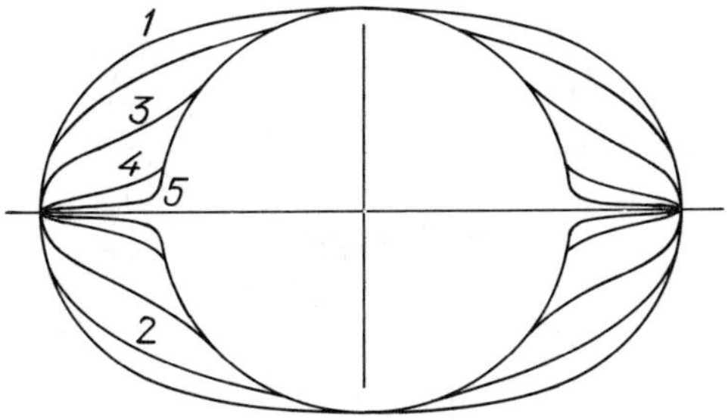

Figure 224. Geometric petal (B)

N	1	2	3	4	5
a	2.00	2.00	2.00	2.00	2.00
b	1.00	1.00	1.00	1.00	1.00
n	1	2	10	100	1000

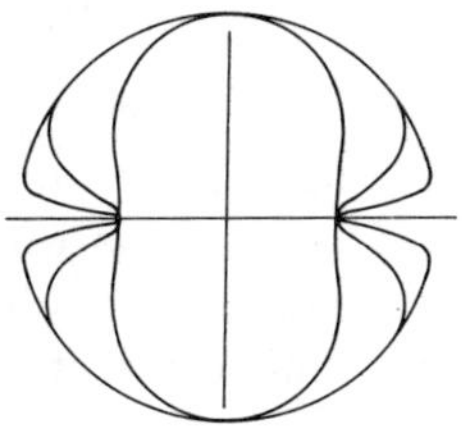

Figure 225. Geometric petal (B)

N	1	2	3
a	2.00	2.00	2.00
b	−1.00	−1.00	−1.00
n	1	10	100

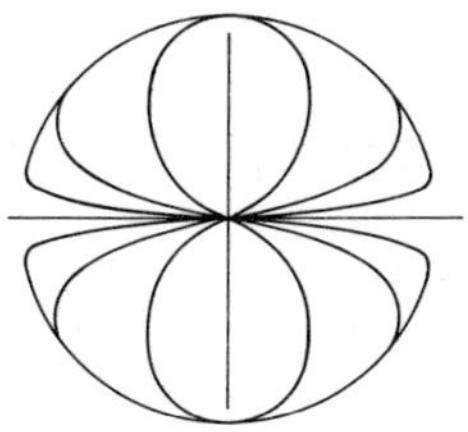

Figure 226. Geometric petal (B)

N	1	2	3
a	2.00	2.00	2.00
b	−2.00	−2.00	−2.00
n	1	10	100

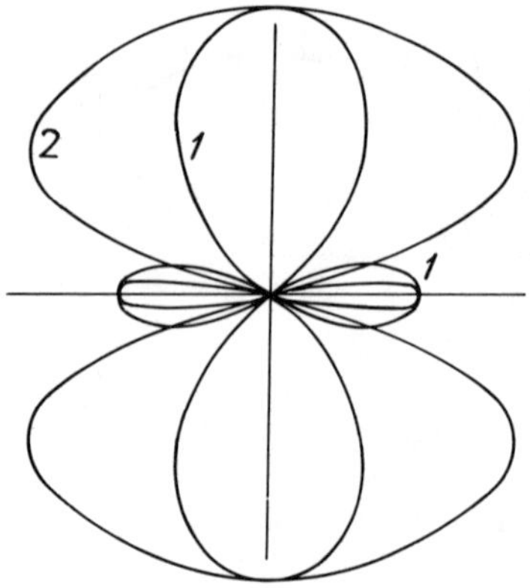

Figure 227. Geometric petal (B)

N	1	2
a	2.00	2.00
b	−3.00	−3.00
n	1	10

Singular points: point $(0,0)$ is a multiple point (for $a+b \leq 0$).
Number of self-intersection points depends on values of parameters a, b and n; for example, the curve has a single self-intersection point for $a = 2$, $b = -3$, $n = 1$, and five of the self-intersection points for $a = 2$, $b = -3$, $n = 10$.
(Figures 223–227).
GEOMETRIC PETAL (B) *is a bounded connected closed central-symmetric curve with two axes of symmetry; number of its singular points depends on values of parameters* a, b *(the curve has a multiple point for* $a+b \leq 0$*) and* n.

Geometric petal (C) (botanic curve (C))
Equations:

$$\rho = a(1 + \cos^{2n+1}\varphi), \quad a > 0, \quad n \in \mathbf{N}$$

It is also called $(n+1)$th heart.

Basic geometric properties:
Boundedness: present.
Closedness: present.
Axial symmetry: straight line $y = 0$ is an axis of symmetry.
Central symmetry: none.
Connectedness: present.
Singular points: point $(0,0)$ is a cuspidal point of the first kind, straight line $y = 0$ is a tangent at the cusp point.

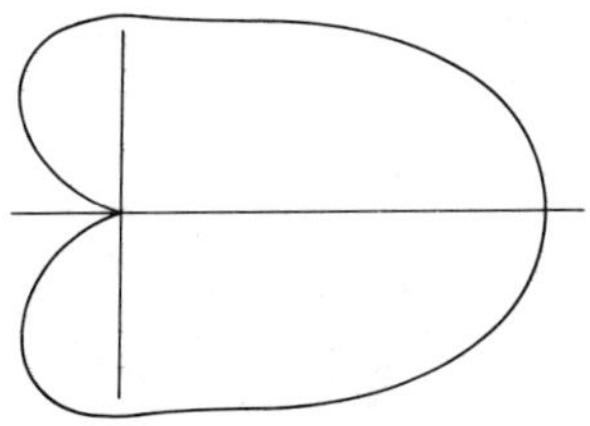

Figure 228. Geometric petal (C)
$a = 1.00, \quad n = 1$

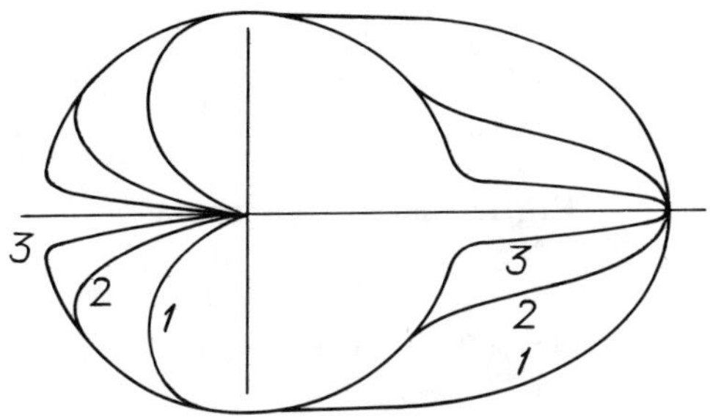

Figure 229. Geometric petal (C)

N	1	2	3
a	1.00	1.00	1.00
n	1	10	100

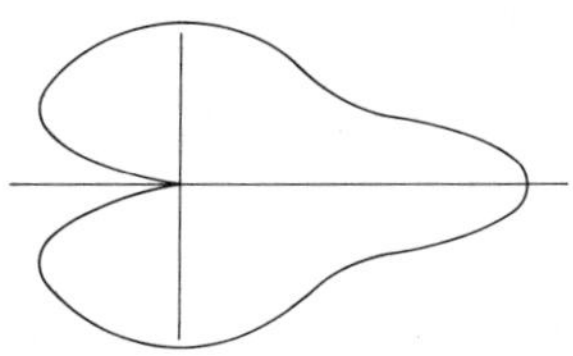

Figure 230. Geometric petal (C)
$a = 1.00, \quad n = 10$

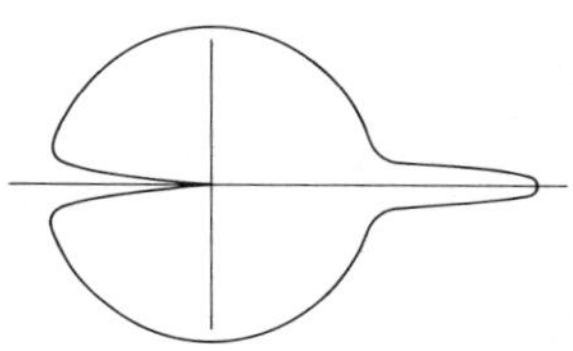

Figure 231. Geometric petal (C)
$a = 1.00, \quad n = 99$

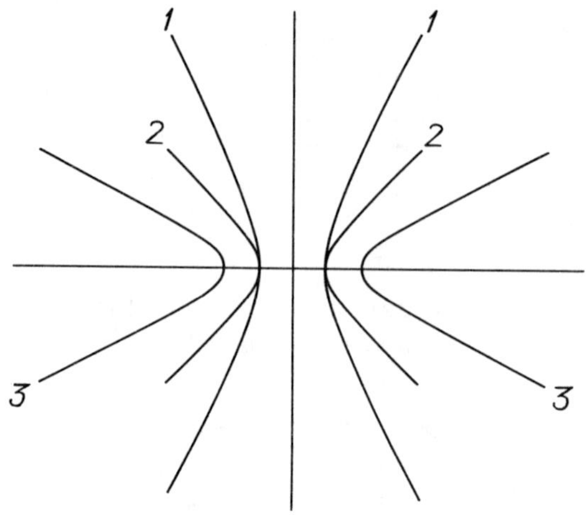

Figure 232. Hyperbola

N	1	2	3
a	1.00	1.00	2.00
n	2.00	1.00	1.00

Self-intersection points: none.
(Figures 228–231).
GEOMETRIC PETAL (C) *is a bounded connected closed curve with a single singular point (cuspidal point of the first kind) and a single axis of symmetry.*

2.2.8 H

Hyperbola
Equations:

$$\frac{x^2}{a^2} - \frac{y^2}{b^2} = 1,$$

$$x = \pm a \cosh t, \quad y = b \sinh t, \quad a > 0, \quad b > 0$$

Additional equations:

$$y^2 = 2px + qx^2, \quad q > 0$$

$$\rho = \frac{p}{1 + \varepsilon \cos \varphi}, \quad \varepsilon > 0$$

Basic geometric properties:
Unboundedness: present.
Asymptotes: straight lines $bx \pm ay = 0$ are inclined asymptotes.
Axial symmetry: straight lines $x = 0$, $y = 0$ are axes of symmetry.

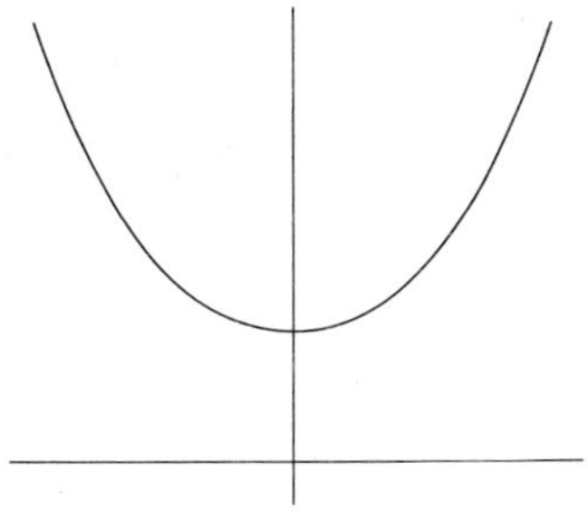

Figure 233. Hyperbolic cosinusoid

Central symmetry: point $(0,0)$ is a center of symmetry.
Connectedness: none.
Singular points: none.
Self-intersection points: none.
Characterized points: points $(\pm a, 0)$ are vertices.
(Figure 232).
HYPERBOLA *is an unbounded unconnected (consisting of two unbounded branches) central-symmetric curve free of singular points with two inclined asymptotes and two axes of symmetry.*

Hyperbolic cosinusoid
Equation:

$$y = \cosh(ax), \quad a > 0$$

Basic geometric properties:
Unboundedness: present.
Asymptotes: none.
Axial symmetry: straight line $x = 0$ is an axis of symmetry.
Central symmetry: none.
Connectedness: present.
Singular points: none.
Self-intersection points: none.
Characterized points: point $(0,1)$ is a vertex.
(Figure 233).
HYPERBOLIC COSINUSOID *is an unbounded connected curve free of singular points with a single axis of symmetry.*

Hyperbolic cotangentoid
Equation:

$$y = \coth(ax), \quad a > 0$$

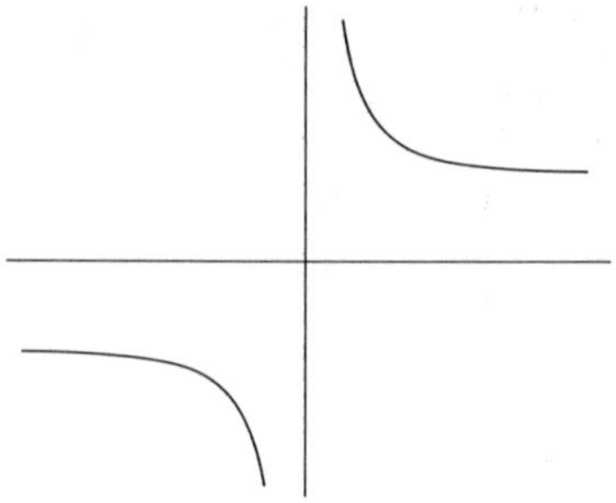

Figure 234.
Hyperbolic cotangentoid

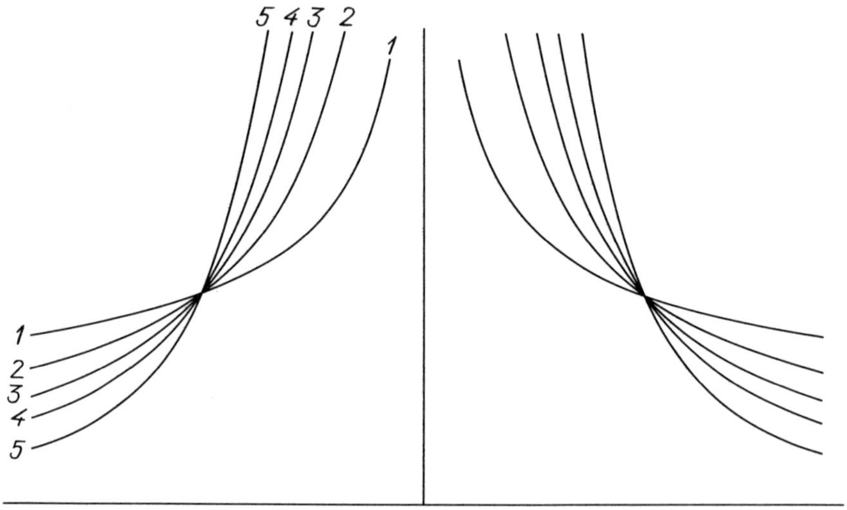

Figure 235. Hyperbolic curve

N	1	2	3	4	5
p	2	4	6	8	12
q	5	5	5	5	5

Basic geometric properties:
Unboundedness: present.
Asymptotes: straight lines $y = \pm 1$ are horizontal asymptotes, straight line $x = 0$ is a vertical asymptote.
Axial symmetry: none.
Central symmetry: point $(0, 0)$ is a center of symmetry.
Connectedness: none.
Singular points: none.
Self-intersection points: none.
(Figure 234).

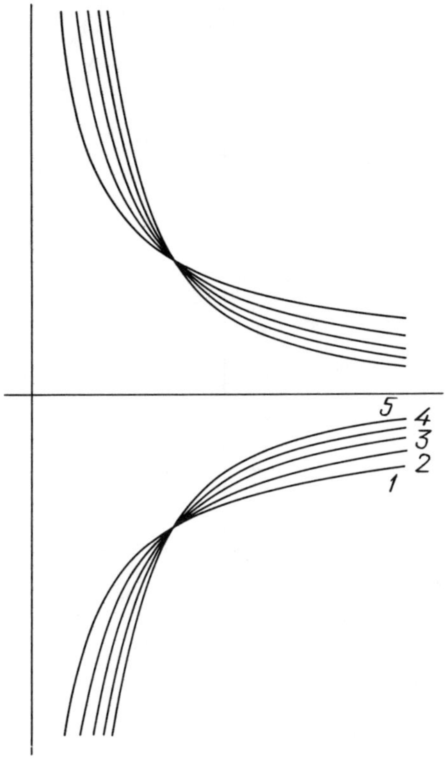

Figure 236. Hyperbolic curve

N	1	2	3	4	5
p	5	7	9	11	13
q	8	8	8	8	8

HYPERBOLIC COTANGENTOID *is an unbounded unconnected (consisting of two unbounded branches) central-symmetric curve free of singular points with two horizontal and a single vertical asymptotes.*

Hyperbolic curve

Equation:

$$x^p y^q = a,$$

$$a > 0, \quad p \in \mathbf{N}, \quad q \in \mathbf{N}, \quad \text{G.C.D.}(p, q) = 1$$

Additional equation:

$$y = cx^{-m}, \quad m > 0$$

Basic geometric properties:

Unboundedness: present.

Asymptotes: straight line $x = 0$ is a vertical asymptote, straight line $y = 0$ is a horizontal asymptote.

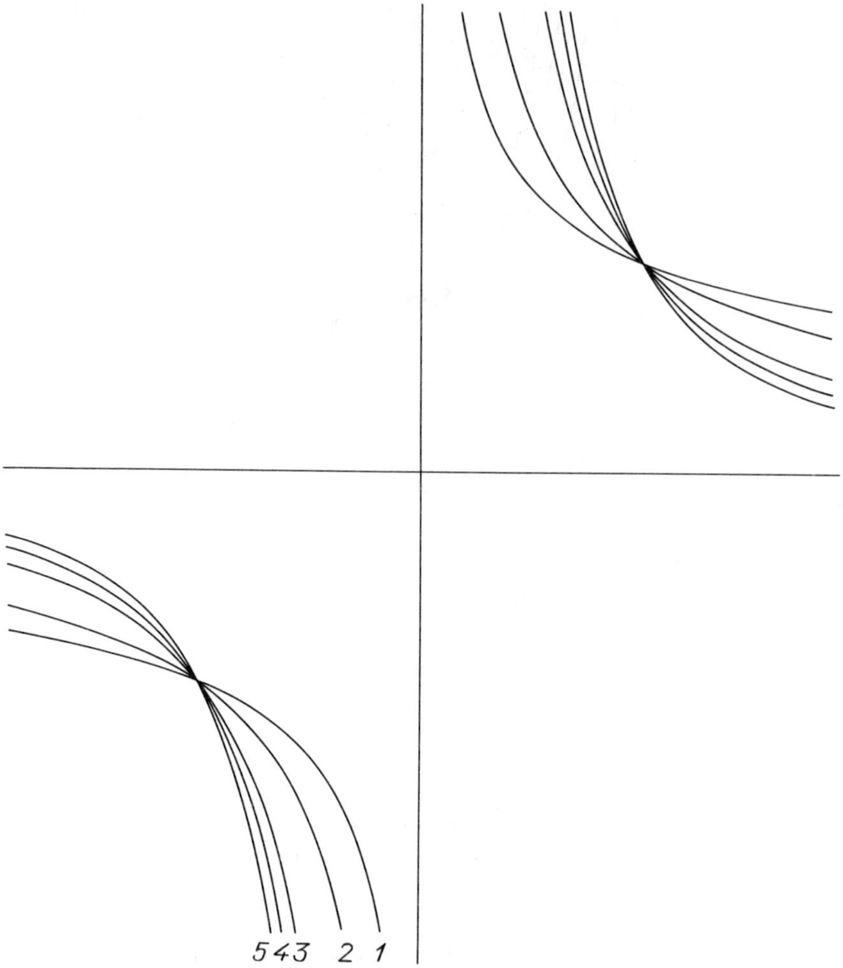

Figure 237. Hyperbolic curve

N	1	2	3	4	5
p	3	5	9	11	13
q	7	7	7	7	7

Axial symmetry: straight line $x = 0$ is an axis of symmetrty (p is even); straight line $y = 0$ is an axis of symmetry (q is even).
Central symmetry: point $(0,0)$ is a center of symmetry (pq is odd).
Connectedness: none.
Singular points: none.
Self-intersection points: none.
(Figures 235–237).
HYPERBOLIC CURVE *is an unbounded unconnected (consisting of two unbounded branches) curve free of singular points with a single horizontal*

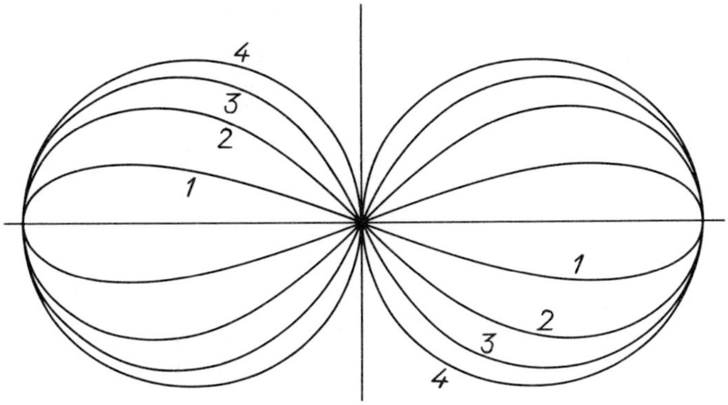

Figure 238. Hyperbolic lemniscate

N	1	2	3	4
a	1.50	1.50	1.50	1.50
b	4.00	1.50	0.50	0.10

and a single vertical asymptotes; the curve is central-symmetric (pq is odd) or axisymmetric (pq is even).

Hyperbolic lemniscate

Equations:

$$(x^2+y^2)^2 = a^2x^2 - b^2y^2,$$

$$\rho = \sqrt{a^2\cos^2\varphi - b^2\sin^2\varphi}, \quad a>0, \quad b>0$$

Basic geometric properties:

Boundedness: present.

Closedness: present.

Axial symmetry: straight lines $x=0$, $y=0$ are axes of symmetry.

Central symmetry: point $(0,0)$ is a center of symmetry.

Connectedness: present.

Singular points: point $(0,0)$ is a node, straight lines $ax \pm by = 0$ are tangents at the node.

Self-intersection points: one, $(0,0)$.

Characterized points: points

$$\left(\pm\frac{a}{2}\sqrt{\frac{2b^2+a^2}{a^2+b^2}}, \pm\frac{a^2}{2\sqrt{a^2+b^2}}\right)$$

are located at maximal distance from x-axis.

Area of domain enclosed by the curve:

$$\frac{a^2-b^2}{2}\arctan\frac{a}{b} + \frac{ab}{2}$$

Special cases: lemniscate of Bernoulli ($a=b$).

(Figure 238).

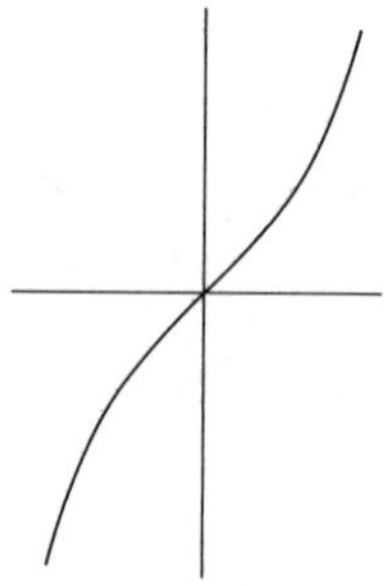

Figure 239. Hyperbolic sinusoid

HYPERBOLIC LEMNISCATE *is a bounded connected closed central-symmetric curve with a single singular point (node) and two axes of symmetry.*

Hyperbolic sinusoid
Equation:

$$y = \sinh(ax), \quad a > 0$$

Basic geometric properties:
Unboundedness: present.
Asymptotes: none.
Axial symmetry: none.
Central symmetry: point $(0,0)$ is a center of symmetry.
Connectedness: present.
Singular points: none.
Self-intersection points: none.
Characterized points: point $(0,0)$ is an inflection point.
(Figure 239).
HYPERBOLIC SINUSOID *is an unbounded connected central-symmetric curve free of singular points.*

Hyperbolic spiral (reciprocal spiral)
Equations:

$$\rho = \frac{a}{\varphi},$$

$$\sqrt{x^2 + y^2} \arctan \frac{y}{x} = a, \quad a > 0$$

Additional equation:

$$\rho = \frac{a}{\rho} + l$$

Basic geometric properties:
Unboundedness: present.

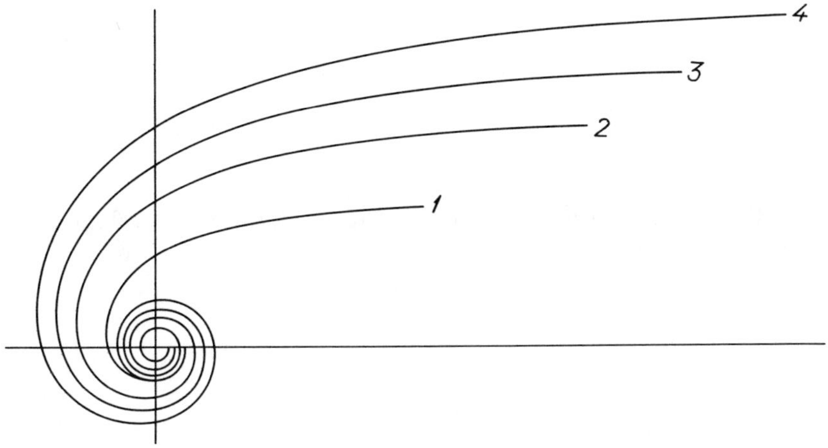

Figure 240. Hyperbolic spiral

N	1	2	3	4
a	0.50	0.80	1.00	1.20

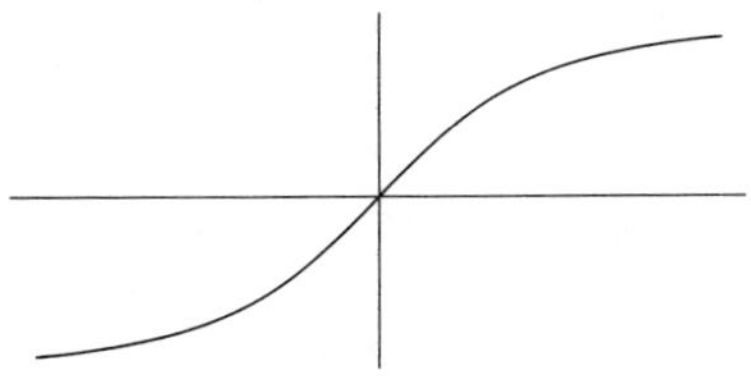

Figure 241. Hyperbolic tangentoid

Asymptotes: straight line $y = a$ is a horizontal asymptote.
Axial symmetry: none.
Central symmetry: none.
Connectedness: present.
Singular points: point $(0, 0)$ is an asymptotic point.
Self-intersection points: none.
Characterized points: point $(0, 0)$ is an asymptotic point.
(Figure 240).
HYPERBOLIC SPIRAL *is an unbounded connected curve with a single singular point (asymptotic point) and a single horizontal asymptote.*

Hyperbolic tangentoid
Equation:

$$y = \tanh(ax), \quad a > 0$$

Basic geometric properties:
Unboundedness: present.

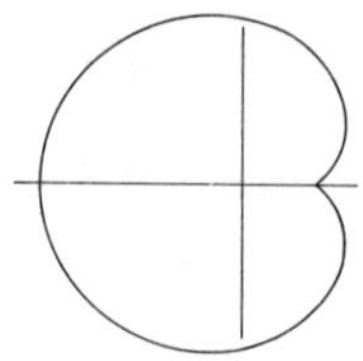

Figure 242. Hypocycloid
$R = 1.00, \quad m = 2.00$

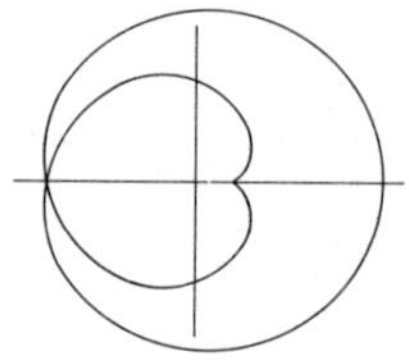

Figure 243. Hypocycloid
$R = 1.00, \quad m = 3.00$

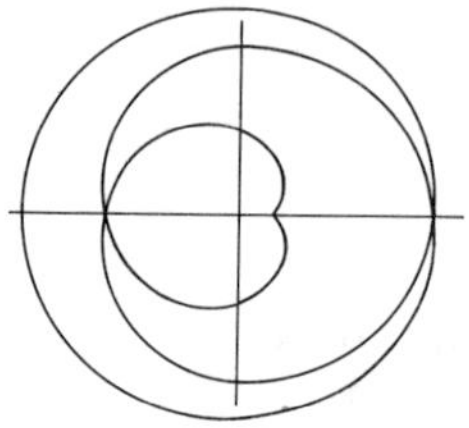

Figure 244. Hypocycloid
$R = 1.00, \quad m = 4.00$

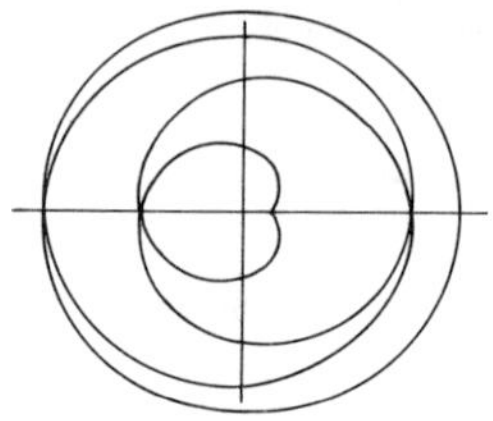

Figure 245. Hypocycloid
$R = 1.00, \quad m = 5.00$

Asymptotes: straight lines $y = \pm 1$ are horizontal asymptotes.
Axial symmetry: none.
Central symmetry: point $(0, 0)$ is a center of symmetry.
Connectedness: present.
Singular points: none.
Self-intersection points: none.
Characterized points: point $(0, 0)$ is an inflection point.
(Figure 241).
HYPERBOLIC TANGENTOID *is an unbounded connected central-symmetric curve free of singular points with two horizontal asymptotes.*

Hypocycloid

Equations:

$$x = (R - mR)\cos mt + m\cos(t - mt), \quad y = (R - mR)\sin mt - m\sin(t - mt),$$

$$\rho^2 = (R - mR)^2 + (mR)^2 + 2(R - mR)mR\cos t,$$

$$R > 0, \quad m = \frac{p}{q}, \quad p \in \mathbf{N}, \quad q \in \mathbf{N}, \quad \text{G.C.D.}(p, q) = 1$$

Basic geometric properties:
Boundedness: present.
Closedness: present.

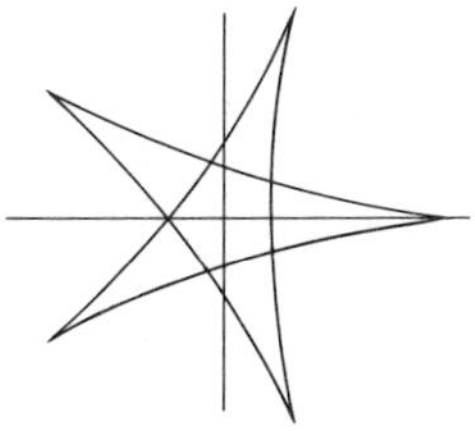

Figure 246. Hypocycloid
$R = 1.00, \quad m = 0.40$

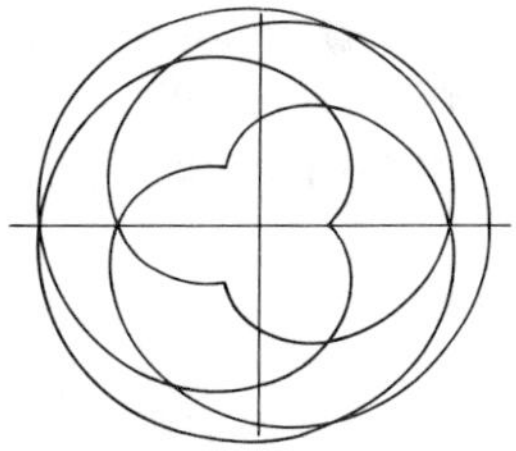

Figure 247. Hypocycloid
$R = 1.00, \quad m = 2.(3)$

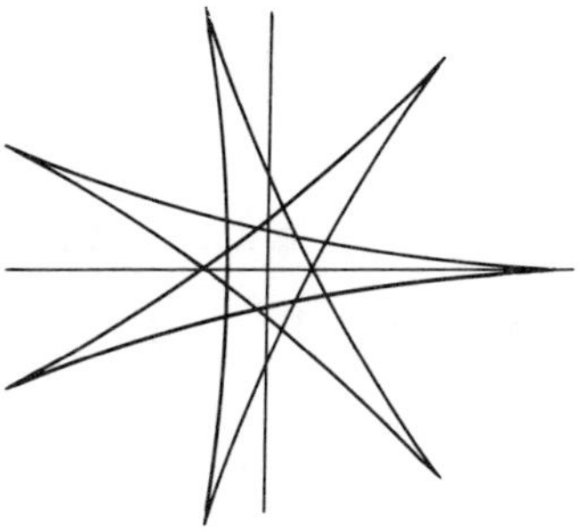

Figure 248. Hypocycloid
$R = 1.00, \quad m = 3/7$

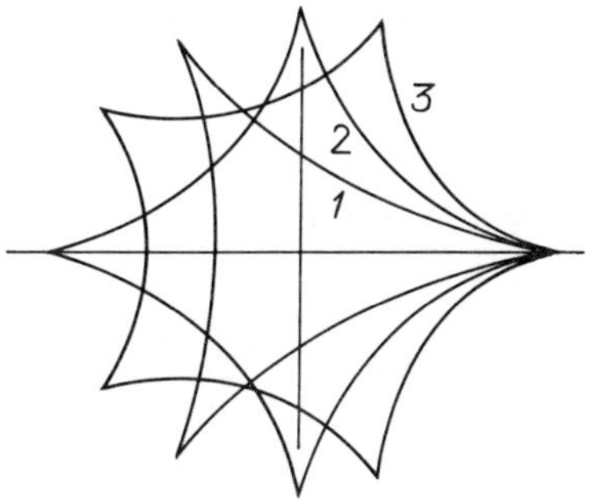

Figure 249. Hypocycloid

N	1	2	3
m	1/3(2/3)	1/4(3/4)	1/5(4/5)

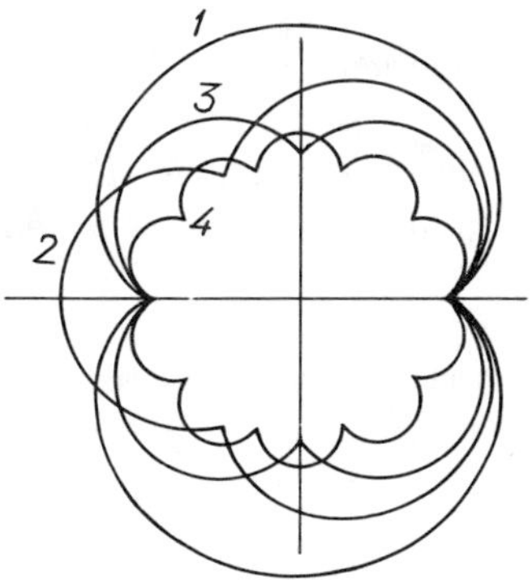

Figure 250. Hypocycloid

N	1	2	3	4
m	3/2	4/3	5/4	11/10

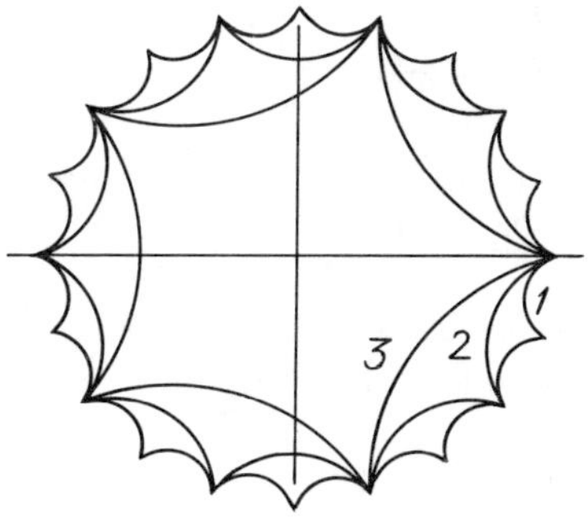

Figure 251. Hypocycloid

N	1	2	3
m	1/20	1/10	1/5

Axial symmetry: straight lines

$$x \sin \frac{\pi}{q} k = y \cos \frac{\pi}{q} k, \quad k = 0, 1, \ldots, q-1$$

are axes of symmetry.

Central symmetry: point $(0,0)$ is a center of symmetry (q is even).
Connectedness: present.
Singular points: points

$$\left(R\cos\frac{2\pi}{q}k,\, R\sin\frac{2\pi}{q}k\right),$$

$$k = 0, 1, \ldots, q-1$$

are cuspidal points of the first kind, straight lines

$$x\sin\frac{2\pi}{q}k = y\cos\frac{2\pi}{q}k,$$

$$k = 0, 1, \ldots, q-1$$

are tangents at the corresponding cusp points; hypocycloid may possess nodes of the form

$$\left(\rho_l\cos\frac{2k+1}{q}\pi,\, \rho_l\sin\frac{2k+1}{q}\pi\right), \quad l = 1, 3, \ldots,$$

$$\left(\rho_n\cos\frac{2k}{q}\pi,\, \rho_n\sin\frac{2k}{q}\pi\right), \quad n = 2, 4, \ldots,$$

$$k = 0, 1, \ldots, q-1$$

Number of self-intersection points is finite and depends on values of parameter m.
Special cases: pericycloid ($m > 1$), straight line segment ($m = 1/2$).
(Figures 242–251).
Hypocycloid *is a bounded connected closed central-symmetric (the last for even q only) curve with q nodes, q cuspidal points of the first kind and q axes of symmetry.*

In case of irrational m the hypocycloid is a transcendent unclosed curve with an infinitely large number of nodes and cusp points.

Hypotrochoid

Equations:

$$x = (R - mR)\cos mt + h\cos(t - mt), \quad y = (R - mR)\sin mt - h\sin(t - mt),$$

$$\rho^2 = (R - mR)^2 + h^2 + 2(R - mR)h\cos t,$$

$$R > 0, \quad m = \frac{p}{q}, \quad p \in \mathbf{N}, \quad q \in \mathbf{N},$$

$$\text{G.C.D.}(p, q) = 1, \quad h > 0, \quad h \neq mR$$

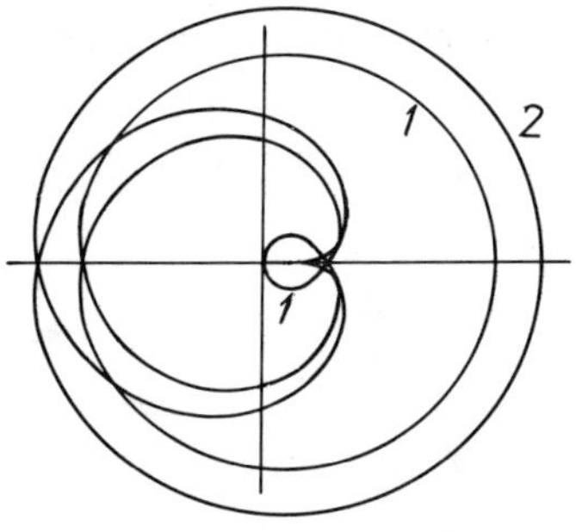

Figure 252. Hypotrochoid

N	1	2
m	3.00	3.00
h	2.00	3.00

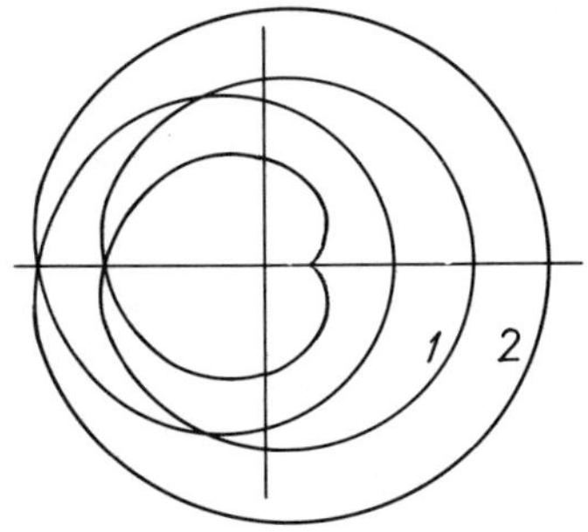

Figure 253. Hypotrochoid

N	1	2
m	3.00	3.00
h	3.00	5.00

Additional equations:

$$x = (R-r)\cos\varphi + h\cos\frac{R-r}{r}\varphi, \quad y = (R-r)\sin\varphi - h\sin\frac{R-r}{r}\varphi,$$

$$\rho^2 = (R-r)^2 + h^2 + 2(R-r)h\cos\frac{R}{r}\varphi$$

Basic geometric properties:
Boundedness: present.
Closedness: present.
Axial symmetry: straight lines

$$x\sin\frac{\pi}{q}k = y\cos\frac{\pi}{q}k,$$
$$k = 0, 1, \ldots, q-1$$

are axes of symmetry.
Central symmetry: point $(0,0)$ is a center of symmetry (q is even).
Connectedness: present.
Singular points: hypotrochoid may possess nodes and cusp points (number of the singular points depends on values of parameters R, m and h), located on the axes of symmetry.
Number of self-intersection points depends on values of parameters R, m and h.
Special cases: prolate hypotrochoid ($h > mR$), curtate hypotrochoid ($h < mR$), hypocycloid ($h = mR$).
(Figures 252–259).
HYPOTROCHOID *is a bounded connected closed central-symmetric (for even q only), q nodes and q axes of symmetry.*

In case of irrational m the hypotrochoid is a transcendent unclosed curve with an infinitely large number of nodes and cusp points.

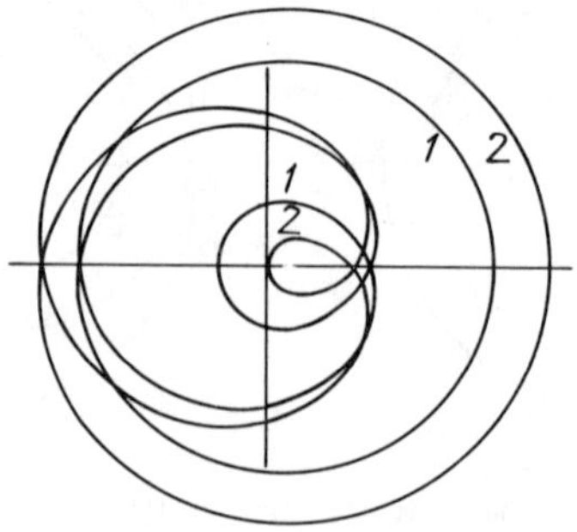

Figure 254. Hypotrochoid

N	1	2
m	3.00	3.00
h	1.25	2.00

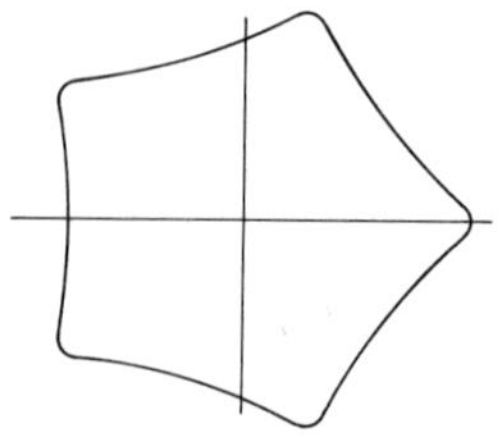

Figure 255. Hypotrochoid
$m = 0.20, \quad h = 0.10$

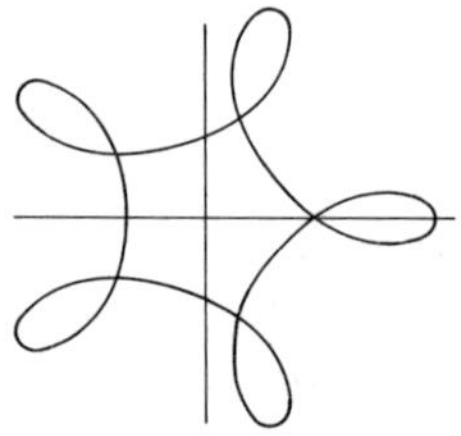

Figure 256. Hypotrochoid
$m = 0.20, \quad h = 0.40$

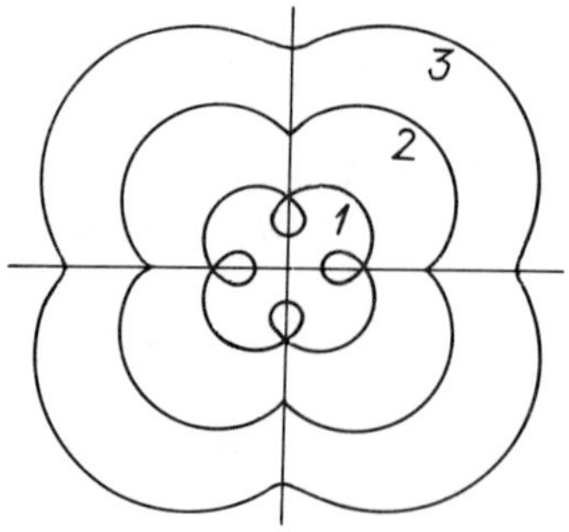

Figure 257. Hypotrochoid

N	1	2	3
m	1.25	1.25	1.25
h	0.50	1.25	2.00

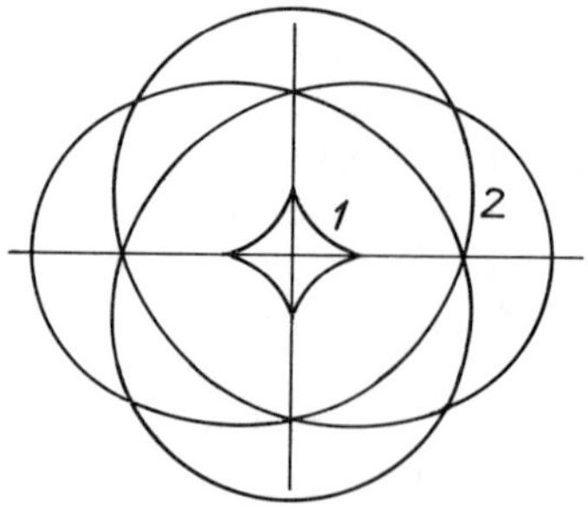

Figure 258. Hypotrochoid

N	1	2
m	0.25	0.25
h	0.25	3.00

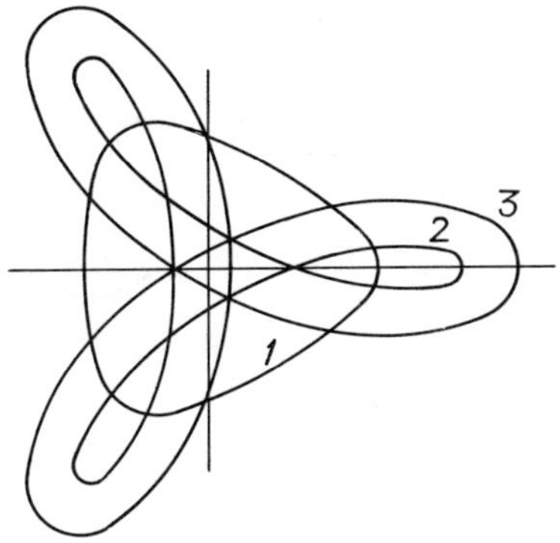

Figure 259. Hypotrochoid

N	1	2	3
m	0.(3)	0.(3)	0.(3)
h	0.10	0.30	0.75

2.2.9 I

Inverse logarithmic curve (reciprocal logarithmic curve)
Equation:

$$y = be^{a/x}, \quad a > 0, \quad b > 0$$

Basic geometric properties:
Unboundedness: present.
Asymptotes: straight line $x = 0$ is a vertical asymptote, straight line $y = b$ is a horizontal asymptote.
Axial symmetry: none.
Central symmetry: none.
Connectedness: none.
Singular points: none.
Self-intersection points: none.
(Figures 260–261).
INVERSE LOGARITHMIC CURVE *is an unbounded unconnected (consisting of two unbounded branches) curve free of singular points with a single vertical asymptote and a single horizontal asymptote.*

Isochrone of Varignon
Equations:

$$\rho = a + (a - c)t^2,$$

$$\varphi = b\sqrt{a - c}(\arctan t - t),$$

$$a > c > 0, \quad b = \frac{2c}{l\sqrt{a}} > 0$$

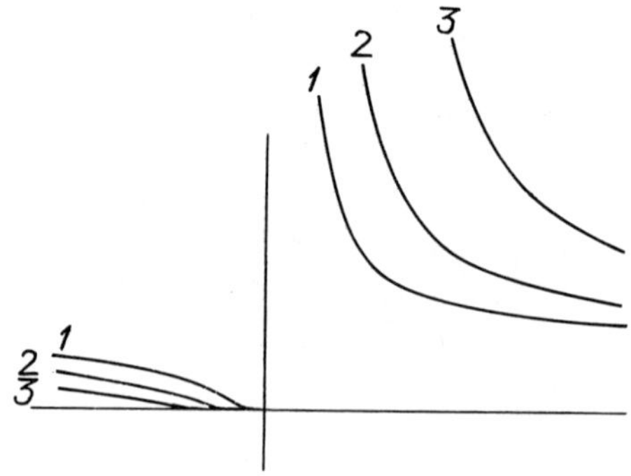

Figure 260. Inverse logarithmic curve

N	1	2	3
a	1.00	2.00	4.00
b	1.00	1.00	1.00

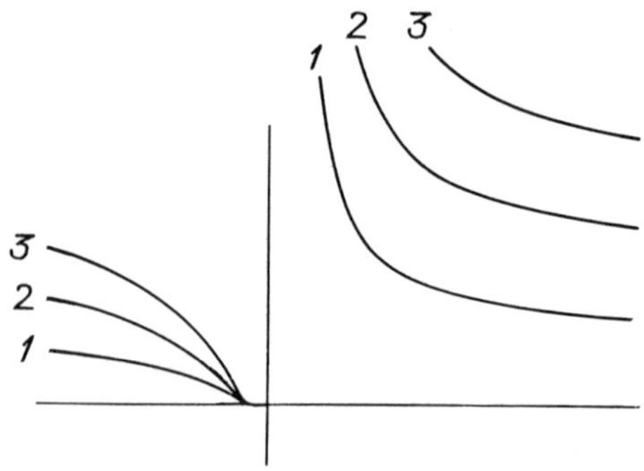

Figure 261. Inverse logarithmic curve

N	1	2	3
b	1.00	2.00	3.00
a	1.00	1.00	1.00

Basic geometric properties:
Unboundedness: present.
Asymptotes: none.
Axial symmetry: straight line $y = 0$ is an axis of symmetry.
Central symmetry: none.
Connectedness: present.
Singular points: point $(a, 0)$ is a cuspidal point of the first kind, straight line $y = 0$ is a tangent at the cusp point; the curve possesses an infinitely large number of nodes (all these nodes are located on the x-axis).
Self-intersection points: infinite.
(Figures 262–266).

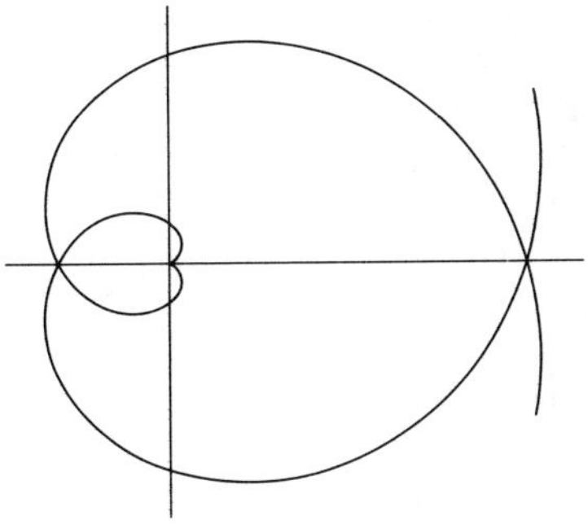

Figure 262.
Isochrone of Varignon
$a = 2.00, \quad b = 0.50, \quad c = 1.00$

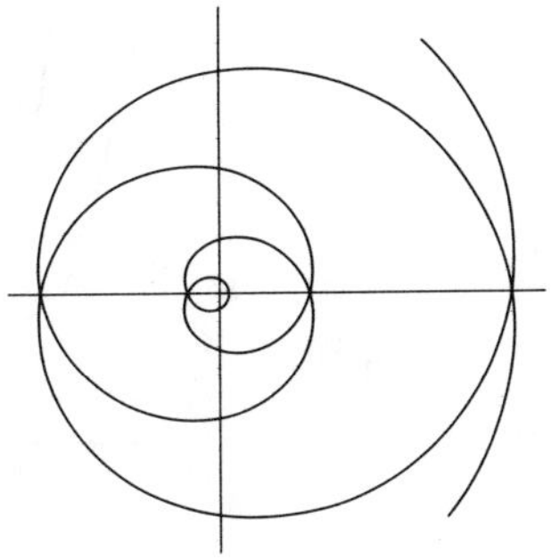

Figure 263.
Isochrone of Varignon
$a = 2.00, \quad b = 1.00, \quad c = 1.00$

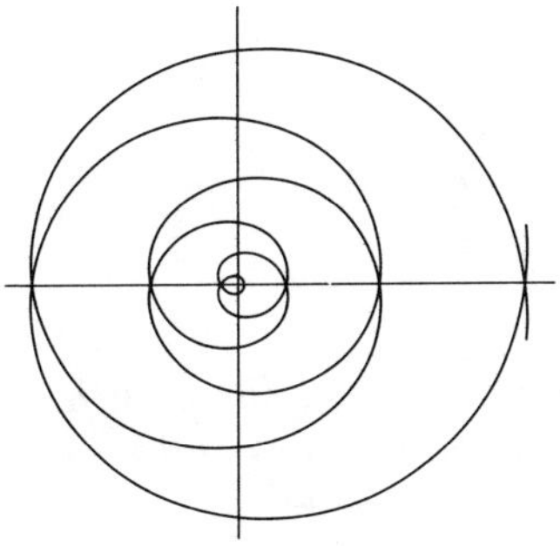

Figure 264.
Isochrone of Varignon
$a = 3.00, \quad b = 1.00, \quad c = 1.00$

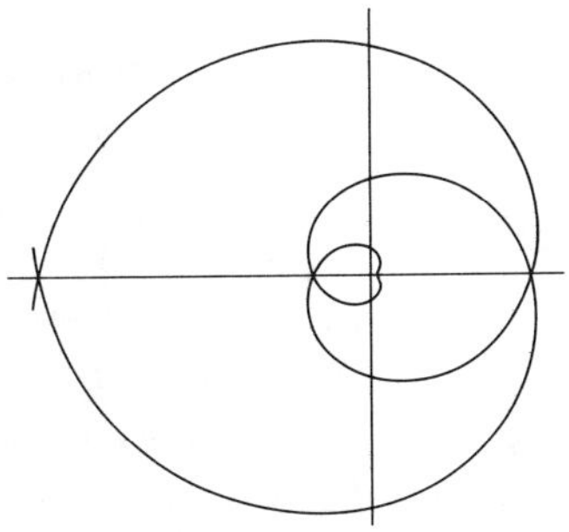

Figure 265.
Isochrone of Varignon
$a = 2.00, \quad b = 1.00, \quad c = 1.50$

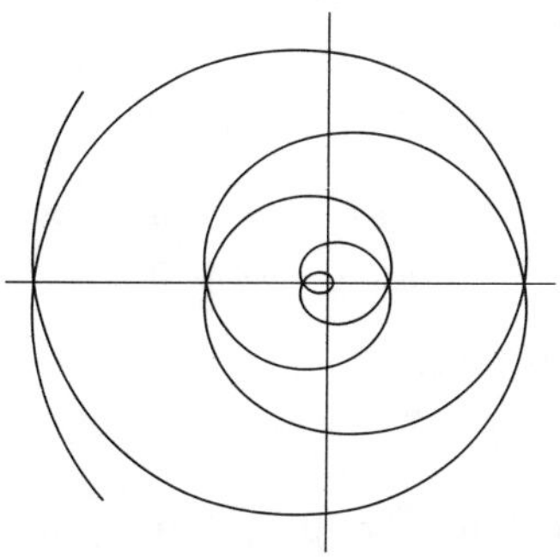

Figure 266. Isochrone of Varignon
$a = 2.00, \quad b = 1.00, \quad c = 0.50$

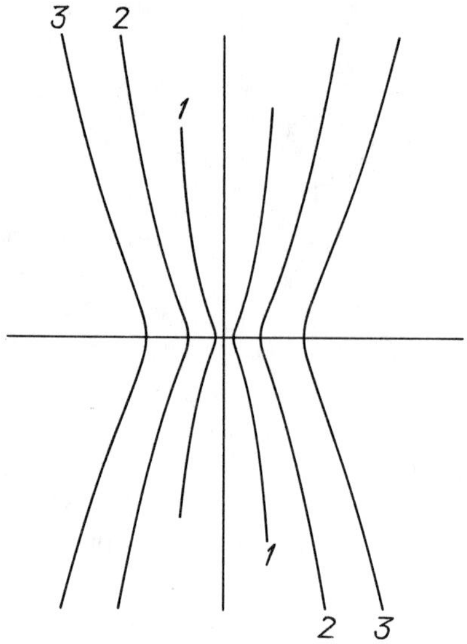

Figure 267. Kampyle

N	1	2	3
a	1.00	1.00	1.00
b	1.00	2.00	3.00

ISOCHRONE OF VARIGNON *is an unbounded connected curve with a single axis of symmetry, a single cuspidal point of the first kind and an infinitely large number of self-intersection points (simple nodes).*

2.2.10 K

Kampyle
Equations:

$$a^2(x^2+y^2) = \frac{a^4}{b^4}x^4, \quad \rho = \frac{b^2}{a\cos^2\varphi},$$

$$x = \frac{b^2}{a}\cdot\frac{1}{\cos t}, \quad y = \frac{b^2}{a}\cdot\frac{\sin t}{\cos^2 t},$$

$$a > 0, \quad b > 0$$

Basic geometric properties:
Unboundedness: present.
Asymptotes: none.
Axial symmetry: straight lines $x = 0,\ y = 0$ are axes of symmetry.
Central symmetry: point $(0,0)$ is a center of symmetry.
Connectedness: none.
Singular points: point $(0,0)$ is an isolated point.

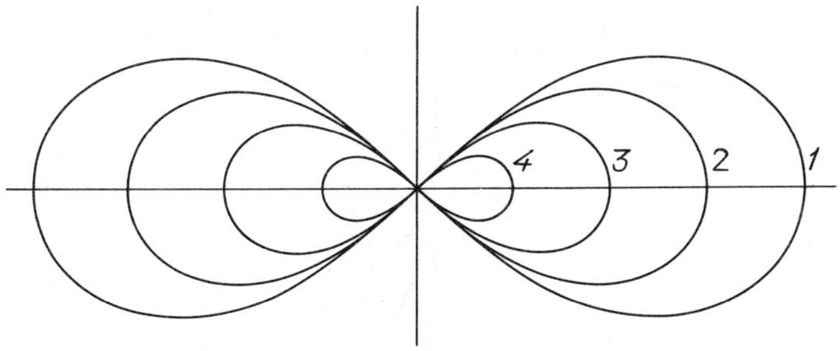

Figure 268.
Lemniscate of Bernoulli

N	1	2	3	4
a	1.00	0.75	0.50	0.25

Self-intersection points: none.
(Figure 267).
KAMPYLE *is an unbounded unconnected (consisting of two unbounded branches and a single isolated point) central-symmetric curve with a single singular point (isolated point) and two axes of symmetry.*

2.2.11 L

Lemniscate of Bernoulli (lemniscate)
Equations:

$$(x^2+y^2)^2 = 2a^2(x^2-y^2),$$

$$\rho = a\sqrt{2\cos 2\varphi},$$

$$x = a\sqrt{2}\frac{t+t^3}{1+t^4}, \quad y = a\sqrt{2}\frac{t-t^3}{1+t^4}, \quad a>0$$

Basic geometric properties:
Boundedness: present.
Closedness: present.
Axial symmetry: straight lines $x=0$, $y=0$ are axes of symmetry.
Central symmetry: point $(0,0)$ is a center of symmetry.
Connectedness: present.
Singular points: point $(0,0)$ is a node, straight lines $x \pm y = 0$ are tangents at the node.
Self-intersection points: one, $(0,0)$.
Characterized points: points $(\pm a\sqrt{2}, 0)$ are located at maximal distance from y-axis; points $(\pm a\sqrt{3/2}, \pm a\sqrt{2})$ are located at maximal distance from x-axis.
Area of domain enclosed by the curve: $2a^2$.
(Figure 268).

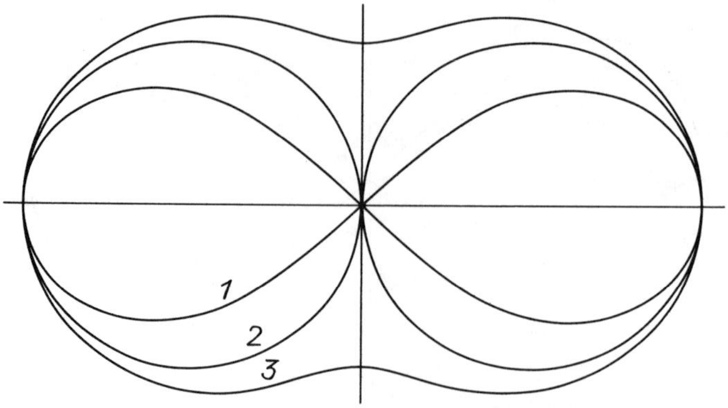

Figure 269. Lemniscate of Booth

N	1	2	3
a	1.00	1.00	1.00
b	0.25	0.00	−1.00
m^2	0.3125	0.2500	0.0000
c	0.375	0.250	1.00

LEMNISCATE OF BERNOULLI *is a bounded connected closed central-symmetric curve with a single singular point (node) and two axes of symmetry.*

Lemniscate of Booth

Equations:

$$(x^2+y^2)^2-(2m^2+c)x^2+(2m^2-c)y^2=0,$$

$$\rho=\sqrt{(2m^2+c)\cos^2\varphi-(2m^2-c)\sin^2\varphi},$$

$$m>0,\quad c\geq 0$$

Basic geometric properties:

Boundedness: present (for $r<a$).

Closedness: present.

Axial symmetry: straight lines $x=0,\ y=0$ are axes of symmetry.

Central symmetry: point $(0,0)$ is a center of symmetry.

Connectedness: present (for $0\leq c\leq 2m^2$), none (for $c>2m^2$).

Singular points: point $(0,0)$ is an isolated point (for $c>2m^2$), or a degenerate node (for $c=2m^2$), straight line $x=0$ is a tangent at the degenerate node, or node (for $c<2m^2$), straight lines

$$x\sqrt{2m^2+c}\pm y\sqrt{2m^2-c}=0$$

are tangents at the node.

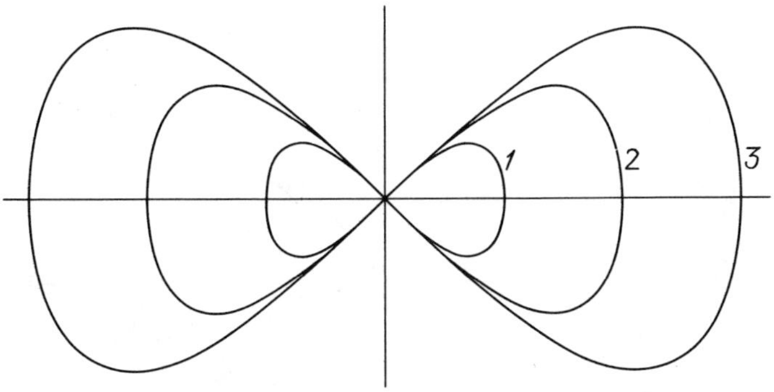

Figure 270. Lemniscate of Gerono

N	1	2	3
r	1.00	2.00	3.00

Self-intersection points: one, $(0,0)$ ($c \leq 0$); none ($c > 2m^2$).
Special cases: elliptic lemniscate ($4c > 2m^2$); hyperbolic lemniscate ($c < 2m^2$); pair of equal circles ($c = 2m^2$); lemniscate of Bernoulli ($c = 0$).
(Figure 269).
LEMNISCATE OF BOOTH ($r < a$) *is a bounded connected closed central-symmetric curve with a single singular point (type of singularity depends on relationship between parameter values) and two axes of symmetry.*

Lemniscate of Gerono (eight curve)
Equations:

$$x^4 = r^2(x^2 - y^2),$$

$$y = \pm\sqrt{x^2 - \frac{x^4}{r^2}},$$

$$r > 0$$

Basic geometric properties:
Boundedness: present.
Closedness: present.
Axial symmetry: straight lines $x = 0$, $y = 0$ are axes of symmetry.
Central symmetry: point $(0,0)$ is a center of symmetry.
Connectedness: none.
Singular points: point $(0,0)$ is a node, straight lines $x \pm y = 0$ are tangents at the node.
Self-intersection points: one, $(0,0)$.
(Figure 270).
LEMNISCATE OF GERONO *is a bounded connected closed central-symmetric curve with a single singular point (node) and two axes of symmetry.*

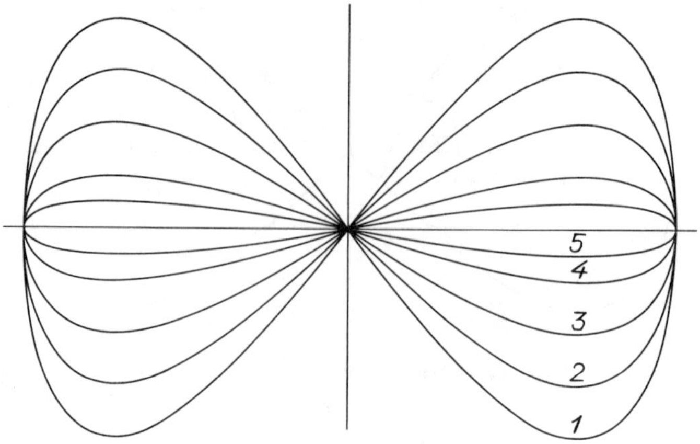

Figure 271.
Lemniscate of Montferrier

N	1	2	3	4	5
b	0.75	1.00	1.50	3.00	6.00
a	1.00	1.00	1.00	1.00	1.00

Lemniscate of Montferrier

Equations:

$$b^2y^2 = x^2(a^2 - x^2),$$

$$x = a\cos t, \quad y = \frac{a^2}{2b}\sin 2t, \quad a > 0, \quad b > 0$$

Additional equations:

$$y = \frac{x\sqrt{a^2 - x^2}}{b},$$

$$x = \sqrt{\frac{a^2}{4} + \frac{by}{2}} + \sqrt{\frac{a^2}{4} - \frac{by}{2}}$$

Basic geometric properties:
Boundedness: present.
Closedness: present.
Axial symmetry: straight lines $x = 0$, $y = 0$ are axes of symmetry.
Central symmetry: point $(0, 0)$ is a center of symmetry.
Connectedness: none.
Singular points: point $(0, 0)$ is a node, straight lines $ax \pm by = 0$ are tangents at the node.
Self-intersection points: one, $(0, 0)$.
(Figures 271–273).

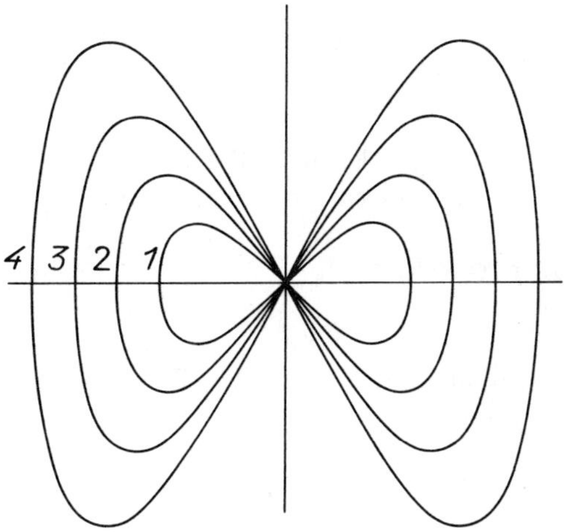

Figure 272. Lemniscate of Montferrier

N	1	2	3	4
a	1.50	2.80	2.50	3.80
b	1.50	1.50	1.50	1.50

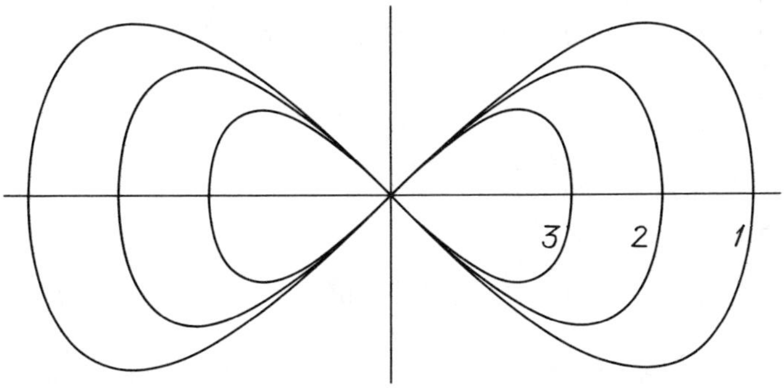

Figure 273. Lemniscate of Montferrier

N	1	2	3
a	1.00	0.75	0.58
b	1.00	0.75	0.50

LEMNISCATE OF MONTFERRIER *is a bounded connected closed central-symmetric curve with a single singular point (node) and two axes of symmetry.*

Lemniscatrix of Ökinghaus

Equation:

$$y = \ln(\cos x + \sqrt{1 + \cos^2 x})$$

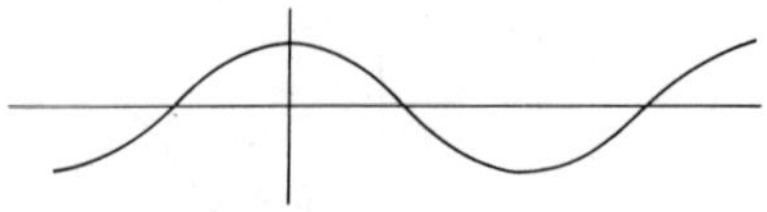

Figure 274.
Lemniscatrix of Ökinghaus

Basic geometric properties:
Unboundedness: present.
Asymptotes: none.
Axial symmetry: straight lines $x = k\pi$, $k \in \mathbf{Z}$, are axes of symmetry.
Central symmetry: points

$$\left(\frac{\pi}{2} + m\pi, 0\right), \quad m \in \mathbf{Z}$$

are centers of symmetry.
Connectedness: present.
Singular points: none.
Self-intersection points: none.
(Figure 274).
LEMNISCATRIX OF ÖKINHAUS *is an unbounded connected central-symmetric curve free of singular points with an infinitely large number of symmetry axes.*

Limaçon (Pascal's limaçon)
Equations:

$$(x^2 + y^2 - 2ax)^2 = l^2(x^2 + y^2), \quad \rho = 2a\cos\varphi + l,$$

$$x = 2a\cos^2 t + l\cos t, \quad y = 2a\cos t\sin t + l\sin t, \quad a > 0, \quad l > 0$$

Additional equations:

$$x = \frac{(1-t^2)(l + 2a + (l-2a)t^2)}{(1+t^2)^2}, \quad y = \frac{2t(l + 2a + (l-2a)t^2)}{(1+t^2)^2}$$

Basic geometric properties:
Boundedness: present.
Closedness: present.
Axial symmetry: straight line $y = 0$ is an axis of symmetry.
Central symmetry: none.
Connectedness: present (for $l \leq 2a$), none (for $l > 2a$).
Singular points: point $(0,0)$ is an isolated point (for $l > 2a$), or a cuspidal point of the first kind (for $l = 2a$), straight line $y = 0$ is a tangent at the

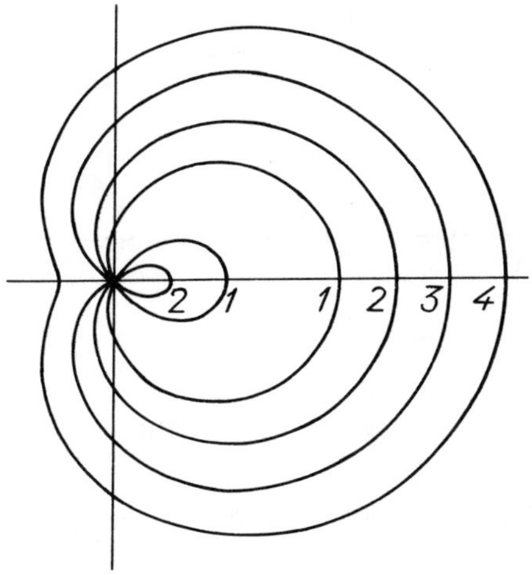

Figure 275. Limaçon

N	1	2	3	4
l	1.00	2.00	3.00	4.00
a	1.50	1.50	1.50	1.50

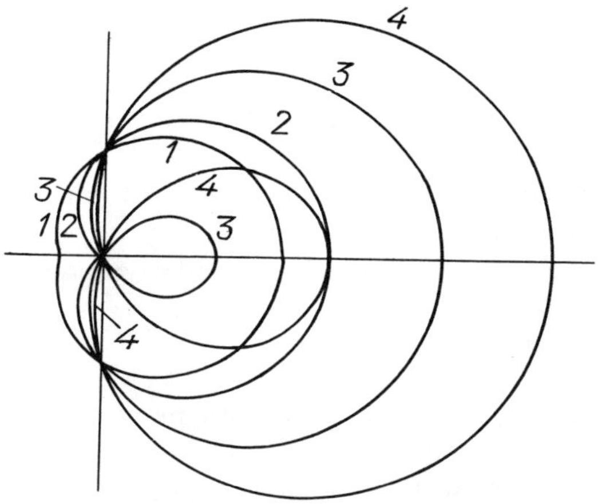

Figure 276. Limaçon

N	1	2	3	4
a	0.30	0.50	1.00	1.50
l	1.00	1.00	1.00	1.00

cusp point, or a node (for $l < 2a$), straight lines

$$y = \pm\frac{\sqrt{4a^2 - l^2}}{l}x$$

are tangents at the node.

Self-intersection points: one, $(0,0)$ (for $l < 2a$), none (for $l \geq 2a$).

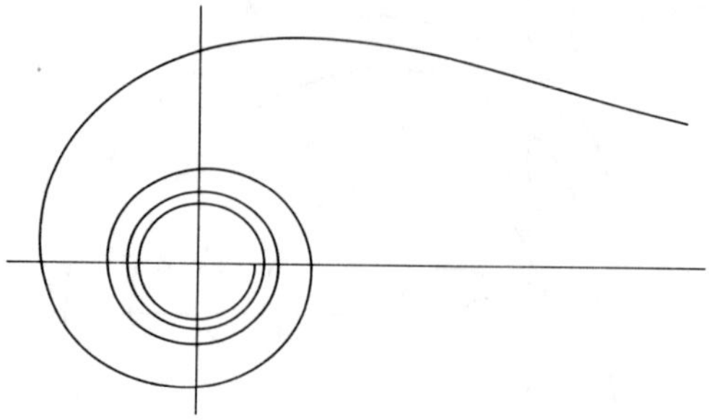

Figure 277. Lituus

Characterized points: points $(0,0), (\pm l + 2a, 0)$ are points of intersection between the curve and x-axis.
Area of domain enclosed by limaçon: $\pi(2a^2 + l^2)$ (for $l > 2a$).
(Figures 275–276).
LIMAÇON *is a bounded connected closed (for $l \leq 2a$ only) curve with a single singular point (type of singularity depends on relationship between parameter values) and a single axis of symmetry.*

Lituus
Equation:

$$\rho = \frac{a}{\varphi}, \quad a > 0$$

Basic geometric properties:
Unboundedness: present.
Asymptotes: straight line $y = 0$ is a horizontal asymptote.
Axial symmetry: none.
Central symmetry: none.
Connectedness: present.
Singular points: point $(0,0)$ is an asymptotic point.
Self-intersection points: none.
Characteristic points: point $(0,0)$ is an asymptotic point.
Generalization:

$$\rho = \frac{a}{\sqrt[n]{\varphi}}$$

(Figure 277).
LITUUS *is an unbounded connected curve with a single singular point (asymptotic point) and a single horizontal asymptote.*

Logarithmic curve
Equation:

$$y = b\log_a x, \quad a > 0, \quad a \neq 1, \quad b \neq 0$$

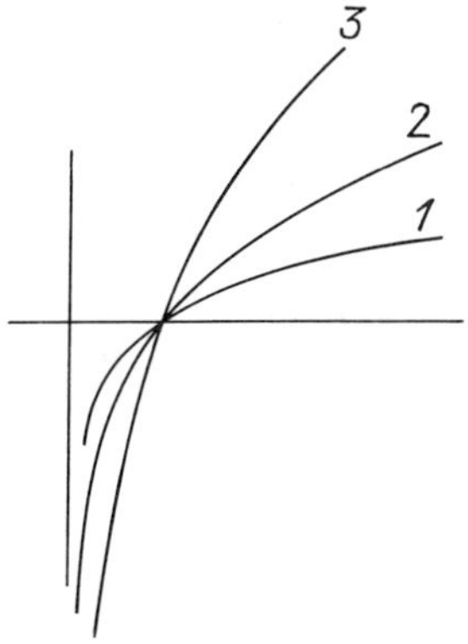

Figure 278. Logarithmic curve

N	1	2	3
b	0.50	1.00	2.00
a	2.00	2.00	2.00

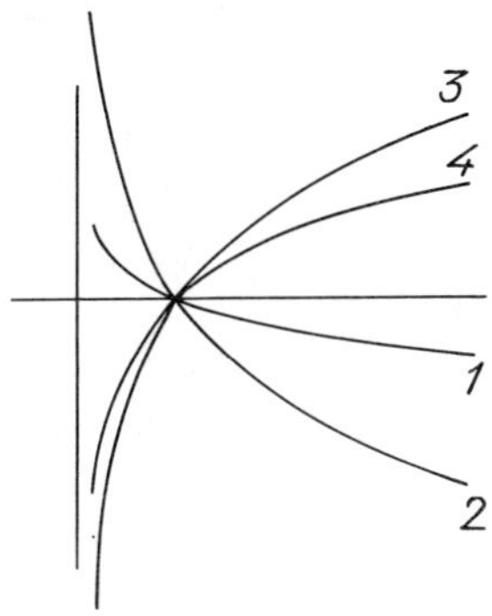

Figure 279. Logarithmic curve

N	1	2	3	4
a	0.10	0.50	2.00	3.00
b	1.00	1.00	1.00	1.00

Basic geometric properties:
Unboundedness: present.
Asymptotes: straight line $x = 0$ is a vertical asymptote.
Axial symmetry: none.
Central symmetry: none.
Connectedness: present.
Singular points: none.
Self-intersection points: none.
(Figures 278–279).
LOGARITHMIC CURVE *is an unbounded connected curve free of singular points with a single vertical asymptote.*

Logarithmic lemniscate
Equation:

$$y^2 = x^2 \ln \frac{a^2}{x^2},$$

$$a > 0$$

Basic geometric properties:
Boundedness: present.
Closedness: none.
Axial symmetry: straight lines $x = 0,\ y = 0$ are axes of symmetry.
Central symmetry: point $(0, 0)$ is a center of symmetry
Connectedness: none.
Singular points: none.

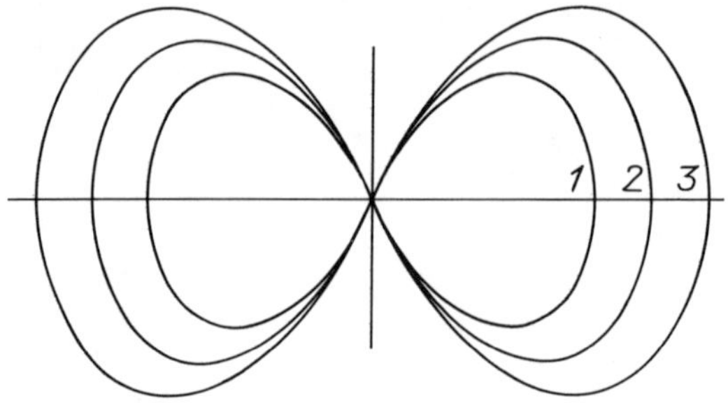

Figure 280.
Logarithmic lemniscate

N	1	2	3
a	0.80	1.00	1.20

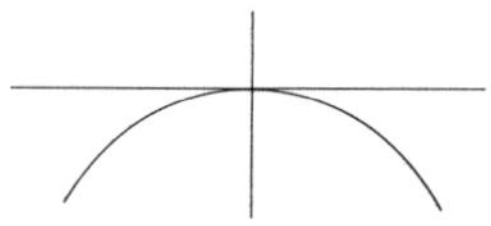

Figure 281.
Logarithmic sinusoid

Self-intersection points: none.
(Figure 280).
LOGARITHMIC LEMNISCATE *is a bounded nonclosed unconnected (consisting of two branches) curve free of singular points with two axes of symmetry.*

Logarithmic sinusoid (logarithmic sine)
Equation:

$$y = \ln\sqrt{1 - x^2}$$

Basic geometric properties:
Unboundedness: present.
Asymptotes: straight lines $x = \pm 1$ are vertical asymptotes.
Axial symmetry: straight line $x = 0$ is an axis of symmetry.
Central symmetry: none.
Connectedness: present.
Singular points: none.
Self-intersection points: none.
(Figure 281).
LOGARITHMIC SINUSOID *is an unbounded connected curve free of singular points with two vertical asymptotes and a single axis of symmetry.*

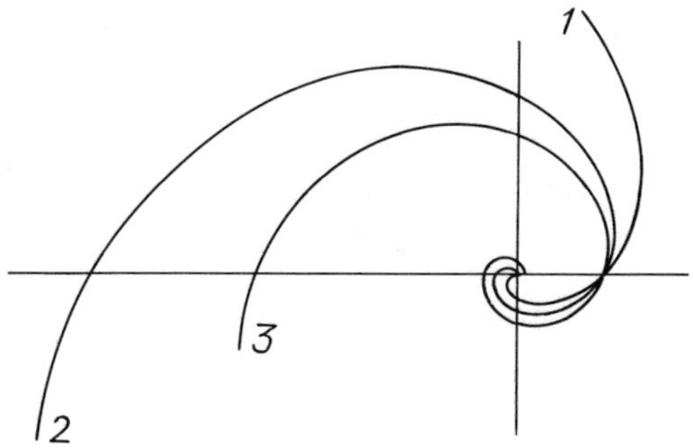

Figure 282. Logarithmic spiral

N	1	2	3
a	1.50	2.20	3.00
c	1.00	1.00	1.00

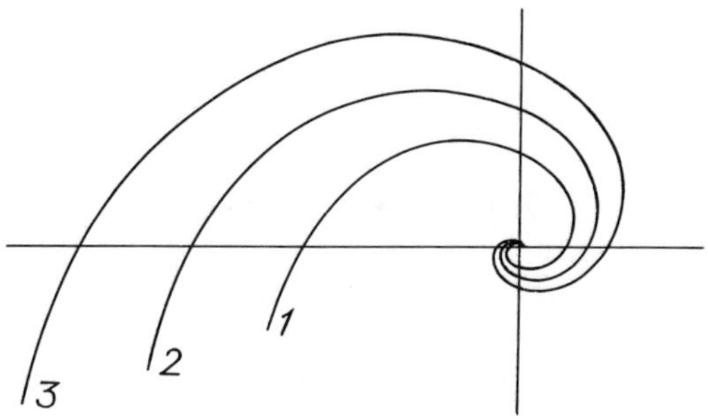

Figure 283. Logarithmic spiral

N	1	2	3
c	0.50	0.75	1.00
a	2.20	2.20	2.20

Logarithmic spiral (equiangular spiral, logistic spiral, logistique)
Equation:

$$\rho = \frac{c}{\sqrt{a^2-1}} e^{\varphi},$$

$$a > 1, \quad c > 0$$

Basic geometric properties:
Unboundedness: present.
Asymptotes: none.
Axial symmetry: none.
Central symmetry: none.
Connectedness: present.
Singular points: point $(0,0)$ is an asymptotic point.
Self-intersection points: none.

Generalization:

$$\rho = ae^{m\varphi} + b$$

(Figures 282–283).
LOGARITHMIC SPIRAL ($r < a$) *is an unbounded connected curve with a single singular point (asymptotic point).*

2.2.12 N

Neoid (stream line, stream-line rankine, swan-like line)
Equation:

$$\rho = a\varphi + b,$$

$$a > 0$$

Basic geometric properties:
Unboundedness: present.
Asymptotes: none.
Axial symmetry: none.
Central symmetry: none.
Connectedness: present.
Singular points: point $(0,0)$ is a point of cessation.
Self-intersection points: none.
Special cases: spiral of Archimedes ($b = 0$).
(Figures 284–285).
NEOID *is an unbounded connected curve with a single singular point (point of cessation).*

Nephroid (nephroid of Freeth)
Equations:

$$(x^2 + y^2)(x^2 + y^2 - a^2)^2 - 4a^2(x^2 + y^2 - ax)^2 = 0,$$

$$\rho = a\left(1 + \sin\frac{\varphi}{2}\right),$$

$$a > 0$$

Additional equation:

$$\rho = 4a \sin\frac{\varphi}{3} \sin\frac{2\varphi}{3}$$

Basic geometric properties:
Boundedness: present.
Closedness: present.
Axial symmetry: straight line $y = 0$ is an axis of symmetry.
Central symmetry: none.
Connectedness: present.

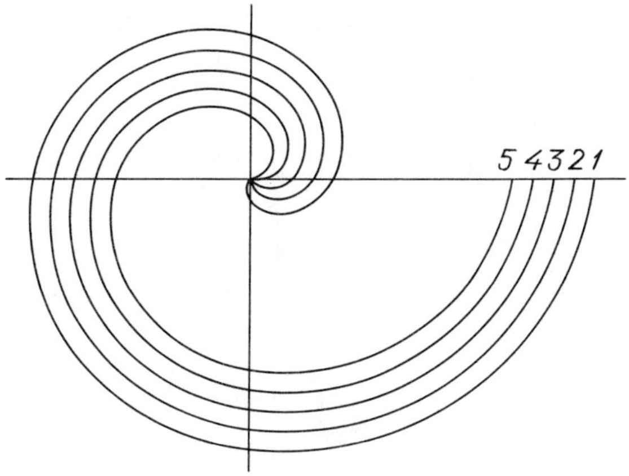

Figure 284. Neoid

N	1	2	3	4	5
b	2.00	1.50	1.00	0.50	0.00
a	1.00	1.00	1.00	1.00	1.00

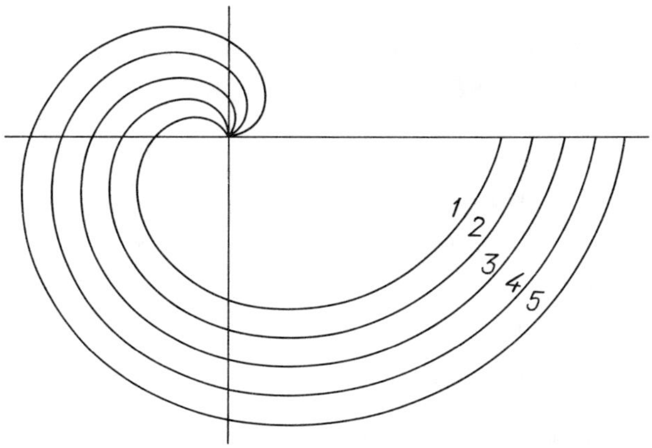

Figure 285. Neoid

N	1	2	3	4	5
b	0.00	−0.50	−1.00	−1.50	−2.00
a	1.00	1.00	1.00	1.00	1.00

Singular points: point $(0,0)$ is a node, straight lines $\sqrt{3}x \pm y = 0$ are tangents at the node; point $(a,0)$ is a triple point, straight lines $x - a = 0$, $x - a \pm \sqrt{3}y = 0$ are tangents at the triple point.
Self-intersection points: two, $(0,0)$, $(a,0)$.
(Figure 286).
NEPHROID *is a bounded connected closed curve with two singular points (node, triple point) and a single axis of symmetry.*

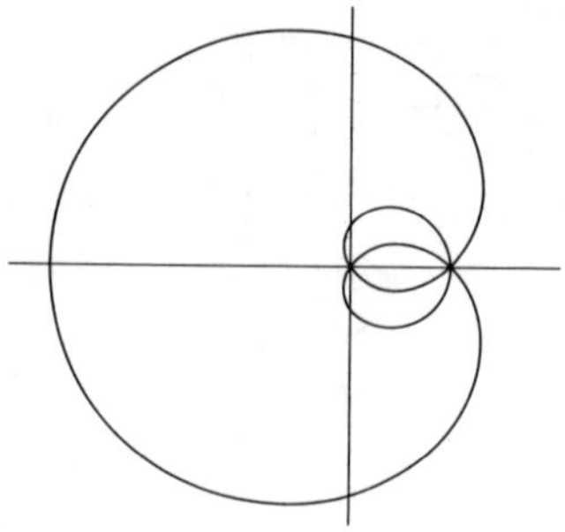

Figure 286. Nephroid

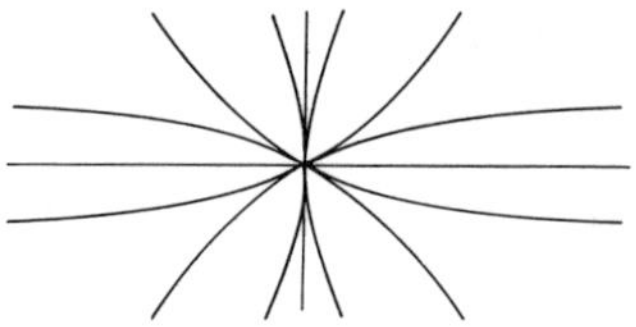

Figure 287. Nodal curve
$a = 1.00, \quad k = 3$

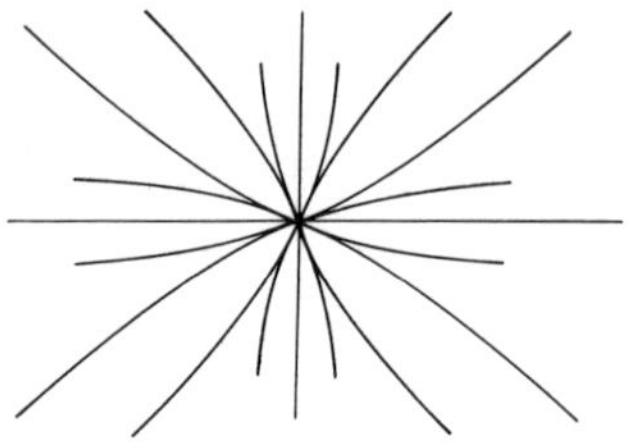

Figure 288. Nodal curve
$a = 1.00, \quad k = 4$

Nodal curve
Equations:

$$\rho = a \cot k\varphi, \quad a > 0, \quad k \in \mathbf{N}$$

Additional equation:

$$\rho = a \tan k\varphi$$

Basic geometric properties:
Unboundedness: present.
Asymptotes: straight lines

$$x \sin \frac{m\pi}{k} + y \cos \frac{m\pi}{k} \pm ka = 0,$$

$$m = 0, 1, \ldots, k - 1$$

are asymptotes.
Axial symmetry: straight lines

$$x \sin \frac{m\pi}{2k} + y \cos \frac{m\pi}{2k} = 0,$$

$$m = 0, 1, \ldots, 2k - 1$$

are axes of symmetry.
Central symmetry: point $(0, 0)$ is a center of symmetry.

Connectedness: present.
Singular points: point $(0,0)$ is a $(4k)$-multiple point, straight lines

$$x \sin \frac{(2l+1)\pi}{2k} + y \cos \frac{(2l+1)\pi}{2k} = 0,$$
$$l = 0, 1, \ldots, k-1$$

are tangents at the multiple point.
Self-intersection points: one, $(0,0)$.
Special cases: cappa ($k = 1$), windmill ($k = 2$).
(Figures 287–288).
NODAL CURVE *is an unbounded connected central-symmetric curve with a single singular point ($4k$-multiple point), $2k$ axes of symmetry and $2k$ asymptotes.*

2.2.13 O

Oblique astroid
Equations:

$$x = \frac{l}{\sin^2 \alpha} \cos t \sin^2(t-\alpha), \quad y = \frac{l}{\sin^2 \alpha} \cos(t-\alpha) \sin^2 t,$$
$$((x^2 + y^2 + 2xy \cos \alpha - l^2)^3 + 27l^2x^2y^2 \sin^2 \alpha) \sin^2 \alpha$$
$$-l^2 \cos \alpha((x^2 + y^2 + 2xy \cos \alpha - l^2)^2 \cos \alpha$$
$$-2xy(9(x^2 + y^2 + 2xy \cos \alpha - l^2) \sin^2 \alpha - 8l^2 \cos^2 \alpha)) = 0,$$
$$l > 0, \quad 0 < \alpha \leq \frac{\pi}{2}$$

Basic geometric properties:
Boundedness: present.
Closedness: present.
Axial symmetry: straight lines $x - y = 0$, $x + y = 0$ are axes of symmetry.
Central symmetry: point $(0,0)$ is a center of symmetry.
Connectedness: present.
Singular points: four cuspidal points of the first kind, for

$$t = \frac{\alpha}{2} \pm \frac{1}{2}\left(\pi - \arccos \frac{\cos \alpha}{3}\right) + \pi n$$

Self-intersection points: none.
Special cases: (right) astroid ($\alpha = \frac{\pi}{2}$).
(Figures 289–290).
OBLIQUE ASTROID *is a bounded connected closed central-symmetric curve with four singular points (cuspidal points of the first kind) and two axes of symmetry.*

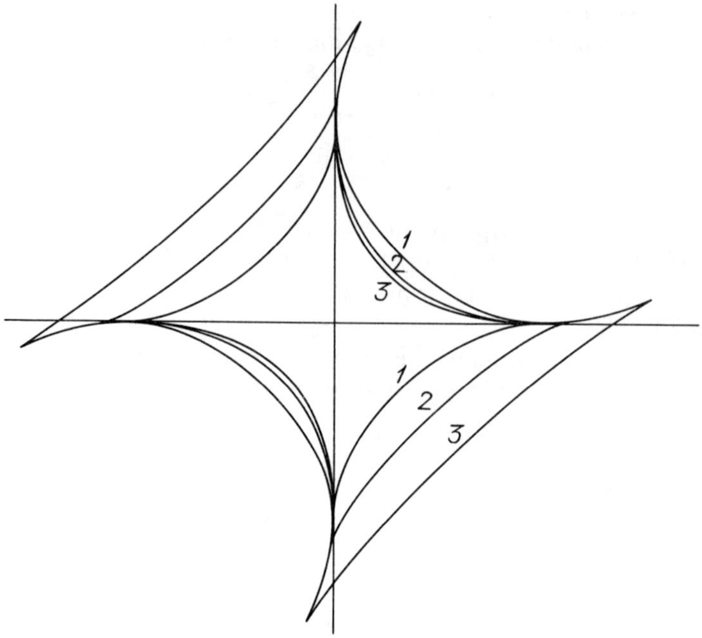

Figure 289. Oblique astroid

N	1	2	3
α	1.57	1.05	0.78
l	1.00	1.00	1.00

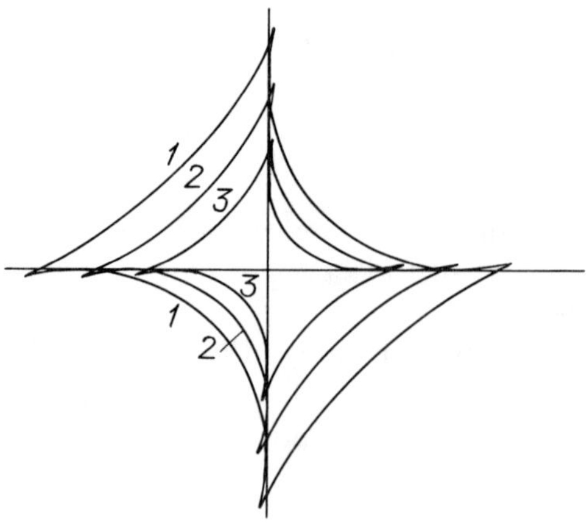

Figure 290. Oblique astroid

N	1	2	3
α	1.05	1.05	1.05
l	1.00	0.75	0.50

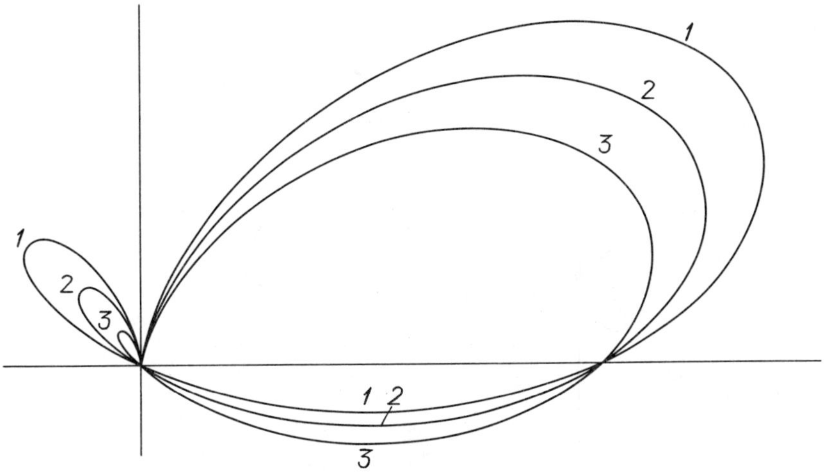

Figure 291. Oblique bifolium

N	1	2	3
b	2.00	1.50	1.00
a	1.00	1.00	1.00

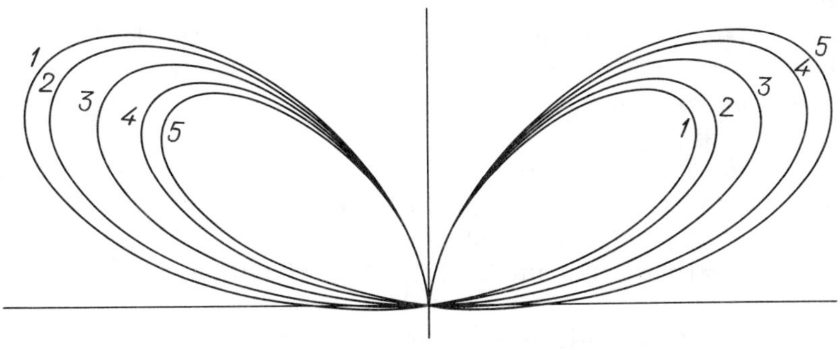

Figure 292. Oblique bifolium

N	1	2	3	4	5
a	−0.18	−0.12	0.00	0.12	0.18
b	1.50	1.50	1.50	1.50	1.50

Oblique bifolium

Equations:

$$(x^2 + y^2)^2 = x^2(ax + by),$$

$$\rho = a\cos^3\varphi + b\sin\varphi\cos^2\varphi, \quad a^2 + b^2 > 0$$

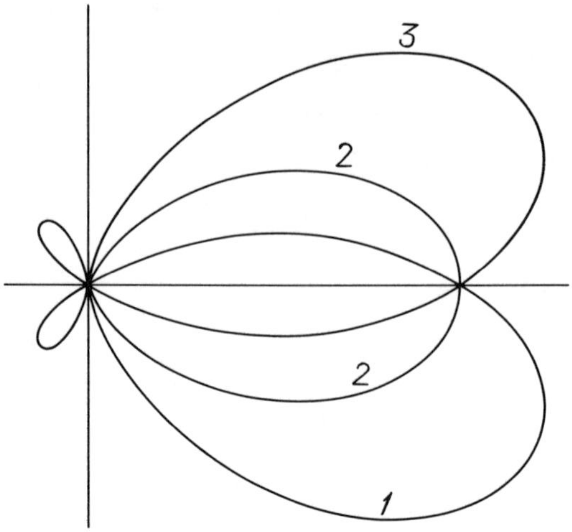

Figure 293. Oblique bifolium

N	1	2	3
b	−1.50	0.00	1.50
a	1.00	1.00	1.00

Basic geometric properties:
Boundedness: present.
Closedness: present.
Axial symmetry: straight line $y = 0$ is an axis of symmetry (for $b = 0$ only); straight line $x = 0$ is an axis of symmetry (for $a = 0$ only).
Central symmetry: none.
Connectedness: present.
Singular points: point $(0,0)$ is a degenerate triple point, straight lines $ax + by = 0$ and $x = 0$ are tangents at the multiple point.
Self-intersection points: one, $(0,0)$ (for $b \neq 0$).
Special cases: (right) bifolium ($a = 0$), folium ($b = 0$).
(Figures 291–293).
OBLIQUE BIFOLIUM *is a bounded connected closed curve with a single singular point (degenerate triple point) and without axes of symmetry (for nonvanishing parameter values).*

Oblique cappa
Equations:

$$(x^2 + y^2)y^2 = a^2(x \sin \alpha + y \cos \alpha)^2,$$

$$\rho = a\frac{\sin(\varphi + \alpha)}{\sin \varphi},$$

$$a > 0, \quad 0 < \alpha \leq \frac{\pi}{2}$$

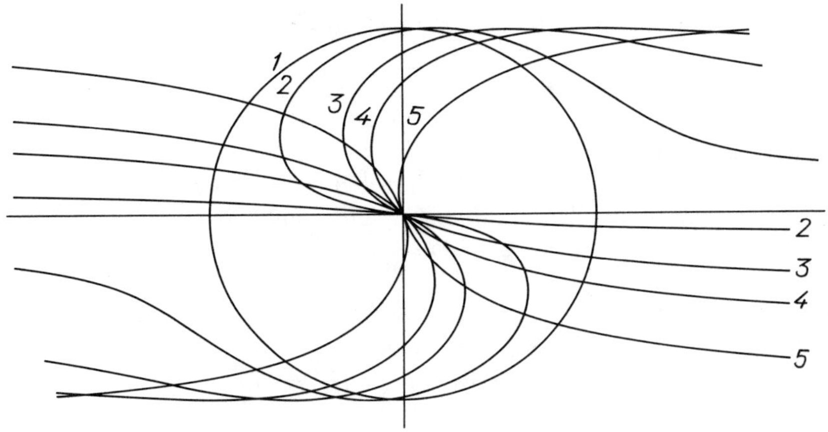

Figure 294. Oblique cappa

N	1	2	3	4	5
α	0.00	0.15	0.50	0.78	1.30
a	1.00	1.00	1.00	1.00	1.00

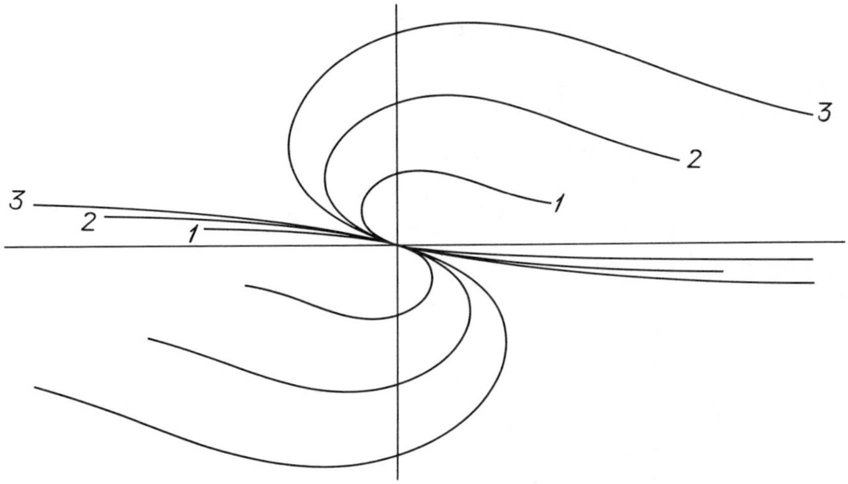

Figure 295. Oblique cappa

N	1	2	3
α	0.30	0.30	0.30
a	0.50	1.00	1.50

Basic geometric properties:
Unboundedness: present.
Asymptotes: straight line $y = 0$ is a horizontal asymptote.
Axial symmetry: none.
Central symmetry: point $(0,0)$ is a center of symmetry.
Connectedness: present.

Singular points: point $(0,0)$ is a degenerate node, straight line $a\sin\alpha + y\cos\alpha = 0$ is a tangent at the node.
Self-intersection points: one, $(0,0)$.
Special cases: (right) cappa ($\alpha = \pi/2$).
(Figures 294–295).
OBLIQUE CAPPA *is an unbounded connected central-symmetric curve with a single singular point (degenerate node) and a single horizontal asymptote.*

Oblique cissoid
Equations:

$$(x^2 + y^2)(x\cos\alpha + y\sin\alpha) = 2ay^2,$$

$$x = \frac{2at}{(1+t^2)(\sin\alpha + t\cos\alpha)}, \quad y = \frac{2a}{(1+t^2)(\sin\alpha + t\cos\alpha)},$$

$$\rho = \frac{2a\sin^2\varphi}{\cos(\varphi - \alpha)}, \quad a > 0, \quad 0 \leq \alpha < \frac{\pi}{2}$$

Additional equation:

$$x(x^2 + y^2) = 2a(x\sin\alpha + y\cos\alpha)^2$$

Basic geometric properties:
Unboundedness: present.
Asymptotes: straight line

$$x\cos\alpha + y\sin\alpha - 2a^2\cos^2\alpha = 0$$

is an inclined asymptote.
Axial symmetry: none.
Central symmetry: none.
Connectedness: present.
Singular points: point $(0,0)$ is a cuspidal point of the first kind, straight line $y = 0$ is a tangent at the cusp point.
Self-intersection points: none.
Characterized points: point $(2a, 2a\cot\alpha)$ is a vertex.
(Figures 296–298).
OBLIQUE CISSOID *is an unbounded connected curve with a single singular point (cuspidal point of the first kind) and a single inclined asymptote.*

Oblique conchoid
Equations:

$$(xy\cos\alpha - (y^2 + ay - l^2)\sin\alpha)^2 = (x\sin\alpha + y\cos\alpha + a\cos\alpha)^2(l^2 - y^2),$$

$$x = a\cot t - l\cos(\alpha + t), \; y = -l\sin(\alpha + t), \quad a > 0, \; l > 0, \; 0 < \alpha < \pi$$

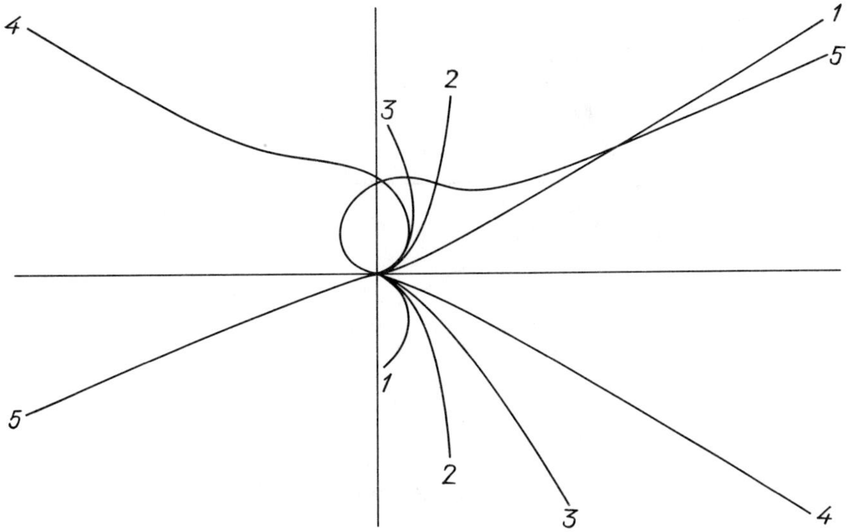

Figure 296. Oblique cissoid

N	1	2	3	4	5
α	−1.00	0.00	0.50	1.00	2.00
a	1.00	1.00	1.00	1.00	1.00

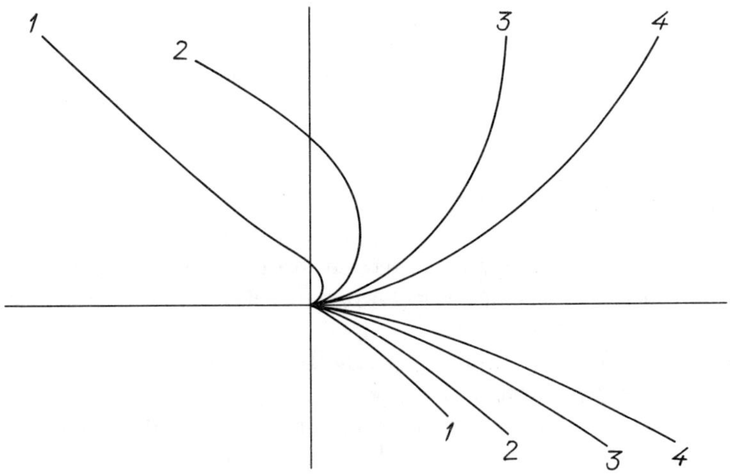

Figure 297. Oblique cissoid

N	1	2	3	4
α	0.80	0.80	0.80	0.80
a	0.50	1.00	2.00	3.00

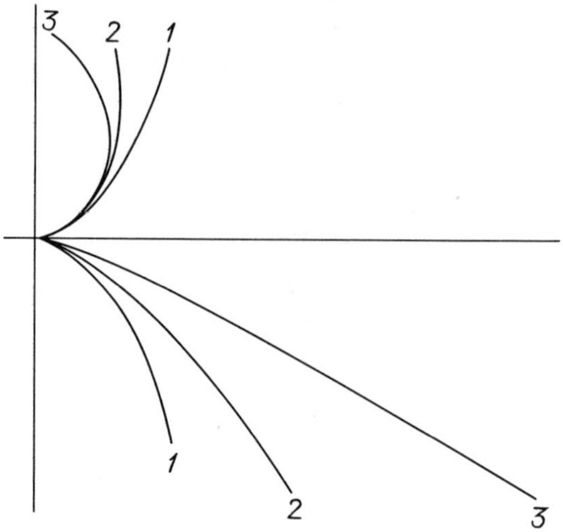

Figure 298. Oblique cissoid

N	1	2	3
α	0.00	0.50	1.00
a	1.00	1.00	1.00

Additional equation: the second branch can be obtained from the first branch with the help of replacement

$$\alpha \longmapsto \alpha + \pi$$

$$x = a \cot t + l \cos(\alpha + t),$$

$$y = l \sin(\alpha + t)$$

Basic geometric properties:

Unboundedness: present.
Asymptotes: straight lines $y = \pm \sin\alpha$ are horizontal asymptotes.
Axial symmetry: straight line $x = 0$ is a symmetry axis (for $\alpha = \pi/2$ only).
Central symmetry: none.
Connectedness: the curve may be connected or not in accordance to the relationship between parameter values.
Singular points: the curve has two singular points, one of them is a node and type of the other depends on relationship between parameter values.
Self-intersection points: one or two.
Characterized points: points $(-a \cot\alpha \pm l, 0)$ are the intersection points of the curve and x-axis.
Special cases: conchoid of Nicomedes ($\alpha = 0,\ \alpha = \pi$), orthoconchoid ($\alpha = \pi/2$).
(Figures 299–301).

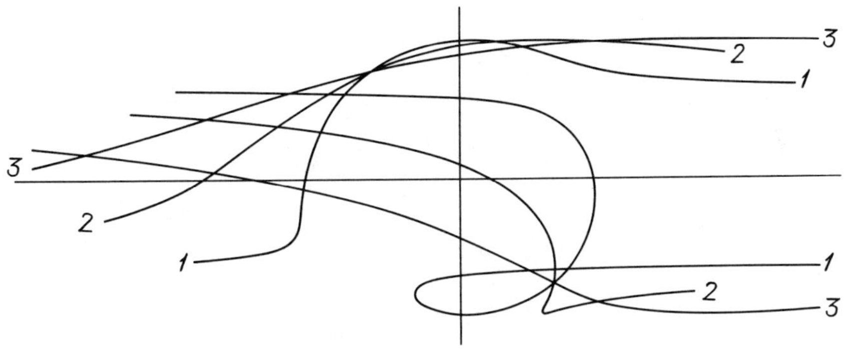

Figure 299. Oblique conchoid

N	1	2	3
a	0.10	1.00	3.00
α	0.70	0.70	0.70
l	1.50	1.50	1.50

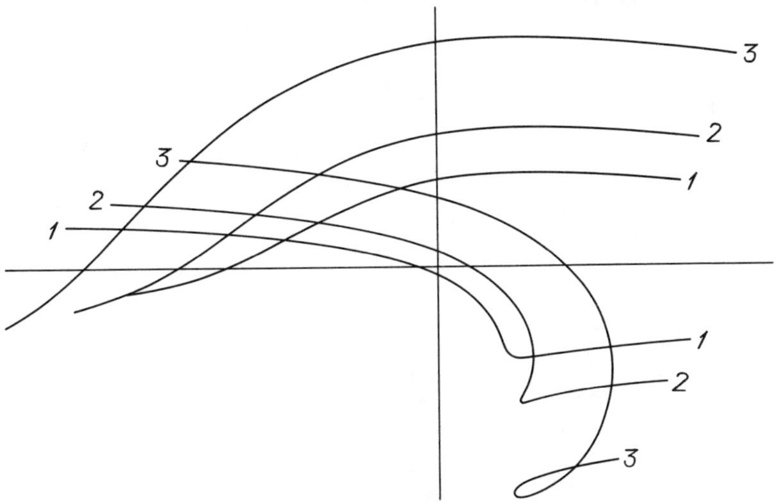

Figure 300. Oblique conchoid

N	1	2	3
l	1.00	1.50	2.50
α	0.70	0.70	0.70
a	1.00	1.00	1.00

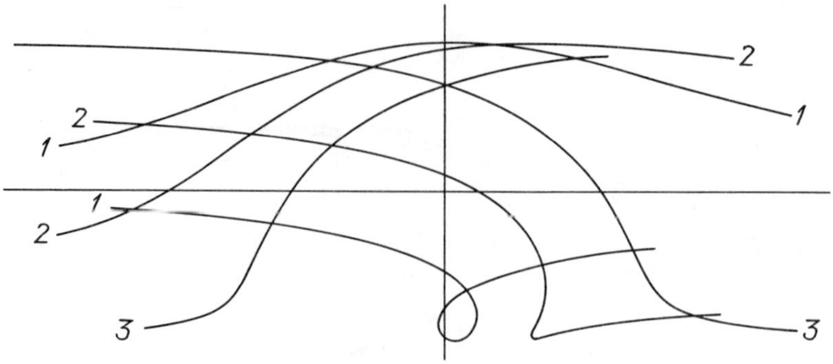

Figure 301. Oblique conchoid

N	1	2	3
α	0.10	0.70	1.50
a	1.00	1.00	1.00
l	1.50	1.50	1.50

OBLIQUE CONCHOID *is an unbounded connected or unconnected curve with two singular points (node, type of the second singular point depends on relationship between parameter values) with two horizontal asymptotes and a single symmetry axis (for $\alpha = \pi/2$ only).*

Oblique pancappa

Equations:

$$(y-a)^2(x^2+y^2) = (bx+cy)^2,$$

$$\rho = \frac{a + b\cos\varphi + c\sin\varphi}{\sin\varphi} \qquad a > 0, \quad b > 0$$

Additional equation:

$$\rho = \frac{a + b\cos(\varphi + \alpha)}{\sin\varphi}$$

Basic geometric properties:

Unboundedness: present.
Asymptotes: straight lines $y = a \pm b$ are horizontal asymptotes.
Axial symmetry: none.
Central symmetry: none.
Connectedness: present.
Singular points: point $(0,0)$ is a node, straight lines

$$(a^2 - c^2)y = (bc \pm \sqrt{a^2b^2 + a^2c^2 - a^4}x$$

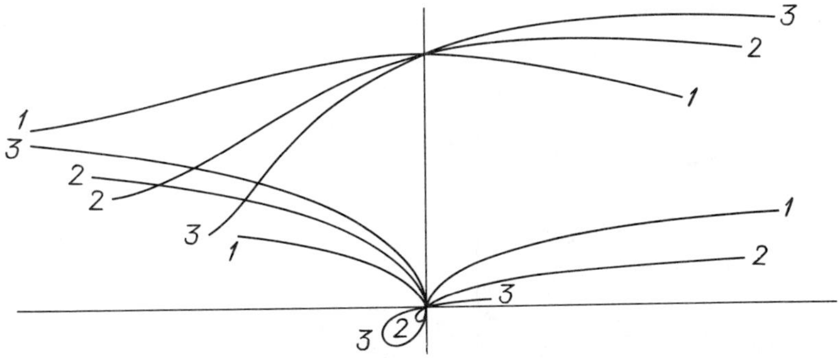

Figure 302. Oblique pancappa

N	1	2	3
b	0.00	0.50	0.80
a	1.00	1.00	1.00
c	1.00	1.00	1.00

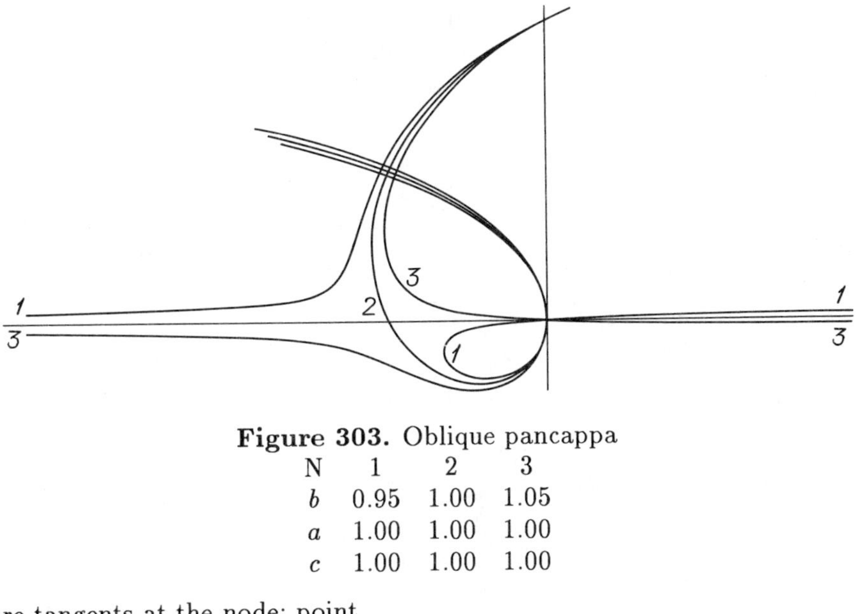

Figure 303. Oblique pancappa

N	1	2	3
b	0.95	1.00	1.05
a	1.00	1.00	1.00
c	1.00	1.00	1.00

are tangents at the node; point

$$\left(-\frac{ac}{a}\right)$$

is a node, straight lines

$$(y-a)\left(-c \pm \frac{a}{b}\sqrt{b^2+c^2}\right) = x + \frac{ac}{b}$$

are tangents at the node.
Self-intersection points: two, $(0,0)$, $(-ac/b, 0)$.
(Figures 302–307).

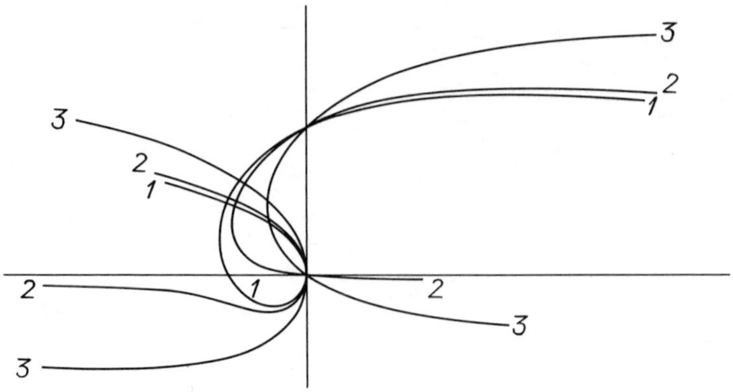

Figure 304.
Oblique pancappa

N	1	2	3
b	1.00	1.10	2.00
a	1.00	1.00	1.00
c	1.00	1.00	1.00

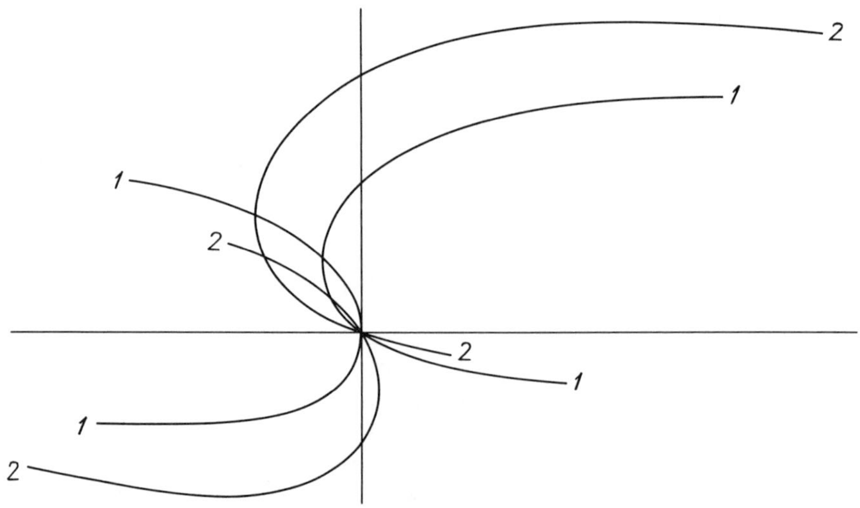

Figure 305.
Oblique pancappa

N	1	2
c	1.00	2.50
a	1.00	1.00
b	2.00	2.00

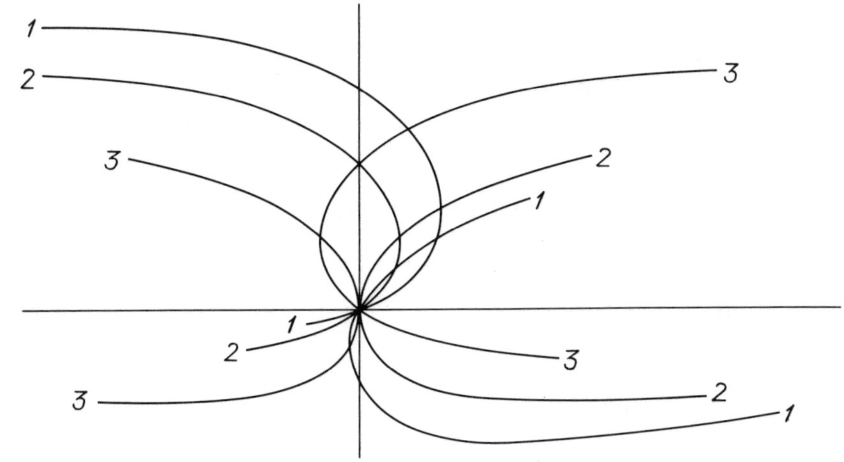

Figure 306. Oblique pancappa

N	1	2	3
c	−2.00	−1.00	1.00
a	1.00	1.00	1.00
b	2.00	2.00	2.00

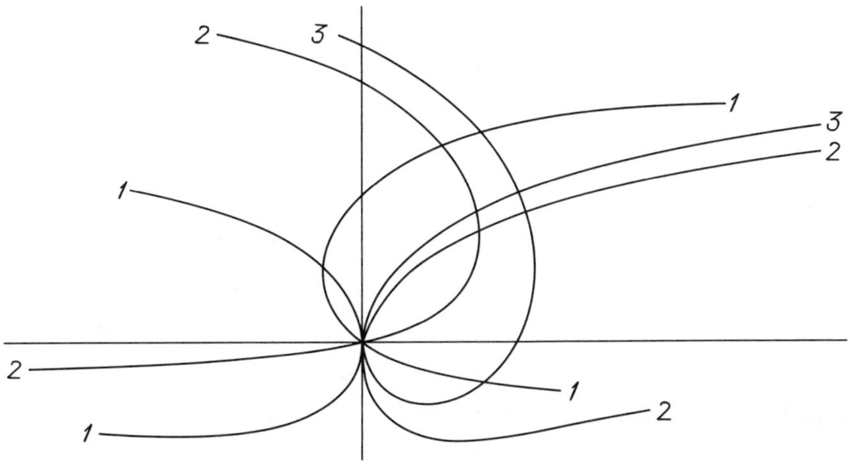

Figure 307. Oblique pancappa

N	1	2	3
a	1.00	1.50	2.00
b	2.00	2.00	2.00
c	−2.00	−2.00	−2.00

OBLIQUE PANCAPPA *is an unbounded connected curve with two singular points (nodes) and two horizontal asymptotes.*

Oblique parabolic folium
Equations:

$$x^3 = a(x^2 - y^2) + bxy,$$

$$x = a(1 - t^2) + bt, \quad y = at(1 - t^2) + bt^2,$$

$$\rho \cos\varphi = a + \tan\varphi(b - a\tan\varphi), \quad a > 0, \quad b \geq 0$$

Basic geometric properties:
Unboundedness: present.
Asymptotes: none.
Axial symmetry: none.
Central symmetry: none.
Connectedness: present.
Singular points: point $(0,0)$ is a node, straight lines

$$y = \frac{b \pm \sqrt{b^2 + 4a^2}}{2a}$$

are tangents at the node.
Self-intersection points: one, $(0,0)$.
Special cases: (right) parabolic folium ($b = 0$).
(Figures 308–309).
OBLIQUE PARABOLIC FOLIUM *is an unbounded connected curve with a single singular point (node).*

Oblique strophoid (strophoid)
Equations:

$$x(x^2 + y^2) = a((x^2 - y^2)\sin\alpha + 2xy\cos\alpha),$$

$$x = a\frac{(1 - t^2)\sin\alpha + 2t\cos\alpha}{1 + t^2}, \quad y = at\frac{(1 - t^2)\sin\alpha + 2t\cos\alpha}{1 + t^2},$$

$$a > 0, \quad 0 < \alpha < \pi$$

Additional equations:

$$(x^2 + y^2)\left(x\cos\frac{\alpha}{2} + y\sin\frac{\alpha}{2}\right) = 2axy,$$

$$x = \frac{2at}{(1 + t^2)}\left(\cos\frac{\alpha}{2} + t\sin\frac{\alpha}{2}\right), \quad y = \frac{2at^2}{(1 + t^2)}\left(\cos\frac{\alpha}{2} + t\sin\frac{\alpha}{2}\right)$$

$$\rho = a\frac{\sin\alpha \pm \sin\varphi}{\sin(\alpha + \varphi)}$$

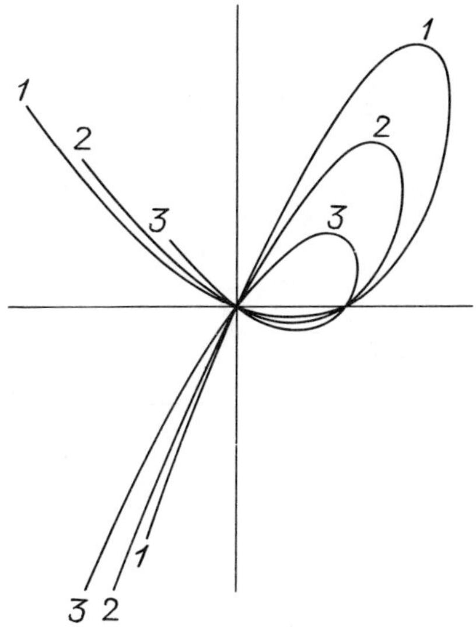

Figure 308. Oblique parabolic folium

N	1	2	3
b	2.00	1.50	0.70
a	1.00	1.00	1.00

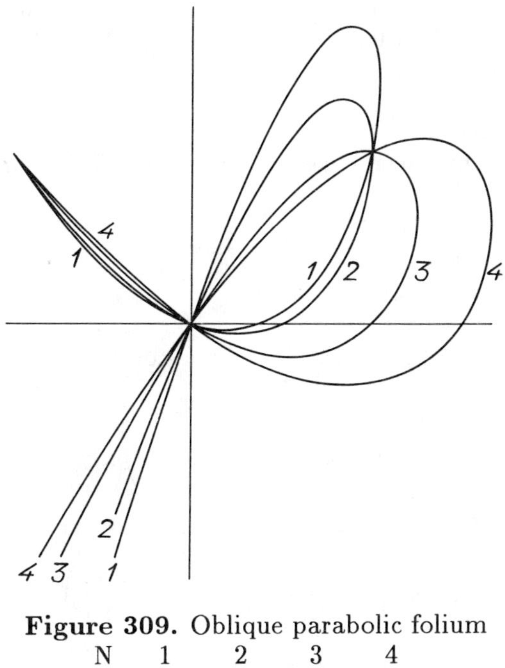

Figure 309. Oblique parabolic folium

N	1	2	3	4
a	0.50	1.00	2.00	3.00
b	2.00	2.00	2.00	2.00

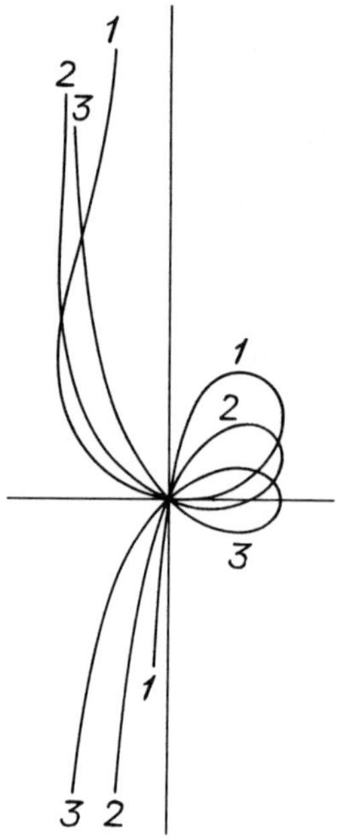

Figure 310. Oblique strophoid

N	1	2	3
α	0.30	0.78	1.57
a	1.00	1.00	1.00

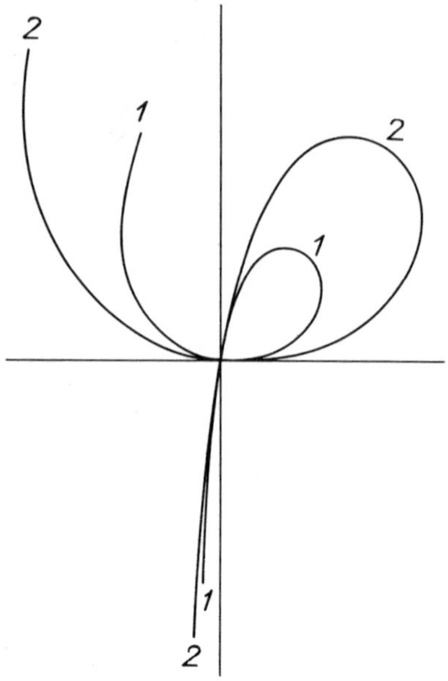

Figure 311. Oblique strophoid

N	1	2
α	0.30	0.30
a	1.00	2.00

Basic geometric properties:
Unboundedness: present.
Asymptotes: straight line $x + a\sin\alpha = 0$ is a vertical asymptote.
Axial symmetry: none.
Central symmetry: none.
Connectedness: present.
Singular points: point $(0, 0)$ is a node,

$$y = -\tan\frac{\alpha}{2}x, \quad y = \cot\frac{\alpha}{2}x$$

are tangents at the node.
Self-intersection points: one, $(0, 0)$.
Characterized points: points $(0, 0)$ and $(a\sin\alpha, 0)$ are intersection points of the curve and x-axis; points

$$\left(-a, \frac{a\cos\alpha}{1-\sin\alpha}\right), \quad \left(a\frac{a\cos\alpha}{1+\sin\alpha}\right),$$

are located at maximal distance from y-axis.
Special cases: (right) strophoid ($\alpha = \pi/2$).
(Figures 310–312).

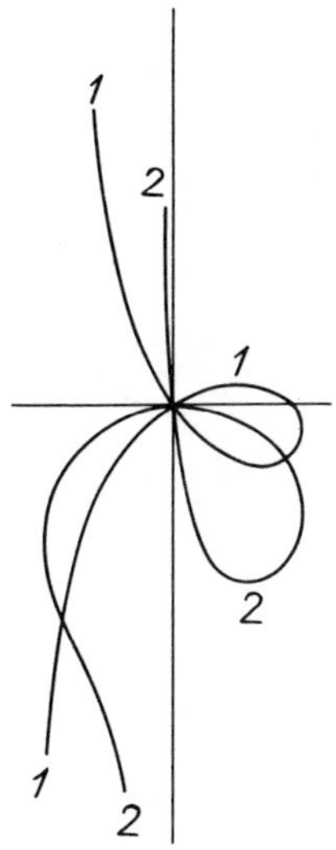

Figure 312. Oblique strophoid

N	1	2
α	2.00	3.00
a	1.00	1.00

OBLIQUE STROPHOID *is an unbounded connected curve with a single singular point (node) and a single vertical asymptote.*

Oblique trifolium (oblique three-leafed rose)
Equations:

$$\frac{(x^2+y^2)^2}{2a} = x(x^2+y^2) + (x^2-3y^2)(x\cos 2\alpha + y\sin 2\alpha),$$

$$\rho = 2a\cos\varphi + 2a\cos(3\varphi - \alpha), \quad a > 0$$

Basic geometric properties:
Boundedness: present.
Closedness: present.
Axial symmetry: straight line $y = 0$ is an axis of symmetry (for $\alpha = 0$ only).
Central symmetry: none.
Connectedness: present.
Singular points: point $(0,0)$ is a triple point.
Self-intersection points: one, $(0,0)$.
Special cases: (right) trifolium ($\alpha = 0$), oblique bifolium ($\alpha = \pi/2$).
(Figures 313–314).

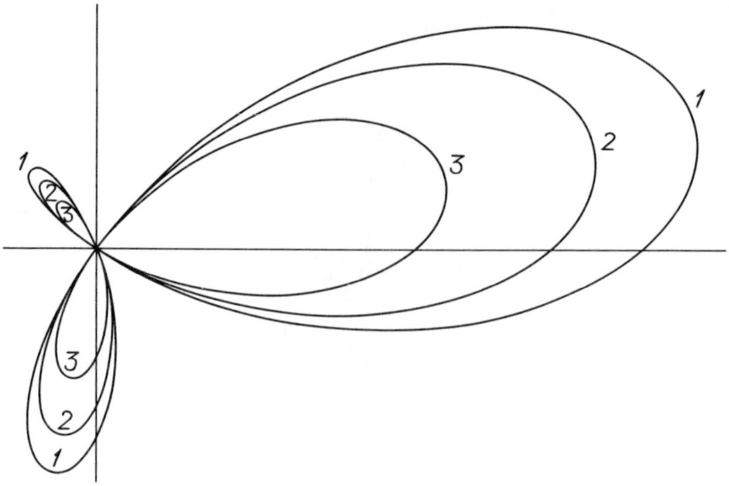

Figure 313. Oblique trifolium

N	1	2	3
α	0.70	0.70	0.70
a	1.20	1.00	0.70

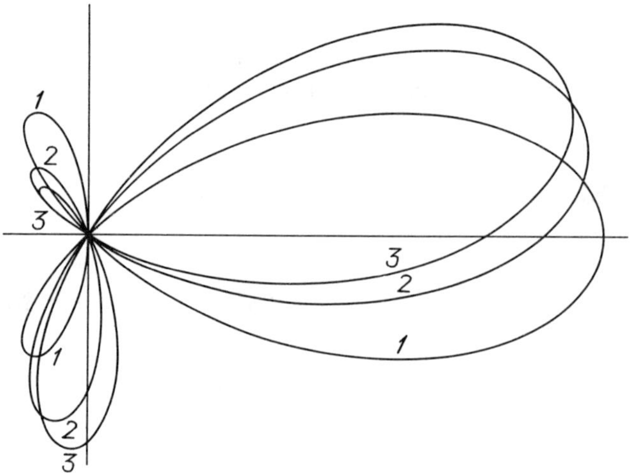

Figure 314. Oblique trifolium

N	1	2	3
α	0.00	0.70	1.00
a	1.00	1.00	1.00

Oblique trifolium *is a bounded connected closed curve with a single singular point (triple point) and a single axis of symmetry (the last for* $\alpha = 0$ *only).*

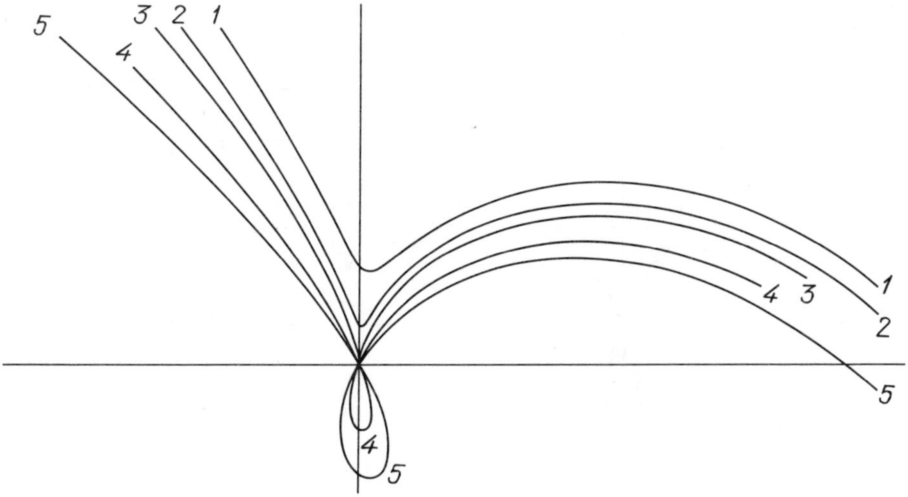

Figure 315. Oblique versiera

N	1	2	3	4	5
α	0.70	0.75	0.78	0.85	0.90
l	1.00	1.00	1.00	1.00	1.00

Oblique versiera
Equations:

$$(x^2 + y^2)(x \cos \alpha + y \sin \alpha - 4 \cos^2 \alpha) + 2ry^2 = 0,$$

$$\rho = 2r \frac{2 \cos^2 \alpha - \sin^2 \varphi}{\cos(\varphi - \alpha)},$$

$$r > 0, \quad 0 < \alpha < \frac{\pi}{2}$$

Basic geometric properties:
Unboundedness: present.
Asymptotes: straight line

$$x \cos \alpha + y \sin \alpha - 2r \cos^2 \alpha = 0$$

are inclined asymptotes.
Axial symmetry: none.
Central symmetry: none.
Connectedness: present (for $\alpha \geq \pi/4$), none (for $\alpha < \pi/4$).
Singular points: point $(0, 0)$ is a node (for $\alpha > \pi/4$), straight lines

$$y = \pm \frac{\sqrt{2} \cos \alpha}{\sqrt{-2 \cos 2\alpha}} x$$

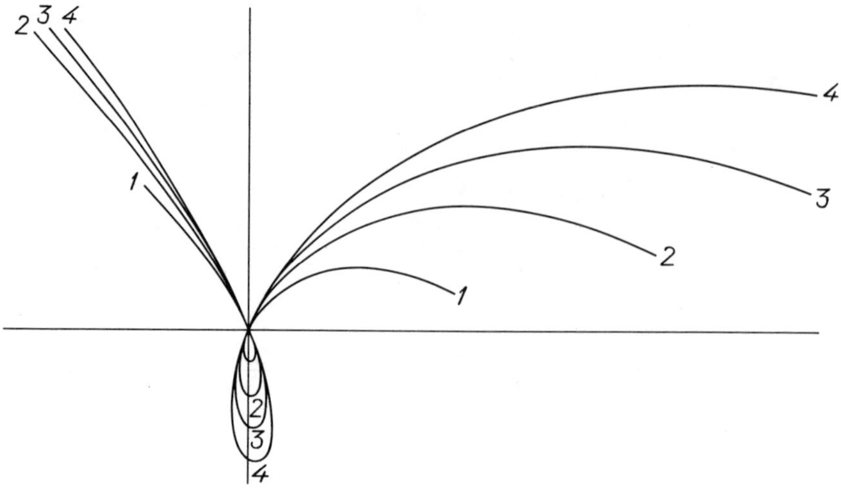

Figure 316. Oblique versiera

N	1	2	3	4
α	0.85	0.85	0.85	0.85
l	0.50	1.00	1.50	2.00

are tangents at the node, or a cuspidal point of the first kind (for $\alpha = \pi/4$), straight line $x = 0$ is a tangent at the cusp point, or an isolated point (for $\alpha < \pi/4$).
Self-intersection points: one, $(0, 0)$ (for $\alpha > \pi/4$ only).
Special cases: cissoid ($\alpha = \pi/4$).
(Figures 315–316).
OBLIQUE VERSIERA *is an unbounded curve with a single inclined asymptote, connectedness and type of singularity of which depends on value of parameter* α.

Ophiuride
Equations:

$$x(x^2 + y^2) = y(cy - bx),$$

$$\rho = (c \sin \varphi - b \cos \varphi) \tan \varphi, \quad b \geq 0, \quad c > 0$$

Additional equation:

$$x(x^2 + y^2) = y(ax - cy)$$

Basic geometric properties:
Unboundedness: present.
Asymptotes: straight line $x = c$ is a vertical asymptote.
Axial symmetry: none.
Central symmetry: none.

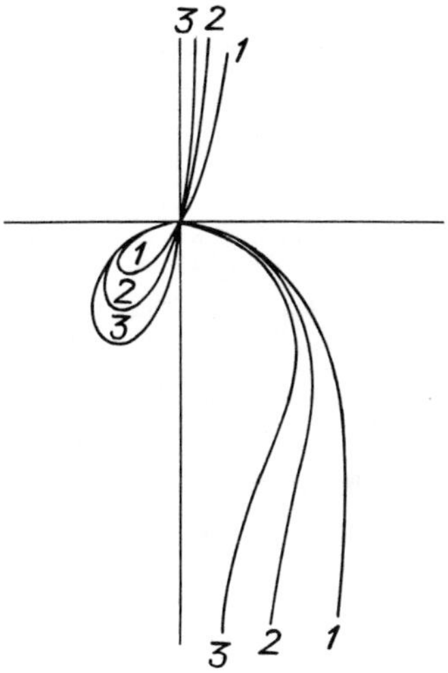

Figure 317. Ophiuride

N	1	2	3
c	2.00	1.00	0.50
b	4.00	4.00	4.00

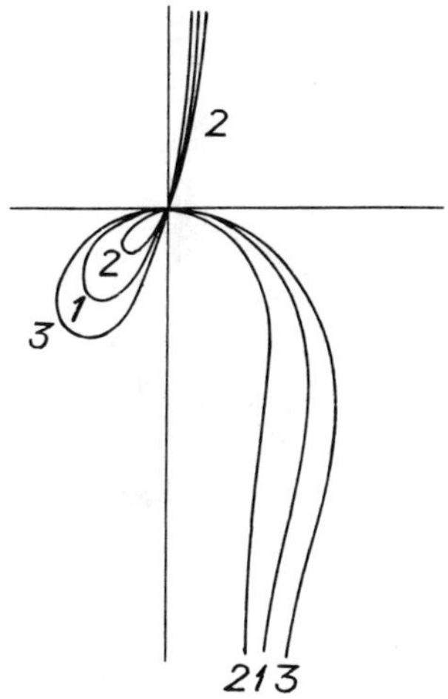

Figure 318. Ophiuride

N	1	2	3
c	1.00	1.00	1.00
b	4.00	2.50	5.00

Connectedness: present.
Singular points: point $(0,0)$ is a node, straight lines

$$y = 0, \quad y = \frac{b}{c}x$$

are tangents at the node.
Self-intersection points: one, $(0,0)$.
Special cases: cissoid (b=0).
(Figure 317–318).
OPHIURIDE *is an unbounded connected curve with a single singular point (node) and a single vertical asymptote.*

Orthoconchoid

Equations:

$$(y^2 + ay - l^2)^2 = x^2(l^2 - y^2),$$

$$x = a\cot t + l\sin t, \quad y = -l\cos t,$$

$$a > 0, \quad l > 0$$

Basic geometric properties:
Unboundedness: present.
Asymptotes: straight lines $y = \pm l$ are horizontal asymptotes.

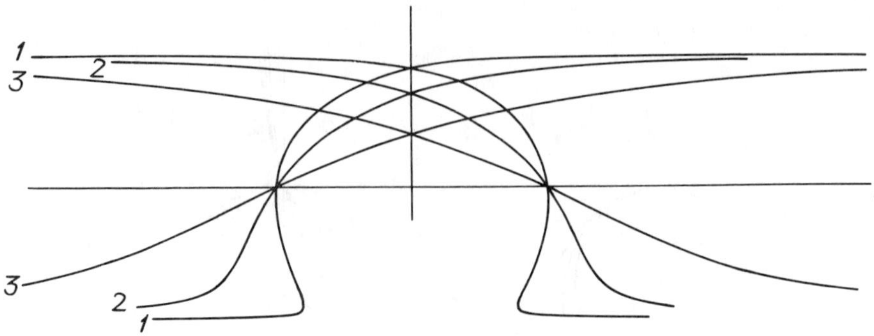

Figure 319. Orthoconchoid

N	1	2	3
a	0.25	1.00	3.00
l	1.50	1.50	1.50

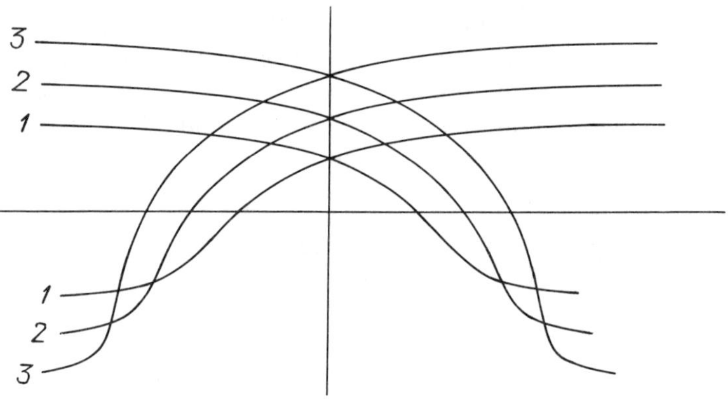

Figure 320. Orthoconchoid

N	1	2	3
l	1.00	1.50	2.00
a	1.00	1.00	1.00

Axial symmetry: straight line $x = 0$ is an axis of symmetry.
Central symmetry: none.
Connectedness: none.
Singular points: point

$$\left(0, -\frac{a + \sqrt{a^2 + 4l^2}}{2}\right)$$

is an isolated point; point

$$\left(0, \frac{\sqrt{a^2 + 4l^2} - a}{2}\right)$$

is a node, straight lines

$$y - \frac{\sqrt{a^2+4l^2}-a}{2} = \pm\sqrt{\frac{a(\sqrt{a^2+4l^2}-a)}{2(a^2+4l^2)}}x$$

are tangents at the node.
Self-intersection points: one,

$$\left(0, \frac{a+\sqrt{a^2+4l^2}-a}{2}\right)$$

Characterized points: points $(\pm l, 0)$ are intersection points of the curve and x-axis; points

$$\left(0, \frac{\pm a+\sqrt{a^2+4l^2}}{2}\right)$$

are intersection points of the curve and y-axis.
(Figures 319–320).
ORTHOCONCHOID *is an unbounded unconnected (consisting of an unbounded connected branch and an isolated point) curve with two singular points (node, isolated point), two horizontal asymptotes and a single axis of symmetry.*

Oval of Münger (egg-formed curve)
Equations:

$$(x^2+y^2)^{2n+1} - 2dx^{2n+1}(x^2+y^2)^n + (d^2-r^2)x^{4n},$$

$$\rho - 2d\rho\cos^{2n+1}\varphi + (d^2-r^2)\cos^{4n}\varphi = 0,$$

$$\rho = \cos^{2n}\varphi\left(d\cos\varphi \pm \sqrt{r^2-d^2\sin^2\varphi}\right), \quad d>0, \quad r \geq 0 \quad n \in \mathbf{N}$$

Basic geometric properties:
Boundedness: present.
Closedness: present.
Axial symmetry: straight line $y = 0$ is an axis of symmetry.
Central symmetry: none.
Connectedness: present (for $d \leq r$), none (for $d > r$).
Singular points: point $(0,0)$ is a degenerate (of $(4n)$th order) node, straight line $x = 0$ is a tangent at the node, or isolated point (for $d > r$).
Self-intersection points: one, $(0,0)$ (for $d < r$); none (for $d \geq r$).
Characterized points: points $(d \pm r, 0)$ and $(0,0)$ are intersection points of the curve and x-axis.
Special cases: double line of Münger ($d = 0$, $n = 1$).
(Figures 321–325).

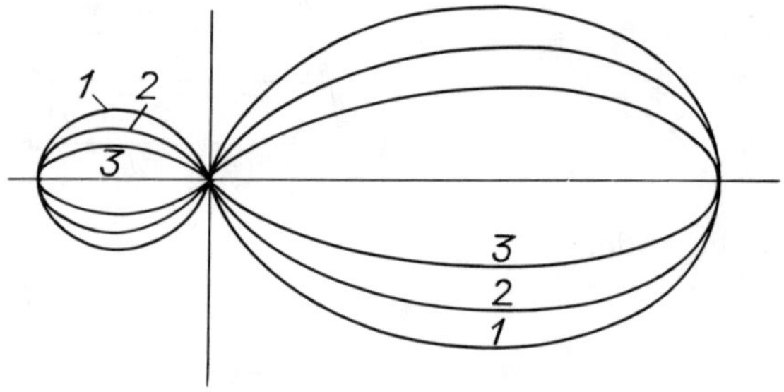

Figure 321. Oval of Münger

N	1	2	3
n	1	2	5
d	1.00	1.00	1.00
r	2.00	2.00	2.00

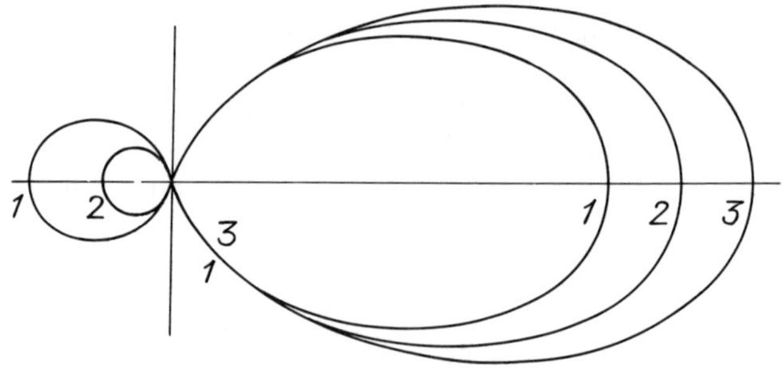

Figure 322. Oval of Münger

N	1	2	3
n	1	1	1
d	1.00	1.50	2.00
r	2.00	2.00	2.00

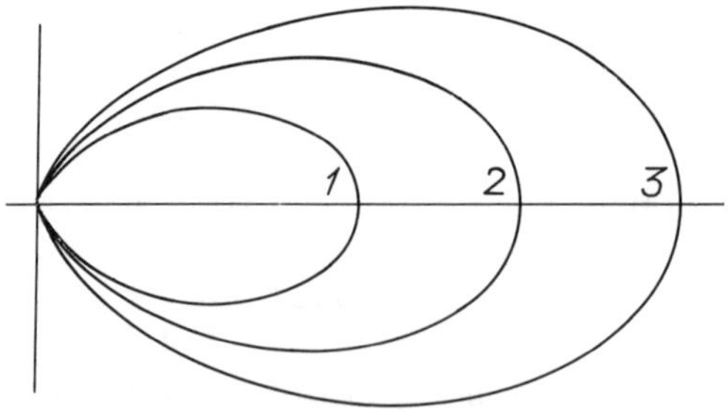

Figure 323. Oval of Münger

N	1	2	3
n	1	1	1
d	1.00	1.50	2.00
r	1.00	1.50	2.00

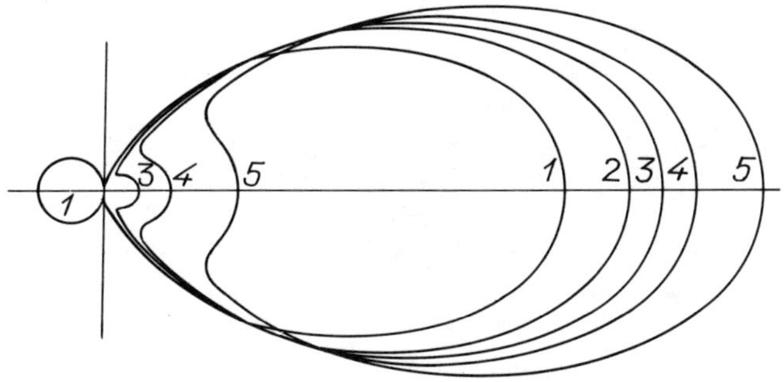

Figure 324. Oval of Münger

N	1	2	3	4	5
d	1.50	2.00	2.25	2.50	3.00
r	2.00	2.00	2.00	2.00	2.00
n	1	1	1	1	1

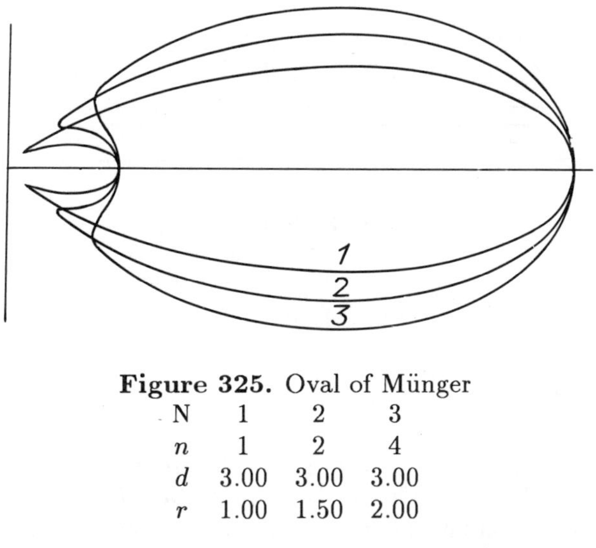

Figure 325. Oval of Münger

N	1	2	3
n	1	2	4
d	3.00	3.00	3.00
r	1.00	1.50	2.00

OVAL OF MÜNGER *is a bounded closed connected (for $d \le r$) or unconnected (consisting of bounded closed curve and isolated point (for $d > r$)) curve with a single singular point (degenerate node ($d < r$), an isolated point ($d > r$)) and a single axis of symmetry.*

2.2.14 P

Pancappa
Equations:

$$(y-a)^2(x^2+y^2) = b^2x^2, \quad \rho = \frac{a + b\cos\varphi}{\sin\varphi}, \quad a > 0, \quad b > 0$$

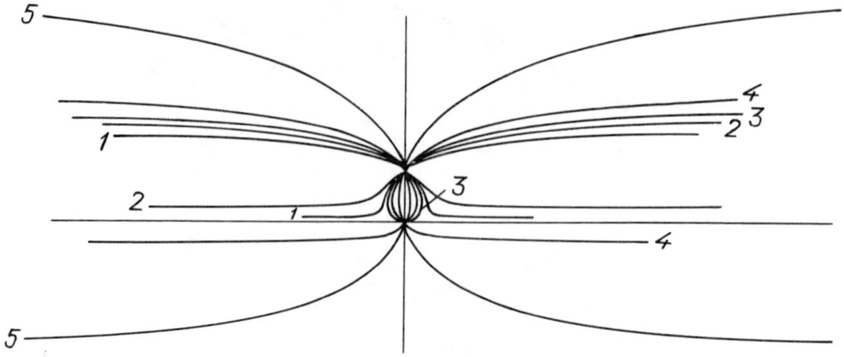

Figure 326. Pancappa

N	1	2	3	4	5
b	1.00	1.30	1.50	2.00	5.00
a	1.50	1.50	1.50	1.50	1.50

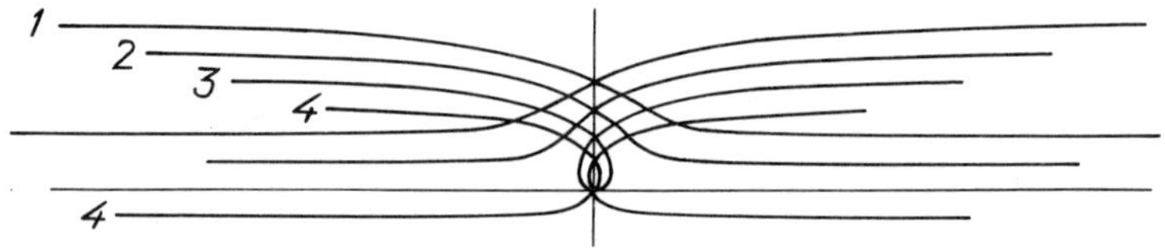

Figure 327. Pancappa

N	1	2	3	4
a	2.00	1.50	1.00	0.50
b	1.00	1.00	1.00	1.00

Basic geometric properties:
Unboundedness: present.
Asymptotes: straight lines $y = a \pm b$ are horizontal asymptotes ($a \neq b$), straight line $y = 2a$ is a horizontal asymptote ($a = b$).
Axial symmetry: straight line $x = 0$ is an axis of symmetry.
Central symmetry: none.
Connectedness: present (for $a \leq b$), none (for $a > b$).
Singular points: point $(0, a)$ is a node, straight lines

$$\pm bx - ay + a^2 = 0$$

are tangents at the node; point $(0, 0)$ is a node (for $a < b$), straight lines

$$y = \pm x\sqrt{\frac{b^2}{a^2} - 1}$$

are tangents at the node, or an isolated point (for $a > b$).

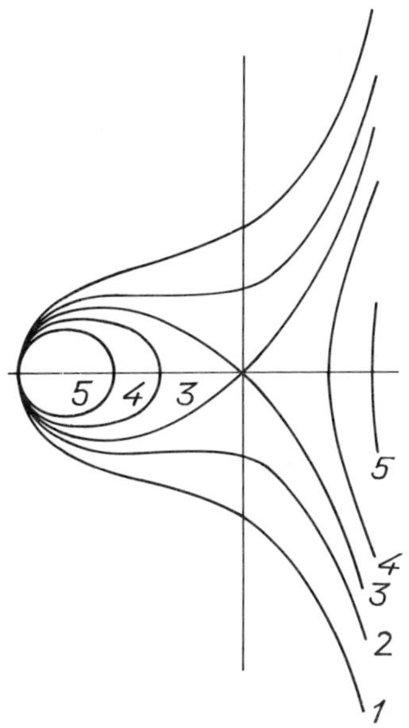

Figure 328. Panstrophoid

N	1	2	3	4	5
k	1.00	0.33	0.00	−0.33	−0.75
g	1.50	1.50	1.50	1.50	1.50

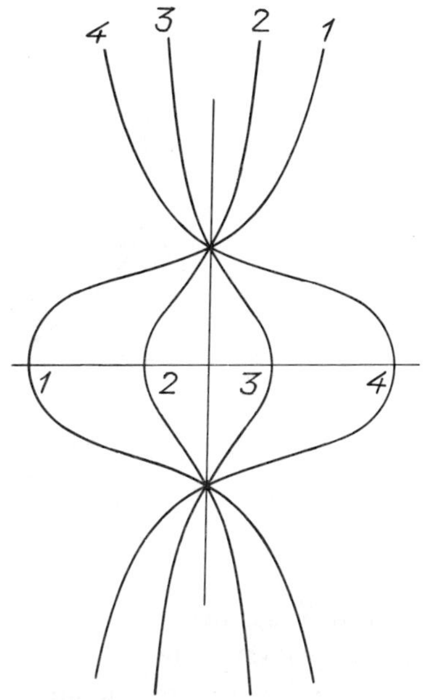

Figure 329. Panstrophoid

N	1	2	3	4
g	1.50	0.50	−0.50	−1.50
k	1.00	1.00	1.00	1.00

Self-intersection points: two, $(0,0)$, $(0,a)$ (for $a < b$): one, $(0,a)$ (for $a \geq b$). (Figures 326–327).

PANCAPPA *is an unbounded curve with a single axis of symmetry; geometric characteristics of the curve (such as number of horizontal asyptotes and self-intersection points, type of singularity and connectedness) depend on relationship between parameter values.*

Panstrophoid

Equations:

$$x(x^2 + y^2) + g(x^2 - y^2) + k(x + g) = 0,$$

$$y = \pm\sqrt{\frac{(x+g)(x^2+k)}{g-x}}, \quad g \neq 0$$

Basic geometric properties:

Unboundedness: present.

Asymptotes: straight line $x - g = 0$ is a vertical asymptote.

Axial symmetry: straight line $y = 0$ is an axis of symmetry.

Central symmetry: none.

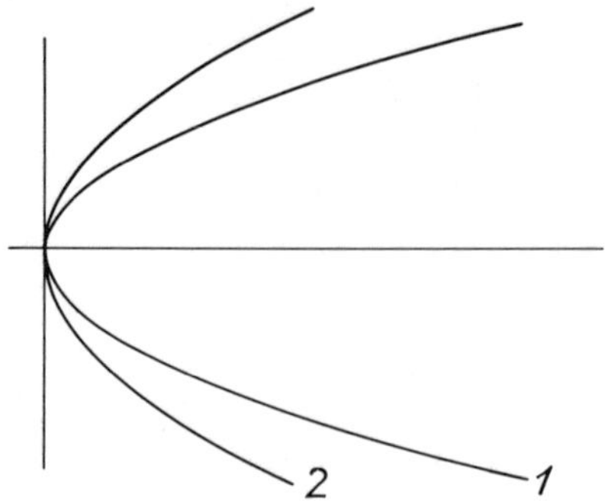

Figure 330. Parabola

N	1	2
p	1.00	0.50

Connectedness: present (for $k > 0$), none (for $k < 0$).
Singular points: none.
Self-intersection points: none.
Characterized points: point $(-g, 0)$ is a vertex.
Special cases: (right) strophoid ($k = 0$).
(Figures 328–329).
PANSTROPHOID *is an unbounded connected (for $k > 0$) or unconnected (consisting of two unbounded branches (for $k < 0$)) free of singular points with a single axis of symmetry and a single vertical asymptote.*

Parabola
Equation:

$$y^2 = 2px, \quad p > 0$$

Additional equation:

$$\rho = \frac{p}{1 + \cos\varphi}$$

Basic geometric properties:
Unboundedness: present.
Asymptotes: none.
Axial symmetry: straight line $y = 0$ is an axis of symmetry.
Central symmetry: none.
Connectedness: present.
Singular points: none.
Self-intersection points: none.
Characterized points: point $(0, 0)$ is a vertex.
(Figure 330).

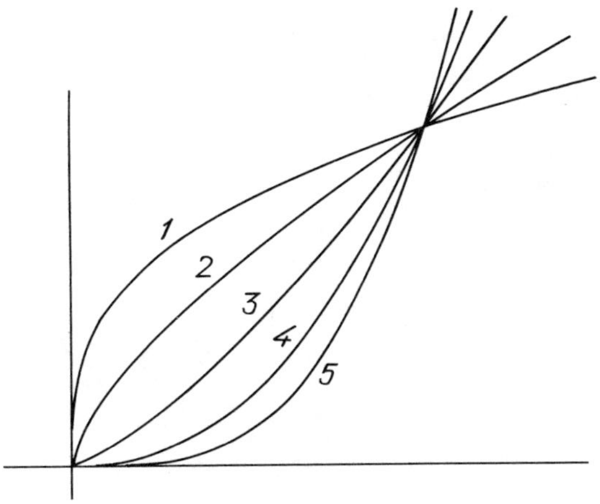

Figure 331. Parabolic curve

N	1	2	3	4	5
p	1	2	4	7	10
q	3	3	3	3	3
a	1.00	1.00	1.00	1.00	1.00

PARABOLA *is an unbounded connected curve free of singular points with a single axis of symmetry.*

Parabolic curve

Equation:

$$y^q = ax^p,$$

$$a > 0, \quad p \in \mathbf{N}, \quad q \in \mathbf{N}, \quad \text{G.C.D.}(p, q) = 1$$

Additional equation:

$$y = cx^m, \quad m > 0$$

Basic geometric properties:

Unboundedness: present.

Asymptotes: none.

Axial symmetry: none (for odd pq), present (for even pq): straight line $x = 0$ is an axis of symmetry (p is even), straight line $y = 0$ is an axis of symmetry (q is even).

Central symmetry: none (for even pq), present (for odd pq): point $(0,0)$ is a center of symmetry.

Connectedness: present.

Singular points: point $(0,0)$ is a cuspidal point of the first kind (straight line $y = 0$ is a tangent at the cusp point (for $p > q$ and even q), straight line $x = 0$ is a tangent at the cusp point (for $p < q$ and odd p).

Self-intersection points: none.

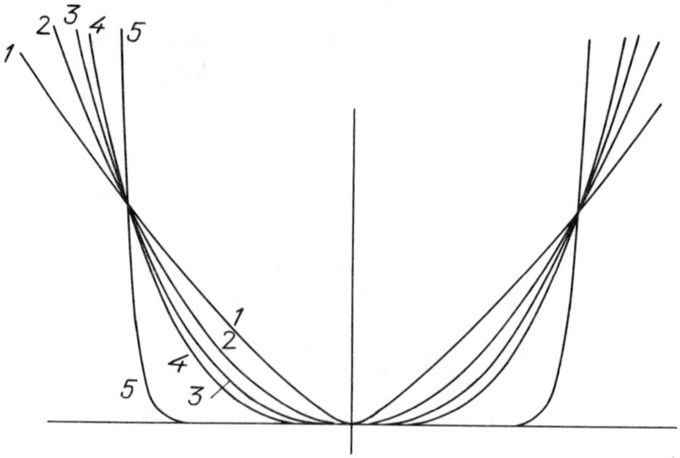

Figure 332. Parabolic curve

N	1	2	3	4	5
p	4	6	8	10	50
q	3	3	3	3	3
a	1.00	1.00	1.00	1.00	1.00

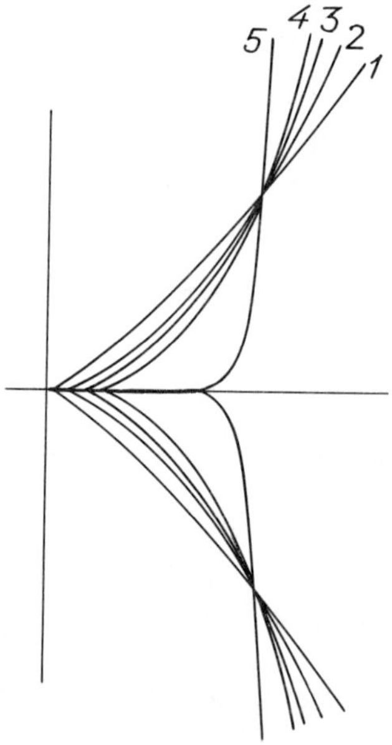

Figure 333. Parabolic curve

N	1	2	3	4	5
p	5	7	9	11	51
q	4	4	4	4	4
a	1.00	1.00	1.00	1.00	1.00

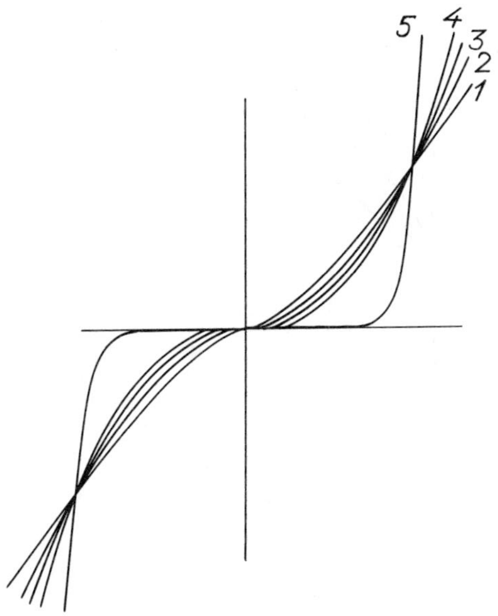

Figure 334. Parabolic curve

N	1	2	3	4	5
p	7	9	11	13	51
q	5	5	5	5	5
a	1.00	1.00	1.00	1.00	1.00

Special cases: semicubical parabola, Neil's parabola ($p = 3$, $q = 2$). (Figures 331–336).

PARABOLIC CURVE *is an unbounded connected central-symmetric (pq is even) or axisymmetric (pq is odd) curve with a single singular point (cusp point of the first kind (for even pq)).*

Parabolic folium (right parabolic folium)

Equations:

$$x^3 = a(x^2 - y^2),$$

$$x = a(1 - t^2), \quad y = at(1 - t^2), \quad a > 0$$

Basic geometric properties:

Unboundedness: present.

Asymptotes: none.

Axial symmetry: straight line $y = 0$ is an axis of symmetry.

Central symmetry: none.

Connectedness: present.

Singular points: point $(0, 0)$ is a node, straight lines $x \pm y = 0$ are tangents at the node.

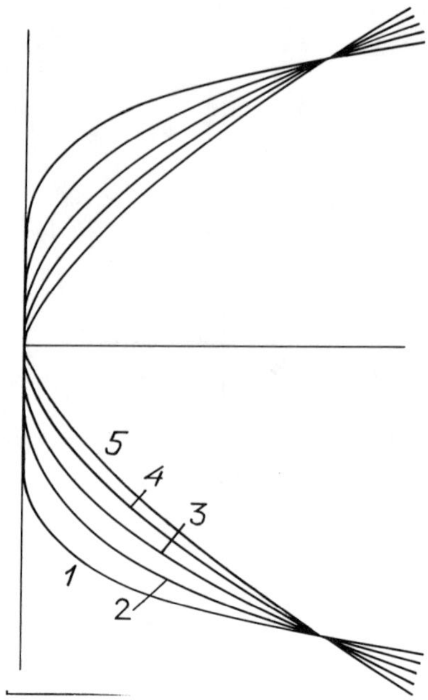

Figure 335. Parabolic curve

N	1	2	3	4	5
p	3	5	7	9	11
q	16	16	16	16	16
a	1.00	1.00	1.00	1.00	1.00

Self-intersection points: one, $(0, 0)$.
(Figure 337).
PARABOLIC FOLIUM *is an unbounded connected curve with a single singular point (node) and a single axis of symmetry.*

Parabolic spiral
Equation:

$$\rho = a\sqrt{\varphi} + l,$$

$$a > 0, \quad l \geq 0$$

Additional equation:

$$(\rho - a)^2 = 2pa\varphi$$

Basic geometric properties:
Unboundedness: present.
Asymptotes: none.
Axial symmetry: none.
Central symmetry: none.

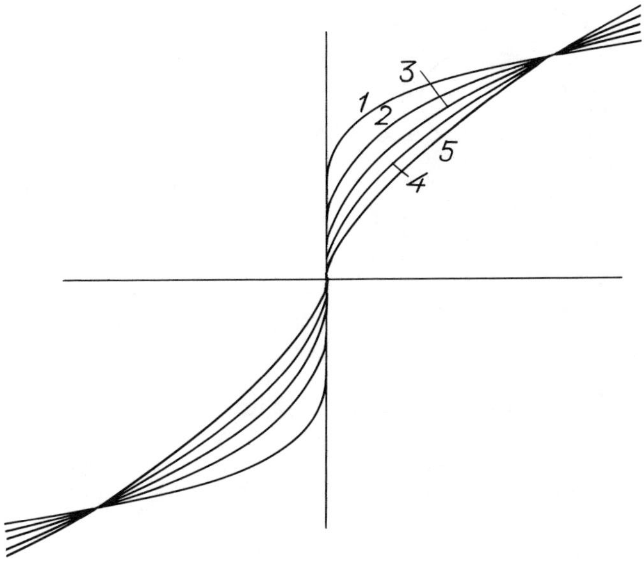

Figure 336. Parabolic curve

N	1	2	3	4	5
p	3	5	7	9	11
q	17	17	17	17	17
a	1.00	1.00	1.00	1.00	1.00

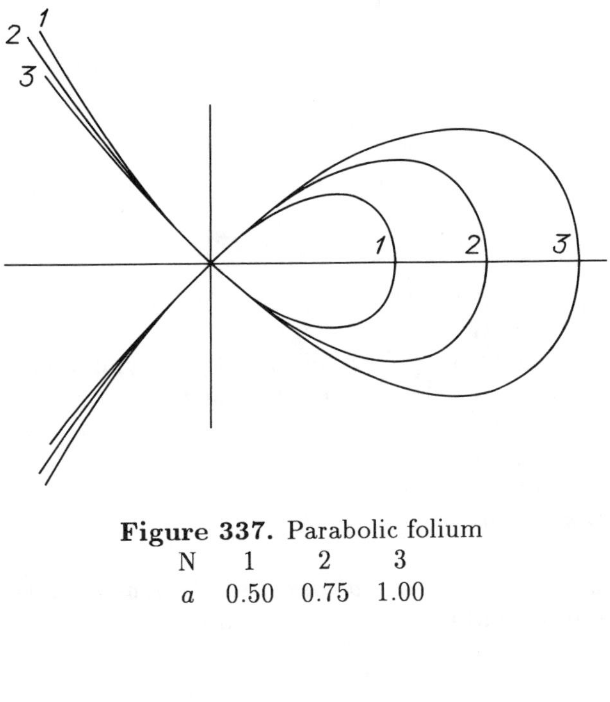

Figure 337. Parabolic folium

N	1	2	3
a	0.50	0.75	1.00

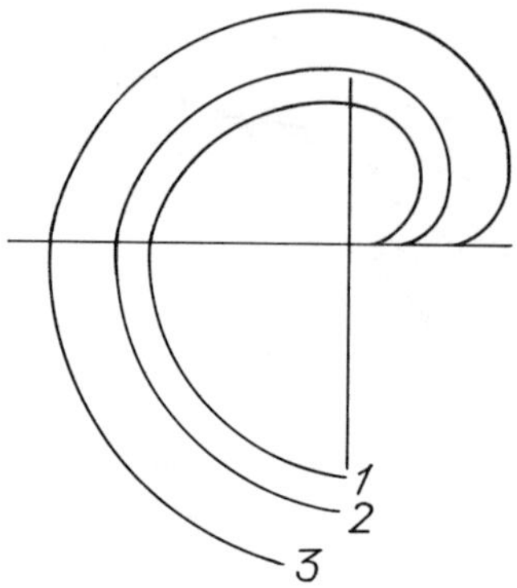

Figure 338. Parabolic spiral

N	1	2	3
l	0.90	0.30	0.00
a	1.00	1.00	1.00

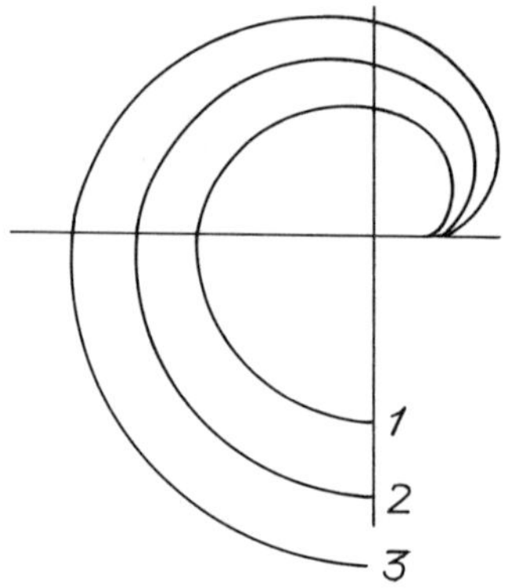

Figure 339. Parabolic spiral

N	1	2	3
l	0.30	0.30	0.30
a	1.00	0.75	0.50

Connectedness: present.
Singular points: point $(0,0)$ is a point of cessation.
Self-intersection points: none.
(Figures 338–339).
PARABOLIC SPIRAL *is an unbounded connected curve with a single singular point (point of cessation).*

Pear
Equations:

$$y^2(x-r)^2 - 2r^2y + r^4 = 0,$$

$$x = r(1+\cos t), \quad y = \frac{r}{1+\sin t},$$

$$r > 0$$

Basic geometric properties:
Unboundedness: present.
Asymptotes: straight line $x - r = 0$ is a vertical asymptote.
Axial symmetry: straight line $x - r = 0$ is an axis of symmetry.
Central symmetry: none.
Connectedness: present.
Singular points: none.
Self-intersection points: none.
(Figure 340).
PEAR *is an unbounded connected curve free of singular points with a single vertical asymptote and a single axis of symmetry.*

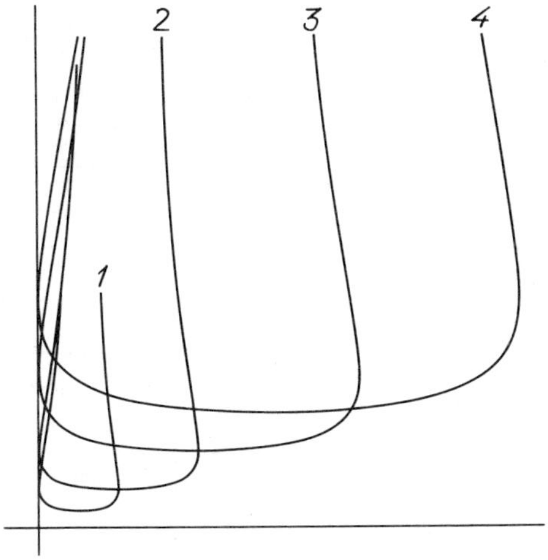

Figure 340. Pear

N	1	2	3	4
r	0.50	1.00	2.00	3.00

Pearformed quartic (piriform)
Equations:

$$x^4 - ax^3 + b^2y^2 = 0,$$

$$y = \pm\frac{x}{b}\sqrt{ax - x^2}, \quad a > 0, \quad b > 0$$

Basic geometric properties:
Boundedness: present.
Closedness: present.
Axial symmetry: straight line $y = 0$ is an axis of symmetry.
Central symmetry: none.
Connectedness: present.
Singular points: point $(0, 0)$ is a cuspidal point of the first kind, straight line $y = 0$ is a tangent at the cusp point.
Self-intersection points: none.
Characterized points: points

$$\left(\frac{3}{4}a, \pm\frac{a^2}{16b}3\sqrt{3}\right)$$

are located at maximal distance from x-axis (vertices).
Area of domain enclosed by the curve: $\pi a^3/(8b)$.
Special cases: quartic of Bonnet ($a = b$), top ($a = 2b$).
(Figures 341–342).
PEARFORMED QUARTIC *is a bounded connected closed curve with a single singular point (cuspidal point of the first kind) and a single axis of symmetry.*

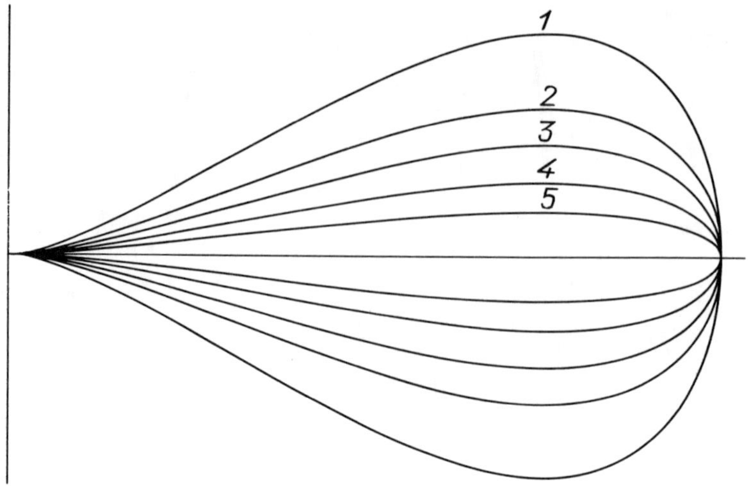

Figure 341. Pearformed quartic

N	1	2	3	4	5
a	1.00	1.00	1.00	1.00	1.00
b	1.00	1.50	2.00	3.00	5.00

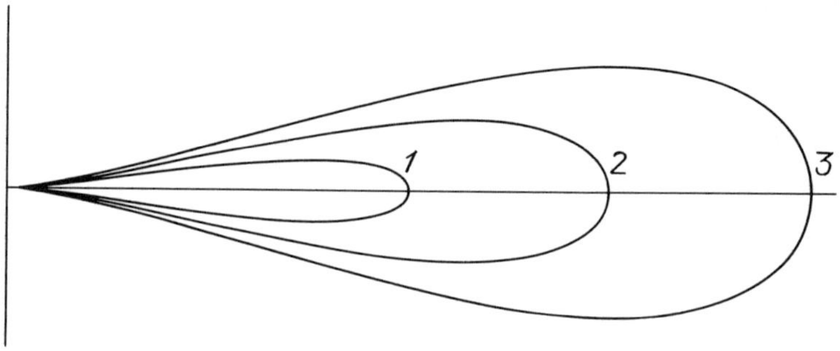

Figure 342. Pearformed quartic

N	1	2	3
a	0.50	0.75	1.00
b	2.00	2.00	2.00

Pearl (teardrop curve)
Equation:

$$x^s(a-x)^r = \frac{y^p}{b^p},$$

$$a > 0, \quad p = 2n, \quad s = 2m-1, \quad r = 2l-1,$$

$$n \in \mathbf{N}, \quad m \in \mathbf{N}, \quad l \in \mathbf{N}$$

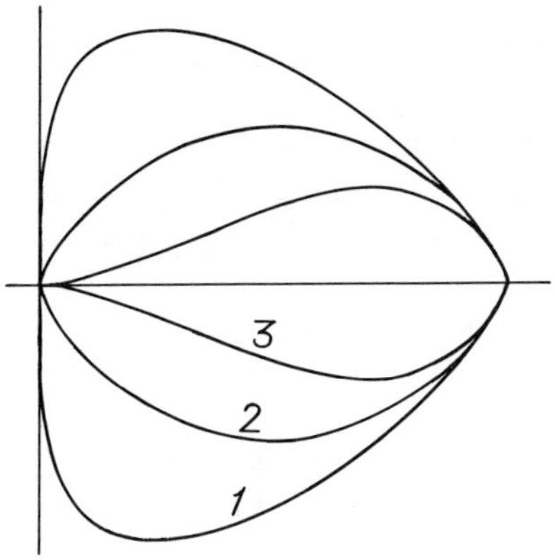

Figure 343. Pearl

N	1	2	3
s	1	3	7
a	1.00	1.00	1.00
b	1.00	1.00	1.00
p	4	4	4
r	3	3	3

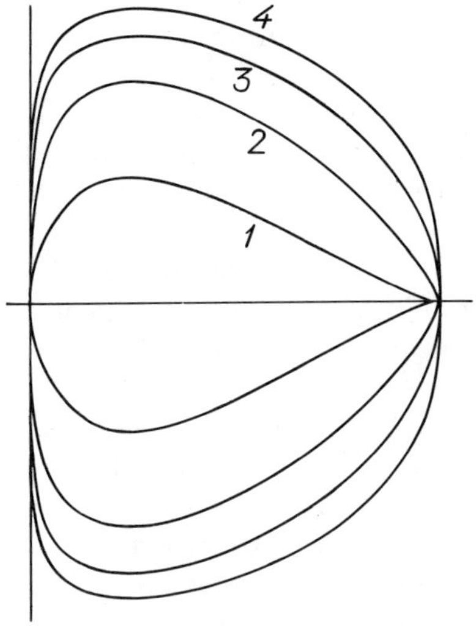

Figure 344. Pearl

N	1	2	3	4
p	2	4	6	8
a	1.00	1.00	1.00	1.00
b	1.00	1.00	1.00	1.00
r	3	3	3	3
s	1	1	1	1

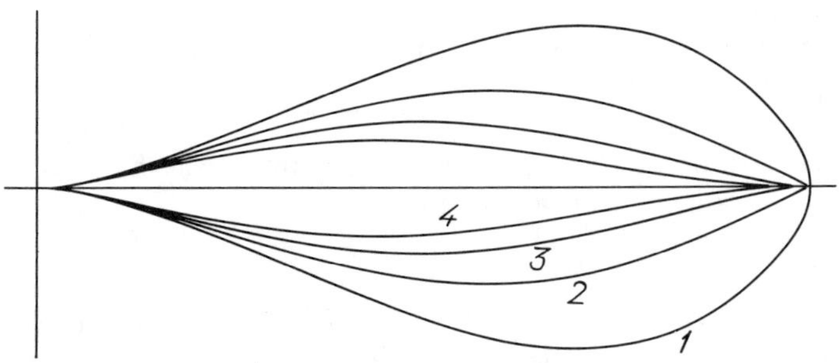

Figure 345. Pearl

N	1	2	3	4
r	3	5	7	9
a	1.00	1.00	1.00	1.00
b	1.00	1.00	1.00	1.00
p	4	4	4	4
s	7	7	7	7

Basic geometric properties:
Boundedness: present.
Closedness: present.
Axial symmetry: straight line $y = 0$ is an axis of symmetry; straight line $2x = a$ is an axis of symmetry (for $s = r$).
Central symmetry: point $(a/2, 0)$ is a center of symmetry (for $s = r$).
Connectedness: present.
Singular points: point $(0,0)$ is a cuspidal point of the first kind (for $s > p$), straight line $y = 0$ is a tangent at the cusp point; point $(a, 0)$ is a cuspidal point of the first kind (for $r > p$), straight line $y = 0$ is a tangent at the cusp point.
Self-intersection points: none.
Special cases: circle of high order ($a = b,\ p = s + r$).
(Figures 343–345).
PEARL *is a bounded connected closed curve with a single axis of symmetry; the curve may possess singular points (cuspidal points of the first kind) for definite relations between parameter values.*

Pearl of the fourth order (A)
Equations:

$$ay^3 - y^4 = a^2x^2,$$

$$x = \pm\frac{1}{a}\sqrt{ay^3 - y^4}, \quad a > 0$$

Basic geometric properties:
Boundedness: present.
Closedness: present.
Axial symmetry: straight line $x = 0$ is an axis of symmetry.
Central symmetry: none.
Connectedness: present.
Singular points: point $(0,0)$ is a cuspidal point of the first kind, straight line $x = 0$ is a tangent at the cusp point.
Self-intersection points: none.
(Figure 346).
PEARL OF THE FOURTH ORDER (A) *is a bounded connected closed curve with a single axis of symmetry and a single singular point (cuspidal point of the first kind).*

Pearl of the fourth order (B)
Equations:

$$ay^3 - y^4 = x^4,$$

$$x = \pm\sqrt[4]{ay^3 - y^4}, \quad a > 0$$

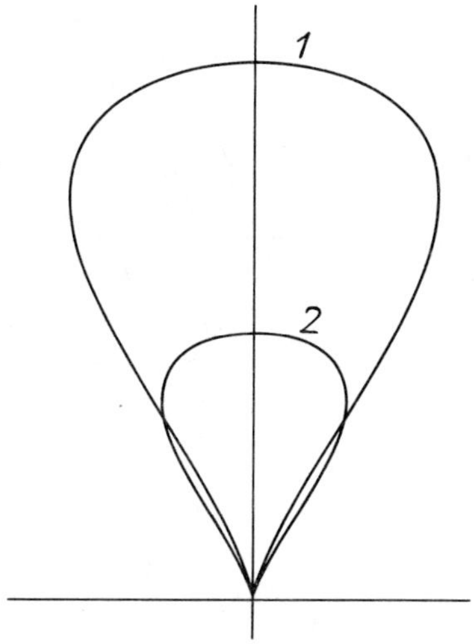

Figure 346.
Pearl of the fourth order (A)

N	1	2
a	1.00	0.50

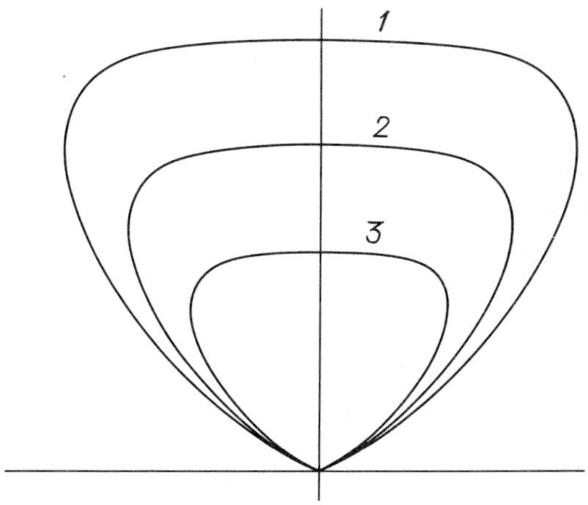

Figure 347.
Pearl of the fourth order (B)

N	1	2	3
a	1.00	0.75	0.50

Basic geometric properties:
Boundedness: present.
Closedness: present.
Axial symmetry: straight line $x = 0$ is an axis of symmetry.

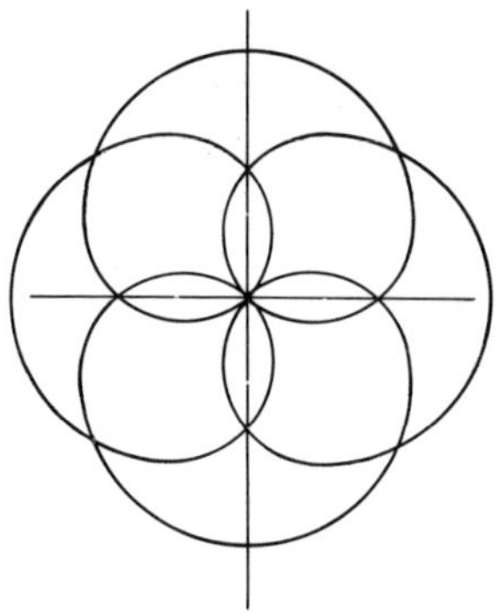

Figure 348.
Pedal of lemniscate

Figure 349.
Pedal of lemniscate

N	1	2
a	1.00	1.30

Central symmetry: none.
Connectedness: present.
Singular points: none.
Self-intersection points: none.
(Figure 347).
PEARL OF THE FOURTH ORDER (B) *is a bounded connected closed curve free of singular points with a single axis of symmetry.*

Pedal of lemniscate (pedal curve of lemniscate, pedal locus of lemniscate, pedal locus line of lemniscate)
Equation:

$$\rho^2 = \frac{a^2}{2} \cos^2 \frac{2\varphi}{3}, \quad a > 0$$

Basic geometric properties:
Boundedness: present.
Closedness: present.
Axial symmetry: straight lines $x = 0,\ y = 0,\ x \pm y = 0$ are axes of symmetry.
Central symmetry: point $(0,0)$ is a center of symmetry.
Connectedness: present.
Singular points: point $(0,0)$ is a four-multiple point, straight lines $x \pm y = 0$ are tangents at the multiple point; points

$$\left(\pm\frac{a\sqrt{3}}{4}, \pm\frac{a\sqrt{3}}{4}\right), \quad \left(\pm\frac{a}{2\sqrt{2}}, 0\right), \quad \left(0, \pm\frac{a}{2\sqrt{2}}\right)$$

are nodes.

Self-intersection points: nine,

$$(0,0), \quad \left(\pm\frac{a\sqrt{3}}{4}, \pm\frac{a\sqrt{3}}{4},\right), \quad \left(\pm\frac{a}{2\sqrt{2}}, 0\right), \quad \left(0, \pm\frac{a}{2\sqrt{2}}\right)$$

(Figures 348–349).

PEDAL OF LEMNISCATE *is a bounded connected closed central-symmetric curve with nine singular points (degenerate four-multiple point, eight nodes) and four axes of symmetry.*

Pericycloid

Equations:

$$x = (R + mR)\cos mt - mR\cos(t + mt),$$

$$y = (R + mR)\sin mt - mR\sin(t + mt),$$

$$R > 0, \quad m = -\frac{p}{q} < -1, \quad p \in \mathbf{N}, \quad q \in \mathbf{N}, \quad \text{G.C.D.}(p,q) = 1$$

Additional equations:

$$x = (R - r)\cos\varphi + r\cos\left(\frac{R-r}{r}\varphi\right), \quad y = -(R - r)\sin\varphi + r\sin\frac{R-r}{r}\varphi,$$

$$\rho^2 = (R-r)^2 + r^2 + 2(R-r)r\cos\frac{R}{r}\varphi$$

Basic geometric properties:

Boundedness: present.

Closedness: present.

Axial symmetry: straight lines

$$x\sin\frac{\pi}{q}k = y\cos\frac{\pi}{q}k, \quad k = 0, 1, \ldots, q-1$$

are axes of symmetry.

Central symmetry: point $(0,0)$ is a center of symmetry (for even q).

Connectedness: present.

Singular points: points

$$\left(R\cos\frac{2\pi}{q}k, R\sin\frac{2\pi}{q}k\right), \quad k = 0, 1, \ldots, q-1$$

are cuspidal points of the first kind, straight lines

$$x\sin\frac{2\pi}{q}k = y\cos\frac{2\pi}{q}k, \quad k = 0, 1, \ldots, q-1$$

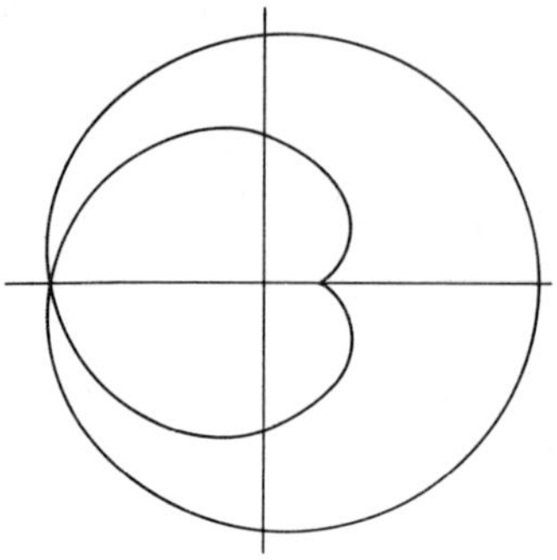

Figure 350. Pericycloid
$m = -3, \quad R = 1.00$

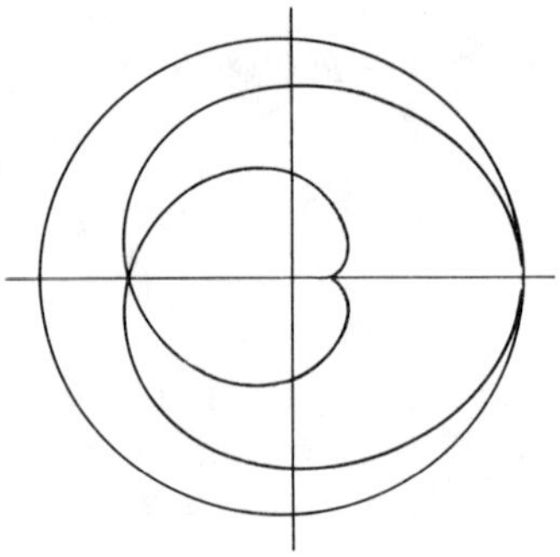

Figure 351. Pericycloid
$m = -4, \quad R = 1.00$

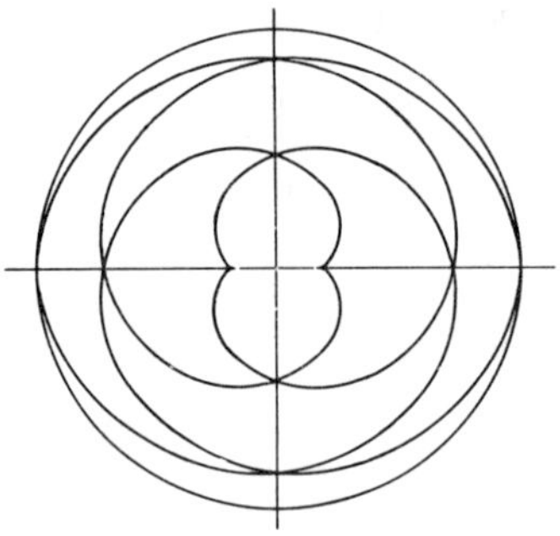

Figure 352. Pericycloid
$m = -7/2, \quad R = 1.00$

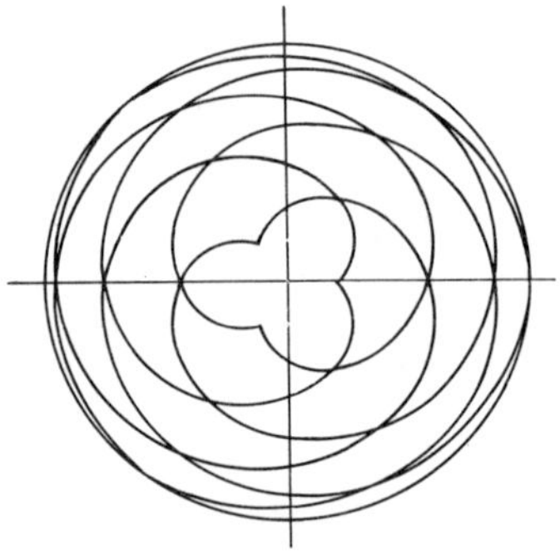

Figure 353. Pericycloid
$m = -10/3, \quad R = 1.00$

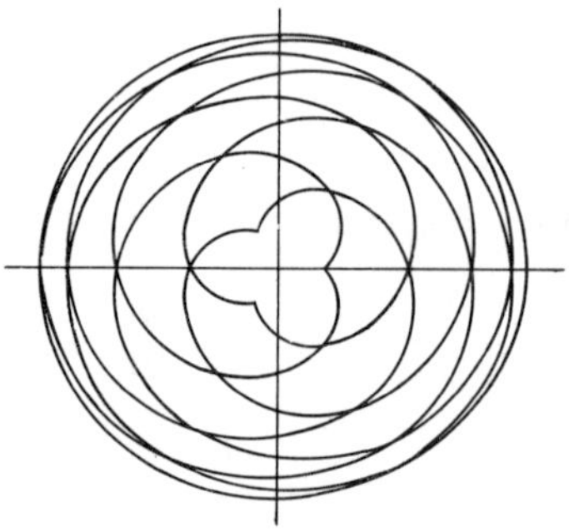

Figure 354. Pericycloid
$m = -11/3, \quad R = 1.00$

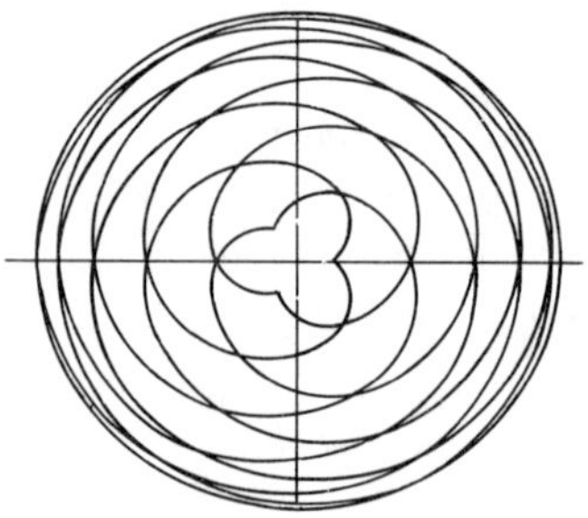

Figure 355. Pericycloid
$m = -13/3, \quad R = 1.00$

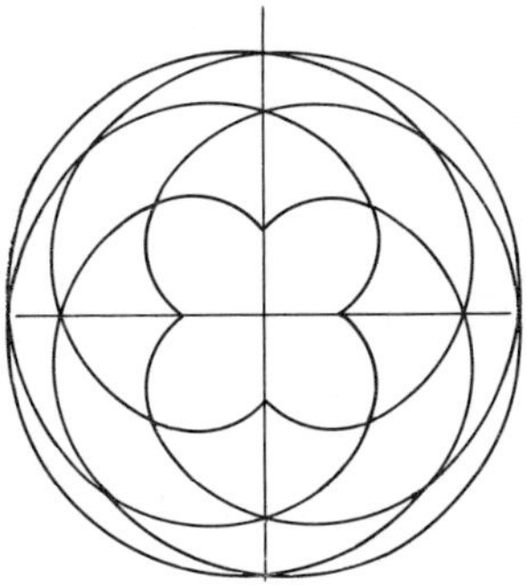

Figure 356. Pericycloid
$m = -9/4, \quad R = 1.00$

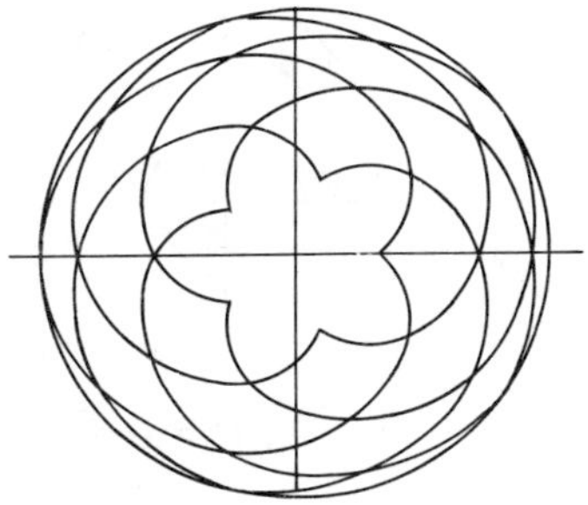

Figure 357. Pericycloid
$m = -11/5, \quad R = 1.00$

are tangents at the corresponding cusp points; the curve has q nodes of the form

$$\left(\rho_l \cos\frac{2k+1}{q}\pi, \rho_l \sin\frac{2k+1}{q}\pi\right), \quad l = 1, 3, \ldots,$$

$$\left(\rho_n \cos\frac{2k}{q}\pi, \rho_n \sin\frac{2k}{q}\pi\right), \quad n = 2, 4, \ldots,$$

$$k = 0, 1, \ldots, q-1$$

Number of self-intersection points is finite and depends on value of parameter m.
(Figures 350–357).
PERICYCLOID *is a bounded connected closed central-symmetric (the last for even q only) curve with q nodes, q cusp points of the first kind and q axes of symmetry.*
In case of irrational m pericycloid is a transcendent nonclosed curve with an infinitely large number of nodes and cusp points.

Polar tractrix
Equations:

$$\rho = a\cos t, \quad \varphi = \tan t - t,$$

$$\varphi = \frac{\sqrt{a^2 - \rho^2}}{\rho} - \arccos\frac{\rho}{a}, \quad a > 0$$

Additional equation:

$$\varphi = \frac{\sqrt{a^2 - \rho^2}}{\rho} - \arcsin\frac{\rho}{a},$$

$$a > 0$$

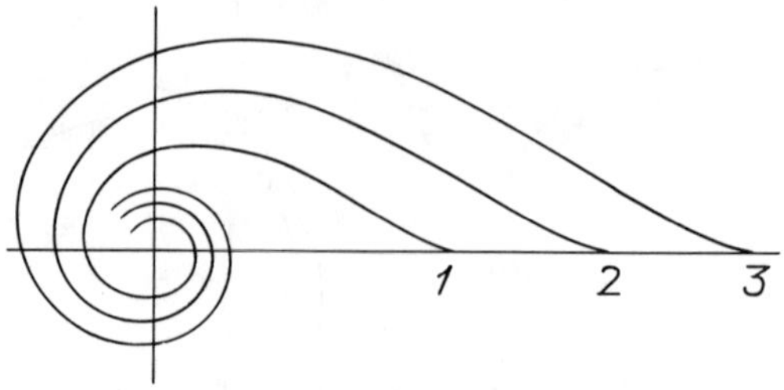

Figure 358. Polar tractrix

N	1	2	3
m	0.50	0.75	1.00

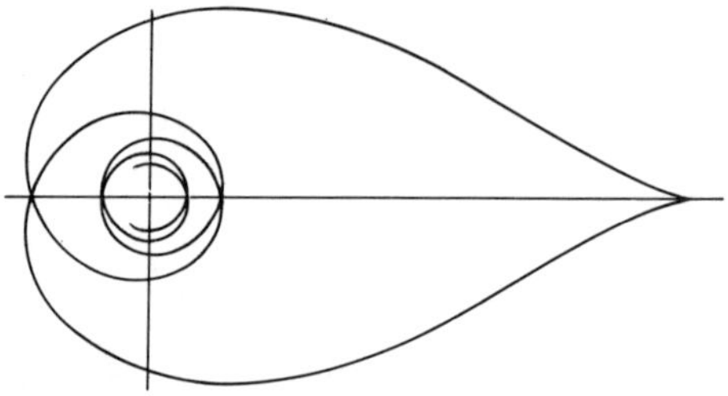

Figure 359. Polar tractrix

Basic geometric properties:
Boundedness: present.
Closedness: none.
Axial symmetry: straight line $y = 0$ is an axis of symmetry.
Central symmetry: none.
Connectedness: present.
Singular points: point $(a, 0)$ is a point of cessation, straight line $y = 0$ is a tangent at the point; $(0, 0)$ is an asymptotic point.
Self-intersection points: none.
(Figures 358–359).
POLAR TRACTRIX *is a bounded nonclosed connected curve with two singular points (point of cessation, asymptotic point) and a single axis of symmetry.*

Polyod of circle
Equations:

$$\rho = 2a \cos\left(2 \arccos \frac{m}{\rho} \mp \varphi\right),$$

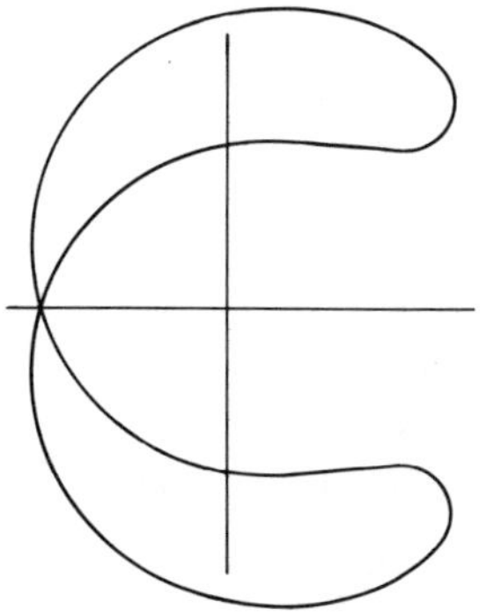

Figure 360. Polyod of circle
$a = 1.00, \quad m = 1.0000$

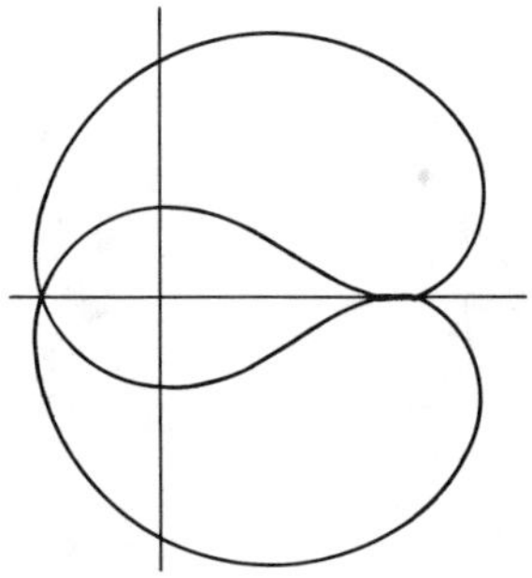

Figure 361. Polyod of circle
$a = 1.00, \quad m = 0.5443$

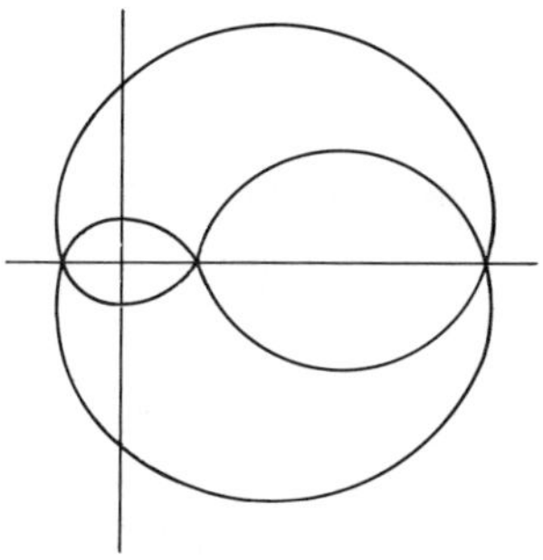

Figure 362. Polyod of circle

$a = 1.00, \quad m = 0.2500$

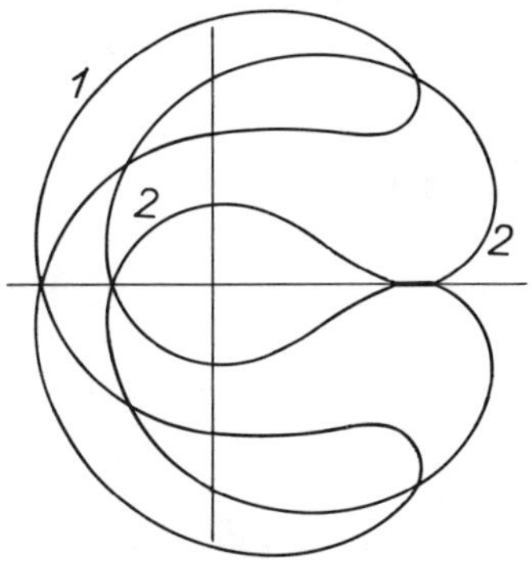

Figure 363. Polyod of circle

N	1	2
a	1.00	1.00
m	1.0000	0.5443

$$(x^2+y^2)\left(\frac{x^2+y^2}{2a}-x\right)^2+4m^4-4m^2(x^2+y^2)+\frac{2m}{a}x(x^2+y^2)=0$$

$$\varphi=\frac{\pi}{2}\arctan\frac{\rho}{\sqrt{4a^2-\rho^2}}\pm a\arctan\frac{m}{\sqrt{\rho^2-m^2}},$$

$$a>0, \quad 0<m<2a$$

Basic geometric properties:

Boundedness: present.
Closedness: present.
Axial symmetry: straight line $y = 0$ is an axis of symmetry.
Central symmetry: none.

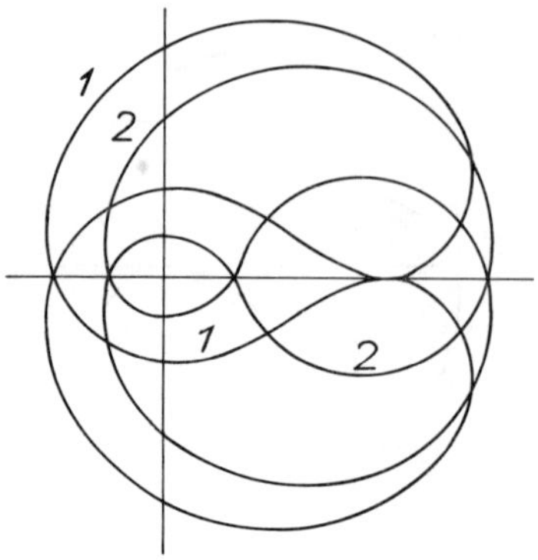

Figure 364. Polyod of circle

N	1	2
a	1.00	1.00
m	0.5443	0.2500

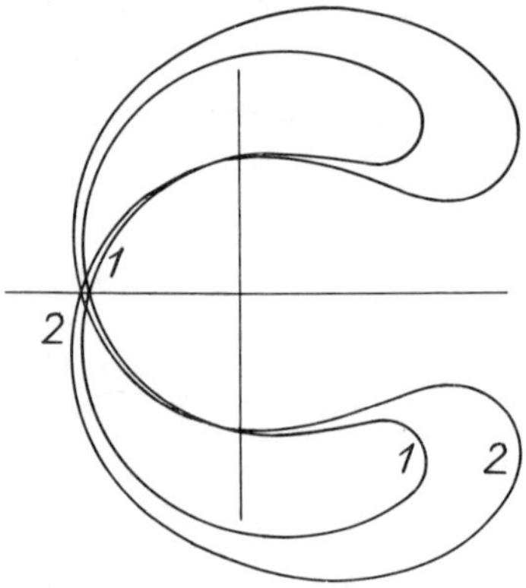

Figure 365. Polyod of circle

N	1	2
a	1.00	1.25
m	1.0000	1.0000

Connectedness: present.
Singular points: nodes located on x-axis: one (for $27m^2 > 8a^2$); two, $(4a/3, 0)$ (degenerate), $(-2a/3, 0)$ (for $27m^2 = 8a^2$); three (for $27m^2 < 8a^2$).
Self-intersection points: three ($27m^2 < 8a^2$), two ($27m^2 = 8a^2$), one, ($27m^2 > 8a^2$).
(Figures 360–365).
POLYOD OF CIRCLE *is a bounded connected closed curve with a single axis of symmetry and one, two or three singular points (nodes); number of singular points depends on relationship between parameter values.*

Polyod of polar axis
Equations:

$$m = \rho \sin \frac{\varphi}{2},$$

$$(x^2 + y^2)x^2 = (x^2 + y^2 - 2m^2)^2,$$

$$m > 0$$

Basic geometric properties:
Unboundedness: present.
Asymptotes: straight lines $y = \pm 2m$ are horizontal asymptotes.
Axial symmetry: straight lines $x = 0$, $y = 0$ are axes of symmetry.
Central symmetry: point $(0, 0)$ is a center of symmetry.
Connectedness: present.

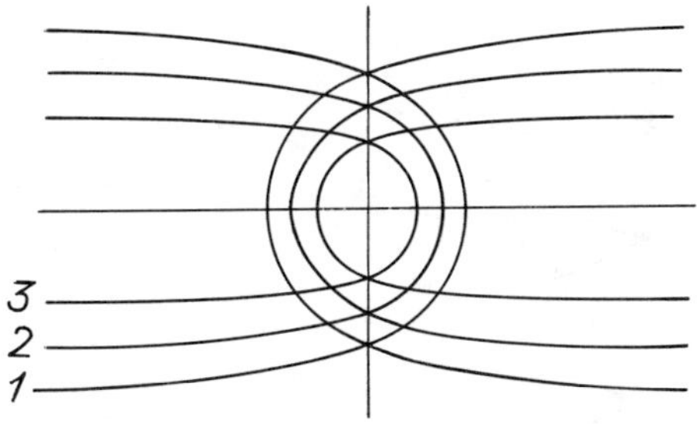

Figure 366. Polyod of polar axis

N	1	2	3
m	1.00	0.75	0.50

Singular points: point $(0, m\sqrt{2})$ is a node, straight lines

$$y - m\sqrt{2} = \pm\frac{1}{2}x$$

are tangents at the node; point $(0, -m\sqrt{2})$ is a node, straight lines

$$y + m\sqrt{2} = \pm\frac{1}{2}x$$

are tangents at the node.
Self-intersection points: two, $(0, \pm m\sqrt{2})$.
(Figure 366).
POLYOD OF POLAR AXIS *is an unbounded connected central-symmetric curve with two singular points (nodes), two axes of symmetry and two horizontal asymptotes.*

Polyod of straight line parallel to polar axis
Equations:

$$\rho \sin\left(\varphi - 2\arcsin\frac{m}{\rho}\right) = k,$$

$$(x^2 + y^2)((k - y)^2 - 4m^2) + 4km^2y + 4m^2 = 0,$$

$$m > 0, \quad k > 0$$

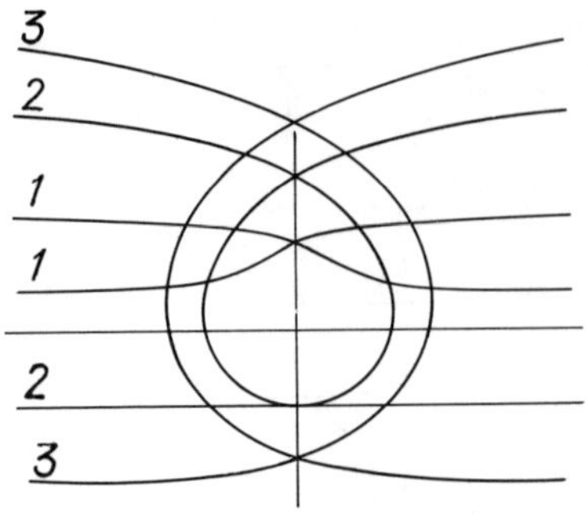

Figure 367.
Polyod of straight line parallel to polar axis

N	1	2	3
m	0.25	1.00	1.50
k	1.00	1.00	1.00

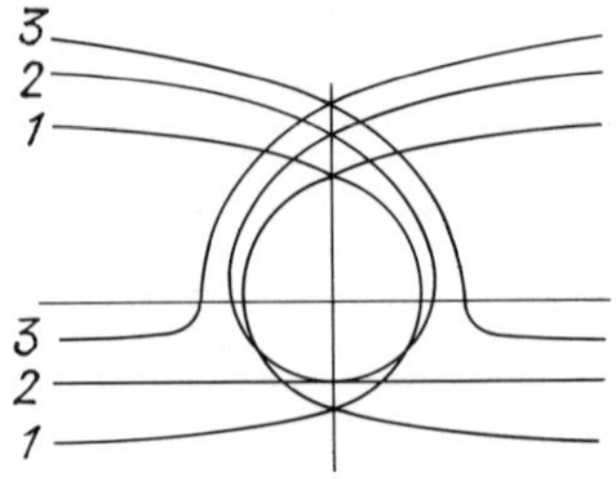

Figure 368.
Polyod of straight line parallel to polar axis

N	1	2	3
k	0.25	1.00	1.50
m	1.00	1.00	1.00

Basic geometric properties:
Unboundedness: present.
Asymptotes: straight lines $y = k \pm 2m$ are horizontal asymptotes.
Axial symmetry: straight lines $x = 0$ is an axis of symmetry.
Central symmetry: none.
Connectedness: present (for $k \leq m$), none (for $k > m$).
Singular points: point

$$\left(0, \frac{1}{2}(k + \sqrt{k^2 + 8m^2})\right)$$

is a node, straight lines

$$y - \frac{1}{2}(k + \sqrt{k^2 + 8m^2}) = \pm\sqrt{\frac{4m^2 - k^2 + \sqrt{k^2 + 8m^2}}{k^2 + 8m^2}}x$$

are tangents at the node; point

$$\left(0, \frac{1}{2}(k - \sqrt{k^2 + 8m^2})\right)$$

is a node ($k \leq m$), straight lines

$$y - \frac{1}{2}(k - \sqrt{k^2 + 8m^2}) = \pm\sqrt{\frac{4m^2 - k^2 - \sqrt{k^2 + 8m^2}}{k^2 + 8m^2}}x$$

are tangents at the node, or isolated point ($k > m$).

Self-intersection points: two,

$$\left(0, \frac{1}{2}(k \pm \sqrt{k^2 + 8m^2})\right)$$

(for $k \leq m$); one,

$$\left(0, \frac{1}{2}(k + \sqrt{k^2 + 8m^2})\right)$$

(for $k > m$).
(Figures 367–368).
POLYOD OF STRAIGHT LINE PARALLEL TO POLAR AXIS *is an unbounded connected or unconnected (consisting of an unbounded branch and an isolated point) curve with a single axis of symmetry, two horizontal asymptotes and two singular points (node, type of the second point depends on relationship between parameter values).*

Polyod of straight line perpendicular to polar axis
Equations:

$$\rho \cos\left(\varphi - 2 \arcsin \frac{m}{\rho}\right) = k,$$

$$y^2(x^2 + y^2 - k^2) = (x^2 + y^2 - kx - 2m^2)^2,$$

$$m > 0, \quad k > 0$$

Basic geometric properties:
Unboundedness: present.
Asymptotes: straight lines $x = k \pm 2m$ are horizontal asymptotes.
Axial symmetry: straight lines $y = 0$ is an axis of symmetry.
Central symmetry: none.
Connectedness: present (for $k \leq m$), none (for $k > m$).
Singular points: point

$$\left(\frac{1}{2}(k + \sqrt{k^2 + 8m^2}), 0\right)$$

is a node, straight lines

$$\pm\sqrt{\frac{4m^2 - k^2 + \sqrt{k^2 + 8m^2}}{k^2 + 8m^2}}\, y = x - \frac{1}{2}(k + \sqrt{k^2 + 8m^2})$$

are tangents at the node; point

$$\left(\frac{1}{2}(k - \sqrt{k^2 + 8m^2}), 0\right)$$

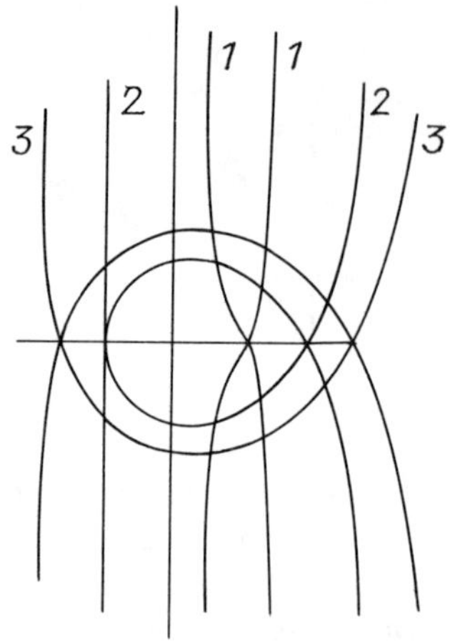

Figure 369.
Polyod of straight line
perpendicular to polar axis

N	1	2	3
m	0.25	1.00	1.50
k	1.00	1.00	1.00

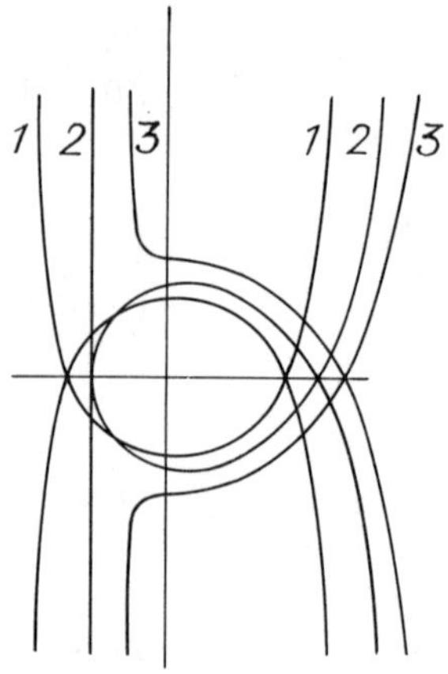

Figure 370.
Polyod of straight line
perpendicular to polar axis

N	1	2	3
k	0.25	1.00	1.50
m	1.00	1.00	1.00

is a node ($k \leq m$), straight lines

$$\pm\sqrt{\frac{4m^2 - k^2 - \sqrt{k^2+8m^2}}{k^2+8m^2}}\,y = x - \frac{1}{2}(k - \sqrt{k^2+8m^2})$$

are tangents at the node, or isolated point ($k > m$).
Self-intersection points: two,

$$\left(\frac{1}{2}(k \pm \sqrt{k^2+8m^2}), 0\right)$$

(for $k \leq m$); one,

$$\left(\frac{1}{2}(k + \sqrt{k^2+8m^2}), 0\right)$$

(for $k > m$).
(Figures 369–370).
POLYOD OF STRAIGHT LINE PERPENDICULAR TO POLAR AXIS *is an unbounded connected or unconnected (consisting of an unbounded branch and an isolated point) curve with a single axis of symmetry, two vertical asymptotes and two singular points (node, type of the second point depends on relationship between values of parameters).*

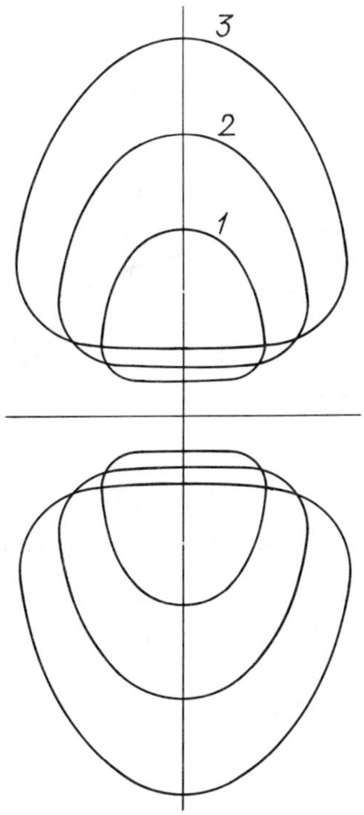

Figure 371.
Polyzomal curve of Bernoulli

N	1	2	3
a	0.50	0.75	1.00

Polyzomal curve of Bernoulli
Equations:

$$y^4 - 6a^2y^2 + 4x^2y^2 + a^4 = 0,$$

$$x = \pm\frac{a}{2}\sqrt{6 - \frac{y^2}{a^2} - \frac{a^2}{y^2}},$$

$$a > 0$$

Additional equations:

$$y = \pm\sqrt{2a^2 - x^2} \pm \sqrt{a^2 - x^2}$$

Basic geometric properties:
Boundedness: present.
Closedness: present.
Axial symmetry: straight lines $x = 0,\ y = 0$ are axes of symmetry.

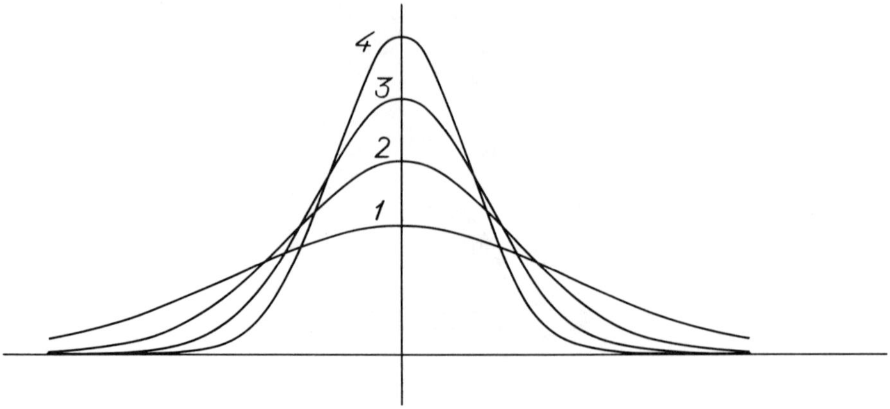

Figure 372.
Probability curve of Gauss

N	1	2	3	4
h	1.00	1.50	2.00	2.50

Central symmetry: point $(0,0)$ is a center symmetry.
Connectedness: none.
Singular points: none.
Self-intersection points: none.
Area of either oval: πa^2.
(Figure 371).

POLYZOMAL CURVE OF BERNOULLI *is a bounded unconnected (consisting of two ovals) central-symmetric closed curve free of singular points with two axes of symmetry.*

Probability curve of Gauss (Gaussian curve, normal curve)
Equation:

$$y = \frac{h}{\sqrt{\pi}} e^{-h^2 x^2}, \quad h > 0$$

Basic geometric properties:
Unboundedness: present.
Asymptotes: straight line $y = 0$ is a horizontal asymptote.
Axial symmetry: straight line $x = 0$ is a symmetry axis.
Central symmetry: none.
Connectedness: present.
Singular points: none.
Self-intersection points: none.

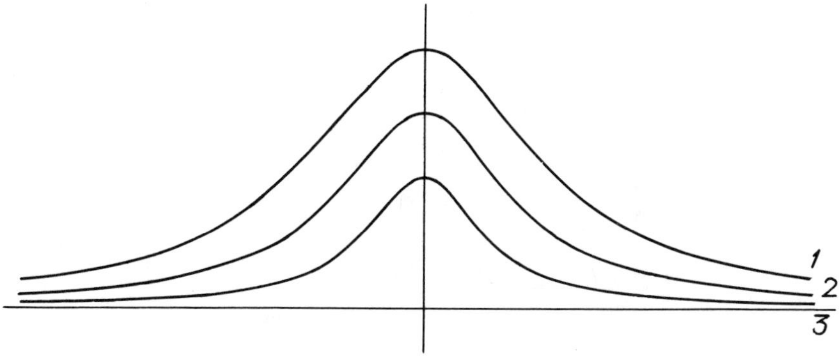

Figure 373. Pseudoversiera

N	1	2	3
a	1.00	0.75	0.50

Characterized points: point $(0, h/\sqrt{\pi})$ is a vertex; points

$$\left(\pm\frac{1}{h\sqrt{2}}, \frac{h}{e\sqrt{\pi}}\right)$$

are inflection points.
Area of domain enclosed by the curve and the asymptote: 1.
(Figure 372).
PROBABILITY CURVE OF GAUSS *is an unbounded connected curve free of singular points with a single horizontal asymptote and a single symmetry axis.*

Pseudoversiera
Equation:

$$(x^Z + a^2)y = 2a^3, \quad a > 0$$

Basic geometric properties:
Unboundedness: present.
Asymptotes: straight line $y = 0$ is a horizontal asymptote.
Axial symmetry: straight line $x = 0$ is a symmetry axis.
Central symmetry: none.
Connectedness: present.
Singular points: none.
Self-intersection points: none.
Characterized points: point $(0, 2a)$ is located at maximal distance from x-axis; points

$$\left(\pm\frac{a}{\sqrt{3}}, \frac{3a}{2}\right)$$

are inflection points.

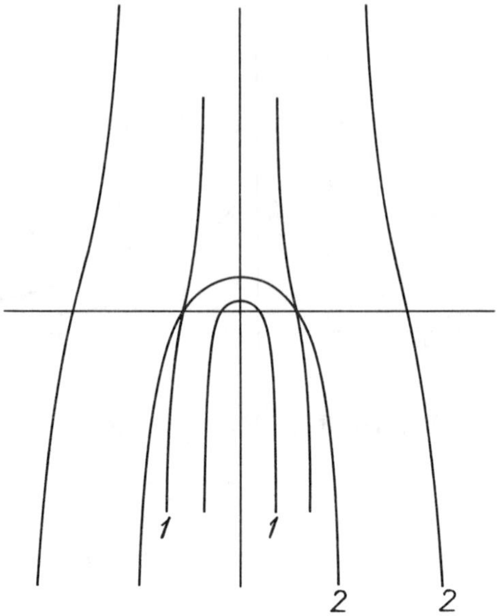

Figure 374.
Quadratrix of Dinostratus

N	1	2
a	1.00	3.00

Area of domain enclosed between the curve and asymptote: $2\pi a^2$.
(Figure 373).
PSEUDOVERSIERA *is an unbounded connected curve free of singular points with a single horizontal asymptote and a single symmetry axis.*

2.2.15 Q

Quadratrix of Dinostratus (Dinostratus' quadratrix, Deinostratos' quadratrix, quadratrix of Hippias)
Equations:

$$y = x \cot \frac{\pi x}{2a}, \quad \rho = \frac{a(\pi - 2\varphi)}{\pi \cos \varphi},$$

$$x = \frac{2a}{\pi} t, \quad y = \frac{2a}{\pi} t \cot t, \quad a > 0$$

Basic geometric properties:
Unboundedness: present.
Asymptotes: straight lines $x = 2ka$, $k \in \mathbf{Z}$ ($k \neq 0$), are vertical asymptotes.
Axial symmetry: straight line $x = 0$ is an axis of symmetry.

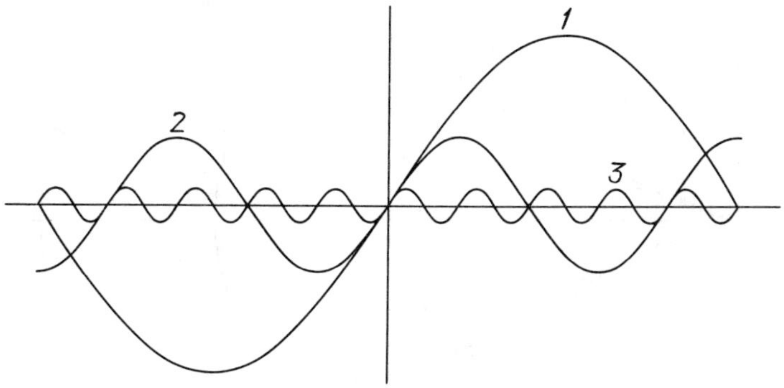

Figure 375.
Quadratrix of Tschirnhausen

N	1	2	3
a	1.00	0.40	0.10

Central symmetry: none.
Connectedness: none.
Singular points: none.
Self-intersection points: none.
(Figure 374).
QUADRATRIX OF DINOSTRATUS *is an unbounded unconnected (consisting of an infinitely large number of unbounded branches) free of singular points with an infinitely large number of vertical asymptotes and a single axis of symmetry.*

Quadratrix of Tschirnhausen
Equation:

$$y = a \sin \frac{\pi x}{2a},$$

$$a > 0$$

Basic geometric properties:
Unboundedness: present.
Asymptotes: none.
Axial symmetry: straight lines $x = (2k+1)$, $k \in \mathbf{Z}$, are axes of symmetry.
Central symmetry: points $(2ka, 0)$, $k \in \mathbf{Z}$, are centers of symmetry.
Connectedness: present.
Singular points: none.
Self-intersection points: none.
(Figure 375).

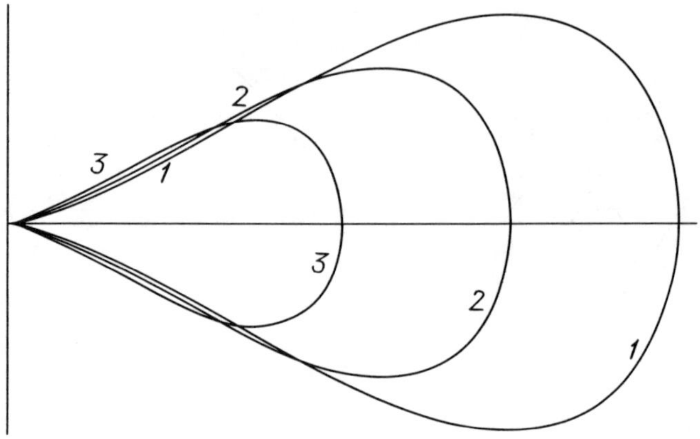

Figure 376. Quartic of Bonnet

N	1	2	3
a	1.00	0.75	0.50

QUADRATRIX OF TSCHIRNHAUSEN *is an unbounded connected central-symmetric curve free of singular points with an infinitely large number of axes and centers of symmetry.*

Quartic of Bonnet

Equation:

$$x^4 - ax^3 + a^2y^2 = 0,$$

$$a > 0$$

Basic geometric properties:

Boundedness: present.

Closedness: present.

Axial symmetry: straight line $y = 0$ is an axis of symmetry.

Central symmetry: none.

Connectedness: present.

Singular points: point $(0,0)$ is a cuspidal point of the first kind, straight line $y = 0$ is a tangent at the cusp point.

Self-intersection points: none.

Characterized points: points

$$\left(\frac{3a}{4}, \pm\frac{a}{16}3\sqrt{3}\right)$$

are located at maximal distance from x-axis.

Area of domain enclosed by the curve: $\pi a^2/8$.

(Figure 376).

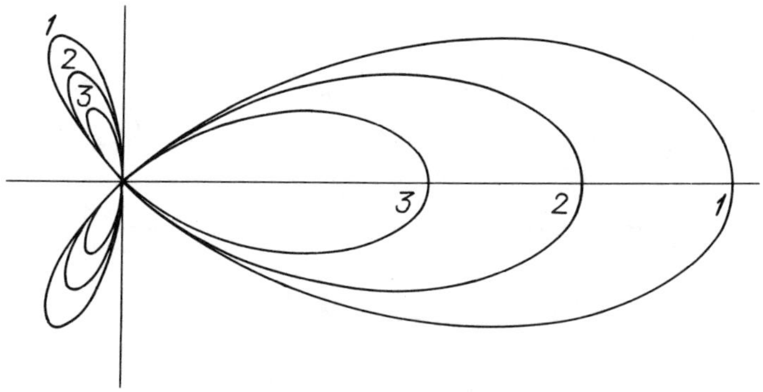

Figure 377. Right trifolium

N	1	2	3
a	1.00	0.75	0.50

QUARTIC OF BONNET *is a bounded closed connected curve with a single singular point (cuspidal point of the first kind) and a single axis of symmetry.*

2.2.16 R

Right trifolium (right three-leafed rose)
Equations:

$$(x^2 + y^2)^2 = ax(x^2 - y^2),$$

$$\rho = a \cos\varphi \cos 2\varphi, \quad a > 0$$

Basic geometric properties:
Boundedness: present.
Closedness: present.
Axial symmetry: straight line $y = 0$ is an axis of symmetry.
Central symmetry: none.
Connectedness: present.
Singular points: point $(0, 0)$ is a triple point, straight lines $x = 0,\ x \pm y = 0$ are tangents at the triple point.
Self-intersection points: one, $(0, 0)$.
Characterized points: points

$$(a, 0), \left(-\frac{a}{9}, \pm\frac{\sqrt{5}}{9}a\right)$$

are vertices; the points farthest from the x-axis of the curve are located on the straight lines $4y \pm a = 0$; the points farthest from the y-axis of the small leaves are located on the straight line $8x + a = 0$.

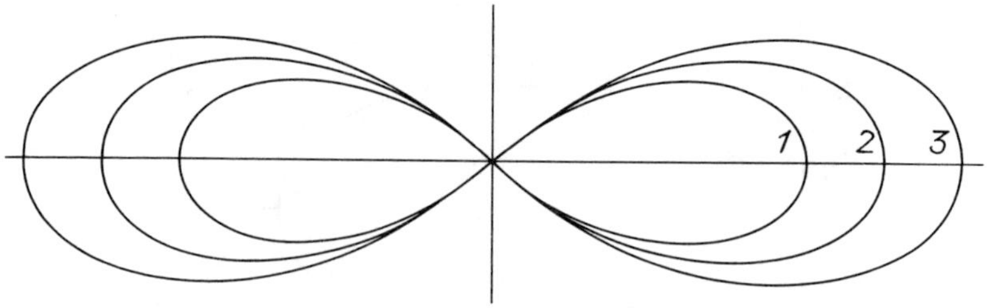

Figure 378. Rose

N	1	2	3
a	0.80	1.00	1.20
k	2	2	2

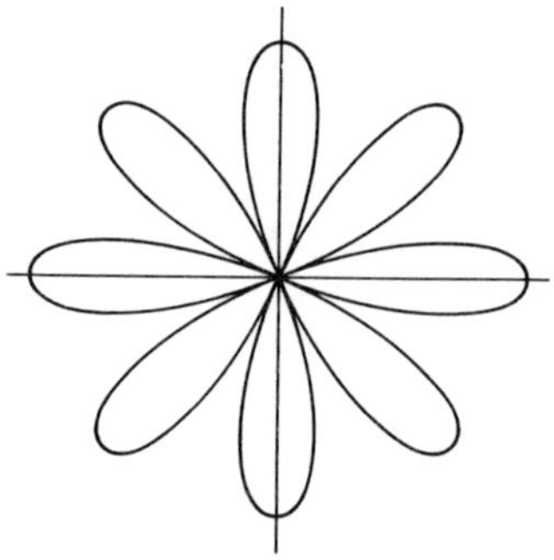

Figure 379. Rose
$k = 4, \quad a = 1.00$

Areas: area of the greater leaf is equal to $a^2(3\pi + 8)/3$, area of the lesser one is equal to $a^2(3\pi - 8)/3$, area enclosed by the curve is $2\pi a^2$. (Figure 377).
RIGHT TRIFOLIUM *is a bounded connected closed curve with a single singular point (triple point) and a single axis of symmetry.*

Rose (rhodonea, rose of Guido Grandi)
Equation:

$$\rho = a \cos k\varphi,$$

$$a > 0, \quad k = \frac{p}{q}, \quad p \in \mathbf{N}, \quad q \in \mathbf{N}, \quad \text{G.C.D.}(p, q) = 1$$

Basic geometric properties:
Boundedness: present.
Closedness: present.

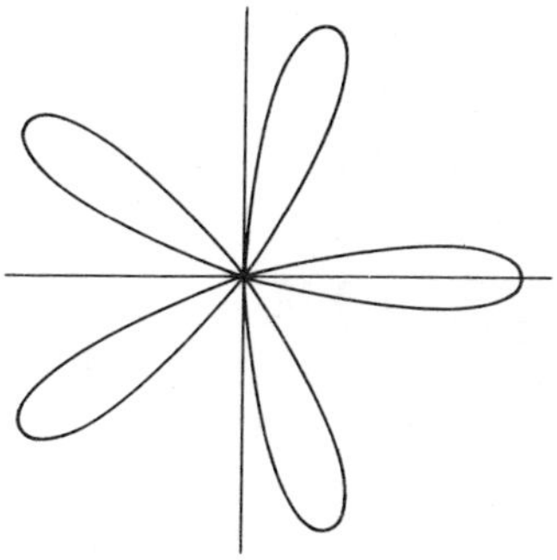

Figure 380. Rose
$k = 5, \quad a = 1.00$

Figure 381. Rose
$k = 5/2, \quad a = 1.00$

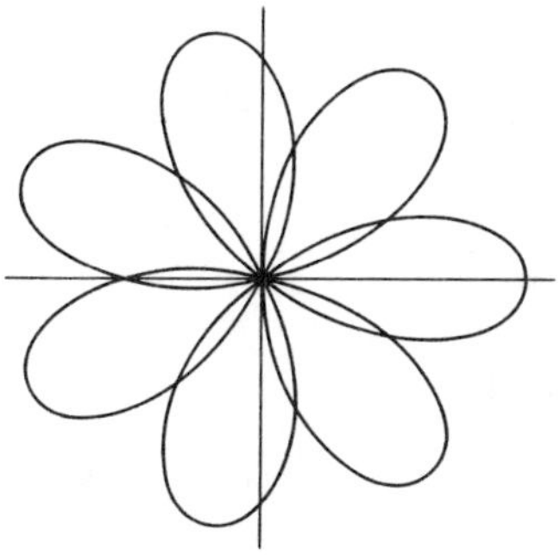

Figure 382. Rose
$k = 7/3, \quad a = 1.00$

Figure 383. Rose
$k = 9/4, \quad a = 1.00$

Figure 384. Rose
$k = 10/3, \quad a = 1.00$

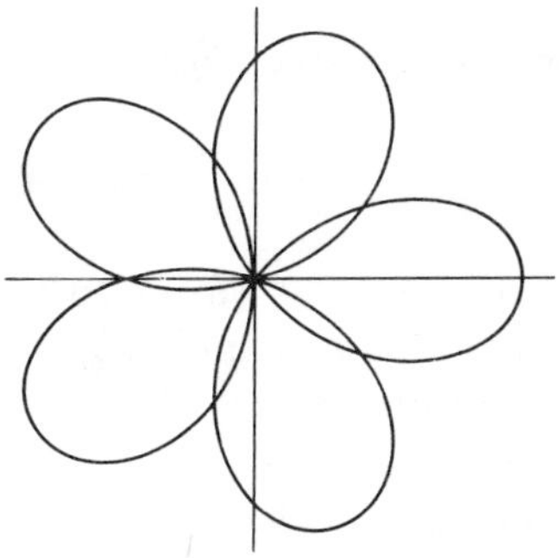

Figure 385. Rose
$k = 5/3, \quad a = 1.00$

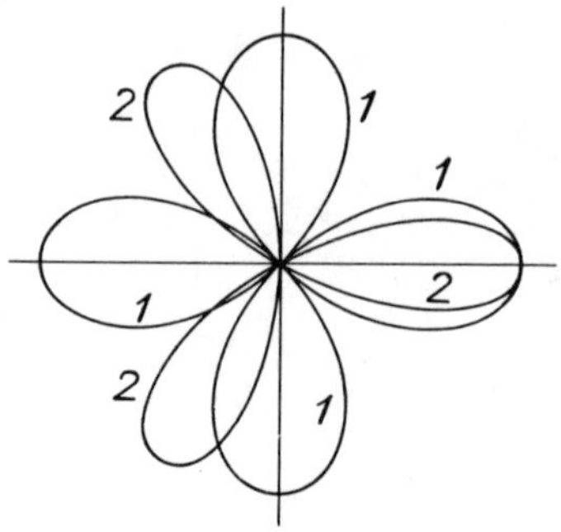

Figure 386. Rose

N	1	2
k	2	3
a	1.00	1.00

Axial symmetry:

k is a positive integer:
straight lines

$$x \sin \frac{2\pi m}{k} = y \cos \frac{2\pi m}{k}, \quad m = 0, 1, \ldots, k-1$$

are axes of symmetry (for odd k); straight lines

$$x \sin \frac{\pi m}{k} = y \cos \frac{\pi m}{k}, \quad m = 0, 1, \ldots, 2k-1$$

are axes of symmetry (for even k);

k is rational:
straight lines

$$x \sin \frac{2\pi l}{p} = y \cos \frac{2\pi l}{p}, \quad l = 0, 1, \ldots, p-1$$

are axes of symmetry (for odd pq); straight lines

$$x \sin \frac{\pi l}{p} = y \cos \frac{\pi l}{p}, \quad l = 0, 1, \ldots, p-1$$

are axes of symmetry (for even pq).
Central symmetry: point $(0, 0)$ is a center of symmetry (for even pq only).
Connectedness: present.
Singular points: point $(0, 0)$ is a multiple point, simple nodes exist for non-integer k only and their number depends on values of p and q.
Self-intersection points: finite but no less than one, $(0, 0)$.
Special cases: two-leafed rose ($k = 1/2$), three-leafed rose ($k = 3$), corolla, Indian pearl ($k = 4$).

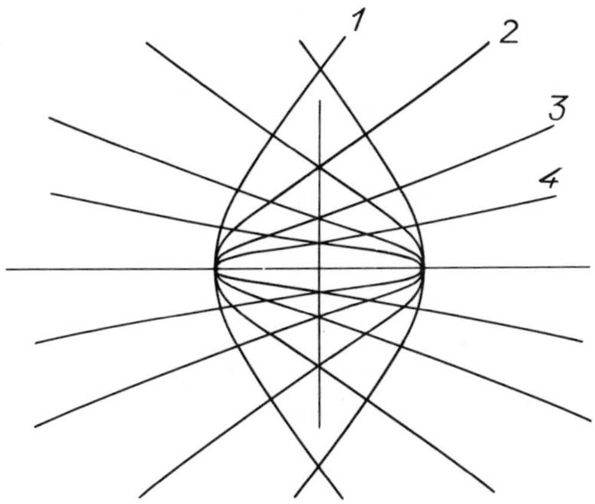

Figure 387. Sand-glass curve

N	1	2	3	4
b	0.25	0.50	1.00	2.00
a	1.00	1.00	1.00	1.00

Area of every leaf: $\pi a^2/(4k)$.
Number of leaves: k ($k \in \mathbf{N}$ and odd), $2k$ ($k \in \mathbf{N}$ and even), p ($k = p/q$ and pq is odd), $2p$ ($k = p/q$ and pq is even).
Order of curve: $p + q$ (pq is odd), $2(p + q)$ (pq is even).
(Figures 378–386).
ROSE *is a bounded connected closed curve with a single point of high multiplicity; properties of symmetry and number of simple nodes are defined by values of parameter* k.

2.2.17 S

Sand-glass curve
Equations:

$$a^2(y^2 - b^2)^2 - 4b^2x^2(y^2 + b^2) = 0,$$

$$x = a\frac{\cos 2t}{\cos t}, \quad y = b\tan t, \quad a > 0, \; b > 0$$

Basic geometric properties:
Unboundedness: present.
Asymptotes: straight lines $2bx \pm ay = 0$ are inclined asymptotes.
Axial symmetry: straight lines $x = 0$, $y = 0$ are axes of symmetry.
Central symmetry: point $(0, 0)$ is a center of symmetry.
Connectedness: present.
Singular points: point $(0, b)$ is a node, straight lines $\pm bx - ay + ab = 0$ are tangents at the node; point $(0, -b)$ is a node, straight lines $\pm bx + ay + ab = 0$ are tangents at the node.

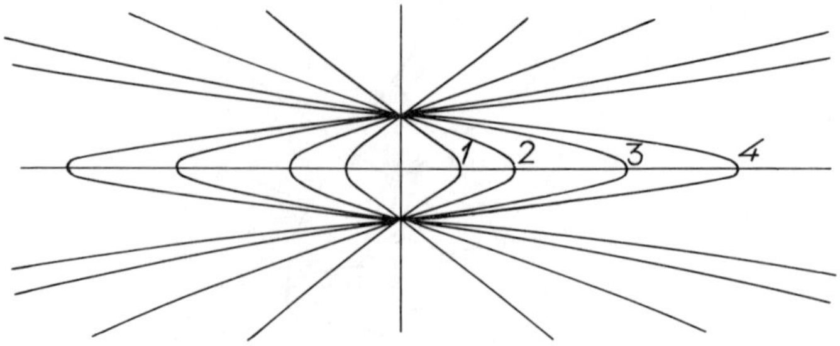

Figure 388. Sand-glass curve

N	1	2	3	4
a	0.50	1.00	2.00	3.00
b	0.50	0.50	0.50	0.50

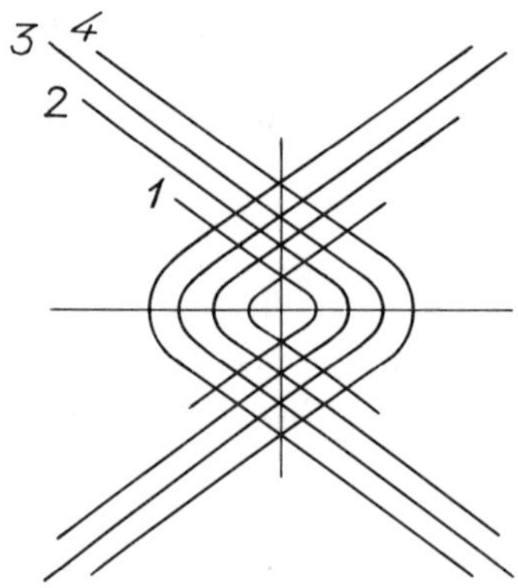

Figure 389. Sand-glass curve

N	1	2	3	4
a	0.50	1.00	1.50	2.00
b	0.50	1.00	1.50	2.00

Self-intersection points: two, $(0, \pm b)$.
(Figures 387–389).
SAND-GLASS CURVE *is an unbounded connected central-symmetric curve with two singular points (nodes), two inclined asymptotes and two axes of symmetry.*

Scarabaeus
Equations:

$$(x^2 + y^2)(x^2 + y^2 + cx)^2 = a^2(x^2 - y^2)^2,$$

$$\rho = a \cos 2\varphi - c \cos \varphi,$$

$$a > 0$$

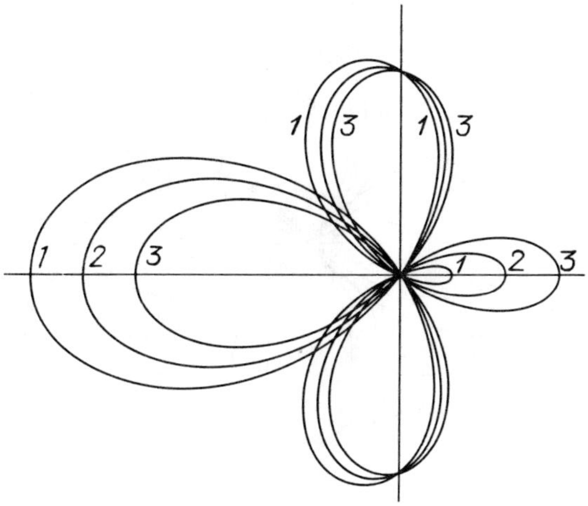

Figure 390. Scarabaeus

N	1	2	3
a	1.00	1.00	1.00
c	0.75	0.50	0.25

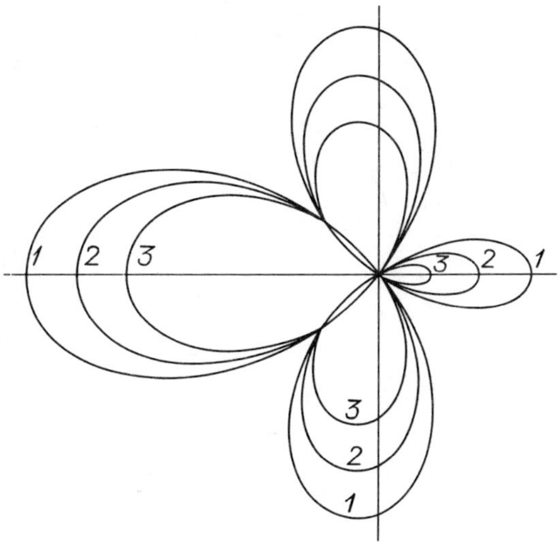

Figure 391. Scarabaeus

N	1	2	3
a	1.25	1.00	0.75
c	0.50	0.50	0.50

Basic geometric properties:
Boundedness: present.
Closedness: present.

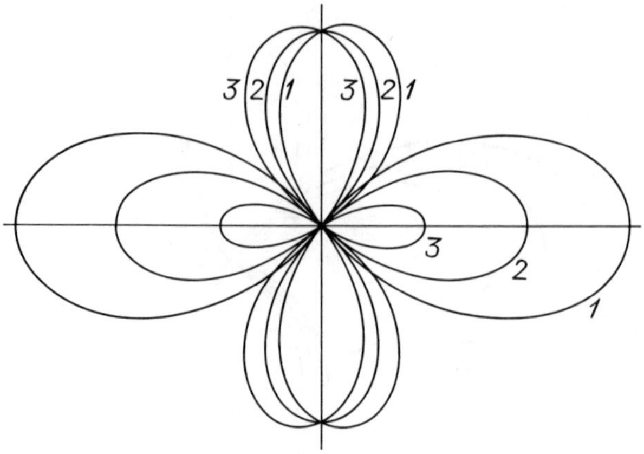

Figure 392. Scarabaeus

N	1	2	3
a	1.00	1.00	1.00
c	−0.50	0.00	0.50

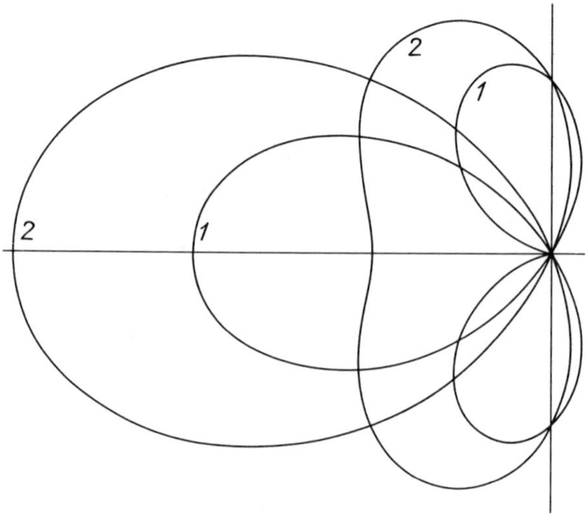

Figure 393. Scarabaeus

N	1	2
a	1.00	1.00
c	1.00	2.00

Axial symmetry: straight line $y = 0$ is an axis of symmetry; straight line $x = 0$ is an axis of symmetry (for $c = 0$ only).
Central symmetry: point $(0, 0)$ is a center of symmetry (for $c = 0$ only).
Connectedness: present.

Singular points: point $(0,0)$ is a multiple point; point

$$\left(-\frac{c}{2},\frac{c}{2}\right)$$

is a node, straight lines

$$\left(-a\pm\frac{c}{\sqrt{2}}\right)\left(y-\frac{c}{2}\right)=a\left(x+\frac{c}{2}\right)$$

are tangents at the node; point

$$\left(-\frac{c}{2},-\frac{c}{2}\right)$$

is a node, straight lines

$$\left(a\pm\frac{c}{\sqrt{2}}\right)\left(y+\frac{c}{2}\right)=a\left(x+\frac{c}{2}\right)$$

are tangents at the node.
Self-intersection points: three,

$$(0,0),\quad\left(-\frac{c}{2},\pm\frac{c}{2}\right)$$

Special cases: four-leafed rose (corolla) $(c=0)$.
(Figures 390–393).
SCARABAEUS $(c\neq 0)$ *is a bounded closed connected curve with three singular points (multiple point, two nodes) and a single axis of symmetry.*
FOUR-LEAFED ROSE $(c=0)$ *is a bounded connected closed central-symmetric curve with a single singular point (multiple point) and two axes of symmetry.*

Secantoid

Equations:

$$y=\sec(ax)=\frac{1}{\cos(ax)},\quad a>0$$

Basic geometric properties:
Unboundedness: present.
Asymptotes: straight lines $2ax=(2k+1)\pi,\ k\in\mathbf{Z}$, are vertical asymptotes.
Axial symmetry: straight lines $ax=k\pi,\ k\in\mathbf{Z}$, are axes of symmetry.
Central symmetry: points

$$\left(\frac{(2k+1)\pi}{a},0\right)$$

are centers of symmetry.

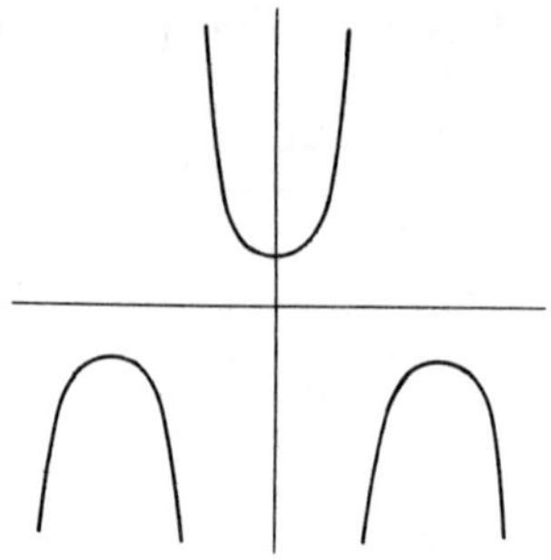

Figure 394. Secantoid

Connectedness: none.
Singular points: none.
Self-intersection points: none.
Characterized points: points

$$\left(\frac{k\pi}{a}, (-1)^k\right)$$

are located at minimal distance from x-axis (vertices).
(Figure 394).
SECANTOID *is an unbounded unconnected (consisting of an infinitely large number of unbounded branches) central-symmetric curve free of singular points with an infinitely large number of symmetry axes and vertical asymptotes.*

Secluded wave
Equation:

$$y = \frac{a}{\cosh \frac{x}{2b}}, \quad a > 0, \quad b > 0$$

Basic geometric properties:
Unboundedness: present.
Asymptotes: straight line $y = 0$ is a horizontal asymptote.
Axial symmetry: straight line $x = 0$ is an axis of symmetry.
Central symmetry: none.
Connectedness: present.
Singular points: none.
Self-intersection points: none.
Characterized points: point $(a, 0)$ is a vertex.
(Figures 395–396).
SECLUDED WAVE *is an unbounded connected curve free of singular points with a single horizontal asymptote and a single axis of symmetry.*

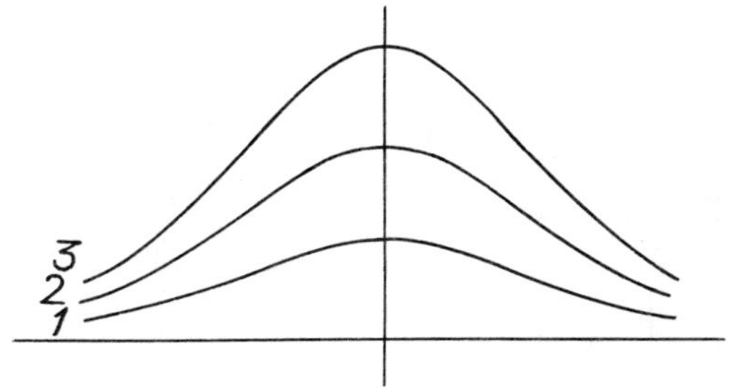

Figure 395. Secluded wave

N	1	2	3
a	1.00	2.00	3.00
b	1.00	1.00	1.00

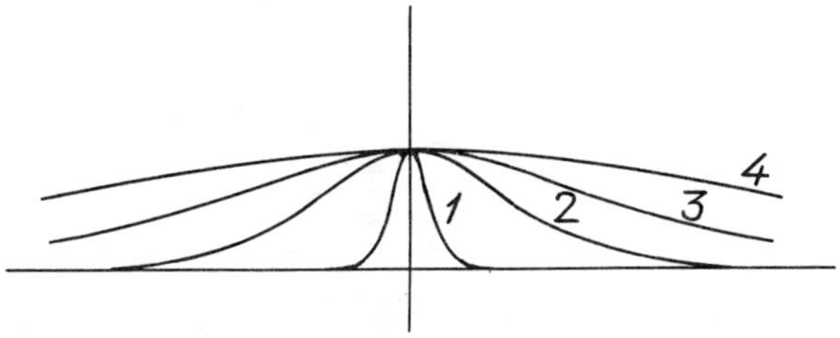

Figure 396. Secluded wave

N	1	2	3	4
b	0.10	0.50	1.00	2.00
a	1.00	1.00	1.00	1.00

Secondary cycloid of Ricci

Equations:

$$x = Rt - r\sin t, \quad y = r - r\cos t,$$

$$R > 0, \quad r > 0$$

Basic geometric properties:

Unboundedness: present.

Asymptotes: none.

Axial symmetry: straight lines $x = k\pi R$, $k \in \mathbf{Z}$, are axes of symmetry.

Central symmetry: none.

Connectedness: present.

Singular points: points $(2\pi mR, r - r\cos t_0)$, $m \in \mathbf{Z}$, are nodes (for $R > r$), t_0 is a least positive solution of equation

$$\frac{R}{r} = \frac{\sin t}{t}$$

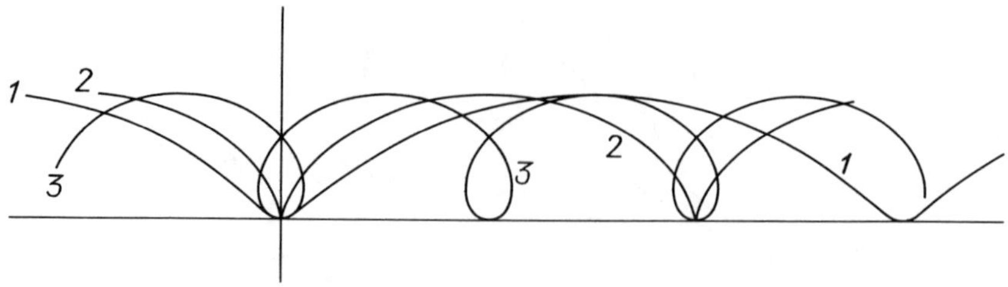

Figure 397.
Secondary cycloid of Ricci

N	1	2	3
r	1.00	1.00	1.00
R	1.50	1.00	0.50

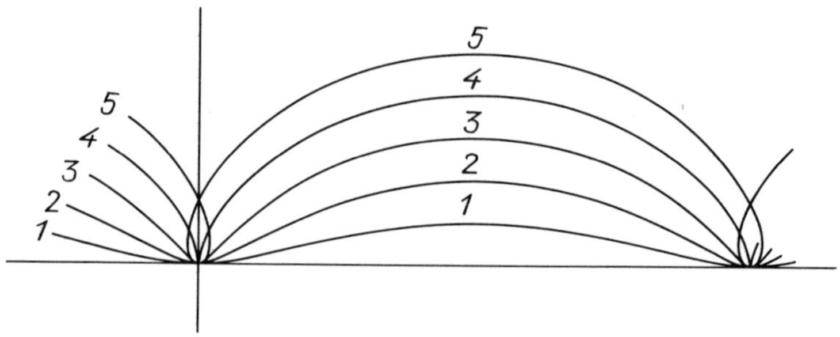

Figure 398.
Secondary cycloid of Ricci

N	1	2	3	4	5
r	0.25	0.50	0.75	1.00	1.25
R	1.00	1.00	1.00	1.00	1.00

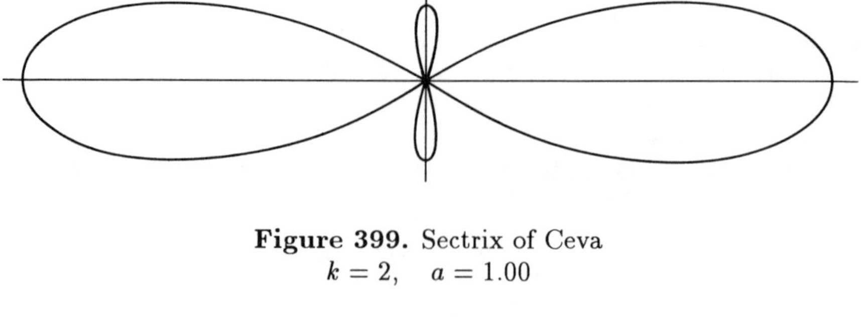

Figure 399. Sectrix of Ceva
$k = 2, \quad a = 1.00$

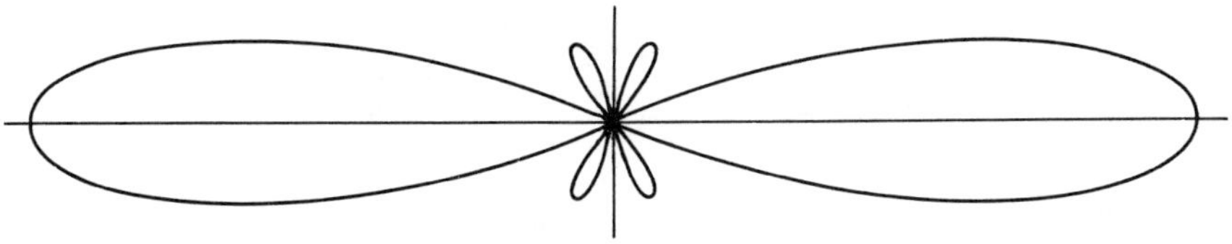

Figure 400. Sectrix of Ceva
$k = 3, \quad a = 1.00$

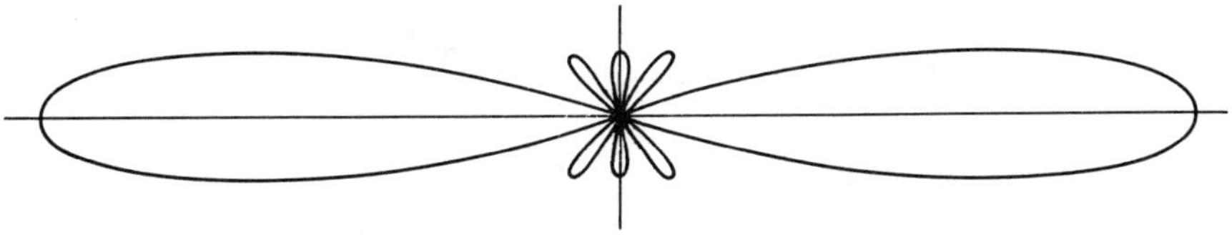

Figure 401. Sectrix of Ceva
$k = 4, \quad a = 1.00$

Self-intersection points: none for $R > r$, infinitely large for $R < r$.
Special cases: cycloid ($R = r$).
Generalization: cycloid of Clairaut

$$x = at + b - r\sin(\alpha t + \beta), \quad y = r - r\cos(\alpha y + \beta)$$

(Figures 397–398).
SECONDARY CYCLOID OF RICCI *is an unbounded connected curve with an infinitely large number of singular points (nodes, for $R < r$ only) and symmetry axes.*

Sectrix of Ceva
Equation:

$$\rho = a + 2a\frac{\sin k\varphi \cos(k+1)\varphi}{\sin\varphi},$$

$$a > 0, \quad k \in \mathbf{N}$$

Basic geometric properties:
Boundedness: present.
Closedness: present.
Axial symmetry: straight lines $x = 0$, $y = 0$ are axes of symmetry.
Central symmetry: point $(0, 0)$ is a center of symmetry.
Connectedness: present.
Singular points: point $(0, 0)$ is a $(2k)$-multiple point.
Self-intersection points: one, $(0, 0)$.
(Figures 399–401).

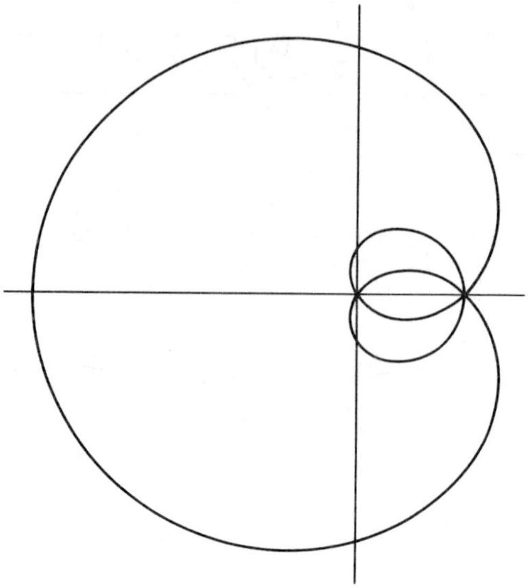

Figure 402. Sectrix of Coppola
$n = 1, \quad a = 1.00$

SECTRIX OF CEVA *is a bounded connected closed curve with a single singular point (multiple point) and two axes of symmetry.*

Sectrix of Coppola
Equations:

$$\rho_1 = a + 2a \sin \frac{\varphi}{2},$$

$$\rho_2 = \rho_1 + 2a \sin \frac{\varphi}{2} \sin \frac{\varphi + \pi}{4},$$

$$\rho_3 = \rho_2 + 2a \sin \frac{\varphi}{2} \sin \frac{\varphi + \pi}{4} \sin \frac{\varphi + 3\pi}{8},$$

$$\rho_n = \rho_{n-1} + 2a \sin \frac{\varphi}{2} \sin \frac{\varphi + \pi}{4} \sin \frac{\varphi + 3\pi}{8} \cdots \sin \frac{\varphi + (2n-1)\pi}{2^n},$$

$$a > 0, \quad n \in \mathbf{N}$$

Basic geometric properties:
Boundedness: present.
Closedness: present.
Axial symmetry: straight line $y = 0$ is a symmetry axis.
Central symmetry: none.
Connectedness: present.
Singular points: nodes and multiple points.

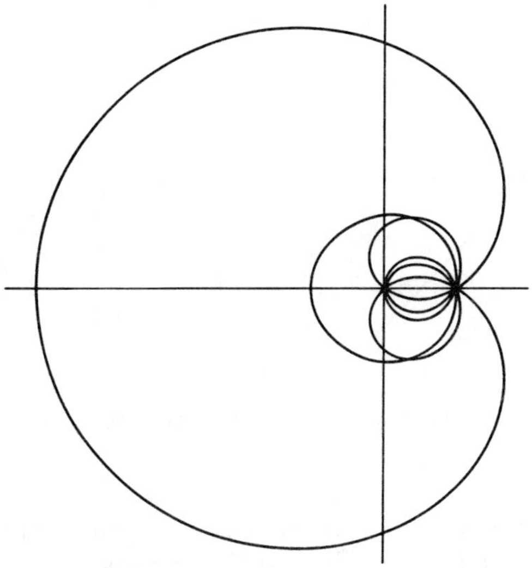

Figure 403. Sectrix of Coppola
$n = 2, \quad a = 1.00$

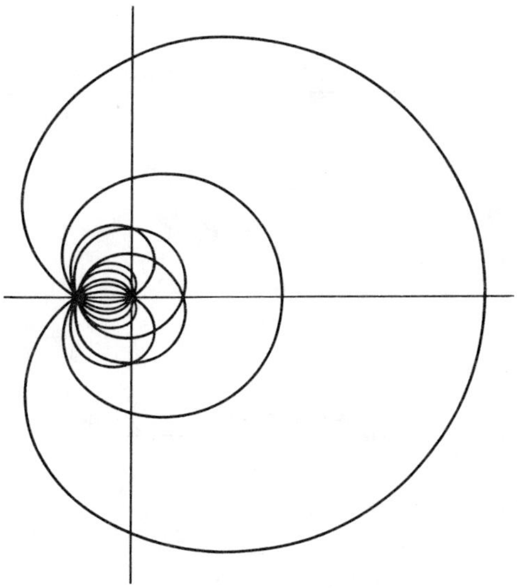

Figure 404. Sectrix of Coppola
$n = 3, \quad a = 1.00$

Number of self-intersection points depends on values of parameter n: two (for $n = 1$), four (for $n = 2$), eleven (for $n = 3$).
(Figures 402–404).

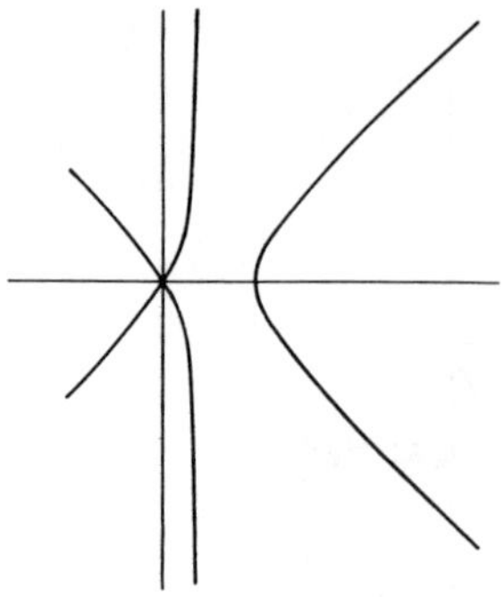

Figure 405. Sectrix of Hesse
$n = 3, \quad a = 1.00$

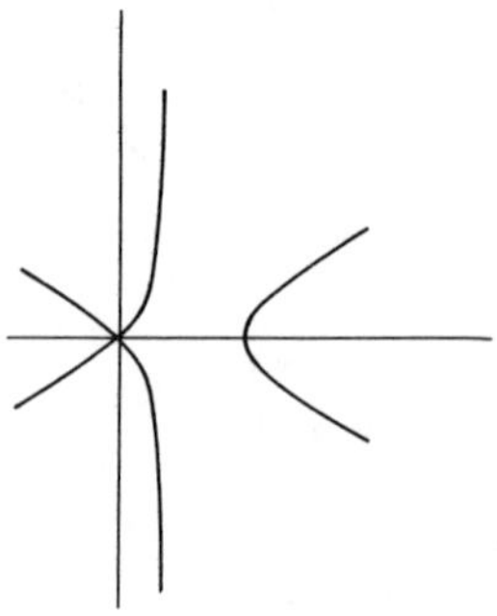

Figure 406. Sectrix of Hesse
$n = 4, \quad a = 1.00$

SECTRIX OF COPPOLA *is a bounded connected closed curve with a single symmetry axis; the number of its multiple points is defined by values of parameter n.*

Sectrix of Hesse
Equations:

$$x = a\frac{\sin\varphi\cos(n-1)\varphi}{\sin n\varphi}, \quad y = a\frac{\cos\varphi\cos(n-1)\varphi}{\sin n\varphi},$$

$$\rho = a\frac{\cos(n-1)\varphi}{\sin n\varphi},$$

$$a > 0, \quad n \in \mathbf{N}, \quad n \geq 3$$

Additional equations:

$$\rho = a\frac{\cos(n-1)(\frac{\pi}{2}-\varphi)}{\sin n(\frac{\pi}{2}-\varphi)},$$

$$\rho = a\frac{\sin(n-1)\varphi}{\sin n\varphi},$$

$$\rho = a\frac{\cos(n-1)\varphi}{\cos n\varphi}$$

Basic geometric properties:
Unboundedness: present.
Asymptotes: number and form of asymptotes depends on values of parameters n and a.

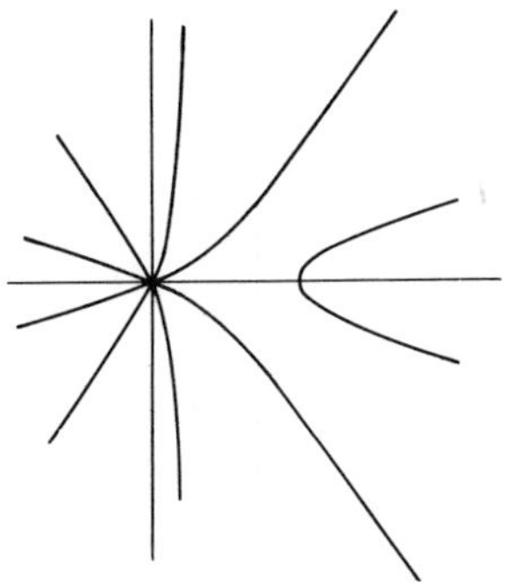

Figure 407. Sectrix of Hesse
$n = 5, \quad a = 1.00$

Axial symmetry: straight line $y = 0$ is an axis of symmetry.
Central symmetry: none.
Connectedness: none.
Singular points: point $(0, 0)$ is a multiple point.
Number of self-intersection points depends on values of parameter n: one, $(0, 0)$ (for $n = 3, 4, 5$).
(Figures 405–407).
SECTRIX OF HESSE *is an unbounded unconnected (consisting of some unbounded branches) curve with a single singular point (multiple point) and a single axis of symmetry.*

Sectrix of Kempe
Equations:

$$\rho_1 = a + 2a \sin\varphi + 2a \cos\left(\frac{\pi}{4} - \frac{\varphi}{2}\right),$$

$$\rho_2 = \rho_1 + 2a \cos\left(\frac{\pi}{4} - \frac{\varphi}{2}\right) \cos\left(\frac{\pi}{8} - \frac{\varphi}{4}\right),$$

$$\rho_n = \rho_{n-1} + 2a \cos\left(\frac{\pi}{4} - \frac{\varphi}{2}\right) \cos\left(\frac{\pi}{8} - \frac{\varphi}{4}\right) \cdots \cos\left(\frac{\pi}{2^{n+1}} - \frac{\varphi}{2^n}\right),$$

$$a > 0, \quad n \in \mathbf{N}$$

Basic geometric properties:
Boundedness: present.
Closedness: present.
Axial symmetry: straight line $x = 0$ is a symmetry axis.
Central symmetry: none.
Connectedness: present.
Singular points: nodes and multiple points.
Number of self-intersection points depends on values of parameter n: two (for $n = 1$), four (for $n = 2$), nine (for $n = 3$).

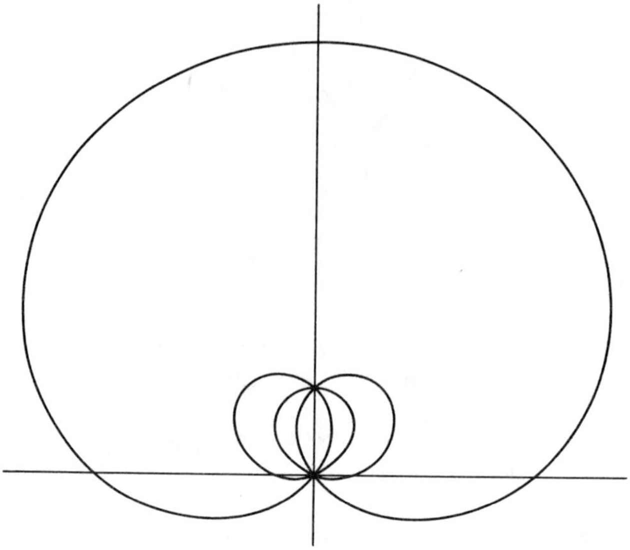

Figure 408. Sectrix of Kempe
$n = 1, \quad a = 1.00$

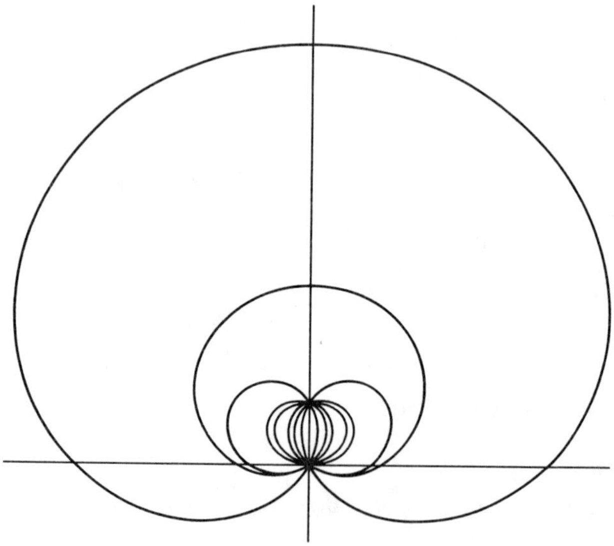

Figure 409. Sectrix of Kempe
$n = 2, \quad a = 1.00$

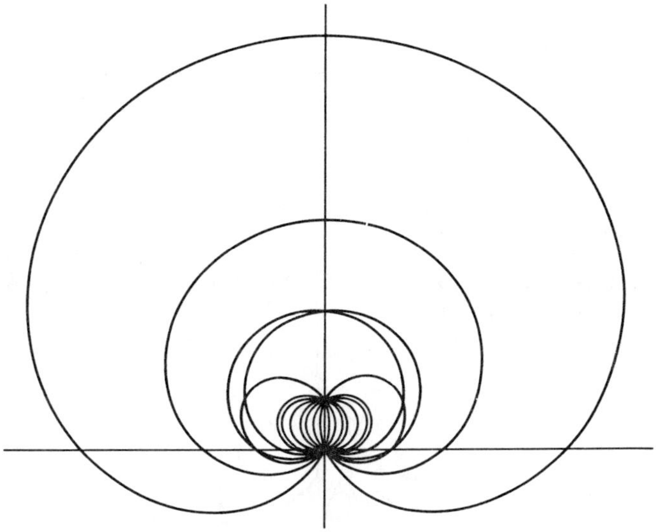

Figure 410. Sectrix of Kempe
$n = 3, \quad a = 1.00$

(Figures 408–410).
SECTRIX OF KEMPE *is a bounded connected closed curve with a single symmetry axis; the number of its multiple points is defined by values of parameter* n.

Sectrix of Ökinghaus (A)
Equation:

$$\rho = a\frac{\sin\left(\frac{n-2}{2}\right)\varphi}{\sin\frac{n}{2}\varphi},$$

$$a > 0, \quad n \in \mathbf{N}, \quad n \geq 3$$

Basic geometric properties:
Unboundedness: present.
Asymptotes: number and form of asymptotes depends on values of parameters n and a.
Axial symmetry: straight line $y = 0$ is an axis of symmetry.
Central symmetry: none.
Connectedness: none.
Singular points: for some values of parameter n the curve has no singular points (in particular, for $n = 3$, $n = 4$), for another values the curve has multiple points (in particular, $(0, 0)$ for $n = 5$).

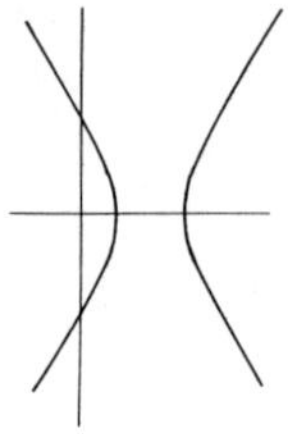

Figure 411.
Sectrix of Ökinghaus (A)
$n = 3, \quad a = 1.00$

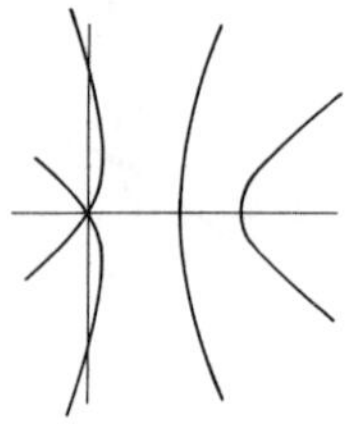

Figure 412.
Sectrix of Ökinghaus (A)
$n = 5, \quad a = 1.00$

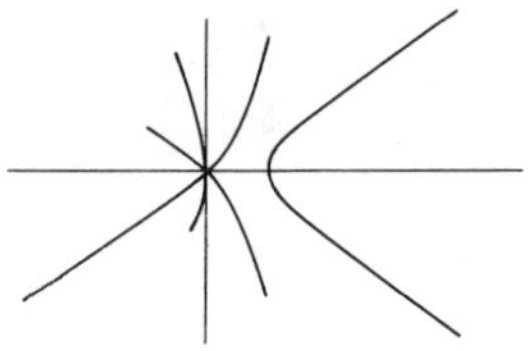

Figure 413.
Sectrix of Ökinghaus (A)
$n = 10, \quad a = 1.00$

Number of self-intersection points depends on values of parameter n: none ($n = 3,\ n = 4$), one ($n = 5$),...
(Figures 411–413).
SECTRIX OF ÖKINGHAUS (A) *is an unbounded unconnected (consisting of some unbounded branches, $n \neq 4$) curve with a single symmetry axis; availability or absence of multiple points is defined by values of parameter n.*

Sectrix of Ökinghaus (B)
Equation:

$$\rho = a\frac{\cos\left(\frac{n-2}{2}\right)\varphi}{\cos\frac{n}{2}\varphi},$$

$$a > 0, \quad n \in \mathbf{N}, \quad n \geq 3$$

Basic geometric properties:
Unboundedness: present.
Asymptotes: number and form of asymptotes depends on values of parameters n and a.
Axial symmetry: straight line $y = 0$ is an axis of symmetry.
Central symmetry: none.
Connectedness: none.

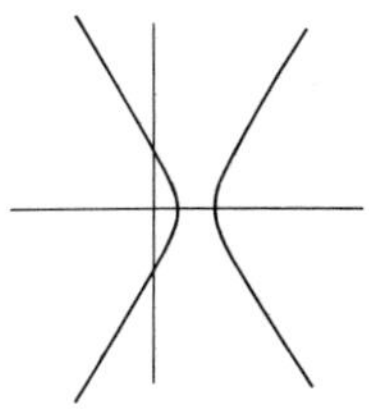

Figure 414.
Sectrix of Ökinghaus (B)
$n = 3, \quad a = 1.00$

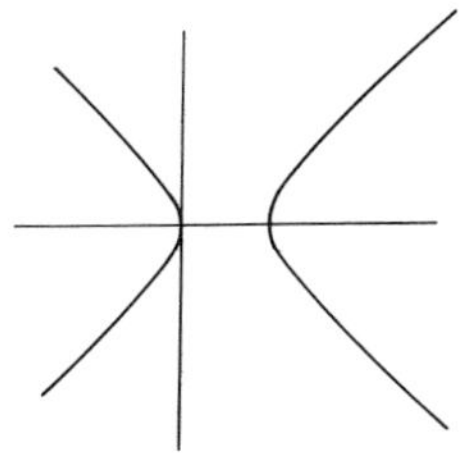

Figure 415.
Sectrix of Ökinghaus (B)
$n = 4, \quad a = 1.00$

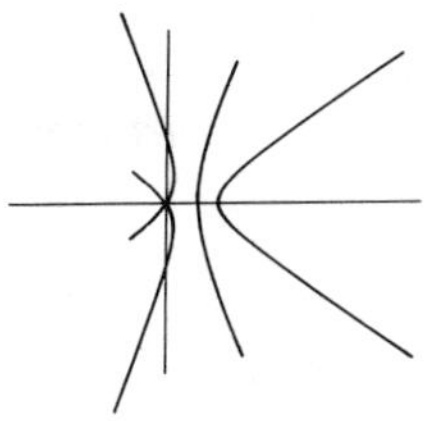

Figure 416.
Sectrix of Ökinghaus (B)
$n = 5, \quad a = 1.00$

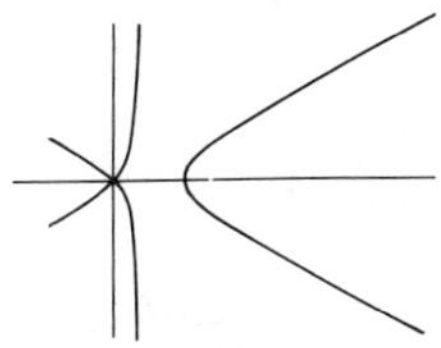

Figure 417.
Sectrix of Ökinghaus (B)
$n = 6, \quad a = 1.00$

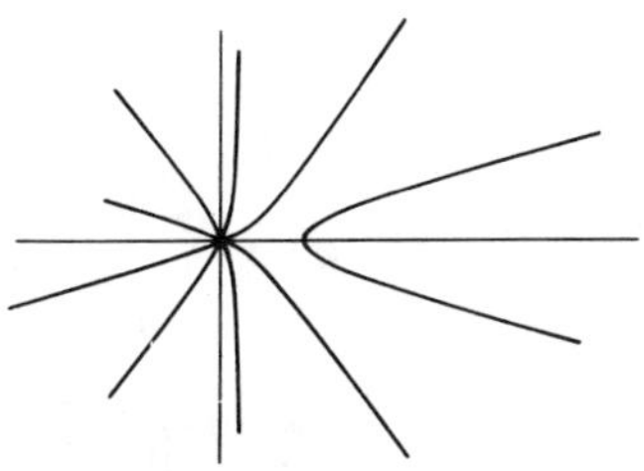

Figure 418.
Sectrix of Ökinghaus (B)
$n = 10, \quad a = 1.00$

Singular points: for some values of parameter n the curve has no singular points (in particular, for $n = 3$, $n = 4$), for other values the curve has multiple points (in particular, $(0,0)$ for $n = 5$, $n = 6$).
Number of self-intersection points depends on values of parameter n: none ($n = 3$, $n = 4$), one ($n = 5$, $n = 6$),...
(Figures 414–418).

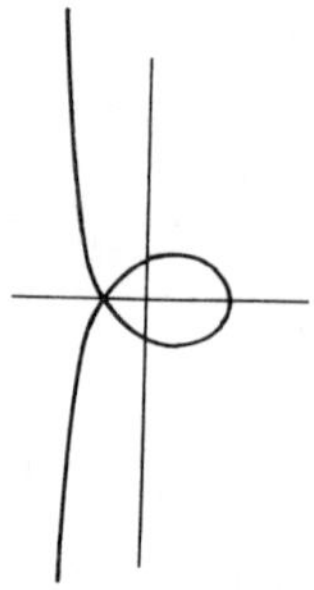

Figure 419. Sectrix of Plateau
$n = 1, \quad n' = 3, \quad a = 1.00$

SECTRIX OF ÖKINGHAUS (B) *is an unbounded unconnected (consisting of some unbounded branches) curve with a single symmetry axis; availability or absence of multiple points is defined by values of parameter n.*

Sectrix of Plateau
Equations:

$$x = a\frac{\sin(n'+n)\varphi}{\sin(n'-n)\varphi}, \quad y = 2a\frac{\sin n'\varphi \sin n\varphi}{\sin(n'-n)\varphi},$$

$$\rho = a\frac{\cos(n-1)\varphi}{\sin n\varphi},$$

$$a > 0, \quad n \in \mathbf{N}, \quad n' \in \mathbf{N},\ n' > n$$

Additional equations:

$$\rho = \sqrt{(x+a)^2 + y^2} = 2a\frac{\sin n'\varphi}{\sin(n'-n)\varphi},$$

$$\rho = \sqrt{(x-a)^2 + y^2} = 2a\frac{\sin n\varphi}{\sin(n'-n)\varphi}$$

Basic geometric properties:
Axial symmetry: straight line $y = 0$ is an axis of symmetry.
Central symmetry: none.
Connectedness: none.
The remaining geometric characteristics *boundedness, closedness, unboundedness, number of asymptotes, number and types of singular points* depend on relationship between values of parameters n and n'.
Special cases: trisectrix of Maclaurin ($n' = 3$, $n = 1$), limaçon ($n' = 3$, $n = 2$).
(Figures 419–425).

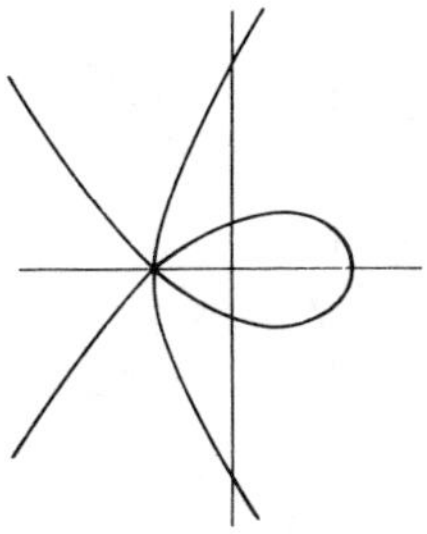

Figure 420. Sectrix of Plateau
$n = 1, \quad n' = 4, \quad a = 1.00$

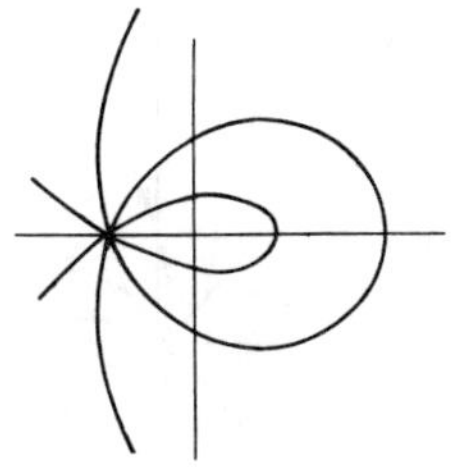

Figure 421. Sectrix of Plateau
$n = 2, \quad n' = 5, \quad a = 1.00$

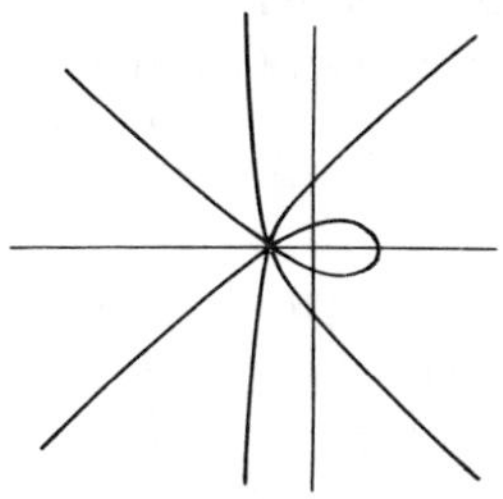

Figure 422. Sectrix of Plateau
$n = 1, \quad n' = 5, \quad a = 1.00$

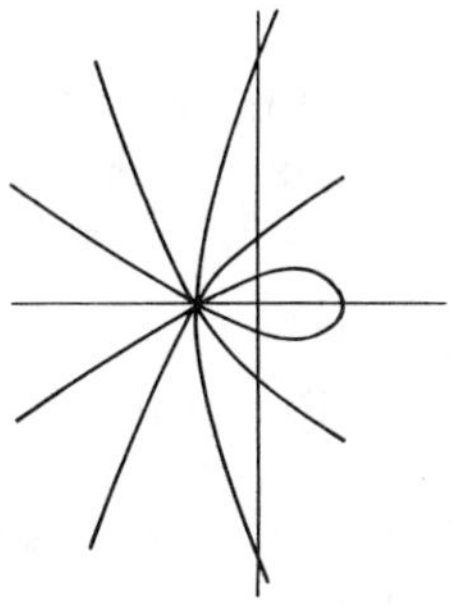

Figure 423. Sectrix of Plateau
$n = 1, \quad n' = 6, \quad a = 1.00$

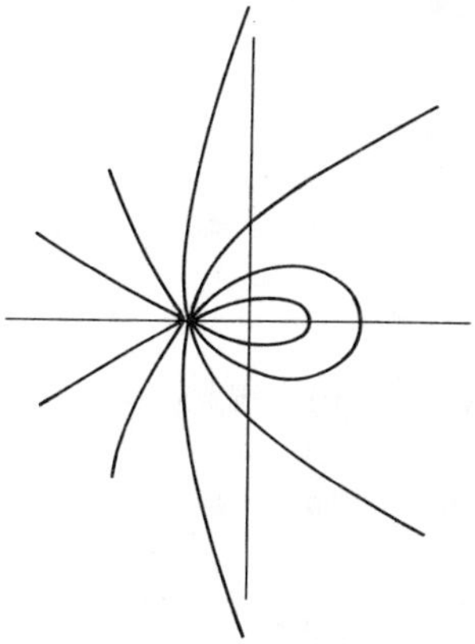

Figure 424. Sectrix of Plateau
$n = 2, \quad n' = 7, \quad a = 1.00$

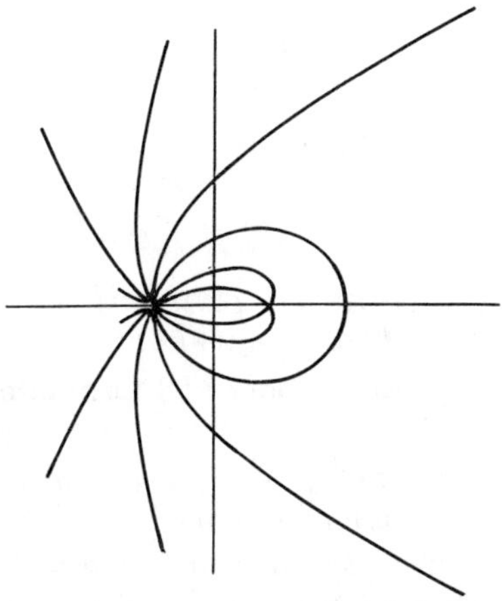

Figure 425. Sectrix of Plateau
$n = 3, \quad n' = 8, \quad a = 1.00$

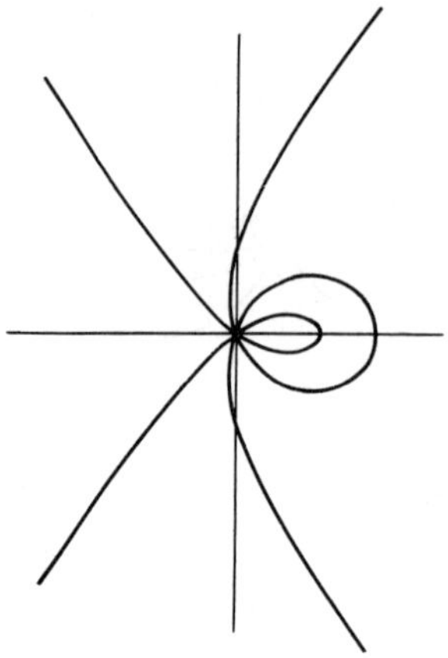

Figure 426.
Sectrix of van Grinten
$n = 4, \quad a = 1.00$

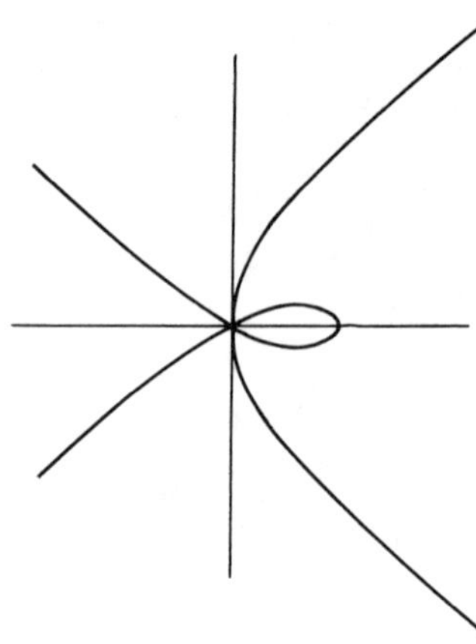

Figure 427.
Sectrix of van Grinten
$n = 5, \quad a = 1.00$

SECTRIX OF PLATEAU *is a connnected curve with a single axis of symmetry; its basic geometric characteristics such as boundedness, closedness, unboundedness, number of asymptotes, number and types of singular points depends on relationship between the integer parameter values.*

Sectrix of van Grinten
Equation:

$$\rho = a\frac{\cos\frac{n+1}{2}\varphi}{\cos\frac{n-1}{2}\varphi}, \quad a > 0, \quad n \in \mathbf{N}, \quad n \geq 3$$

Additional equation:

$$\rho = a\frac{\sin n\varphi - \sin\varphi}{\sin(n-1)\varphi}$$

Basic geometric properties:
Unboundedness: present.
Asymptotes: number and form of asymptotes depends on values of parameters n and a.
Axial symmetry: straight line $y = 0$ is an axis of symmetry.
Central symmetry: none.
Connectedness: present (for $n = 3, 4, 5, \ldots$).
Singular points: point $(0, 0)$ is a multiple point.
Number of self-intersection points depends on values of parameter n: one $(0, 0)$ (for $n = 3, 4, 5$).
(Figures 426–427).

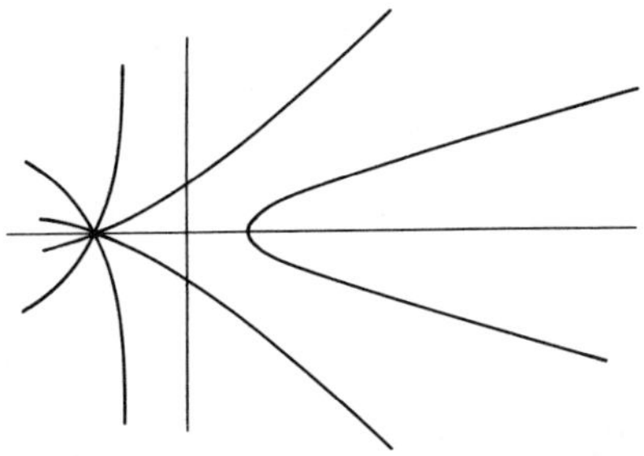

Figure 428. Sectrix of Wagner
$n = 6, \quad r = 5, \quad a = 1.00$

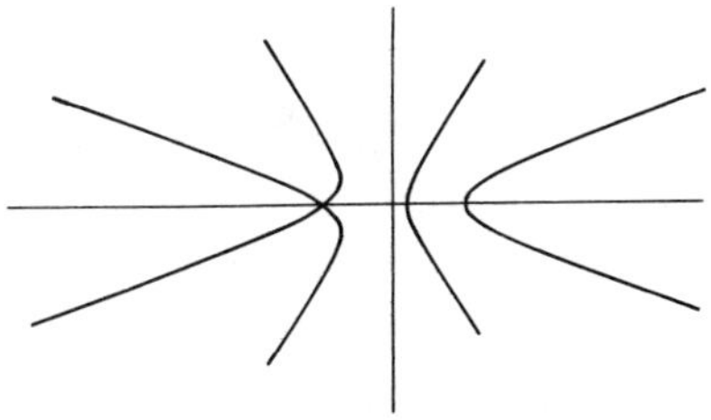

Figure 429. Sectrix of Wagner
$n = 5, \quad r = 3, \quad a = 1.00$

SECTRIX OF VAN GRINTEN *is an unbounded connected (for* $n = 3, 4, 5, \ldots$*) curve with a single singular point (multiple point) and a single axis of symmetry.*

Sectrix of Wagner
Equations:

$$x = a\frac{\sin(2r-n)\varphi}{\sin n\varphi}, \quad y = a\frac{\sin r\varphi \sin(n-1)\varphi}{\sin n\varphi},$$

$$a > 0, \quad r \in \mathbf{N}, \quad n \in \mathbf{N}, \quad r < n < 2r$$

Basic geometric properties:
Unboundedness: present.
Asymptotes: number and form of asymptotes depends on relationship between values of parameters r and n.
Axial symmetry: straight line $y = 0$ is an axis of symmetry.
Central symmetry: none.
Connectedness: none.
Singular points: for some values of parameters r and n the curve has no singular points (in particular, for $r = 2, \ n = 3$), for other values the curve has a multiple point (in particular, for $r = 3, \ n = 4$, or $r = 3, \ n = 5$, or $r = 4, \ n = 5$, or $r = 5, \ n = 6$).
Number of self-intersection points depends on relationship between values of parameters r and n: none ($r = 2, \ n = 3$), one ($r = 3, \ n = 4$; $r = 3, \ n = 5$; $r = 4, \ n = 5$; $r = 5, \ n = 6$).
(Figures 428–432).
SECTRIX OF WAGNER *is an unbounded unconnected (consisting of some unbounded branches) curve with a single symmetry axis; availability or absence of multiple points is defined by relationship between parameters* r *and* n.

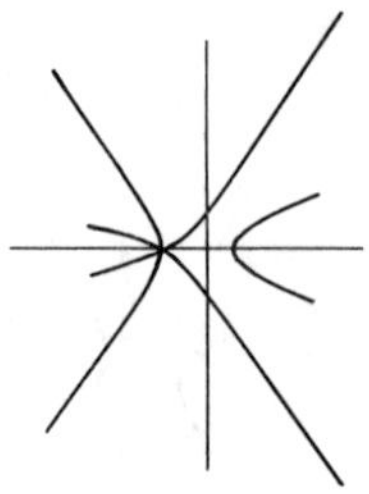

Figure 430. Sectrix of Wagner
$n = 5, \quad r = 4,$
$a = 1.00$

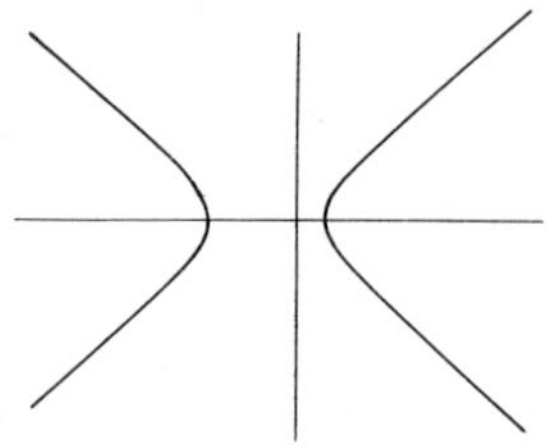

Figure 431. Sectrix of Wagner
$n = 3, \quad r = 2,$
$a = 1.00$

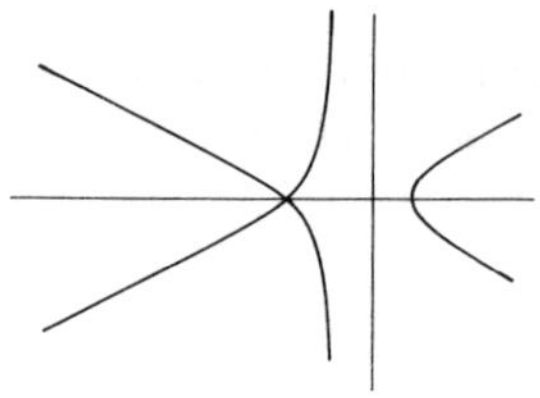

Figure 432. Sectrix of Wagner
$n = 4, \quad r = 3,$
$a = 1.00$

Serpentine of Newton (anguinea, serpentine, serpentine curve)
Equation:

$$y(x^2 + a^2) = acx,$$

$$a > 0, \quad c > 0$$

Basic geometric properties:
Unboundedness: present.
Asymptotes: straight line $y = 0$ is a horizontal asymptote.
Axial symmetry: none.
Central symmetry: point $(0, 0)$ is a center of symmetry.
Connectedness: present.
Singular points: none.
Self-intersection points: none.
Characterized points: points $(a, c/2), (-a, -c/2)$ are located at maximal distance from x-axis (vertices).
(Figures 433–434).
SERPENTINE OF NEWTON *is an unbounded connected central-symmetric curve free of singular points with a single horizontal asymptote.*

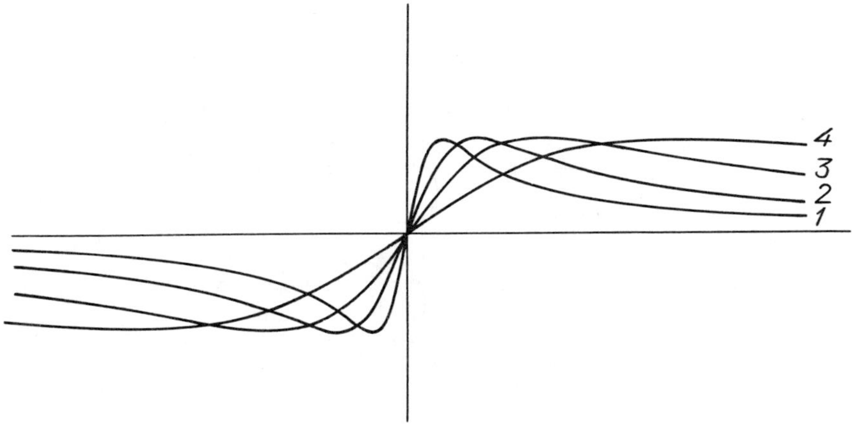

Figure 433. Serpentine of Newton

N	1	2	3	4
a	0.25	0.50	1.00	2.00
c	1.50	1.50	1.50	1.50

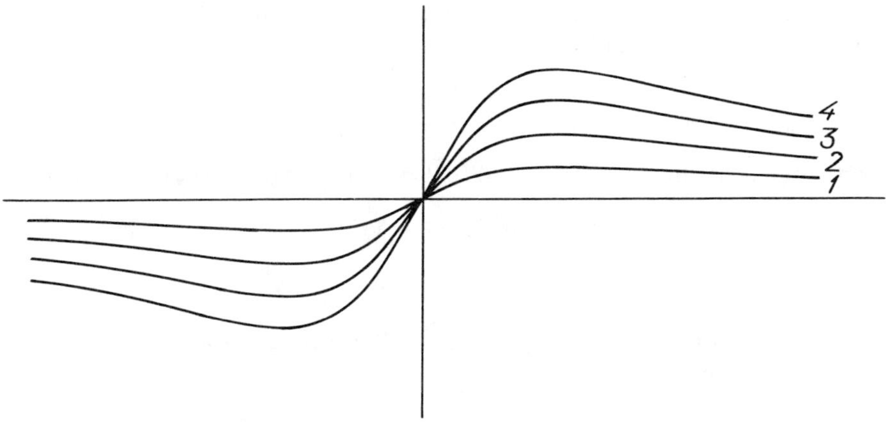

Figure 434. Serpentine of Newton

N	1	2	3	4
c	0.50	1.00	1.50	2.00
a	1.00	1.00	1.00	1.00

Sextic of Cayley (Cayley's sextet)
Equation:

$$\rho = l \cos \frac{\varphi}{4},$$

$$l > 0$$

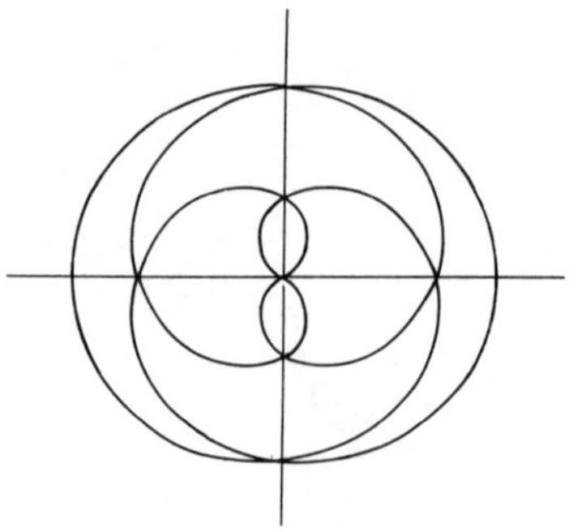

Figure 435. Sextic of Cayley

Basic geometric properties:
Boundedness: present.
Closedness: present.
Axial symmetry: straight lines $x = 0,\ y = 0$ are axes of symmetry.
Central symmetry: point $(0, 0)$ is a center of symmetry.
Connectedness: present.
Singular points: point $(0, 0)$ is a degenerate node, straight line $y = 0$ is a tangent at the node; point

$$\left(\frac{l\sqrt{2}}{2}, 0\right)$$

is a node, straight lines $4x \pm y - 2l\sqrt{2} = 0$ are tangents at the node; point

$$\left(-\frac{l\sqrt{2}}{2}, 0\right)$$

is a node, straight lines $4x \pm y + 2l\sqrt{2} = 0$ are tangents at the node.
Self-intersection points: three,

$$(0, 0), \quad \left(\pm l\frac{\sqrt{2}}{2}, 0\right)$$

(Figure 435).
Sextic of Cayley *is a bounded connected closed central-symmetric curve with three singular points (degenerate node, nodes) and two axes of symmetry.*

Sinusoid
Equation:

$$y = \sin(ax), \quad a > 0$$

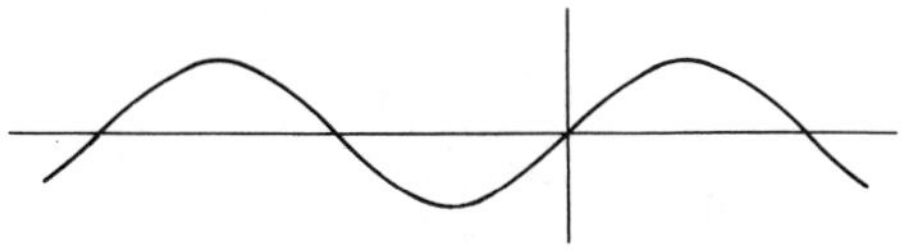

Figure 436. Sinusoid

Basic geometric properties:
Unboundedness: present.
Asymptotes: none.
Axial symmetry: straight lines $2ax = (2k+1)\pi$, $k \in \mathbf{Z}$, are axes of symmetry.
Central symmetry: points $(k\pi/a, 0)$, $k \in \mathbf{Z}$, are centers of symmetry.
Connectedness: present.
Singular points: none.
Self-intersection points: none.
Characterized points: points

$$\left(\frac{(2k+1)\pi}{2a}, (-1)^k\right),$$

$$k \in \mathbf{Z}$$

are located at maximal distance from x-axis (vertices); points $(k\pi/a, 0)$, $k \in \mathbf{Z}$, are inflection points.
(Figure 436).
SINUSOID *is an unbounded connected central-symmetric curve free of singular points with an infinitely large number of the symmetry axes.*

Sinusoidal spiral (A)
Equation:

$$\rho^m = a^m \cos m\varphi,$$

$$a > 0, \quad m = \frac{p}{q},$$

$$p \in \mathbf{N}, \quad q \in \mathbf{N},$$

$$\text{G.C.D.}(p,q) = 1$$

Additional equation:

$$\rho^m = a^m \sin m\varphi$$

Basic geometric properties:
Boundedness: present.
Closedness: present.

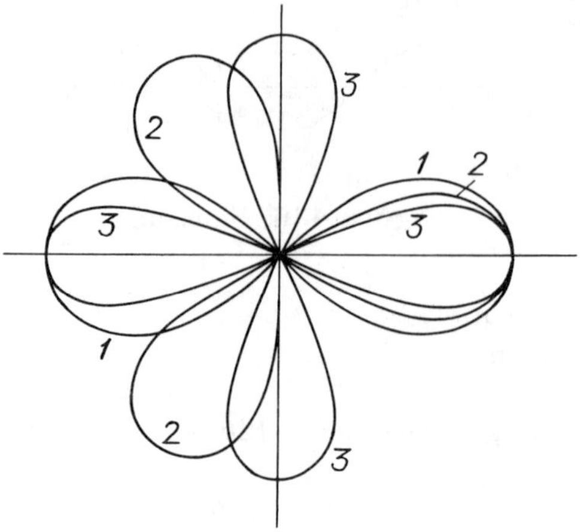

Figure 437. Sinusoidal spiral (A)

N	1	2	3
m	2	3	4
a	1.00	1.00	1.00

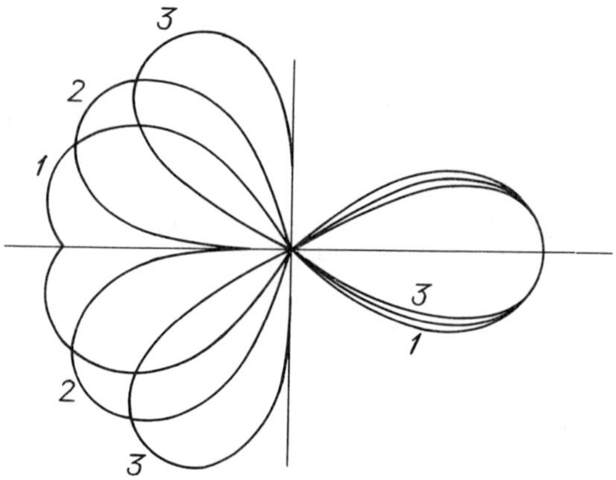

Figure 438. Sinusoidal spiral (A)

N	1	2	3
m	11/10	5/2	3
a	1.00	1.00	1.00

Axial symmetry:
m is a positive integer:
straight lines

$$x \sin \frac{2\pi k}{m} = y \cos \frac{2\pi k}{m}, \quad k = 0, 1, \ldots, m - 1$$

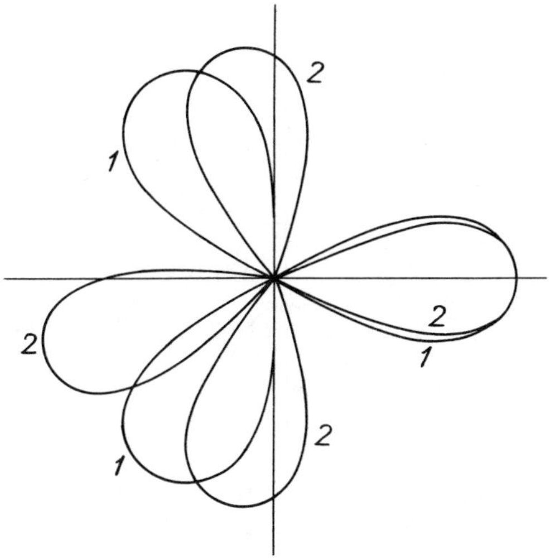

Figure 439. Sinusoidal spiral (A)

N	1	2
m	3	18/5
a	1.00	1.00

are axes of symmetry (for odd m); straight lines

$$x \sin \frac{\pi k}{m} = y \cos \frac{\pi k}{m}, \quad k = 0, 1, \ldots, 2m - 1$$

are axes of symmetry (for even m);

m is rational:
straight lines

$$x \sin \frac{2\pi l}{p} = y \cos \frac{2\pi l}{p}, \quad l = 0, 1, \ldots, p - 1$$

are axes of symmetry (for odd pq); straight lines

$$x \sin \frac{\pi l}{p} = y \cos \frac{\pi l}{p}, \quad l = 0, 1, \ldots, 2p - 1$$

are axes of symmetry (for even pq).
Central symmetry: point $(0, 0)$ is a center of symmetry (for even pq only).
Connectedness: present.
Singular points: point $(0, 0)$ is a multiple point (a single singular point for $q = 1$), simple nodes exist for non-integer m only and their number depends on values of p and q.
Self-intersection points: finite but no less than one, $(0, 0)$.

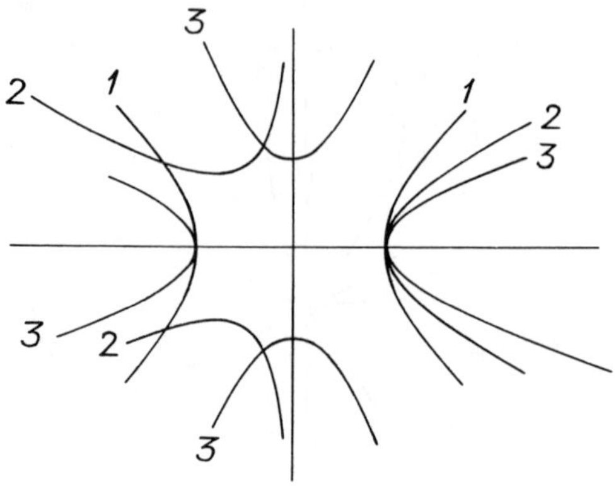

Figure 440. Sinusoidal spiral (B)

N	1	2	3
m	-2	-3	-4
a	1.00	1.00	1.00

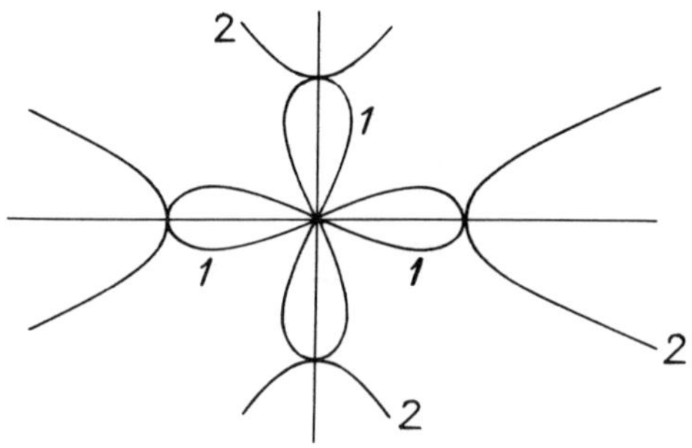

Figure 441. Sinusoidal spiral (A)(B)

N	1	2
m	4	-4
a	1.00	1.00

Number of leaves: m ($m \in \mathbf{N}$ and odd), $2m$ ($m \in \mathbf{N}$ and even), p ($m = p/q$ and pq is odd), $2p$ ($m = p/q$ and pq is even).
(Figures 437–439).
SINUSOIDAL SPIRAL (A) *is a bounded connected closed curve with* p *leaves (self-intersected* ($q > 1$) *or not* ($q = 1$)) *and a single point of high multiplicity; properties of symmetry and number of simple nodes are defined by values of parameter* m.

In case of irrational m sinusoidal spiral (A) is a transcendent nonclosed curve with an infinitely large number of nodes.

Sinusoidal spiral (B)
Equation:

$$\rho^m = a^m \cos m\varphi,$$

$$a > 0, \quad m = -\frac{p}{q}, \quad p \in \mathbf{N}, \quad q \in \mathbf{N}, \quad \text{G.C.D.}(p, q) = 1$$

Additional equation:

$$\rho^m = a^m \sin m\varphi$$

Basic geometric properties:
Unboundedness: present.
Asymptotes: straight lines

$$x \sin \frac{2l+1}{2p}\pi = y \cos \frac{2l+1}{2p}\pi, \quad l = 0, 1, \ldots, p-1$$

are asymptotes.
Axial symmetry:

m is a positive integer:
straight lines

$$x \sin \frac{2\pi k}{m} = y \cos \frac{2\pi k}{m}, \quad k = 0, 1, \ldots, m-1$$

are axes of symmetry (for odd m); straight lines

$$x \sin \frac{\pi k}{m} = y \cos \frac{\pi k}{m}, \quad k = 0, 1, \ldots, 2m-1$$

are axes of symmetry (for even m);

m is rational:
straight lines

$$x \sin \frac{2\pi l}{p} = y \cos \frac{2\pi l}{p}, \quad l = 0, 1, \ldots, p-1$$

are axes of symmetry (for odd pq); straight lines

$$x \sin \frac{\pi l}{p} = y \cos \frac{\pi l}{p}, \quad l = 0, 1, \ldots, 2p-1$$

are axes of symmetry (for even pq).
Central symmetry: point $(0, 0)$ is a center of symmetry (for even pq only).
Connectedness: none (for $q = 1$), present (for $q > 1$).
Singular points: none (for $q = 1$), nodes (for $q > 1$).
Self-intersection points: finite (for $q > 1$), none (for $q = 1$).
(Figures 440–441).

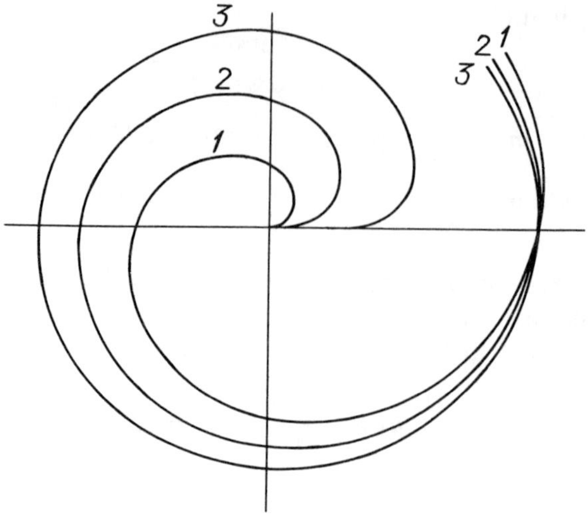

Figure 442. Spiral of higher order

N	1	2	3
k	1	2	5
a	1.00	1.00	1.00

SINUSOIDAL SPIRAL (B) *is an unbounded connected* ($q > 1$) *or unconnected* ($q = 1$) *curve; properties of symmetry and number of asymptotes and simple nodes are defined by values of parameter* m.

In case of irrational m sinusoidal spiral (B) is a transcendent connected curve with an infinitely large number of nodes.

Spiral of higher order
Equation:

$$\rho^k = a^k \frac{\varphi}{2\pi}, \quad a > 0, \quad k \in \mathbf{N}$$

Basic geometric properties:
Unboundedness: present.
Asymptotes: none.
Axial symmetry: none.
Central symmetry: none.
Connectedness: present.
Singular points: point $(0, 0)$ is a cessation point.
Self-intersection points: none.
Special cases: Archimedean spiral ($k = 1$), miraculous (strange) curve ($k = 2$).
(Figure 442).

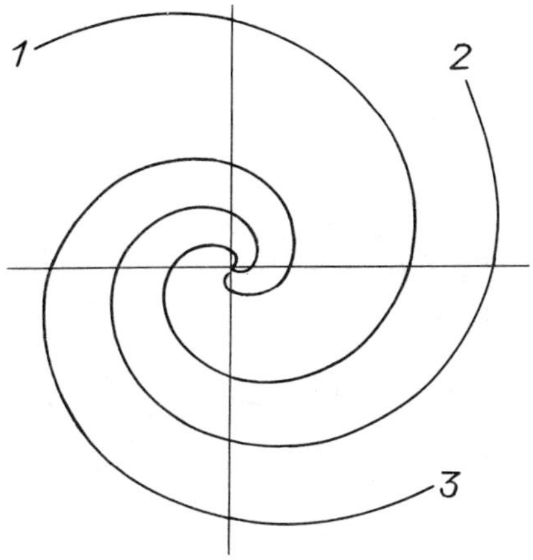

Figure 443. Spiral of Norwich

N	1	2	3
α	−1.57	0.00	1.57
a	1.00	1.00	1.00

SPIRAL OF HIGHER ORDER *is an unbounded connected curve with a single singular point (cessation point).*

Spiral of Norwich

Equation:

$$\varphi + \alpha = \sqrt{\frac{\rho - a}{a}} - 2\arccos\sqrt{\frac{a}{\rho}}, \quad a > 0$$

Additional equations:

$$\varphi + \alpha = \sqrt{\frac{\rho - a}{a}} - \pi + 2\arctan\sqrt{\frac{a}{\rho - a}},$$

$$\varphi + \alpha = \sqrt{\frac{2\rho - c}{c}} - 2\,\mathrm{arccot}\sqrt{\frac{2\rho - c}{c}},$$

Basic geometric properties:

Unboundedness: present.

Asymptotes: none.

Axial symmetry: none.

Central symmetry: none.

Connectedness: present.

Singular points: point $(a\cos\alpha, -a\sin\alpha)$ is a cessation point.

Self-intersection points: none.

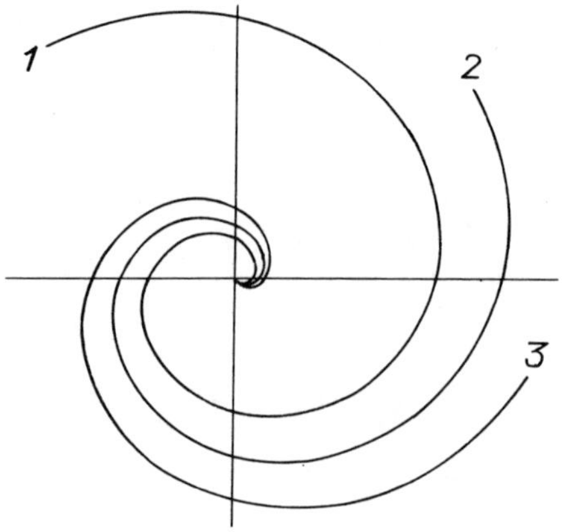

Figure 444. Spiral of Norwich

N	1	2	3
α	0.00	0.00	0.00
a	0.75	1.00	0.50

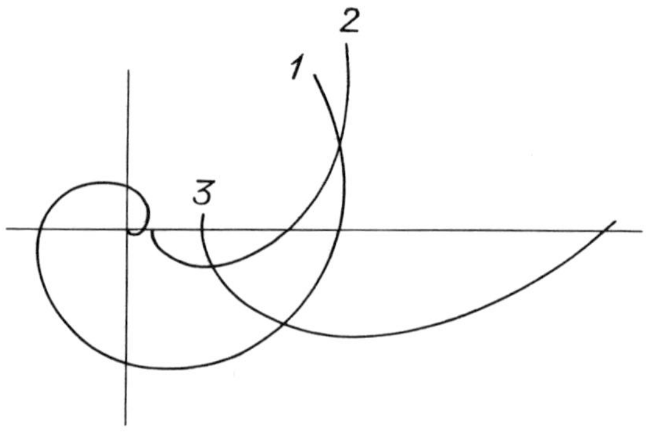

Figure 445. Spiral of Norwich

N	1	2	3
α	0.00	0.00	0.00
a	1.00	10.00	30.00

(Figures 443–445).
SPIRAL OF NORWICH *is an unbounded connected curve with a single singular point (cessation point).*

Spiril curve
Equations:

$$(x^2 + y^2 + p^2 + d^2 - R^2)^2 = 4d^2(x^2 + p^2),$$

$$y = \pm\sqrt{R^2 - (d \pm \sqrt{x^2 + p^2})}, \quad d > 0, \quad R > 0, \quad p \geq 0$$

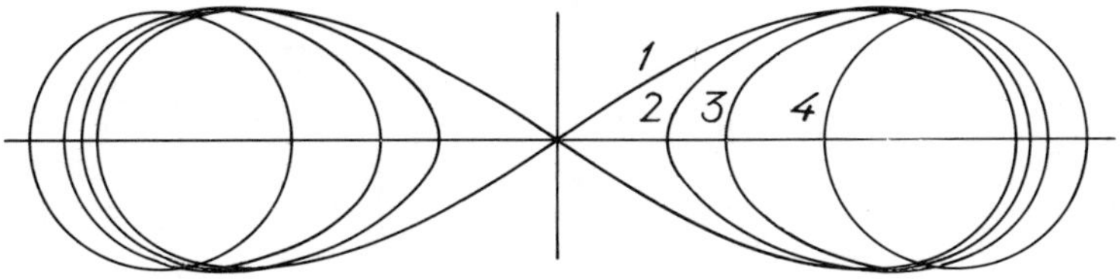

Figure 446. Spiril curve

N	1	2	3	4
p	2.00	1.80	1.50	0.00
d	3.00	3.00	3.00	3.00
R	1.00	1.00	1.00	1.00

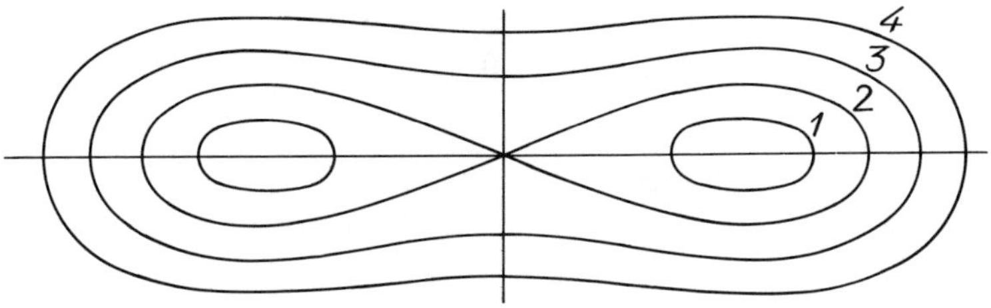

Figure 447. Spiril curve

N	1	2	3	4
R	0.25	0.50	0.75	1.00
d	3.00	3.00	3.00	3.00
p	2.50	2.50	2.50	2.50

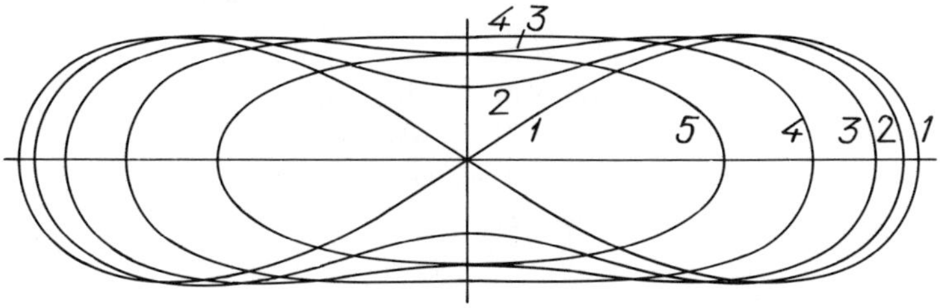

Figure 448. Spiril curve

N	1	2	3	4	5
p	2.00	2.20	2.50	3.00	3.50
d	3.00	3.00	3.00	3.00	3.00
R	1.00	1.00	1.00	1.00	1.00

Additional equation:

$$((x-m)^2+y^2)((x+m)^2+y^2)=c(x^2+y^2)+c_1,$$

$$c_1=-((p+d)^2-R^2)((p-d)^2-R^2)+m^4, \quad c=2(R^2-p^2), \quad m=d$$

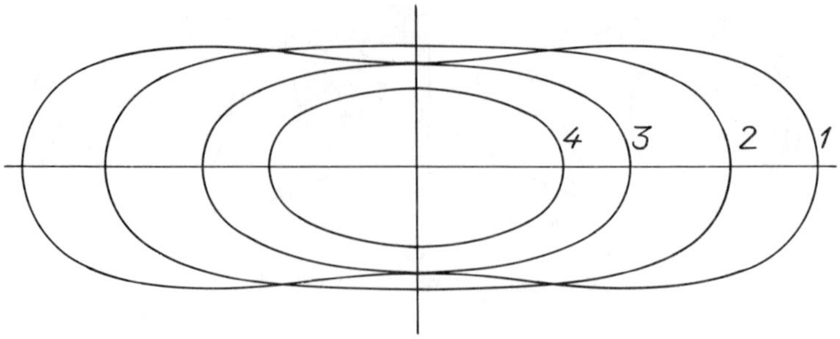

Figure 449. Spiril curve

N	1	2	3	4
d	3.00	2.50	2.00	1.75
p	2.50	2.50	2.50	2.50
R	1.00	1.00	1.00	1.00

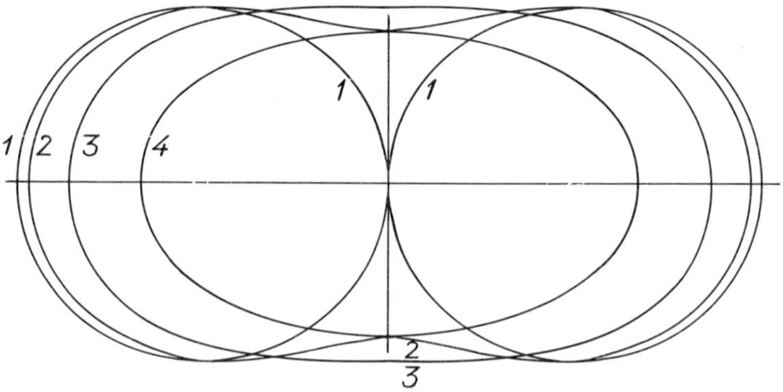

Figure 450. Spiril curve

N	1	2	3	4
p	0.00	0.50	1.00	1.50
d	1.00	1.00	1.00	1.00
R	1.00	1.00	1.00	1.00

Basic geometric properties:
Boundedness: present.
Closedness: present.
Axial symmetry: straight lines $x = 0,\ y = 0$ are axes of symmetry.
Central symmetry: point $(0, 0)$ is a center of symmetry.
Connection: none (for $d - R > p \geq 0$ or for $R > d + p$).
Singular points: point $(0, 0)$ is a node ($p = d - R \geq 0$); straight lines $x \pm y = 0$ are tangents at the node; point $(0, \sqrt{R^2 - d^2})$ is a node, straight lines

$$y = \pm \frac{d}{\sqrt{R^2 - d^2}}(x - \sqrt{R^2 - d^2})$$

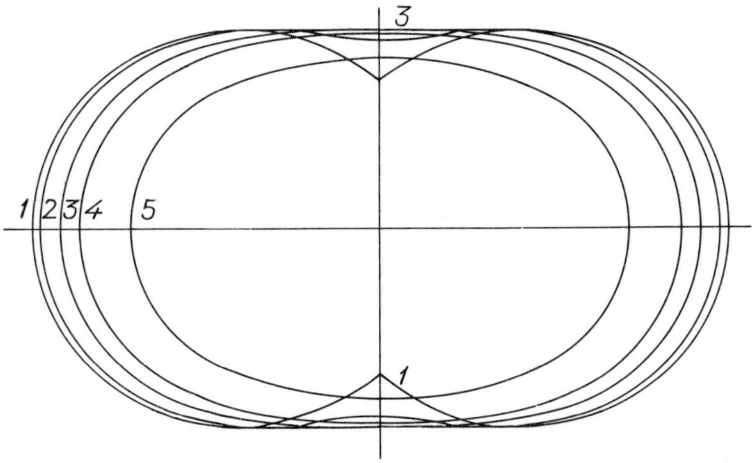

Figure 451. Spiril curve

N	1	2	3	4	5
p	0.00	0.50	1.00	1.25	1.75
d	1.00	1.00	1.00	1.00	1.00
R	1.50	1.50	1.50	1.50	1.50

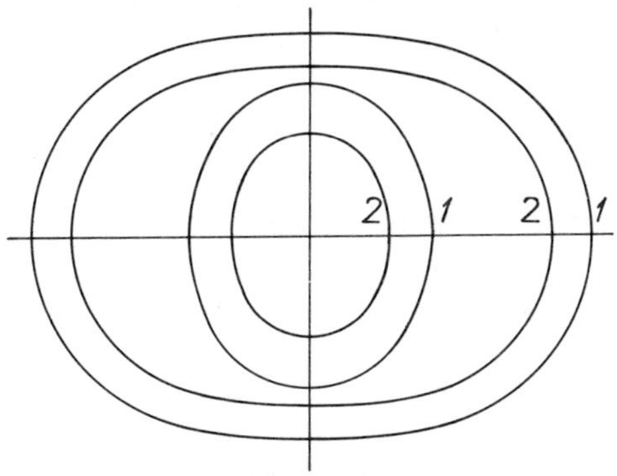

Figure 452. Spiril curve

N	1	2
R	3.00	2.50
d	1.00	1.00
p	1.00	1.00

are tangents at the node; point $(0, -\sqrt{R^2 - d^2}$ is a node ($d < R$, $p = 0$), straight lines

$$y = \pm \frac{d}{\sqrt{R^2 - d^2}}(x + \sqrt{R^2 - d^2})$$

are tangents at the node.

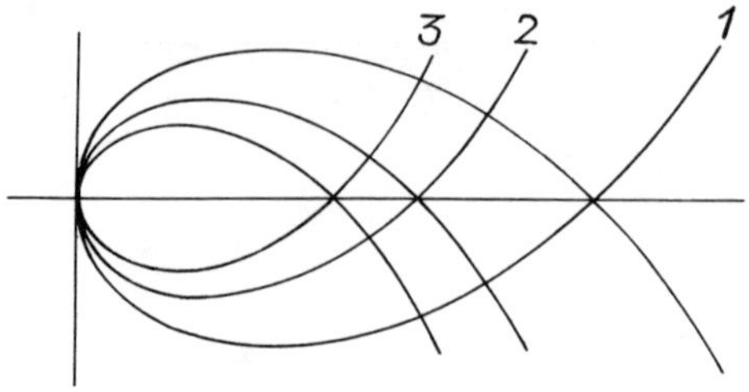

Figure 453. Strophoid

N	1	2	3
a	0.50	0.33	0.25

Number of self-intersection points depends on relationship between values of parameters p, d and R: one, $(0,0)$ (for $p = d - R \geq 0$); two, $(0, \pm\sqrt{R^2 - d^2})$ (for $d < R$, $p = 0$); none (in all other cases).
Special cases (for additonal equation): oval of Cassini ($c = 0$), lemniscate of Bernoulli ($c = 0, c_1 = m$), lemniscate of Booth ($c_1 = m$).
(Figures 446–452).
SPIRIL CURVE (SECTION OF TORUS BY PLANE PARALLEL TO THE ROTATIONAL AXIS) *is a bounded connected or unconnected (the last for $d - R > p \geq 0$ or $R > d + p$ only) closed central-symmetric curve with two axes of symmetry and singular points (nodes: one ($p = d - R \geq 0$), two ($d < R, p = 0$)).*

Strophoid
Equations:

$$(x^2 + y^2)(x - 2a) + a^2 x = 0,$$

$$x = \frac{2at^2}{1 + t^2},$$

$$y = \frac{at(t^2 - 1)}{1 + t^2}, \quad a > 0$$

Additional equations:

$$y = \pm(x - a)\sqrt{\frac{x}{2a - x}},$$

$$\rho = a\frac{1 + \sin\varphi}{\cos\varphi}, \quad \rho = a\tan\left(\frac{\pi}{4} + \frac{\varphi}{2}\right)$$

Basic geometric properties:
Unboundedness: present.

Asymptotes: straight line $x = 2a$ is a vertical asymptote.
Axial symmetry: straight line $y = 0$ is an axis of symmetry.
Central symmetry: none.
Connectedness: present.
Singular points: point $(a, 0)$ is a node, straight lines $x \pm y = 0$ are tangents at the node.
Self-intersection points: one, $(a, 0)$.
Characterized points: points

$$\left(\frac{3-\sqrt{5}}{2}a, \pm\frac{3-\sqrt{5}}{2\sqrt{\sqrt{5}+1}}a\right)$$

are loop points located at maximal distance from x-axis.
Areas: area of the loop is equal to

$$2a^2 - \frac{\pi}{2}a^2$$

area of domain enclosed by the curve and the asymptote is equal to

$$2a^2 + \frac{\pi}{2}a^2$$

(Figure 453).
STROPNOID *is an unbounded connected curve with a single singular point (node), a single vertical asymptote and a single axis of symmetry.*

2.2.18 T

Tangentoid
Equation:

$$y = \tan(ax), \quad a > 0$$

Basic geometric properties:
Unboundedness: present.
Asymptotes: straight lines

$$2ax = (2k+1)\pi, \quad k \in \mathbf{Z}$$

are vertical asymptotes.
Axial symmetry: none.
Central symmetry: points

$$\left(\frac{k\pi}{a}, 0\right), \quad k \in \mathbf{Z}$$

are centers of symmetry.

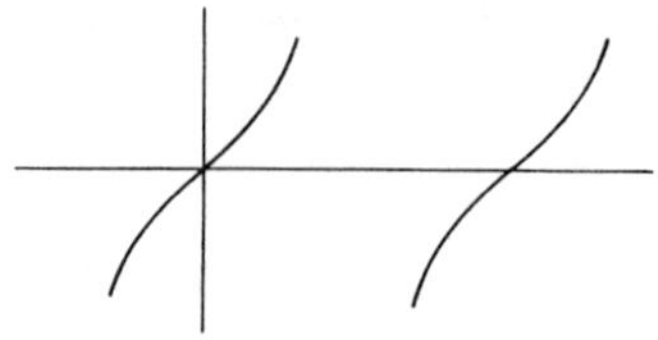

Figure 454. Tangentoid

Connectedness: none.
Singular points: none.
Self-intersection points: none.
(Figure 454).
TANGENTOID *is an unbounded unconnected (consisting of an infinitely large number of unbounded branches) central-symmetric curve free of singular points with an infinitely large number of symmetry axes and vertical asymptotes.*

Three-cuspid hypocycloid (detloid, hypocycloid of Steiner, Steiner's three-cuspid hypocycloid, tricuspid, tricuspid curve)
Equations:

$$(x^2 + y^2)^2 + 8rx(3y^2 - x^2) + 18r^2(x^2 + y^2) - 27r^4 = 0,$$

$$x = 2r\cos t + r\cos 2t, \quad y = 2r\sin t - r\sin 2t,$$

$$r > 0$$

Basic geometric properties:
Boundedness: present.
Closedness: present.
Axial symmetry: straight lines

$$y = 0, \quad \sqrt{3}x \pm y = 0$$

are axes of symmetry.
Central symmetry: none.
Connectedness: present.
Singular points: point $(3r, 0)$ is a cuspidal point of the first kind, straight line $y = 0$ is a tangent at the cusp point; point $(-3r/2, 3\sqrt{3}r)$ is a cuspidal point of the first kind, straight line $\sqrt{3}x + y = 0$ is a tangent at the cusp point; point $(-3r/2, -3\sqrt{3}r)$ is a cuspidal point of the first kind, straight line $\sqrt{3}x - y = 0$ is a tangent at the cusp point.
Self-intersection points: none.
Area of domain enclosed by the curve: $2\pi r^2$.
(Figure 455).

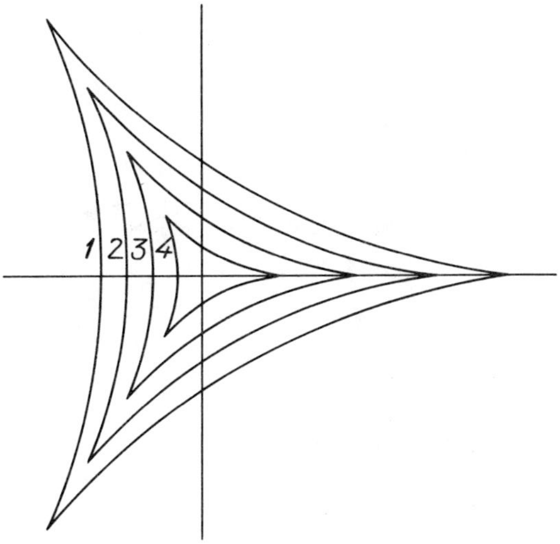

Figure 455.
Three-cuspid hypocycloid

N	1	2	3	4
r	1.00	0.75	0.50	0.25

THREE-CUSPID HYPOCYCLOID *is a bounded connected closed curve with three singular points (cuspidal points of the first kind) and three axes of symmetry.*

Three-leafed rose
Equations:

$$(x^2 + y^2)^2 + rx(3y^2 - x^2) = 0,$$

$$\rho = r \cos 3\varphi,$$

$$x = r\frac{1 - 3t^2}{(1 + t^2)^2}, \quad y = r\frac{1 - 3t^2}{(1 + t^2)^2}t,$$

$$r > 0$$

Basic geometric properties:
Boundedness: present.
Closedness: present.
Axial symmetry: straight lines $y = 0$, $\sqrt{3}x \pm y = 0$ are axes of symmetry.
Central symmetry: none.
Connectedness: none.
Singular points: point $(0, 0)$ is a triple point, straight lines $x = 0$, $x \pm \sqrt{3}y = 0$ are tangents at the point.
Self-intersection points: one, $(0, 0)$.

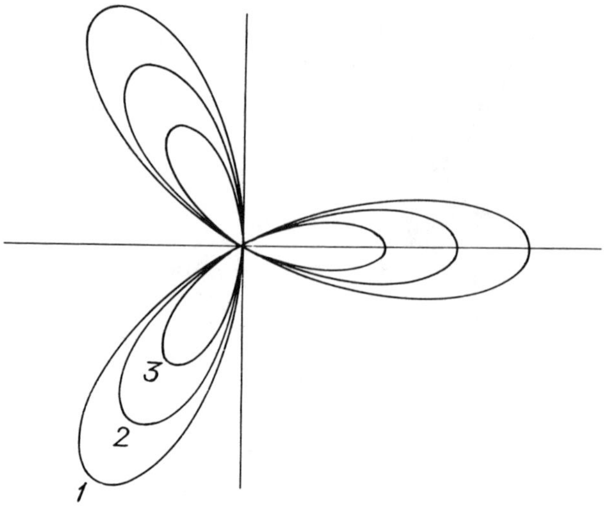

Figure 456. Three-leafed rose

N	1	2	3
r	1.00	0.75	0.50

Area of every leaf: $\pi r^2/12$.
(Figure 456).
THREE-LEAFED ROSE *is a bounded connected closed curve with a single singular point (triple point) and three axes of symmetry.*

Top
Equation:

$$x^4 - ax^3 + \frac{a^2}{4}y^2 = 0,$$

$$a > 0$$

Basic geometric properties:
Boundedness: present.
Closedness: present.
Axial symmetry: straight line $y = 0$ is an axis of symmetry.
Central symmetry: none.
Connectedness: present.
Singular points: point $(0, 0)$ is a cuspidal point of the first kind, straight line $y = 0$ is a tangent at the point.
Self-intersection points: none.

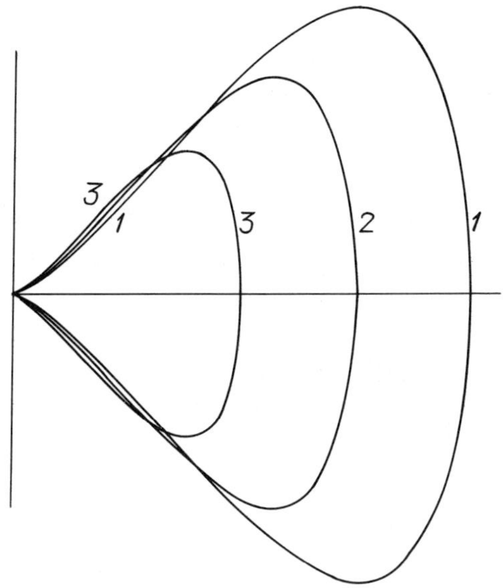

Figure 457. Top

N	1	2	3
a	2.00	1.50	1.00

Characterized points: points

$$\left(\frac{3}{4}a, \pm\frac{3\sqrt{3}}{8}a\right)$$

are located at maximal distance from x-axis.
Area of the domain enclosed by the curve: $\pi a^2/4$.
(Figure 457).
Top *is a bounded connected closed curve with a single singular point (cusp point of the first kind) and a single axis of symmetry.*

Tractrix
Equations:

$$x = a \ln \tan \frac{t}{2} + a \cos t, \quad y = a \sin t,$$

$$x = \ln \frac{a - \sqrt{a^2 - y^2}}{y} + \sqrt{a^2 - y^2}, \quad a > 0$$

Additional equations:

$$x = -\frac{y}{2} \cos\theta \pm \left(\frac{c}{q}\sqrt{q^2 - y^2} - c \ln \frac{q + \sqrt{q^2 - y^2}}{y}\right), \quad q^2 = \frac{4c^2}{4 - \cos^2\theta}$$

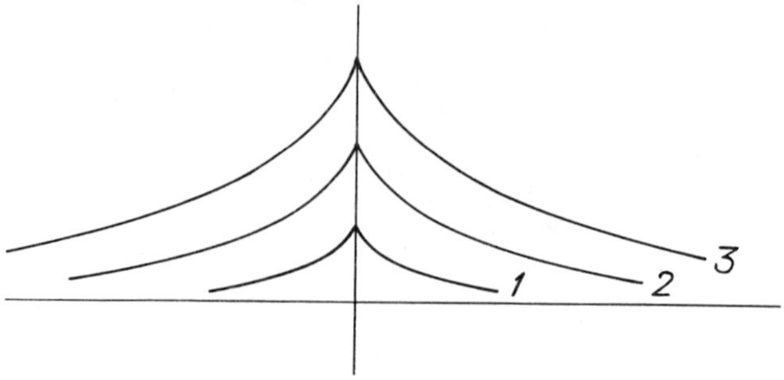

Figure 458. Tractrix

N	1	2	3
a	0.50	1.00	1.50

Basic geometric properties:
Unboundedness: present.
Asymptotes: straight line $y = 0$ is a horizontal asymptote.
Axial symmetry: straight line $x = 0$ is an axis of symmetry.
Central symmetry: none.
Connectedness: present.
Singular points: point $(0, a)$ is a cuspidal point of the first kind, straight line $x = 0$ is a tangent at the point.
Self-intersection points: none.
Area of domain enclosed by the curve and the asymptote is equal to $\pi a^2/2$. (Figure 458).
TRACTRIX *is an unbounded connected curve with a single singular point (cusp point of the first kind), a single horizontal asymptote and a single axis of symmetry.*

Trefoil (trifolium)
Equations:

$$x = r(2\cos 2t - \cos t), \quad y = r(2\sin 2t + \sin t),$$

$$r > 0$$

Basic geometric properties:
Boundedness: present.
Closedness: present.
Axial symmetry: straight lines $y = 0,\ y = \pm\sqrt{3}$ are axes of symmetry.
Central symmetry: none.
Connectedness: present.
Singular points: point

$$\left(-\frac{3}{2}r, 0\right)$$

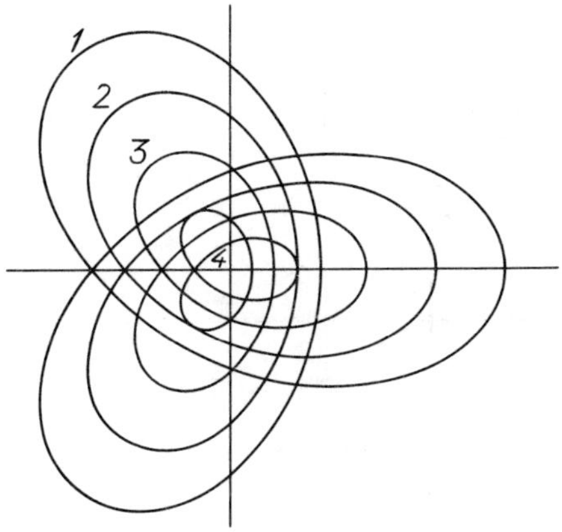

Figure 459. Trefoil

N	1	2	3	4
r	1.00	0.75	0.50	0.25

is a node, straight lines

$$y = \pm\sqrt{15}\left(x + \frac{3}{2}\right)$$

are tangents at the node; point

$$\left(\frac{3}{4}r, \frac{3\sqrt{3}}{4}\right)$$

is a node, straight lines

$$(\sqrt{3} \mp \sqrt{15})x - (1 \pm 3\sqrt{5})y \pm 3\sqrt{15}r = 0$$

are tangents at the node; point

$$\left(\frac{3}{4}r, -\frac{3\sqrt{3}}{4}r\right)$$

is a node, straight lines

$$-(\sqrt{3} \pm \sqrt{15})x + (1 \pm 3\sqrt{5})y \pm 3\sqrt{15}r = 0$$

are tangents at the node.
Self-intersection points: three,

$$\left(-\frac{3}{2}r, 0\right), \quad \left(\frac{3}{4}r, \pm\frac{3\sqrt{3}}{4}r\right)$$

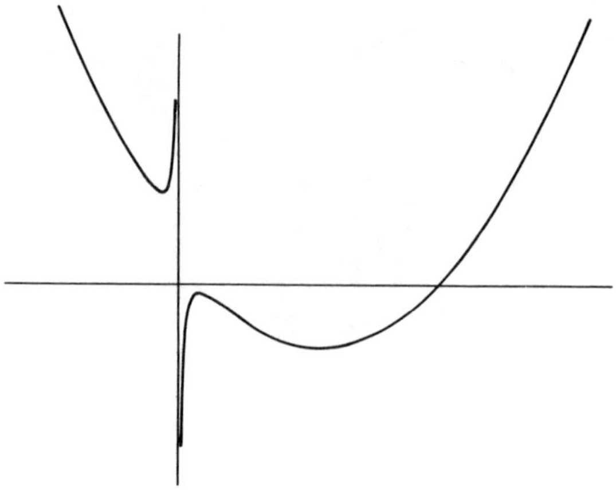

Figure 460. Trident of Newton
$a = 0.1 \quad b = -1.4, \quad c = 1.9, \quad d = -1.2$

Characterized points: points

$$(r, 0), \quad (3r, 0), \quad \left(-\frac{3}{2}r, \pm\frac{3\sqrt{3}}{2}r\right), \quad \left(-\frac{r}{2}, \pm\frac{\sqrt{3}}{2}r\right)$$

are vertices.
(Figure 459).
TREFOIL *is a bounded connected closed curve with three singular points (nodes) and three axes of symmetry.*

Trident of Newton (Cartesian parabola, Descartes' parabola, parabola of Descartes, trident)
Equation:

$$xy = ax^3 + bx^2 + cx + d,$$

$$a \neq 0, \quad d \neq 0$$

Basic geometric properties:
Unboundedness: present.
Asymptotes: straight line $x = 0$ is a vertical asymptote.
Axial symmetry: none.
Central symmetry: none.
Connectedness: none.
Singular points: none.
Self-intersection points: none.
(Figures 460–461).

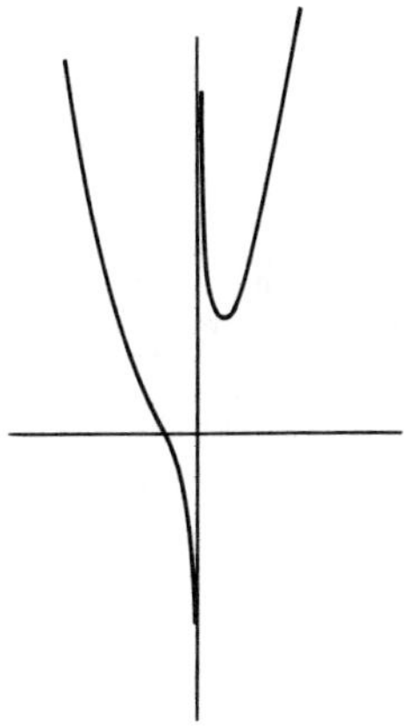

Figure 461. Trident of Newton
$a = b = c = d = 1.0$

TRIDENT OF NEWTON *is an unbounded unconnected (consisting of two unbounded branches) curve free of singular points with a single vertical asymptote.*

Trifolium (three-leafed figure)
Equations:

$$(x+a)y^2 = kx^2(x-3a),$$

$$y = \pm x\sqrt{k\frac{x-3a}{x+a}},$$

$$\rho = a\frac{\sin^2\varphi + 3k\cos^2\varphi}{\cos\varphi(k\cos^2\varphi - \sin^2\varphi)}, \quad a > 0, \quad k \neq 0$$

Basic geometric properties:
Unboundedness: present.
Asymptotes: straight line $x + a = 0$ is a vertical asymptote, straight lines $y = \pm x\sqrt{k}(k > 0)$ are inclined asymptotes.
Axial symmetry: straight lines

$$y = 0, \quad y = \pm\frac{1}{\sqrt{3}}x\left(k = \frac{1}{3}\right)$$

are axes of symmetry.
Central symmetry: none.
Connectedness: present (for $k < 0$), none (for $k > 0$).
Singular points: point $(0,0)$ is a node ($k < 0$), straight lines $y = \pm\sqrt{-3k}x$ are tangents at the node, or isolated point ($k > 0$).

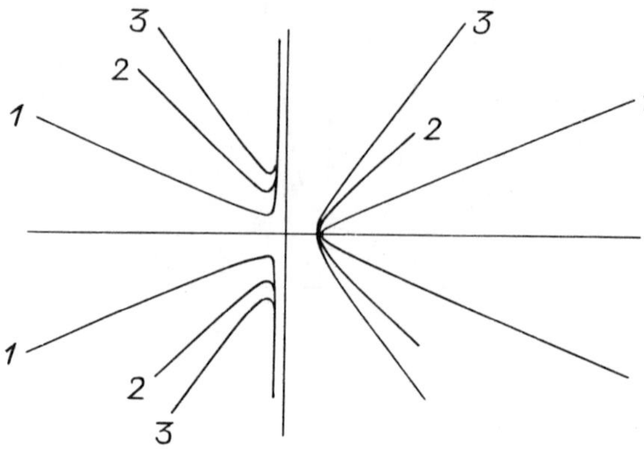

Figure 462. Trifolium

N	1	2	3
k	0.20	1.00	2.00
a	1.00	1.00	1.00

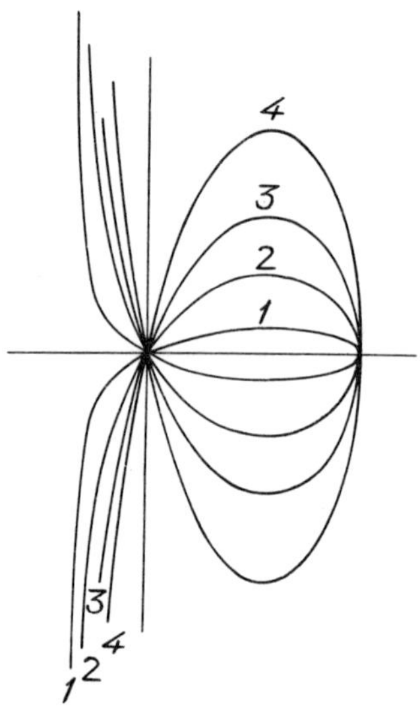

Figure 463. Trifolium

N	1	2	3	4
k	−0.10	−1.00	−3.00	−8.00
a	1.00	1.00	1.00	1.00

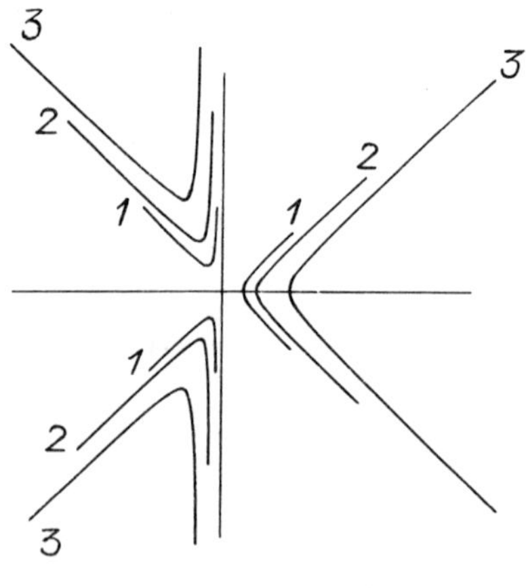

Figure 464. Trifolium

N	1	2	3
a	0.50	1.00	2.00
k	1.00	1.00	1.00

Self-intersection points: one, $(0,0)$ (for $k<0$); none (for $k>0$).
Special cases: Cartesian folium ($k=-1/3$), trisectrix of Maclaurin ($k=-1$), trisectrix of Longchamps ($k=1/3$), trifolium of Marie ($k>0$).
(Figures 462–466).

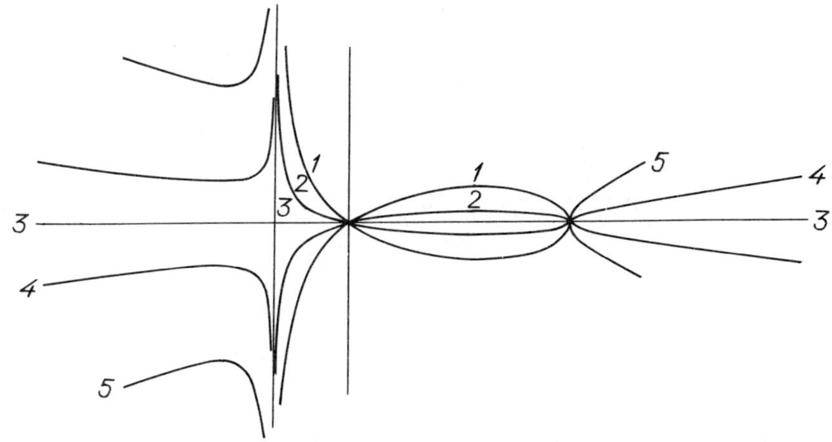

Figure 465. Trifolium

N	1	2	3	4	5
k	−0.20	−0.02	0.00	0.02	0.20
a	1.00	1.00	1.00	1.00	1.00

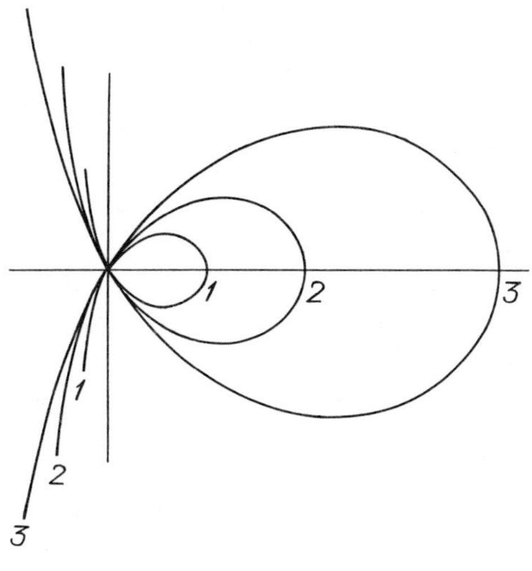

Figure 466. Trifolium

N	1	2	3
a	0.50	1.00	2.00
k	−1.00	−1.00	−1.00

TRIFOLIUM *is an unbounded connected (for $k < 0$) or unconnected (for $k > 0$) curve with a single singular point (node or isolated point), a single (for $k \neq 1/3$) axis of symmetry, a single vertical and two inclined (the last for $k > 0$ only) asymptotes.*

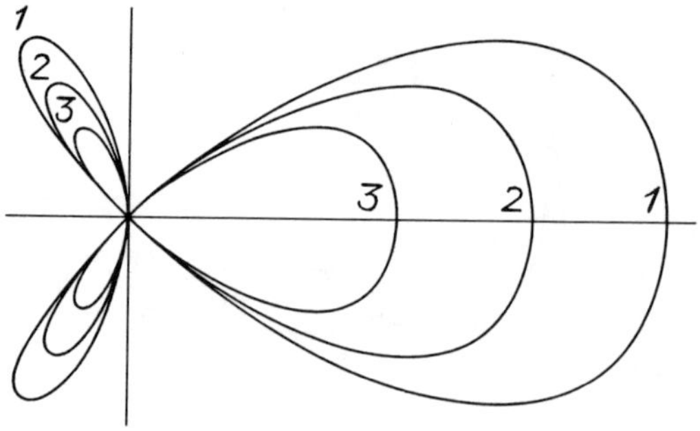

Figure 467. Trifolium of Cramer

N	1	2	3
a	1.00	0.75	0.50

Trifolium of Cramer
Equations:

$$x^4 + y^4 - 2ax(x^2 - y^2) = 0,$$

$$\rho = a\frac{2a\cos\varphi\cos 2\varphi}{\cos^4\varphi + \sin^4\varphi},$$

$$a > 0$$

Basic geometric properties:
Boundedness: present.
Closedness: present.
Axial symmetry: straight line $y = 0$ is an axis of symmetry.
Central symmetry: none.
Connectedness: present.
Singular points: point $(0, 0)$ is a triple point, straight lines $x = 0$, $x \pm y = 0$ are tangents at the point.
Self-intersection points: one, $(0, 0)$.
(Figure 467).
TRIFOLIUM OF CRAMER *is a bounded connected closed curve with a single singular point (triple point) and a single axis of symmetry.*

Trisectrix of Delanges
Equations:

$$(a^2 - y^2)(x^2 + y^2) - \frac{a^4}{4} = 0,$$

$$x = \pm\frac{\frac{a^2}{2} - y^2}{\sqrt{a^2 - y^2}},$$

$$a > 0$$

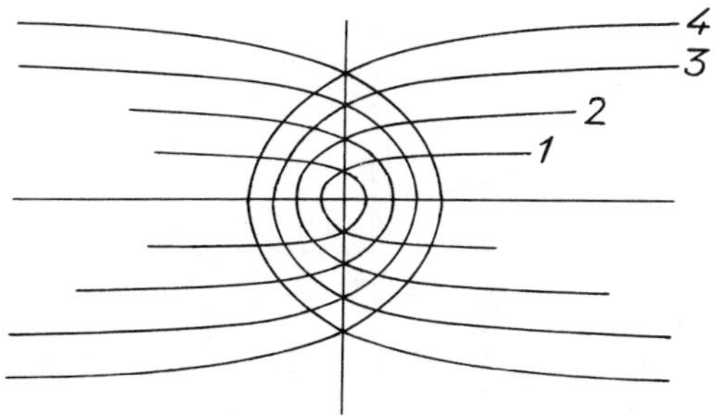

Figure 468. Trisectrix of Delanges

N	1	2	3	4
a	0.25	0.50	0.75	1.00

Basic geometric properties:
Unboundedness: present.
Asymptotes: straight lines $y = \pm a$ are horizontal asymptotes.
Axial symmetry: straight lines $x = 0,\ y = 0$ are axes of symmetry.
Central symmetry: point $(0,0)$ is a center of symmetry.
Connectedness: present.
Singular points: point $(0, a/\sqrt{2})$ is a node, straight lines

$$y = \pm\frac{1}{2}\left(x - \frac{a}{\sqrt{2}}\right)$$

are tangents at the node; point $(0, -a/\sqrt{2})$ is a node, straight lines

$$y = \pm\frac{1}{2}\left(x + \frac{a}{\sqrt{2}}\right)$$

are tangents at the node.
Self-intersection points: two, $(0, \pm a/\sqrt{2})$.
Area of domain enclosed between branches of the curve: a^2.
(Figure 468).
TRISECTRIX OF DELANGES *is an unbounded connected central-symmetric curve with two singular points (nodes), two horizontal asymptotes and two axes of symmetry.*

Trisectrix of Longchamps
Equations:

$$x(x^2 - 3y^2) - r(x^2 + y^2) = 0,$$

$$y = \pm x\sqrt{\frac{x - r}{3x + r}}, \quad \rho = \frac{r}{\cos 3\varphi}, \quad r > 0$$

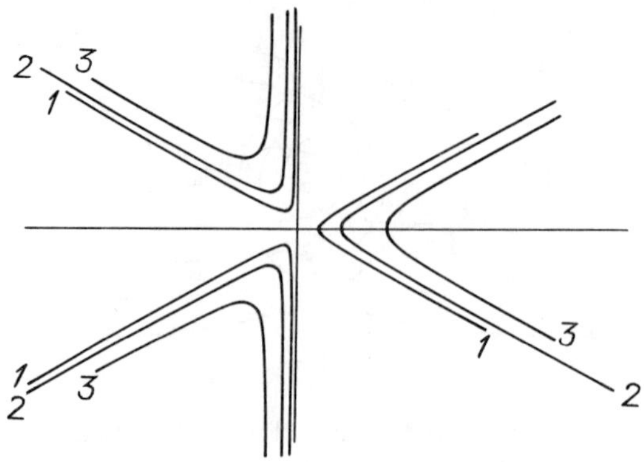

Figure 469.
Trisectrix of Longchamps

N	1	2	3
r	0.50	1.00	2.00

Additional equations:

$$x(x^2 - 3y^2) + r(x^2 + y^2) = 0,$$

$$\rho = -\frac{r}{\cos 3\varphi},$$

Basic geometric properties:
Unboundedness: present.
Asymptotes: straight line $3x + r = 0$ is a vertical asymptote, straight lines $3x \pm 3\sqrt{3} - 2r = 0$ are inclined asymptotes.
Axial symmetry: straight lines $y = 0$, $\sqrt{3}x \pm y = 0$ are axes of symmetry.
Central symmetry: none.
Connectedness: none.
Singular points: point $(0, 0)$ is an isolated point.
Self-intersection points: none.
Characterized points: points

$$(r, 0), \quad \left(-\frac{r}{2}, \pm r\frac{\sqrt{3}}{2}\right)$$

are vertices.
(Figure 469).
TRISECTRIX OF LONGCHAMPS *is an unbounded unconnected (consisting of three unbounded branches and an isolated point) curve with a single singular point (isolated point), three axes of symmetry, a single vertical asymptote and two inclined asymptotes.*

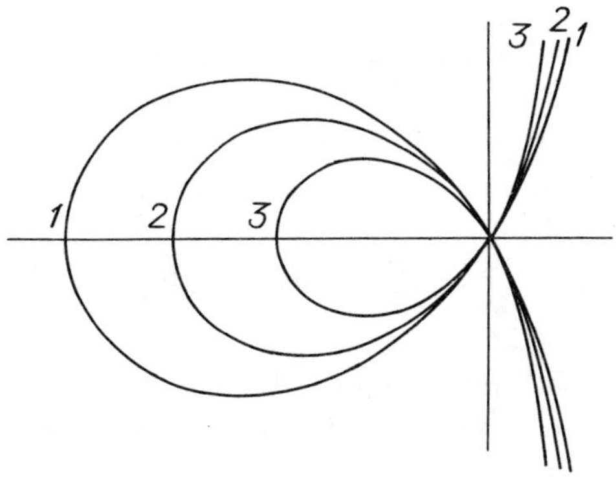

Figure 470.
Trisectrix of Maclaurin

N	1	2	3
a	1.00	0.75	0.50

Trisectrix of Maclaurin
Equations:

$$x(x^2 + y^2) = a(y^2 - 3x^2) = 0,$$

$$y = \pm x\sqrt{\frac{3a + x}{a - x}},$$

$$\rho = a\frac{1 - 4cos^2\varphi}{\cos\varphi},$$

$$x = a(\sin^2 t - 3\cos^2 t), \quad y = a(\sin^2 t - 3\cos^2 t)\tan t,$$

$$x = a\frac{t^3 - 3}{t^2 + 1}, \quad y = at\frac{t^3 - 3}{t^2 + 1}, \quad a > 0$$

Additional equations:

$$(x - 3a)(x^2 + y^2) + 4a^2 = 0,$$

$$\rho = \frac{a}{\cos\frac{\varphi}{3}},$$

$$x(x^2 + y^2) = a(3x^2 - y^2),$$

$$x = a\frac{\sin 3t}{\sin t}, \quad y = a\frac{\sin 3t}{\cos t}$$

Basic geometric properties:
Unboundedness: present.
Asymptotes: straight line $3x - a = 0$ is a vertical asymptote.
Axial symmetry: straight line $y = 0$ is an axis of symmetry.
Central symmetry: none.

Connectedness: present.
Singular points: point $(0,0)$ is a node, straight lines $\sqrt{3}x \pm y = 0$ are tangents at the node.
Self-intersection points: one, $(0,0)$.
Characterized points: points

$$\left(-a\sqrt{3}, \pm a\sqrt{6\sqrt{3}-9}\right)$$

are the loop points located at maximal distance from x-axis; point $(-3a,0)$ is a loop point located at maximal distance from origin $(0,0)$ (vertex).
Area of the loop: $3\sqrt{3}a^2$.
(Figure 470).
TRISECTRIX OF MACLAURIN *is an unbounded connected curve with a single singular point (node), a single axis of symmetry and a single vertical asymptote.*

Trochoidal rose
Equations:

$$x = (R + mR)\cos mt - (R + mR)\cos(t + mt),$$

$$y = (R + mR)\sin mt + (R + mR)\sin(t + mt),$$

$$x = 2(R+mR)\sin\left(mt + \frac{t}{2}\right)\sin\frac{t}{2}, \quad y = -2(R+mR)\cos\left(mt + \frac{t}{2}\right)\sin\frac{t}{2},$$

$$R > 0, \quad m = \pm\frac{p}{q}, \quad p \in \mathbf{N}, \quad q \in \mathbf{N}, \quad \text{G.C.D.}(p,q) = 1, \quad m \neq -1$$

Additional equations:

$$\rho = 2(R + mR)\sin\frac{\varphi}{2m+1}, \quad m \neq -1, -\frac{1}{2}, \quad \rho = a\sin\mu\varphi$$

Basic geometric properties:
Boundedness: present.
Closedness: present.
Axial symmetry: straight lines

$$x\sin\frac{\pi}{q}k = y\cos\frac{\pi}{q}k, \quad k = 0, 1, \ldots, q-1$$

are axes of symmetry.
Central symmetry: point $(0,0)$ is a center of symmetry (q is even).
Connectedness: present.

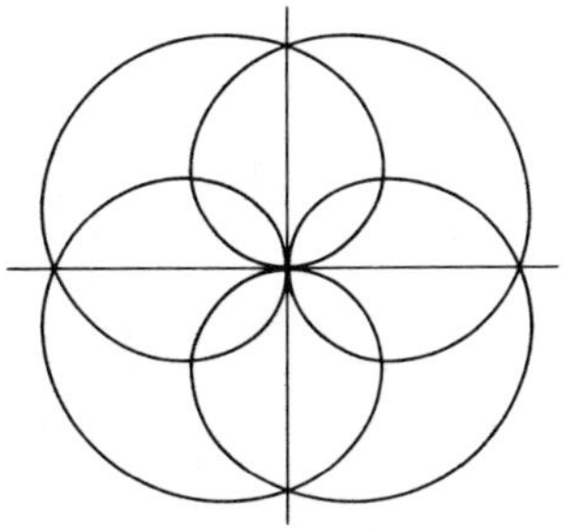

Figure 471. Trochoidal rose
$m = 1/4, \quad R = 1.00$

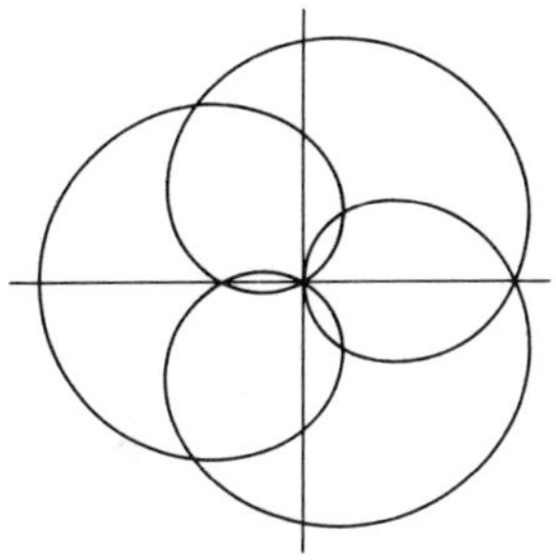

Figure 472. Trochoidal rose
$m = 1/3, \quad R = 1.00$

Singular points: point $(0, 0)$ is a multiple point; trochoidal rose has nodes of the following form:

$$\left(\rho_l \cos\frac{2k+1}{q}\pi, \rho_l \sin\frac{2k+1}{q}\pi\right), \quad l = 1, 3, \ldots,$$

$$\left(\rho_n \cos\frac{2k}{q}\pi, \rho_n \sin\frac{2k}{q}\pi\right), \quad n = 2, 4, \ldots,$$

$$k = 0, 1, \ldots, q-1$$

Number of self-intersection points is finite and depends on values of parameter m.
Special cases: epitrochoidal rose ($m > 0R$), hypotrochoidal rose ($-1/2 < m < 0$).
(Figures 471–484).
TROCHOIDAL ROSE *is a bounded connected closed central-symmetric (for even q only), a single multiple point (for $q \geq 2$) and q axes of symmetry; number of nodes is a number divisible by q.*
In case of irrational m the trochoidal rose is a transcendent unclosed curve with an infinitely large number of nodes.

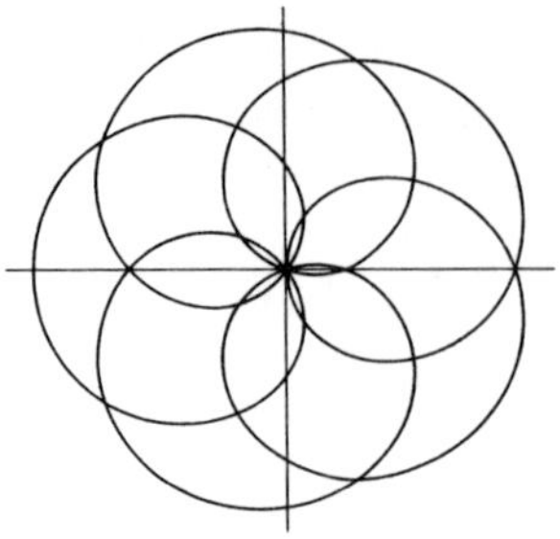

Figure 473. Trochoidal rose
$m = 1/5, \quad R = 1.00$

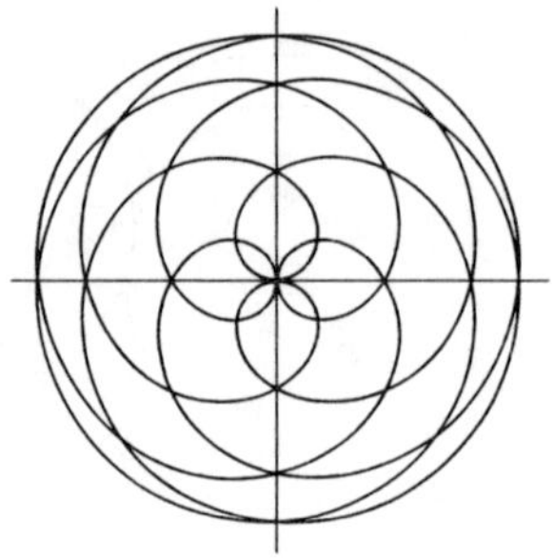

Figure 474. Trochoidal rose
$m = 5/4, \quad R = 1.00$

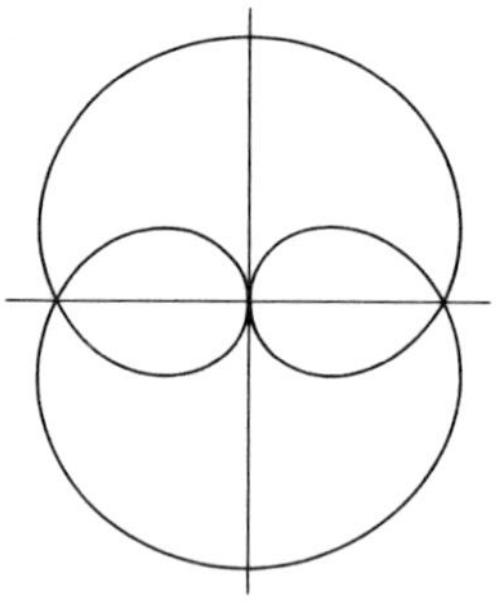

Figure 475. Trochoidal rose
$m = 1/2, \quad R = 1.00$

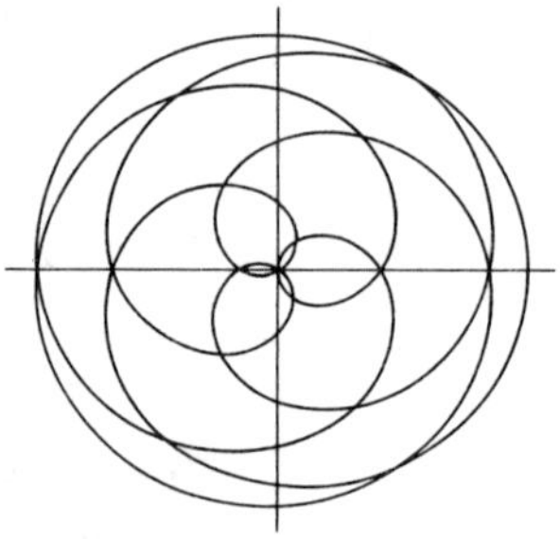

Figure 476. Trochoidal rose
$m = 4/3, \quad R = 1.00$

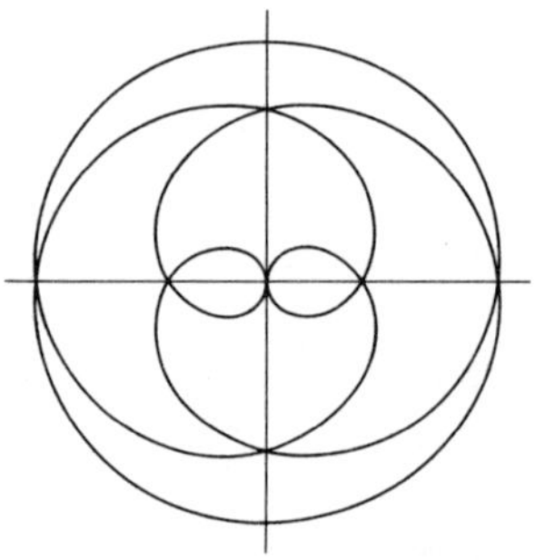

Figure 477. Trochoidal rose
$m = 3/2, \quad R = 1.00$

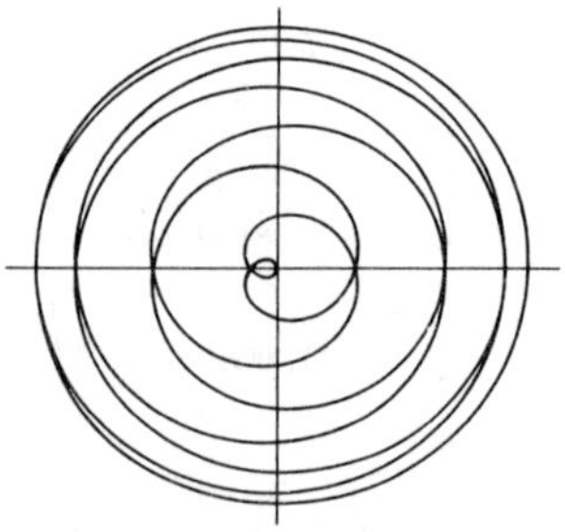

Figure 478. Trochoidal rose
$m = 7, \quad R = 1.00$

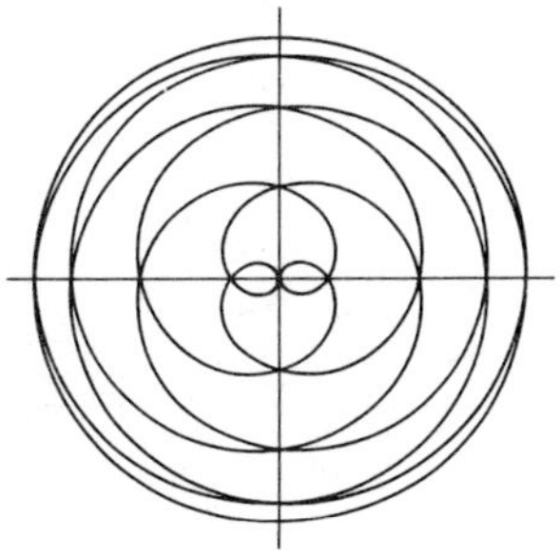

Figure 479. Trochoidal rose
$m = 7/2, \quad R = 1.00$

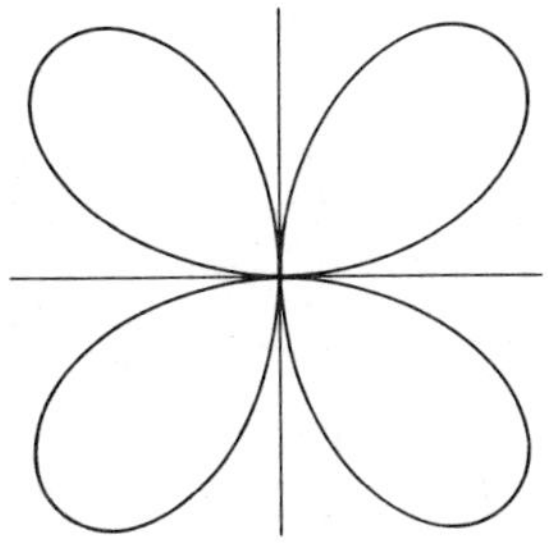

Figure 480. Trochoidal rose
$m = -1/4, \quad R = 1.00$

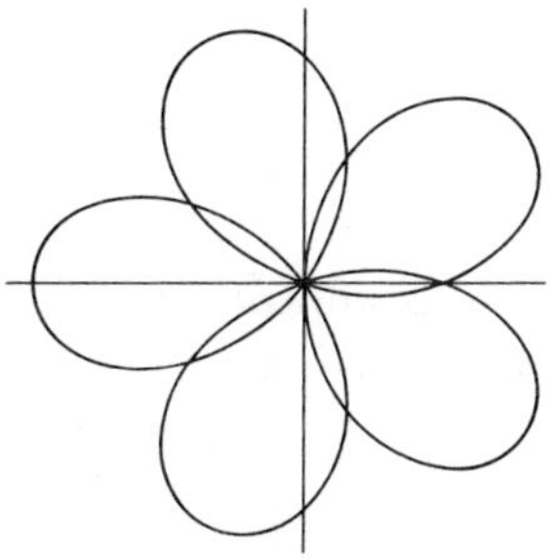

Figure 481. Trochoidal rose
$m = -1/5, \quad R = 1.00$

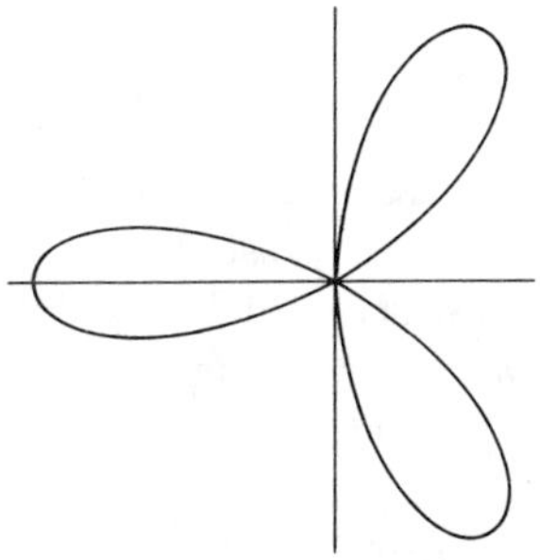

Figure 482. Trochoidal rose
$m = -1/3, \quad R = 1.00$

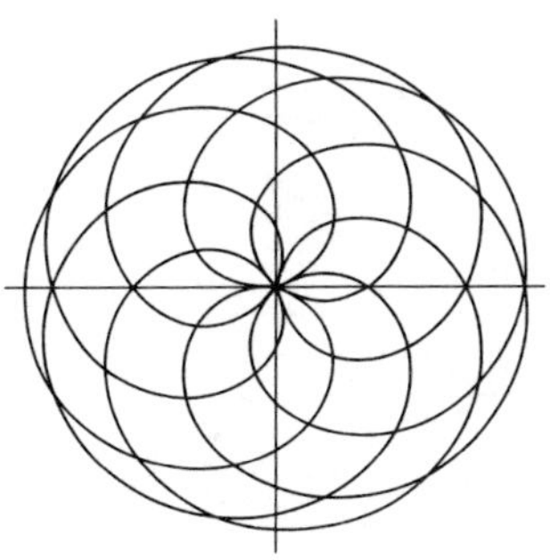

Figure 483. Trochoidal rose
$m = 3/7, \quad R = 1.00$

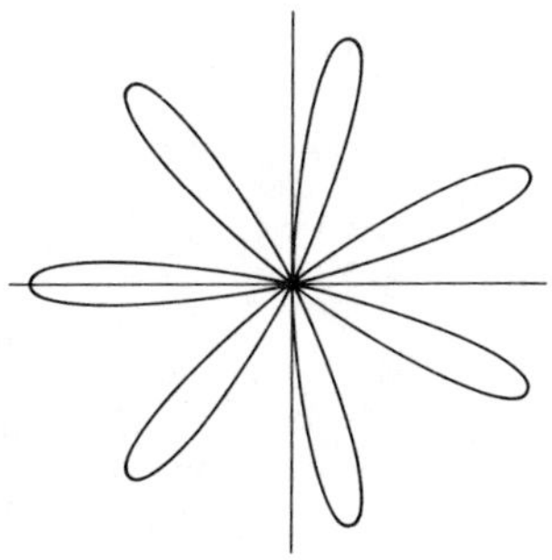

Figure 484. Trochoidal rose
$m = -3/7, \quad R = 1.00$

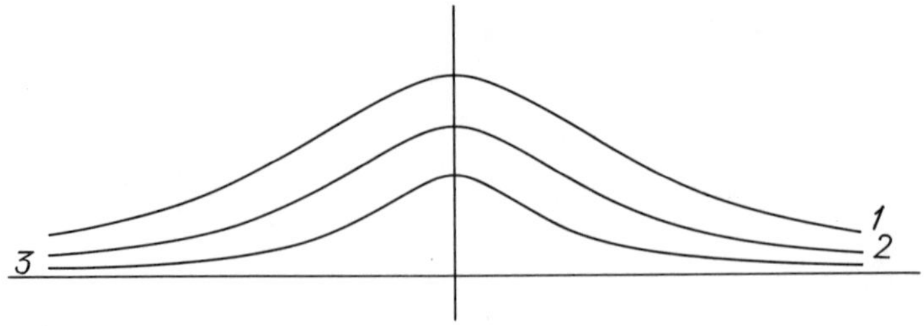

Figure 485. Versiera

N	1	2	3
a	1.00	0.75	0.50

2.2.19 V

Versiera (Agnesi's versiera, witch, witch of Agnesi)
Equation:

$$(x^2 + a^2)y = a^3, \quad a > 0$$

Basic geometric properties:
Unboundedness: present.
Asymptotes: straight line $y = 0$ is a horizontal asymptote.
Axial symmetry: straight line $x = 0$ is an axis of symmetry.
Central symmetry: none.
Connectedness: present.
Singular points: none.
Self-intersection points: none.
Characterized points: point $(0, a)$ is located at maximal distance from x-axis; points $(\pm a/\sqrt{3}, 3a/4)$ are inflection points.
Area of domain enclosed by the curve and the asymptote: πa^2.
(Figure 485).
VERSIERA *is an unbounded connected curve free of singular points with a single axis of symmetry and a single horizontal asymptote.*

Virtual parabola (A)
Equations:

$$d^2(y^2 - x^2) = y^4,$$

$$y = \sqrt{\frac{d^2}{4} + \frac{da}{2}} + \sqrt{\frac{d^2}{4} - \frac{da}{2}}, \quad d > 0$$

Basic geometric properties:
Boundedness: present.
Closedness: present.
Axial symmetry: straight lines $x = 0$ and $y = 0$ are axes of symmetry.
Central symmetry: point $(0, 0)$ is a center of symmetry.

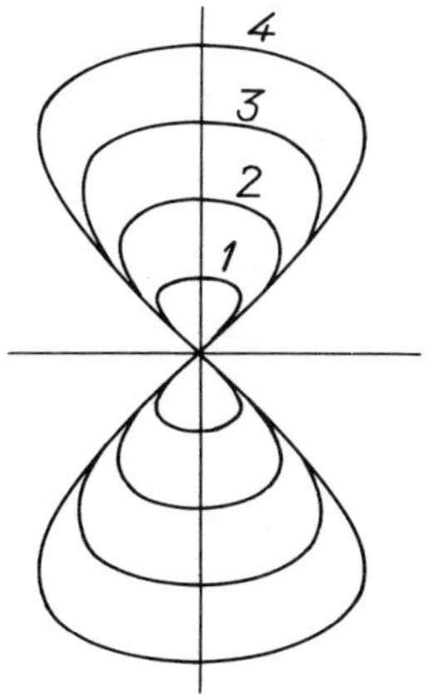

Figure 486. Virtual parabola (A)

N	1	2	3	4
a	0.50	1.00	1.50	2.00

Connectedness: present.
Singular points: point $(0,0)$ is a node.
Self-intersection points: one, $(0,0)$.
(Figure 486).
VIRTUAL PARABOLA (A) *is a bounded connected closed central-symmetric curve with a single singular point (node) and two axes of symmetry.*

Virtual parabola (B)
Equations:

$$x = 2a\cos^2\varphi, \quad y = 2(a\cos\varphi + b\sin\varphi),$$

$$y = \sqrt{2ax} + \sqrt{ab^2 - \frac{2b^2}{a}x},$$

$$a > 0, \quad b > 0$$

Basic geometric properties:
Boundedness: present.
Closedness: present.
Axial symmetry: straight line $y = 0$ is an axis of symmetry, straight line $x = a$ is an axis of symmetry (for $a = b$ only).
Central symmetry: point $(a, 0)$ is a center of symmetry (for $a = b$ only).
Connectedness: present.
Singular points: point

$$\left(\frac{2ab^2}{a^2+b^2}, 0\right)$$

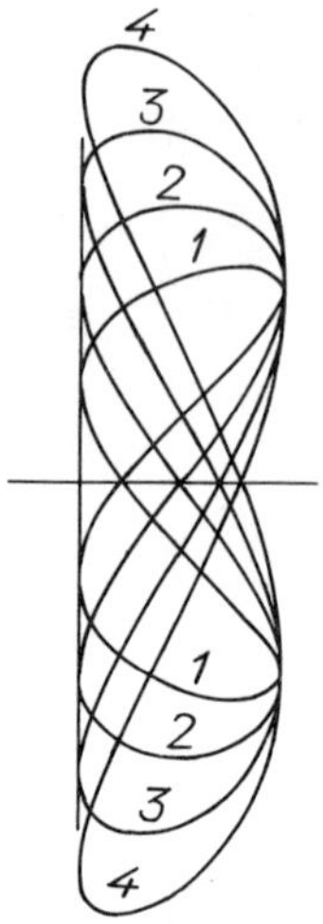

Figure 487. Virtual parabola (B)

N	1	2	3	4
a	1.00	1.00	1.00	1.00
b	0.50	1.00	1.50	2.00

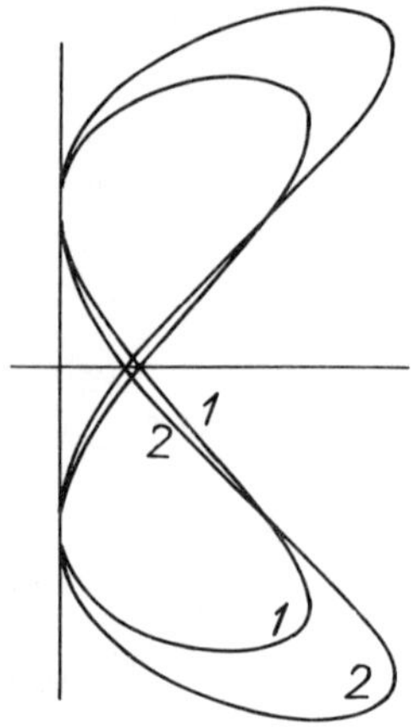

Figure 488. Virtual parabola (B)

N	1	2
a	0.75	1.00
b	0.50	0.50

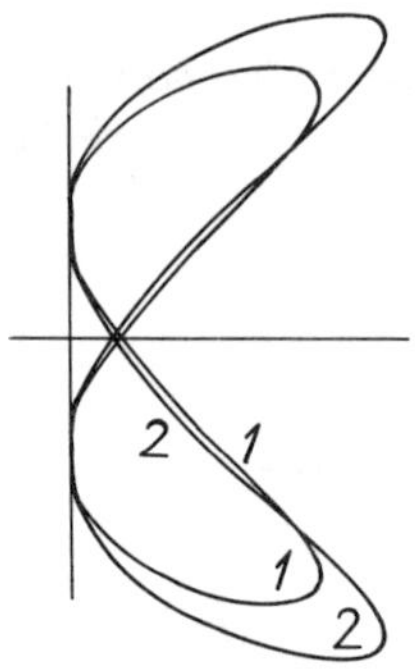

Figure 489. Virtual parabola (B)

N	1	2
a	1.00	1.25
b	0.50	0.50

is a node, straight lines

$$y = \pm \frac{\sqrt{a^2+b^2}(a^2+b^2)}{2a^2b}\left(a - \frac{2ab^2}{a^2+b^2}\right)$$

are tangents at the point.

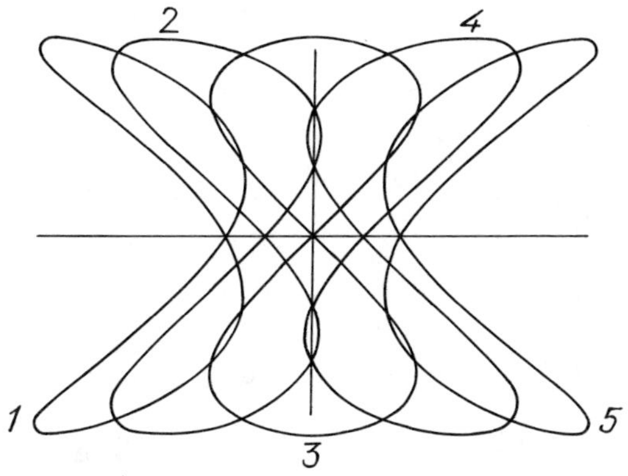

Figure 490. Virtual parabola (C)

N	1	2	3	4	5
α	−1.00	−0.50	0.00	0.50	1.00
r	1.00	1.00	1.00	1.00	1.00

Self-intersection points: one,

$$\left(\frac{2ab^2}{a^2+b^2}, 0\right)$$

(Figures 487–489).

VIRTUAL PARABOLA (B) *is a bounded connected closed curve with a single singular point (node) and a single axis of symmetry (for $a \neq b$).*

Virtual parabola (C)

Equations:

$$x = r\sin\alpha + 2r\sin\varphi\cos(\alpha-\varphi), \quad y = 2r\sin\varphi,$$

$$\frac{y}{r} = \sqrt{\cos^2\alpha - \frac{1}{r}(1-\sin\alpha)(x-r\sin\alpha)}$$
$$+\sqrt{\cos^2\alpha - \frac{1}{r}(1+\sin\alpha)(x-r\sin\alpha)},$$
$$r > 0, \quad -\frac{\pi}{2} < \alpha < \frac{\pi}{2}$$

Basic geometric properties:

Boundedness: present.

Closedness: present.

Axial symmetry: straight line $y = 0$ is an axis of symmetry; straight line $x = 0$ is an axis of symmetry (for $\alpha = 0$).

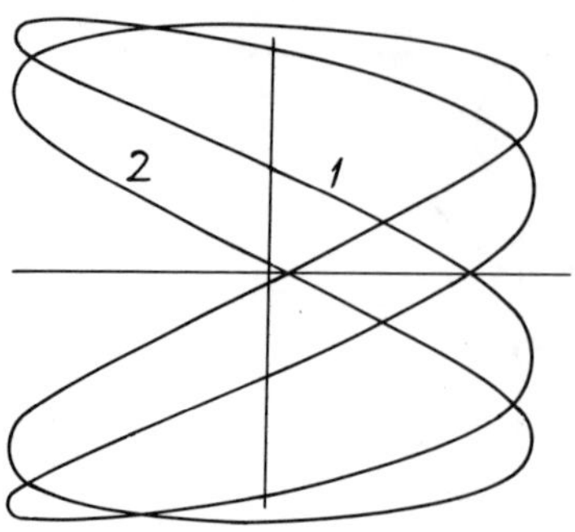

Figure 491.
Virtual parabola (D)

N	1	2
α	0.70	1.50
r	1.00	1.00

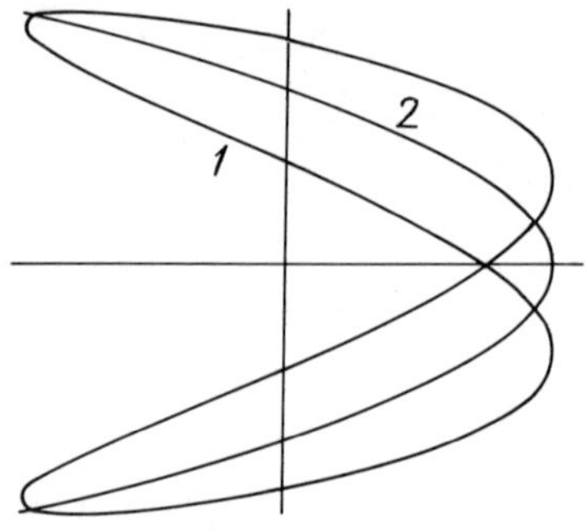

Figure 492.
Virtual parabola (D)

N	1	2
α	0.70	0.00
r	1.00	1.00

Central symmetry: point $(0, 0)$ is a center of symmetry.
Connectedness: present.
Singular points: point $(r \sin\alpha, 0)$ is a node, straight lines

$$y = \pm \frac{1}{\cos\alpha}(x - r\sin\alpha)$$

are tangents at the node.
Self-intersection points: one, $(r, \sin\alpha, 0)$.
(Figure 490).
VIRTUAL PARABOLA (C) *is a bounded connected closed curve with a single singular point (node) and a single axis of symmetry* $(\alpha \neq 0)$.

Virtual parabola (D)
Equations:

$$x = \cos(\alpha - 2\varphi), \quad y = r\sin\varphi,$$

$$\sqrt{2}\frac{y}{r} = \sin\frac{\alpha}{2}\sqrt{r^2 + rx} + \cos\frac{\alpha}{2}\sqrt{r^2 - rx},$$

$$r > 0, \quad 0 < \alpha < \frac{\pi}{2}$$

Basic geometric properties:
Boundedness: present.
Closedness: present.
Axial symmetry: straight line $y = 0$ is an axis of symmetry.
Central symmetry: none.

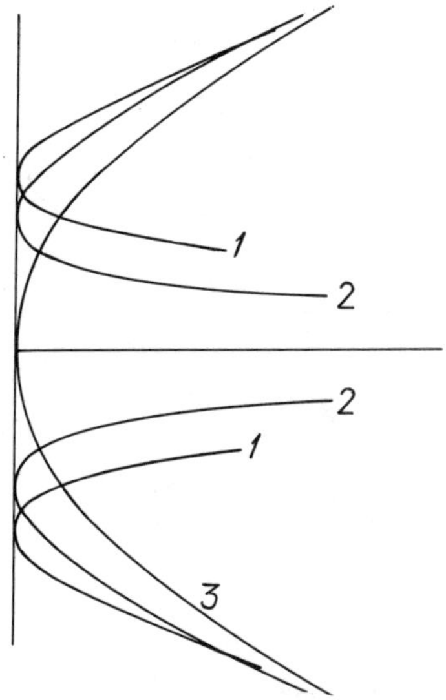

Figure 493.
Virtual parabola (E)

N	1	2	3
r	1.00	1.00	1.00
R	0.50	0.75	1.00

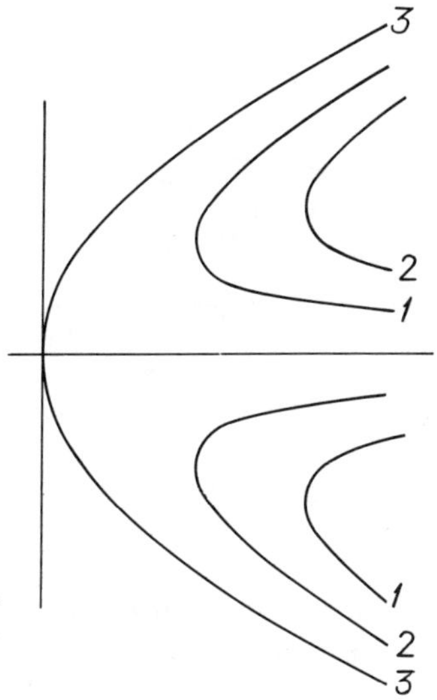

Figure 494.
Virtual parabola (E)

N	1	2	3
r	0.50	0.75	1.00
R	1.00	1.00	1.00

Connectedness: present.
Singular points: point $(r, \cos\alpha, 0)$ is a node, straight lines

$$y = \pm\frac{x - r\cos\alpha}{2\sin\alpha}$$

are tangents at the node.
Self-intersection points: one, $(r\cos\alpha, 0)$.
Special cases: parabola $(\alpha = 0)$.
(Figures 491–492).
VIRTUAL PARABOLA (D) *is a bounded connected closed curve with a single singular point (node) and a single axis of symmetry.*

Virtual parabola (E)
Equations:

$$x = 2R\cos^2\varphi, \quad y = R\cos\varphi + \sqrt{r^2 - R^2\sin^2\varphi},$$

$$y = \sqrt{\frac{Rx}{2}} + \sqrt{\frac{Rx}{2} + r^2 - R^2},$$

$$r > 0, \quad R > 0, \quad r \neq R$$

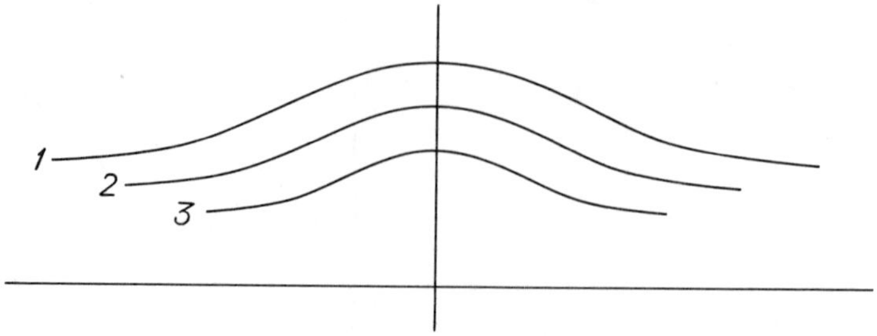

Figure 495. Visiera

N	1	2	3
a	1.00	0.80	0.60

Basic geometric properties:
Unboundedness: present.
Asymptotes: straight line $y = 0$ is a horizontal asymptote.
Axial symmetry: straight line $y = 0$ is an axis of symmetry.
Central symmetry: none.
Connectedness: none.
Singular points: none.
Self-intersection points: none.
Special cases: parabola ($r = R$).
(Figures 493–494).
VIRTUAL PARABOLA (E) *is an unbounded unconnected (consisting of two unbounded branches) curve free of singular points with a single axis of symmetry and a single horizontal asymptote.*

Visiera (anguinea, cissoidal of cubic, visiera Peano)
Equations:

$$(2y - a)(x^2 + y^2) = ay^2,$$

$$\rho = \frac{a}{2}\left(\sin\varphi + \frac{1}{\sin\varphi}\right),$$

$$x = at\frac{t^2+2}{2(t^2+1)}, \quad y = a\frac{t^2+1}{2(t^2+1)}, \quad a > 0$$

Basic geometric properties:
Unboundedness: present.
Asymptotes: straight line $2y - a = 0$ is a horizontal asymptote.
Axial symmetry: straight line $x = 0$ is an axis of symmetry.
Central symmetry: none.
Connectedness: none.

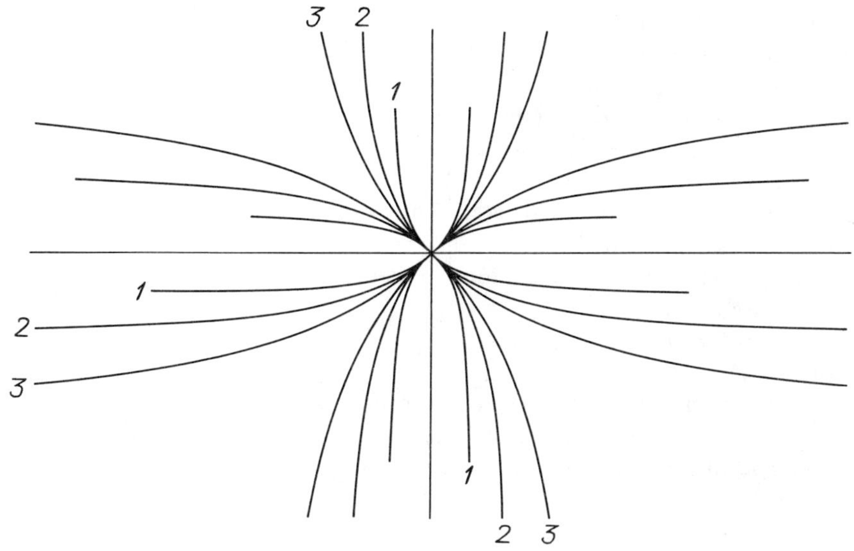

Figure 496. Windmill

N	1	2	3
a	1.00	2.00	4.00

Singular points: point $(0,0)$ is an isolated point.
Self-intersection points: none.
Characterized points: point $(0,a)$ is located at maximal distance from x-axis; points

$$\left(\pm\frac{4a}{5}\sqrt{\frac{2}{3}},\frac{4a}{5})\right)$$

are inflection points.
Area of domain enclosed by the curve and the asymptote: $5\pi a^2$.
(Figure 495).
VISIERA *is an unbounded unconnected curve with a single singular point (isolated point), a single horizontal asymptote and a single axis of symmetry.*

2.2.20 W

Windmill
Equations:

$$4x^2y^2(x^2+y^2)=a^2(x^2-y^2)^2,$$

$$\rho=a\cot 2\varphi,$$

$$a>0$$

Basic geometric properties:
Unboundedness: present.
Asymptotes: straight lines $2x \pm a = 0$ are vertical asymptotes, straight lines $2y \pm a = 0$ are horizontal asymptotes.
Axial symmetry: straight lines $x = 0,\ y = 0,\ x - y = 0,\ x + y = 0$ are axes of symmetry.
Central symmetry: point $(0,0)$ is a center of symmetry.
Connectedness: present.
Singular points: point $(0,0)$ is a four-multiple point, straight lines $x \pm y = 0$ are tangents at the point.
Self-intersection points: one, $(0,0)$.
(Figure 496).
WINDMILL *is an unbounded connected central-symmetric curve with a single singular point (degenerate four-multiple point), four axes of symmetry, two vertical and two horizontal asymptotes.*

2.3 Geometric indices for Atlas of curves

In this section the curves from the Atlas are classified into groups, each being determined by definite geometric property; in particular, the property defines a name of the group.

2.3.1 Algebraic curves

ALGEBRAIC CURVES OF THE SECOND ORDER

Hyperbola

$$\frac{x^2}{a^2} - \frac{y^2}{b^2} = 1$$

Circle

$$(x-a)^2 + (y-b)^2 = R^2$$

Parabola

$$y^2 = 2px$$

Ellipse

$$\frac{x^2}{a^2} + \frac{y^2}{b^2} = 1$$

ALGEBRAIC CURVES OF THE THIRD ORDER

Agnesiana

$$y(b^2 + x^2) = ab^2$$

Cartesian folium

$$x^3 + y^3 = 3axy$$

Cissoid

$$x(x^2 + y^2) = 2ay^2$$

Cissoid of the second order curves

$$(x - k)(y^2 - qx^2) + 2px^2 = 0$$

Conchoid of Slüse (+)

$$a(x - a)(x^2 + y^2) = k^2x^2$$

Conchoid of Slüse (−)

$$a(x - a)(x^2 + y^2) = -k^2x^2$$

Cubic of Tschirnhausen

$$2(2p + x)^3 = 27p(x^2 + y^2)$$

Curtate and prolate cissoids

$$(a - x)y^2 = (2r - a + x)x^2$$

Curve of Cramer

$$x(x^2 + y^2) = -(r + l)x^2 + (r - l)y^2$$

Curve of Rolle

$$xy^2 = a(x + y)^2$$

Duplicatrix cubic

$$x^3 = l(x^2 + y^2)$$

First pearl

$$x^2(a - x) = b^2y$$

Oblique cissoid

$$(x^2 + y^2)(x \cos\alpha + y \sin\alpha) = 2ay^2$$

Oblique parabolic folium

$$x^3 = a(x^2 - y^2) + bxy$$

Oblique strophoid

$$x(x^2 + y^2) = a((x^2 - y^2)\sin\alpha + 2xy\cos\alpha)$$

Oblique versiera

$$(x^2 + y^2)(x \cos\alpha + y \sin\alpha - 4r\cos^2\alpha) + 2ry^2 = 0$$

Ophiuride

$$x(x^2 + y^2) = y(cy - bx)$$

Panstrophoid

$$x(x^2 + y^2) + g(x^2 - y^2) + k(x + g) = 0$$

Parabolic folium

$$x^3 = a(x^2 - y^2)$$

Pseudoversiera

$$(x^2 + a^2)y = 2a^3$$

Serpentine of Newton

$$y(x^2 + a^2) = acx$$

Strophoid

$$(x^2 + y^2)(x - 2a) + a^2 x = 0$$

Trident of Newton

$$xy = ax^3 + bx^2 + cx + d$$

Trifolium

$$(x + a)y^2 = kx^2(x - 3a)$$

Trisectrix of Longchamps

$$x(x^2 - 3y^2) = r(x^2 + y^2)$$

Trisectrix of Maclaurin

$$x(x^2 + y^2) = a(y^2 - 3x^2)$$

Versiera

$$(x^2 + y^2)y = a^3$$

Visiera

$$(2y - a)(x^2 + y^2) = ay^2$$

ALGEBRAIC CURVES OF THE FOURTH ORDER

Antiversiera

$$x^4 - 2rx^3 + 4r^2y^2 = 0$$

Besace

$$\begin{aligned} x^4 - 4bxy + 4(a^2 + b^2)y^2 - 2(a^2 - b^2 - r^2)x^2 \\ -4b(a^2 + b^2 - r^2)y + (a^2 + b^2 - r^2)^2 = 0 \end{aligned}$$

Bicorn curve

$$(x^2 + 2ay - a^2)^2 = y^2(a^2 - x^2)$$

Bifolium

$$(x^2+y^2)^2+4rxy^2=0$$

Biquadratic cubic (parabola)

$$(x+y)^4=ax^3$$

Bullet-nose curve

$$a^2y^2-b^2x^2=x^2y^2$$

Capricornoid

$$g^2x^2(x^2+y^2)=c^2(gy-x^2-y^2)^2$$

Cardioid

$$(x^2+y^2+2rx)^2=4r^2(x^2+y^2)$$

Cartesian oval

$$m\sqrt{x^2+y^2}+n\sqrt{(x-d)^2+y^2}=c$$

Cassinian oval

$$(x^2+y^2)^2-2c^2(x^2-y^2)=a^4-c^4$$

Cissoid of the fourth order

$$y^2(k^2-(a+x)^2)=(x^2+k^2-a^2)^2$$

Conchal

$$(x+a)^2((x-a)^2+y^2)a^2k^2$$

Conchoid of Dürer

$$(xy+b^2-y^2)^2=(x+y-a)^2(b^2-y^2)$$

Conchoid of Külp

$$x^2y^2=a^2(a^2-x^2)$$

Conchoid of Nicomedes

$$(x^2+y^2)(y-a)^2-l^2y^2=0$$

Cruciform curve

$$\frac{a^2}{x^2}+\frac{b^2}{y^2}=1$$

Curve of Clairaut

$$y^4=a^2(x^2+y^2)$$

Curve of Gutschoven

$$l^2x^2(x^2+y^2)=a^2(x^2+y^2-ly)^2$$

Curve of Jeřabek

$$r^2(x^2+y^2-ax)^2=a^2(x^2+y^2)(x-a)^2$$

Curve of Mascheroni

$$x^2(x^2 + y^2) = (lx + ay)^2$$

Curve of Slüse

$$x^4 - r^2x^2 + a^2y^2 = 0$$

Double-heart curve

$$(x^2 + y^2)^2 - 6axy^2 - ax^3 + a^2x^2 = 0$$

Elliptic lemniscate

$$(x^2 + y^2)^2 = a^2x^2 + b^2y^2$$

Folium

$$(x^2 + y^2)^2 = 4rx^3$$

Generalized kampyle (A)

$$\frac{x^2}{a^2} + \frac{y^2}{b^2} = \frac{x^4}{a^4}$$

Hyperbolic lemniscate

$$(x^2 + y^2)^2 = a^2x^2 - b^2y^2$$

Kampyle

$$a^2(x^2 + y^2) = \frac{a^4}{b^4}x^4$$

Lemniscate of Bernoulli

$$(x^2 + y^2)^2 = 2a^2(x^2 - y^2)$$

Lemniscate of Booth

$$(x^2 + y^2)^2 - (2m^2 + c)x^2 + (2m^2 - c)y^2 = 0$$

Lemniscate of Gerono

$$x^4 = r^2(x^2 - y^2)$$

Lemniscate of Montferrier

$$b^2y^2 = x^2(a^2 - x^2)$$

Limaçon

$$(x^2 + y^2 - 2ax)^2 = l^2(x^2 + y^2)$$

Oblique bifolium

$$(x^2 + y^2)^2 = x^2(ax + by)^2$$

Oblique cappa

$$(x^2 + y^2)y^2 = a^2(x \sin\alpha + y \cos\alpha)^2$$

Oblique conchoid

$$x = a\cot t + l\cos(\alpha + t), \quad y = -\sin(\alpha + t)$$

Oblique pancappa

$$(y-a)^2(x^2+y^2) = (bx+cy)^2$$

Oblique trifolium

$$\frac{(x^2+y^2)^2}{2a} = x(x^2+y^2) + (x^2-3y^2)\cos 2\alpha + y(x^2-3y^2)\sin 2\alpha$$

Orthoconchoid

$$(y^2+ay-l^2)^2 = x^2(l^2-y^2)$$

Pancappa

$$(y-a)^2(x^2+y^2) = b^2x^2$$

Pear

$$y^2(x-r)^2 - 2r^3y + r^4 = 0$$

Pearformed quartic

$$x^4 - ax^3 + b^2y^2 = 0$$

Pearl of the fourth order (A)

$$ay^3 - y^4 = a^2x^2$$

Pearl of the fourth order (B)

$$ay^3 - y^4 = x^4$$

Polyzomal curve of Bernoulli

$$y^4 - 6a^2y^2 + 4x^2y^2 + a^4 = 0$$

Quartic of Bonnet

$$x^4 - ax^3 + a^2y^2 = 0$$

Sand-glass curve

$$a^2(y^2-b^2) - 4b^2x^2(y^2+b^2) = 0$$

Polyod of polar axis

$$(x^2+y^2)x^2 = (x^2+y^2-2m^2)^2$$

Polyod of straight line parallel to polar axis

$$(x^2+y^2)((k-y)^2-4m^2) + 4km^2y + 4m^4$$

Polyod of straight line perpendicular to polar axis

$$y^2(x^2+y^2-k^2)=(x^2+y^2-kx-2m^2)^2$$

Right trifolium

$$(x^2+y^2)^2=ax(x^2-y^2)$$

Spiril curve

$$(x^2+y^2+p^2+d^2-R^2)^2=4d^2(x^2+p^2)$$

Three-cuspid hypocycloid

$$x=2r\cos t+r\cos 2t,$$

$$y=2r\sin t-r\sin 2t$$

Three-leafed rose

$$(x^2+y^2)^2+rx(3y^2-x^2)=0$$

Top

$$x^4-ax^3+\frac{a^2}{4}y^2=0$$

Trifolium of Cramer

$$x^4+y^4-2ax(x^2-y^2)=0$$

Trisectrix of Delanges

$$(a^2-y^2)(x^2+y^2)-\frac{a^4}{4}=0$$

Virtual parabola (A)

$$d^2(y^2-x^2)=y^4$$

Virtual parabola (B)

$$x=2a\cos^2\varphi,\quad y=2(a\cos\varphi+b\sin\varphi)$$

Virtual parabola (C)

$$x=r\sin\alpha+2r\sin\varphi\cos(\alpha-\varphi),\quad y=2r\sin\varphi$$

Virtual parabola (D)

$$x=r\cos(\alpha-2\varphi),\quad y=r\sin\varphi$$

Virtual parabola (E)

$$x=2R\cos^2\varphi,\quad y=R\cos\varphi+\sqrt{r^2-R^2\sin^2\varphi}$$

Algebraic curves of the sixth order

Astroid

$$x = b\cos^3 t, \quad y = b\sin^3 t$$

Atrophtaloid (A)

$$x^4(x^2+y^2) = (hx^2-k^3)^2$$

Atrophtaloid (B)

$$(x^2+y^2)^2x^2 = (hx^2-k^3)^2$$

Cappa

$$y^2(x^2+y^2+cx)^2 = a^2x^2$$

Cornoid

$$x = r\cos t(1-2\sin^2 t), \quad y = r\sin t(1+2\cos^2 t)$$

Cranioid

$$\rho = a\cos\varphi + b\sqrt{1-m^2\sin^2\varphi} + c\sqrt{1-k^2\sin^2\varphi}$$

Curve of Kiepert

$$\rho^3 = l^3\cos 3\varphi$$

Curve of Legendre

$$x = bh\sin t\left(1+\frac{m}{3}\sin^2 t\right), \quad y = bh\cos t\left(1+m-\frac{m}{3}\cos^2 t\right)$$

Curve of Watt

$$(x^2+y^2)(x^2+y^2-a^2-b^2+c^2)^2 + 4a^2y^2(x^2+y^2-b^2) = 0$$

Generalized kampyle (B)

$$b^2x^2y^2 = (x^2-a^2)(x^2-a^2+b^2)^2$$

Nephroid

$$(x^2+y^2)(x^2+y^2-a^2)^2 = 4a^2(x^2+y^2-ax)^2$$

Oblique astroid

$$x = \frac{l}{\sin^2\alpha}\cos t\sin^2(t-\alpha), \quad y = \frac{l}{\sin^2\alpha}\cos(t-\alpha)\sin^2 t$$

Polyod of circle

$$\rho = 2a\cos\left(2\arcsin\frac{m}{\rho}\pm\varphi\right)$$

Scarabaeus

$$(x^2+y^2)(x^2+y^2+cx)^2 = a^2(x^2-y^2)^2$$

Trefoil

$$x = r(2\cos 2t - \cos t),$$

$$y = r(2\sin 2t + \sin t)$$

Windmill

$$4x^2y^2(x^2 + y^2) = a^2(x^2 - y^2)^2$$

ALGEBRAIC CURVES OF THE SEVENTH ORDER

Acceleration curve

$$(x - y)(x^2 + y^2)^3 = 3a^2l^2x^3$$

ALGEBRAIC CURVES OF THE EIGHTH ORDER

Sextic of Cayley

$$\rho = l\cos\frac{\varphi}{4}$$

ALGEBRAIC CURVES OF AN ARBITRARY ORDER

Arachnida (A)

$$\rho = 2a\frac{\sin n\varphi}{\sin(n-1)\varphi}$$

Arachnida (B)

$$\rho = 2a\frac{\sin n\varphi}{\sin(n+1)\varphi}$$

Conchoid of higher-order circle

$$\rho = a\frac{\sin\frac{m+1}{m}\varphi}{\sin\frac{\varphi}{m}}$$

Curve of Lamet

$$\left(\frac{x}{a}\right)^m + \left(\frac{y}{b}\right)^m = 1$$

Curve of Lissajous

$$x = a\sin(mt + \gamma), \quad y = b\sin(nt + \delta)$$

(m, n, γ, δ are rational numbers)
Curve with convexities

$$\rho = \frac{p}{1 + \varepsilon\cos m\varphi}$$

Epicycloid

$$x = (R + mR)\cos mt - mR\cos(t + mt),$$

$$y = (R + mR)\sin mt - mR\sin(t + mt)$$

(m is a rational number)
Epitrochoid

$$x = (R + mR)\cos mt - h\cos(t + mt),$$

$$y = (R + mR)\sin mt - h\sin(t + mt)$$

(m is a rational number)
Geometric petal (A)

$$\rho = a + b\cos 2n\varphi$$

Geometric petal (B)

$$\rho = a + b\cos^{2n}\varphi$$

Geometric petal (C)

$$\rho = a(1 + \cos^{2n+1}\varphi)$$

Generalized hyperbola

$$\left(\frac{x}{a}\right)^m - \left(\frac{y}{b}\right)^m = 1$$

Hyperbolic curve

$$x^p y^q = a$$

Hypocycloid

$$x = (R - mR)\cos mt + mR\cos(t - mt),$$

$$y = (R - mR)\sin mt - mR\sin(t - mt)$$

(m is a rational number)
Hypotrochoid

$$x = (R - mR)\cos mt + h\cos(t - mt),$$

$$y = (R - mR)\sin mt - h\sin(t - mt)$$

(m is a rational number)
Nodal curve

$$\rho = a\cot k\varphi$$

Oval of Münger

$$(x^2 + y^2)^{2n+1} - 2dx^{2n+1}(x^2 + y^2)^n + (d^2 - r^2)x^{4n} = 0$$

Parabolic curve

$$y^q = ax^p$$

Pearl

$$x^s(a - x)^r = \frac{y^p}{b^p}$$

Pericycloid

$$x = (R + mR)\cos mt + mR\cos(t + mt),$$

$$y = (R + mR)\sin mt - mR\sin(t + mt)$$

(m is a rational number, $m < -1$)
Rose

$$\rho = a\cos k\varphi$$

Sectrix of Ceva

$$\rho = a + 2a\frac{\sin k\varphi\cos(k+1)\varphi}{\sin\varphi}$$

Sectrix of Coppola

$$\rho = a + 2a\sin\frac{\varphi}{2} + 2a\sin\frac{\varphi}{2}\sin\frac{\varphi+\pi}{4} + 2a\sin\frac{\varphi}{2}\sin\frac{\varphi+\pi}{4}\sin\frac{\varphi+3\pi}{8} + \dots$$

$$+2a\sin\frac{\varphi}{2}\sin\frac{\varphi+\pi}{4}\sin\frac{\varphi+3\pi}{8}\cdots\sin\frac{\varphi+(2n-1)\pi}{2^{n+1}}$$

Sectrix of Hesse

$$\rho = a\frac{\cos(n-1)\varphi}{\sin n\varphi}$$

Sectrix of Kempe

$$\rho = a + 2a\sin\varphi + 2a\cos\left(\frac{\pi}{4} - \frac{\varphi}{2}\right) + 2a\cos\left(\frac{\pi}{4} - \frac{\varphi}{2}\right)\cos\left(\frac{\pi}{8} - \frac{\varphi}{4}\right) + \cdots$$

$$+2a\cos\left(\frac{\pi}{4} - \frac{\varphi}{2}\right)\cos\left(\frac{\pi}{8} - \frac{\varphi}{4}\right)\cdots\cos\left(\frac{\pi}{2^{n+1}} - \frac{\varphi}{2^n}\right)$$

Sectrix of Ökinghaus (A)

$$\rho = a\frac{\sin\frac{n-2}{2}\varphi}{\sin\frac{n}{2}\varphi}$$

Sectrix of Ökinghaus (B)

$$\rho = a\frac{\cos\frac{n-2}{2}\varphi}{\cos\frac{n}{2}\varphi}$$

Sectrix of Plateau

$$x = a\frac{\sin(n'+n)\varphi}{\sin(n'-n)\varphi}, \quad y = 2a\frac{\sin n'\varphi\sin n\varphi}{\sin(n'-n)\varphi}$$

Sectrix of van Grinten

$$\rho = a\frac{\cos\frac{n+1}{2}\varphi}{\cos\frac{n-1}{2}\varphi}$$

Sectrix of Wagner

$$x = a\frac{\sin(2r-n)\varphi}{\sin n\varphi}, \quad y = a\frac{\sin r\varphi\sin(n-r)\varphi}{\sin n\varphi}$$

Sinusoidal spiral

$$\rho^m = a^m \cos m\varphi$$

(m is rational number)
Trochoidal rose

$$x = (R + mR)\cos mt - (R + mR)\cos(t + mt),$$

$$y = (R + mR)\sin mt - (R + mR)\sin(t + mt)$$

(m is a rational number)

2.3.2 Transcendental curves

Arc cosinusoid

$$y = \mathrm{Arccos}(bx)$$

Arc cotangentoid

$$y = \mathrm{Arccot}(bx)$$

Arc sinusoid

$$y = \mathrm{Arcsin}(bx)$$

Arc tangentoid

$$y = \mathrm{Arctan}(bx)$$

Archimedean spiral

$$\rho = a\varphi$$

Area-hyperbolic cosinusoid

$$y = \mathrm{arcosh}(ax)$$

Area-hyperbolic cotangentoid

$$y = \mathrm{arcoth}(ax)$$

Area-hyperbolic sinusoid

$$y = \mathrm{arsinh}(ax)$$

Area-hyperbolic tangentoid

$$y = \mathrm{artanh}(ax)$$

Bellshaped curve

$$y = e^{-x^4}$$

Catenary curve

$$y = a \cosh \frac{x}{a}$$

Catenary of equal resistance

$$e^{y/a}\left|\cos\frac{x}{a}\right| = 1$$

Circular tractrix

$$\rho = 2a\cos t, \quad \varphi = t - \tan t$$

Cochleoid

$$\rho = a\frac{\sin\varphi}{\varphi}$$

Conchoid of hyperbolic spiral

$$\rho = \frac{a}{\varphi} + l$$

Convolute of circle

$$\varphi = \frac{1}{a}\sqrt{\rho^2 - \frac{a^2}{4}} + \arcsin\frac{a}{2\rho}$$

Cosecantoid

$$y = \csc(ax)$$

Cosinusoid

$$y = \cos(ax)$$

Cotangentoid

$$y = \cot(ax)$$

Curtate and prolate cycloids

$$x = rt - d\sin t, \quad y = r - d\cos t$$

Curtate and prolate quadratrix

$$\rho = a\frac{\varphi \pm \alpha}{\sin\varphi}$$

Curve of Baum

$$\frac{y^2}{b^2} = \frac{x^2}{a^2} - c\sin\frac{\pi x}{2a} + 1$$

Curve of Caraccioli

$$\rho\frac{\cos\varphi}{a} = 1 - \left(\frac{2\varphi}{\pi}\right)^{m/n}$$

Curve of damped oscillations

$$y = Ae^{-kx}\sin(\omega x + a)$$

Curve of Debeaune

$$x = \frac{y}{a} - \frac{b}{a} + ce^{-y/b}$$

Curve of Lissajous

$$x = a\sin(mt+\gamma), \quad y = b\sin(nt+\delta)$$

Curve of Ozanam

$$x = 2a\sin\frac{y}{2a}$$

Cycloid

$$x = rt - r\sin t, \quad y = r - r\cos t$$

Double sinusoid

$$y = a\sin(\sin x)$$

Epicycloid

$$x = (R+mR)\cos mt - mR\cos(t+mt),$$

$$y = (R+mR)\sin mt - mR\sin(t+mt)$$

(m is an irrational number)
Epitrochoid

$$x = (R+mR)\cos mt - h\cos(t+mt),$$

$$y = (R+mR)\sin mt - h\sin(t+mt)$$

(m is an irrational number)
Exponential curve

$$y = a^x$$

Fermat's spiral

$$\rho = a\sqrt{\varphi}$$

Galileo's spiral

$$\rho = a - b\varphi^2$$

Hyperbolic cosinusoid

$$y = \cosh(ax)$$

Hyperbolic cotangentoid

$$y = \coth(ax)$$

Hyperbolic sinusoid

$$y = \sinh(ax)$$

Hyperbolic spiral

$$\rho = \frac{a}{\varphi}$$

Hyperbolic tangentoid

$$y = \tanh(ax)$$

Hypocycloid

$$x = (R-mR)\cos mt + mR\cos(t-mt),$$

$$y = (R - mR)\sin mt - mR\sin(t - mt)$$

(m is an irrational number)
Hypotrochoid

$$x = (R - mR)\cos mt + h\cos(t - mt),$$

$$y = (R - mR)\sin mt - h\sin(t - mt)$$

(m is an irrational number)
Isochrone of Varignon

$$\rho = a + (a - c)\varphi^2,$$

$$\varphi = b\sqrt{a - c}(\arctan t - t)$$

Inverse logarithmic curve

$$y = be^{a/x}$$

Lemniscatrix of Ökinghaus

$$y = \ln(\cos x + \sqrt{1 + \cos^2 x})$$

Lituus

$$\rho = \frac{a}{\sqrt{\varphi}}$$

Logarithmic curve

$$y = b\log_a x$$

Logarithmic lemniscate

$$y^2 = x^2 \ln\frac{a^2}{x^2}$$

Logarithmic sinusoid

$$y = \ln\sqrt{1 - x^2}$$

Logarithmic spiral

$$\rho = ce^{\varphi/\sqrt{a^2-1}}$$

Neoid

$$\rho = a\varphi + b$$

Parabolic spiral

$$\rho = a\sqrt{\varphi} + l$$

Pedal curve of lemniscate

$$\rho^2 = \frac{a^2}{2}\cos^2\frac{2\varphi}{3}$$

Pericycloid

$$x = (R + mR)\cos mt + mR\cos(t + mt),$$

$$y = (R + mR)\sin mt - mR\sin(t + mt)$$

(m is an irrational number, $m < -1$)
Polar tractrix

$$\rho = a\cos t, \quad \varphi = \tan t - t$$

Probability curve of Gauss

$$y = \frac{h}{\sqrt{\pi}} e^{-h^2x^2}$$

Quadratrix of Dinostratus

$$y = x\cot\frac{\pi x}{2a}$$

Quadratrix of Tschirnhausen

$$y - a\sin\frac{\pi x}{2a}$$

Secantoid

$$y = \sec(ax)$$

Secluded wave

$$y = \frac{a}{\cosh^2\frac{x}{2b}}$$

Secondary cycloid of Ricci

$$x = Rt - r\sin t,$$

$$y = r - r\cos t$$

Sinusoid

$$y = \sin(ax)$$

Sinusoidal spiral

$$\rho^m = a^m\cos m\varphi$$

(m is an irrational number)
Spiral of higher order

$$\rho^k = a^k\frac{\varphi}{2\pi}$$

Spiral of Norwich

$$\varphi + \alpha = \sqrt{\frac{\rho}{a} - 1} - 2\arccos\sqrt{\frac{a}{\rho}}$$

Tangentoid

$$y = \tan(ax)$$

Tractrix

$$x = a \ln \tan \frac{t}{2} + a \cos t, \quad y = a \sin t$$

Trochoidal rose

$$x = (R + mR)\cos mt - (R + mR)\cos(t + mt),$$

$$y = (R + mR)\sin mt - (R + mR)\sin(t + mt)$$

(m is an irrational number)

2.3.3 Bounded curves

The classes of curves labelled with an asterisk have the property for certain of the parameter values only.

antiversiera
astroid
besace
bicorn curve
bifolium
capricornoid *
cardioid
Cartesian oval
Cassinian oval
circle
circular tractrix
cochleoid
conchoid of higher-order circle
conchoid of hyperbolic spiral *
cornoid
cranioid
curve of Gutschoven *
curve of Jeřabek *
curve of Kiepert
curve of Lamet (A) *
curve of Legendre
curve of Lissajous
curve of Slüse
curve of Watt
curve with convexities
double-heart curve
ellipse
elliptic lemniscate
epicycloid
epitrochoid
folium
Galileo's spiral *
geometric petal (A)
geometric petal (B)
geometric petal (C)
hyperbolic lemniscate
hypocycloid
hypotrochoid
lemniscate of Bernoulli
lemniscate of Booth
lemniscate of Gerono
lemniscate of Montferrier
limaçon
logarithmic lemniscate
nephroid
oblique astroid
oblique bifolium
oblique trifolium
oval of Münger
pearformed quartic
pearl
pearl of the fourth order (A)
pearl of the fourth order (B)
pedal curve of lemniscate
pericycloid
polar tractrix

polyod of circle
polyzomal curve of Bernoulli
quartic of Bonnet
right trifolium
rose
scarabaeus
sectrix of Ceva
sectrix of Coppola
sectrix of Kempe
sectrix of Plateau *
sextic of Cayley
sinusoidal spiral (A)
spiral of higher order
spiril curve
three-cuspid hypocycloid
three-leafed rose
top
trefoil
trifolium of Cramer
trochoidal rose
virtual parabola (A)
virtual parabola (B)
virtual parabola (C)
virtual parabola (D)

2.3.4 Closed curves

The classes of curves labelled with an asterisk have the property for certain of the parameter values only.

antiversiera
astroid
besace
bicorn curve
bifolium
capricornoid *
cardioid
Cartesian oval
Cassinian oval
circle
conchoid of higher-order circle
cornoid
cranioid
curve of Gutschoven *
curve of Jeřabek *
curve of Kiepert
curve of Lamet (A) *
curve of Legendre
curve of Lissajous
curve of Slüse
curve of Watt
curve with convexities
double-heart curve
ellipse
elliptic lemniscate
epicycloid *
epitrochoid *
folium
Galileo's spiral *
geometric petal (A)
geometric petal (B)
geometric petal (C)
hyperbolic lemniscate
hypocycloid *
hypotrochoid *
lemniscate of Bernoulli
lemniscate of Booth
lemniscate of Gerono
lemniscate of Montferrier
limaçon
nephroid
oblique astroid
oblique bifolium
oblique trifolium
oval of Münger
pearformed quartic
pearl
pearl of the fourth order (A)
pearl of the fourth order (B)
pedal curve of lemniscate
pericycloid *
polyod of circle
polyzomal curve of Bernoulli
quartic of Bonnet

right trifolium
rose
scarabaeus
sectrix of Ceva
sectrix of Coppola
sectrix of Kempe
sectrix of Plateau *
sextic of Cayley
sinusoidal spiral (A) *
spiril curve
three-cuspid hypocycloid
three-leafed rose
top
trefoil
trifolium of Cramer
trochoidal rose *
virtual parabola (A)
virtual parabola (B)
virtual parabola (C)
virtual parabola (D)

2.3.5 Unbounded curves

The classes of curves labelled with an asterisk have the property for certain of the parameter values only.

acceleration curve
agnesiana
arachnida (A)
arachnida (B)
arc cosinusoid
arc cotangentoid
arc sinusoid
arc tangentoid
Archimedean spiral
area-hyperbolic cosinusoid
area-hyperbolic cotangentoid
area-hyperbolic sinusoid
area-hyperbolic tangentoid
atriphtaloid (A)
atriphtaloid (B)
bellshaped curve
biquadratic cubic (parabola)
bullet-nose curve
cappa
capricornoid *
Cartesian folium
catenary curve
catenary of equal resistance
cissoid
cissoid of the fourth order
cissoid of second-order curves
conchal
conchoid of Dürer
conchoid of hyperbolic spiral *
conchoid of Külp
conchoid of Nicomedes
conchoid of Slüse (+)
conchoid of Slüse (−)
convolute of circle
cosecantoid
cosinusoid
cotangentoid
Côtes' spiral
cruciform curve
cubic of Tschirnhausen
curtate and prolate cissoids
curtate and prolate cycloids
curtate and prolate quadratrices
curve of Baum
curve of Caraccioli
curve of Clairaut
curve of Cramer
curve of damped oscillations
curve of Debeaune
curve of Gutschoven *
curve of Jeřabek *
curve of Lamet (A) *
curve of Lamet (B)
curve of Mascheroni
curve of Ozanam
curve of Rolle
cycloid
double sinusoid

duplicatrix cubic
exponential curve
Fermat's spiral
first pearl
Galileo's spiral
generalized hyperbola (A)
generalized hyperbola (B)
generalized kampyle (A)
generalized kampyle (B)
hyperbola
hyperbolic cosinusoid
hyperbolic cotangentoid
hyperbolic curve
hyperbolic sinusoid
hyperbolic spiral
hyperbolic tangentoid
isochrone of Varignon
inverse logarithmic curve
kampyle
lemniscatrix of Ökinghaus
lituus
logarithmic curve
logarithmic sinusoid
logarithmic spiral
neoid
nodal curve
oblique cappa
oblique cissoid
oblique conchoid
oblique pancappa
oblique parabolic folium
oblique strophoid
oblique versiera
ophiuride
orthoconchoid
pancappa
panstrophoid
parabola
parabolic curve
parabolic folium
parabolic spiral
pear
polyod of polar axis
polyod of straight line
(parallel to polar axis)
polyod of straight line
(perpendicular to polar axis)
probability curve of Gauss
pseudoversiera
quadratrix of Dinostratus
quadratrix of Tschirnhausen
sand-glass curve
secantoid
secluded wave
secondary cycloid of Ricci
sectrix of Hesse
sectrix of Ökinghaus (A)
sectrix of Ökinghaus (B)
sectrix of Plateau *
sectrix of van Grinten
sectrix of Wagner
serpentine of Newton
sinusoid
sinusoidal spiral (B)
spiral of higher order
spiral of Norwich
strophoid
tangentoid
tractrix
trident of Newton
trifolium
trisectrix of Delanges
trisectrix of Longchamps
trisectrix of Maclaurin
versiera
virtual parabola (E)
visiera
windmill

2.3.6 Curves with asymptotes

Type (inclined, vertical, horizontal) and number of asymptotes of appropriate type are indicated; when the number of asymptotes depends on relationship between parameter values the curve is labelled with an asterisk and the maximal permissible number of asymptotes is indicated (but no more than four); the curves with more than four asymptotes are not labelled; the curves with an infinitely large number of asymptotes are labelled with the simbol ∞.

acceleration curve — 1 (inclined)
agnesiana — 1 (horizontal)
arachnida (A)
arachnida (B)
arc cotangentoid — ∞ (horizontal)
arc tangentoid — ∞ (horizontal)
area-hyperbolic cotangentoid — 2 (vertical) + 1 (horizontal)
area-hyperbolic tangentoid — 2 (vertical)
atriphtaloid (A) — 1 (vertical)
atriphtaloid (B) — 1 (vertical)
bellshaped curve — 1 (horizontal)
bullet-nose curve— 2 (vertical)
cappa — 2 (horizontal)
capricornoid — 2* (inclined) + 1* (horizontal)
Cartesian folium — 1 (inclined)
catenary of equal resistance — ∞ (vertical)
cissoid — 1 (vertical)
cissoid of second-order curves — 2* (inclined) + 1 (vertical)
conchal — 1 (vertical)
conchoid of Dürer — 2 (horizontal)
conchoid of hyperbolic spiral — 1* (horizontal)
conchoid of Külp — 1 (vertical)
conchoid of Nicomedes — 1 (horizontal)
conchoid of Slüse (+) — 1 (vertical)
conchoid of Slüse (−) — 1 (vertical)
cosecantoid — ∞ (vertical)
cotangentoid — ∞ (vertical)
Côtes' spiral
cruciform curve — 2 (vertical) + 2 (horizontal)
curtate cissoid — 1 (vertical)
curtate and prolate quadratrices — 1 (horizontal)
curve of Caraccioli — 1 (vertical)
curve of Cramer — 1 (vertical)
curve of damped oscillations — 1 (horizontal)
curve of Debeaune — 1 (inclined)
curve of Gutschoven — 2* (inclined)
curve of Jeřabek — 2* (inclined)
curve of Lamet (A) — 1* (inclined)
curve of Lamet (B) — 2 (vertical) + 2 (horizontal)
curve of Mascheroni — 1 (vertical)

curve of Rolle — 1 (vertical)
exponential curve —1 (horizontal)
generalized hyperbola (A) — 2* (inclined)
generalized hyperbola (B) — 2* (vertical) + 1* (horizontal)
hyperbola — 2 (inclined)
hyperbolic cotangentoid — 1 (vertical) + 2 (horizontal)
hyperbolic curve — 1 (vertical) + 1 (horizontal)
hyperbolic spiral — 1 (horizontal)
hyperbolic tangentoid — 2 (horizontal)
inverse logarithmic curve — 1 (vertical) + 1 (horizontal)
lituus — 1 (horizontal)
logarithmic curve — 1 (vertical)
logarithmic sinusoid — 2 (vertical)
nodal curve
oblique cappa — 1 (horizontal)
oblique cissoid — 1 (inclined)
oblique conchoid — 2 (horizontal)
oblique pancappa — 2 (horizontal)
oblique strophoid — 1 (vertical)
oblique versiera — 1 (inclined)
ophiuride — 1 (vertical)
orthoconchoid — 2 (horizontal)
pancappa — 2* (horizontal)
panstrophoid — 1 (vertical)
pear — 1 (vertical)
polyod of polar axis straight line — 2 (horizontal)
polyod of straight line parallel to polar axis — 2 (horizontal)
polyod of straight line perpendicular to polar axis — 2 (vertical)
probability curve of Gauss — 1 (horizontal)
pseudoversiera — 1 (horizontal)
quadratrix of Dinostratus — ∞ (vertical)
sand-glass curve — 2 (inclined)
secantoid — ∞ (vertical)
secluded wave — 1 (horizontal)
sectrix of Hesse
sectrix of Kempe
sectrix of Ökinghaus (A)
sectrix of Ökinghaus (B)
sectrix of Plateau
sectrix of van Grinten
sectrix of Wagner
serpentine of Newton — 1 (horizontal)
sinusoidal spiral (B)
strophoid — 1 (vertical)
tangentoid — ∞ (vertical)

tractrix — 1 (horizontal)
trident of Newton — 1 (vertical)
trifolium — 2* (inclined) + 1 (vertical)
trifolium of Delanges — 2 (horizontal)
trisectrix of Longchamps — 2 (inclined) + 1 (vertical)
trisectrix of Maclaurin — 1 (vertical)
versiera — 1 (horizontal)
virtual parabola (E) — 1 (horizontal)
visiera — 1 (horizontal)
windmill — 2 (vertical) + 2 (horizontal)

2.3.7 Axisymmetric curves

Number of the symmetry axes is indicated; when the number depends on relationship between parameter values the curve is labelled with an asterisk and the maximal permissible number of axes is indicated (but no more than four); the curves with more than four axes are not labelled; the curves with an infinitely large number of axes are labelled with the simbol ∞.

agnesiana — 1
antiversiera — 1
arachnida (A) — 1
arachnida (B) — 1
arc cosinusoid — ∞
arc sinusoid — ∞
area-hyperbolic cosinusoid — 1
astroid — 4
atriphtaloid (A) — 2
atriphtaloid (B) — 2
bellshaped curve — 1
besace — 1
bicorn curve — 1
bifolium — 1
bullet-nose curve — 2
cappa — 2
capricornoid — 1
cardioid — 1
Cartesian folium — 1
Cartesian oval — 1
Cassinian oval — 2
catenary curve — 1
catenary of equal resistance — ∞
circle — ∞
circular tractrix — 1
cissoid — 1
cissoid of the fourth order — 1
cissoid of second-order curves — 1*
cochleoid — 1
conchal — 1
conchoid of higher-order circle — 1
conchoid of Külp — 2
conchoid of Nicomedes — 1
conchoid of Slüse (+) — 1
conchoid of Slüse (−) — 1
cornoid — 2
cosecantoid — ∞
cosinusoid — ∞
Côtes' spiral
cranioid — 1
cruciform curve — 2
cubic of Tschirnhausen — 1
curtate and prolate cissoids — 1
curtate and prolate cycloids — ∞
curtate and prolate quadratrices — 1
curve of Baum — 1
curve of Clairaut — 2
curve of Cramer — 1
curve of Gutschoven — 1
curve of Jeřabek — 1

curve of Kiepert — 3
curve of Lamet (A) — 2*
curve of Lamet (B) — 2*
curve of Legendre — 2
curve of Lissajous *
curve of Ozanam — ∞
curve of Slüse — 2
curve of Watt — 2
curve with convexities
cycloid — ∞
double-heart curve — 1
double sinusoid — ∞
duplicatrix cubic — 1
ellipse — 2
elliptic lemniscate — 2
epicycloid
epitrochoid
folium — 1
Galileo's spiral — 1
generalized hyperbola (A) — 2*
generalized hyperbola (B) — 2*
generalized kampyle (A) — 2
generalized kampyle (B) — 2
geometric petal (A)
geometric petal (B) — 2
geometric petal (C) — 1
hyperbola — 2
hyperbolic cosinusoid — 1
hyperbolic curve — 1*
hyperbolic lemniscate — 2
hypocycloid
hypotrochoid
isochrone of Varignon — 1
kampyle — 2
lemniscate of Bernoulli — 2
lemniscate of Booth — 2
lemniscate of Gerono — 2
lemniscate of Montferrier — 2
lemniscatrix of Ökinghaus — ∞
limaçon — 1
logarithmic lemniscate — 2
logarithmic sinusoid — 1
nephroid — 1
nodal curve
oblique astroid — 2
oblique bifolium — 1*
oblique cissoid — 1*
oblique conchoid — 1*
oblique strophoid — 1*
oblique trifolium — 1*
orthoconchoid — 1
oval of Münger — 1
pancappa — 1
panstrophoid — 1
parabola — 1
parabolic curve — 1*
parabolic folium — 1
pear — 1
pearformed quartic — 1
pearl — 2*
pearl of the fourth order (A) — 1
pearl of the fourth order (B) — 1
pedal curve of lemniscate — 4
pericycloid
polar tractrix — 1
polyod of circle — 1
polyod of polar axis — 2
polyod of straight line
(parallel to polar axis) — 1
polyod of straight line
(perpendicular to polar axis) — 1
polyzomal curve of Bernoulli — 2
probability curve of Gauss — 1
pseudoversiera — 1
quadratrix of Dinostratus — 1
quadratrix of Tschirnhausen — ∞
quartic of Bonnet — 1
right trifolium — 1
rose
sand-glass curve — 2
scarabaeus — 2*
secantoid — ∞
secluded wave — 1
secondary cycloid of Ricci — ∞
sectrix of Ceva — 2
sectrix of Coppola — 1
sectrix of Hesse — 1
sectrix of Kempe — 1
sectrix of Ökinghaus (A) — 1
sectrix of Ökinghaus (B) — 1

sectrix of Plateau — 1
sectrix of van Grinten — 1
sectrix of Wagner — 1
sextic of Cayley — 2
sinusoid — ∞
sinusoidal spiral (A)
sinusoidal spiral (B)
spiril curve — 2
strophoid — 1
three-cuspid hypocycloid — 3
three-leafed rose — 3
top — 1
tractrix — 1
trefoil — 3
trifolium — 3*
trifolium of Cramer — 1
trisectrix of Delanges — 2
trisectrix of Longchamps — 3
trisectrix of Maclaurin — 1
trochoidal rose
versiera — 1
virtual parabola (A) — 2
virtual parabola (B) — 2*
virtual parabola (C) — 2*
virtual parabola (D) — 1
virtual parabola (E) — 1
visiera — 1
windmill — 4

2.3.8 Centrally symmetric curves

The classes of curves labelled with an asterisk have the property for certain of the parameter values only.

acceleration curve
arc cosinusoid
arc cotangentoid
arc sinusoid
arc tangentoid
area-hyperbolic cotangentoid
area-hyperbolic sinusoid
area-hyperbolic tangentoid
astroid
atriphtaloid (A)
atriphtaloid (B)
bullet-nose curve
cappa
Cassinian oval
circle
conchoid of Külp
cornoid
cosecantoid
cosinusoid
cotangentoid
Côtes' spiral *
cruciform curve
curve of Clairaut
curve of Lamet (A) *
curve of Lamet (B) *
curve of Legendre
curve of Lissajous *
curve of Mascheroni
curve of Slüse
curve of Watt
curve with convexities *
double sinusoid
ellipse
elliptic lemniscate
epicycloid *
epitrochoid *
generalized hyperbola (A) *
generalized hyperbola (B) *
generalized kampyle (A)
generalized kampyle (B)
geometric petal (A)
geometric petal (B)
hyperbola
hyperbolic cotangentoid
hyperbolic curve *
hyperbolic lemniscate
hyperbolic sinusoid
hyperbolic tangentoid
hypocycloid *
hypotrochoid *

kampyle
lemniscate of Bernoulli
lemniscate of Booth
lemniscate of Gerono
lemniscate of Montferrier
lemniscatrix of Ökinghaus
logarithmic lemniscate
nodal curve
oblique astroid
oblique cappa
oval of Münger *
parabolic curve *
pearl *
pedal curve of lemniscate
pericycloid *
polyod of polar axis
polyzomal curve of Bernoulli
quadratrix of Tschirnhausen
rose *
sand-glass curve
scarabaeus *
secantoid
secluded wave
secondary cycloid of Ricci
sectrix of Ceva
serpentine of Newton
sextic of Cayley
sinusoid
sinusoidal spiral (A) *
sinusoidal spiral (B) *
spiril curve
tangentoid
trisectrix of Delanges
trochoidal rose *
virtual parabola (A)
virtual parabola (B) *
virtual parabola (C) *
windmill

2.3.9 Connected curves

The classes of curves labelled with an asterisk have the property for certain of the parameter values only.

acceleration curve
agnesiana
antiversiera
arachnida (A)
arc cosinusoid
arc sinusoid
Archimedean spiral
area-hyperbolic cosinusoid
area-hyperbolic sinusoid
area-hyperbolic tangentoid
astroid
bellshaped curve
besace *
bicorn curve
bifolium
biquadratic cubic (parabola)
bullet-nose curve
cappa
capricornoid
cardioid
Cartesian folium
Cartesian oval *
Cassinian oval *
catenary curve
circle
circular tractrix
cissoid
cissoid of the fourth order *
cissoid of second-order curves *
cochleoid
conchoid of Dürer *
conchoid of higher order circle
conchoid of hyperbolic spiral
conchoid of Slüse (–) *
convolute of circle
cornoid
cosinusoid
Côtes' spiral *
cranioid
cubic of Tschirnhausen

curtate and prolate cissoids
curtate and prolate cycloids
curve of Baum *
curve of Caraccioli
curve of Cramer
curve of damped oscillations
curve of Debeaune
curve of Gutschoven
curve of Jeřabek *
curve of Kiepert
curve of Lamet (A)
curve of Lamet (B) *
curve of Legendre
curve of Lissajous
curve of Mascheroni
curve of Ozanam
curve of Slüse
curve of Watt *
curve with convexities
cycloid
double-heart curve
double sinusoid
ellipse
epicycloid
epitrochoid
exponential curve
Fermat's spiral
first pearl
folium
Galileo's spiral
generalized hyperbola (A) *
generalized hyperbola (B) *
geometric petal (A)
geometric petal (B)
geometric petal (C)
hyperbolic cosinusoid
hyperbolic lemniscate
hyperbolic sinusoid
hyperbolic spiral
hyperbolic tangentoid
hypocycloid
hypotrochoid
isochrone of Varignon
lemniscate of Bernoulli
lemniscate of Booth *
lemniscate of Gerono
lemniscate of Montferrier
lemniscatrix of Ökinghaus
limaçon *
lituus
logarithmic curve
logarithmic sinusoid
logarithmic spiral
nephroid
neoid
nodal curve
oblique astroid
oblique bifolium
oblique cappa
oblique cissoid
oblique conchoid *
oblique pancappa
oblique parabolic folium
oblique strophoid
oblique trifolium
oblique versiera *
ophiuride
oval of Münger *
pancappa *
panstrophoid *
parabola
parabolic curve
parabolic folium
parabolic spiral
pear
pearformed quartic
pearl
pearl of the fourth order (A)
pearl of the fourth order (B)
pedal curve of lemniscate
pericycloid
polar tractrix
polyod of circle
polyod of polar axis
polyod of straight line
(parallel to polar axis) *
polyod of straight line
(perpendicular to polar axis) *
probability curve of Gauss
pseudoversiera

quadratrix of Tschirnhausen
quartic of Bonnet
right trifolium
rose
sand-glass curve
scarabaeus
secluded wave
secondary cycloid of Ricci
sectrix of Ceva
sectrix of Coppola
sectrix of Kempe
sectrix of Plateau
sectrix of van Grinten
serpentine of Newton
sextic of Cayley
sinusoid
sinusoidal spiral (A)
spiral of higher order
spiral of Norwich
spiril curve *
strophoid
three-cuspid hypocycloid
three-leafed rose
top
tractrix
trefoil
trifolium *
trifolium of Cramer
trisectrix of Delanges
trisectrix of Maclaurin
trochoidal rose
versiera
virtual parabola (A)
virtual parabola (B)
virtual parabola (C)
virtual parabola (D)
windmill

2.3.10 Disconnected curves

The classes of curves labelled with an asterisk have the property for certain of the parameter values only.

arachnida (B)
arc cotangentoid
arc tangentoid
area-hyperbolic cotangentoid
atriphtaloid (A)
atriphtaloid (B)
besace *
Cartesian oval *
Cassinian oval *
catenary of equal resistance
cissoid of the fourth order *
cissoid of second-order curves *
conchal
conchoid of Dürer *
conchoid of Külp
conchoid of Nicomedes
conchoid of Slüse (+)
conchoid of Slüse (−) *
cosecantoid
cotangentoid
Côtes' spiral *
cruciform curve
curtate cissoid
curtate and prolate quadratrices
curve of Baum *
curve of Clairaut
curve of Jeřabek *
curve of Lamet (B) *
curve of Rolle
curve of Watt *
duplicatrix cubic
elliptic lemniscate
generalized hyperbola (A) *
generalized hyperbola (B) *
generalized kampyle (A) *
generalized kampyle (B) *
hyperbola
hyperbolic cotangentoid

hyperbolic curve *
inverse logarithmic curve
kampyle
lemniscate of Booth *
limaçon *
logarithmic lemniscate
oblique conchoid *
oblique versiera *
orthoconchoid
oval of Münger *
pancappa *
panstrophoid *
polyod of straight line
(parallel to polar axis) *
polyod of straight line
(perpendicular to polar axis) *
polyzomal curve of Bernoulli
quadratrix of Dinostratus
secantoid
sectrix of Hesse
sectrix of Ökinghaus (A)
sectrix of Ökinghaus (B)
sectrix of Wagner
sinusoidal spiral (B) *
spiril curve *
tangentoid
trident of Newton
trifolium *
trisectrix of Longchamps
virtual parabola (E)
visiera

2.3.11 Curves free of singular points

The classes of curves labelled with an asterisk have the property for certain of the parameter values only.

acceleration curve
agnesiana
arachnida (B) *
arc cosinusoid
arc cotangentoid
arc sinusoid
arc tangentoid
area-hyperbolic cosinusoid
area-hyperbolic cotangentoid
area-hyperbolic sinusoid
area-hyperbolic tangentoid
atriphtaloid (A) *
atriphtaloid (B) *
bellshaped curve
besace *
biquadratic cubic (parabola)
cardioid
Cartesian oval *
Cassinian oval *
catenary curve
catenary of equal resistance
circle
cissoid of the fourth order *
conchal *
conchoid of hyperbolic spiral *
conchoid of Külp
cosecantoid
cosinusoid
cotangentoid
Côtes' spiral *
cranioid
cruciform curve
curtate cycloid
curve of Baum *
curve of damped oscillations
curve of Debeaune
curve of Lamet (A) *
curve of Lamet (B)
curve of Legendre
curve of Ozanam
curve with convexities
double sinusoid
ellipse
epitrochoid *

exponential curve
first pearl
folium
generalized hyperbola (A) *
generalized hyperbola (B) *
geometric petal (A) *
geometric petal (B) *
hyperbola
hyperbolic cosinusoid
hyperbolic cotangentoid
hyperbolic curve
hyperbolic lemniscate
hyperbolic sinusoid
hyperbolic spiral
hyperbolic tangentoid
hypocycloid *
hypotrochoid *
inverse logarithmic curve
kampyle
lemniscatrix of Ökinghaus
lituus
logarithmic curve
logarithmic lemniscate
logarithmic sinusoid
logarithmic spiral
panstrophoid
parabola
parabolic curve *
pear
pearl *
pearl of the fourth order (B)
polyzomal curve of Bernoulli
probability curve of Gauss
pseudoversiera
quadratrix of Dinostratus
quadratrix of Tschirnhausen
quartic of Bonnet
secantoid
secluded wave
secondary cycloid of Ricci *
sectrix of Ökinghaus (A) *
sectrix of Ökinghaus (B) *
sectrix of Wagner *
serpentine of Newton
sinusoid
sinusoidal spiral (B) *
spiril curve *
tangentoid
trident of Newton
versiera
virtual parabola (E)

2.3.12 Curves with asymptotic points

circular tractrix
hyperbolic spiral
lituus
logarithmic spiral
polar tractrix

2.3.13 Curves with cuspidal points

Number of the cuspidal points is indicated; when the number depends on relationship between parameter values the curve is labelled with an asterisk and the maximal permissible number of the cusp points is indicated (but no more than four); the curves with more than four axes are not labelled.

antiversiera — 1
astroid — 4
bicorn curve — 2
cardioid — 1
circular tractrix — 1
cissoid — 1

cissoid of the fourth order — 1*
cissoid of second-order curves — 1*
conchoid of Dürer — 1*
conchoid of Nicomedes — 1*
conchoid of Slüse (−) — 1
curtate and prolate cissoids — 1*
curve of Lamet (A) — 4*
curve of Rolle — 1
cycloid
epicycloid
Galileo's spiral — 1*
generalized hyperbola (A) — 2*
geometric petal (C) — 1
hypocycloid
hypotrochoid
isochrone of Varignon — 1
limaçon — 1*
oblique astroid — 4
oblique cissoid — 1
oblique conchoid — 1*
oblique versiera — 1*
parabolic curve — 1*
pearformed quartic — 1
pearl — 2*
pearl of the fourth order (A) — 1
pericycloid
polar tractrix — 1
quartic of Bonnet
secondary cycloid of Ricci *
three-cuspid hypocycloid — 3
top — 1
tractrix — 1

2.3.14 Curves with isolated points

Number of the isolated points is indicated; when the number depends on relationship between parameter values the curve is labelled with an asterisk and the maximal permissible number of isolated points is indicated.

atriphtaloid (A) — 2*
atriphtaloid (B) — 2*
cissoid of the fourth order — 1*
cissoid of second-order curves — 1*
conchoid of Dürer — 1*
conchoid of Nicomedes — 1*
conchoid of Slüse (+) — 1
conchoid of Slüse (−) — 1*
curtate cissoid — 1
curve of Clairaut — 1
curve of Jeřabek — 1*
curve of Watt — 3*
duplicatrix cubic — 1
elliptic lemniscate — 1
generalized kampyle (A) — 1
generalized kampyle (B) — 2*
kampyle — 1
lemniscate of Booth — 1*
limaçon — 1*
oblique conchoid — 1*
oblique versiera — 1*
orthoconchoid — 1
oval of Münger — 1*
pancappa — 1*
polyod of straight line (parallel to polar axis) — 1*
polyod of straight line (perpendicular to polar axis) — 1*
trifolium — 1*
trisectrix of Longchamps — 1
visiera — 1

2.3.15 Curves with nodes

Number of the nodes is indicated; when the number depends on relationship between parameter values the curve is labelled with an asterisk and the maximal permissible number of nodes is indicated (but no more than four);

the curves with more than four nodes are not labelled; the curves with an infinitely large number of nodes are labelled with the simbol ∞.

besace — 1*
bifolium — 1
bullet-nose curve — 1
cappa — 1
capricornoid — 2
Cartesian folium — 1
Cartesian oval — 1*
Cassinian oval — 1
cissoid of the fourth order — 1*
cissoid of second-order curves — 1*
conchal — 1*
conchoid of Dürer — 2*
conchoid of higher-order circle
conchoid of Nicomedes — 1*
conchoid of Slüse (–) — 1*
cornoid — 2
Côtes' spiral
cubic of Tschirnhausen — 1
prolate cissoid
prolate cycloid
curve of Baum — 1*
curve of Cramer — 1
curve of Gutschoven — 2
curve of Jeřabek — 2*
curve of Kiepert — 1
curve of Lissajous
curve of Mascheroni — 1
curve of Slüse — 1
curve of Watt — 3*
double-heart curve — 1
epicycloid
epitrochoid
Galileo's spiral — ∞
geometric petal (A) — 1*
hyperbolic lemniscate — 1
hypocycloid
hypotrochoid
isochrone of Varignon — ∞
lemniscate of Bernoulli — 1
lemniscate of Booth — 1*
lemniscate of Gerono — 1
lemniscate of Montferrier — 1

limaçon — 1*
nephroid — 1
nodal curve
oblique bifolium — 1
oblique cappa — 1
oblique conchoid — 2*
oblique pancappa — 2
oblique parabolic folium — 1
oblique strophoid — 1
oblique trifolium — 1
oblique versiera — 1*
ophiuride — 1
orthoconchoid — 1
oval of Münger — 1*
pancappa — 2*
pedal curve of lemniscate — 8
pericycloid
polyod of circle — 3*
polyod of polar axis — 2
polyod of straight line
(parallel to polar axis) — 2*
polyod of straight line
(perpendicular to polar axis) — 2*
prolate cissoid — 1
prolate cycloid — ∞
right trifolium — 1
rose
sand-glass curve — 2
scarabaeus — 3*
secondary cycloid of Ricci — ∞ *
sectrix of Ceva
sectrix of Coppola
sectrix of Hesse
sectrix of Kempe
sectrix of Ökinghaus (A) — 1*
sectrix of Ökinghaus (B) — 1*
sectrix of Plateau
sectrix of van Grinten
sectrix of Wagner — 1*
sextic of Cayley — 2
sinusoidal spiral (A)
sinusoidal spiral (B)

spiril curve — 2*
strophoid — 1
trefoil — 3
trifolium — 1*
trifolium of Cramer — 1
trisectrix of Delanges — 2
trisectrix of Maclaurin — 1
trochoidal rose
virtual parabola (A) — 1
virtual parabola (B) — 1
virtual parabola (C) — 1
virtual parabola (D) — 1
windmill — 1

2.3.16 Curves with multiple points

Number of the multiple points with an order of multiplicity greater than two is indicated; when the number depends on relationship between parameter values the curve is labelled with an asterisk and the maximal permissible number of the points is indicated.

arachnida (A) — 1*
arachnida (B) — 1*
bifolium — 1
conchoid of higher-order circle — 1*
curve of Kiepert — 1
geometric petal (A) — 1*
geometric petal (B) — 1*
nephroid — 1
nodal curve — 1
oblique bifolium — 1
oblique trifolium — 1
pedal curve of lemniscate — 1
right trifolium — 1
rose — 1
scarabaeus — 1*
sectrix of Ceva — 1
sectrix of Coppola *
sectrix of Hesse — 1
sectrix of Kempe *
sectrix of Ökinghaus (A) — 1*
sectrix of Ökinghaus (B) — 1*
sectrix of Plateau — 1*
sectrix of van Grinten — 1
sectrix of Wagner — 1*
sinusoidal spiral (A) — 1*
three-leafed rose — 1
trifolium of Cramer — 1
trochoidal rose — 1*
windmill — 1

2.3.17 Curves with self-osculation points

Number of the self-osculation points is indicated; when the number depends on relationship between parameter values the curve is labelled with an asterisk and the maximal permissible number of the self-osculation points is indicated.

cappa — 1
capricornoid — 1
cochleoid — 1
cornoid — 2
curve of Gutschoven — 1
curve of Jeřabek — 1
curve of Mascheroni — 1
curve of Watt — 1
double-heart curve — 1
geometric petal (A) — 1*
geometric petal (B) — 1*
lemniscate of Booth — 1*

oblique cappa — 1
oval of Münger — 1*
polyod of circle — 1*
polyod of straight line (parallel to polar axis) — 1*
polyod of straight line (perpendicular to polar axis) — 1*
sextic of Cayley — 1
trochoidal rose — 1*

2.3.18 Curves with cessation points

Number of cessation points is indicated; when the number depends on relationship between parameter values the curve is labelled with an asterisk and the maximal permissible number of cessation points is indicated.

Archimedean spiral — 1
convolute of circle — 1
conchoid of hyperbolic spiral — 1*
curtate and prolate quadratrices — 1
curve of Caraccioli — 1
curve of Lamet (A) — 2*
Fermat's spiral — 1
generalized hyperbola (A) — 1*
generalized hyperbola (B) — 1*
neoid — 1
parabolic spiral — 1
spiral of higher order — 1
spiral of Norwich — 1

2.3.19 Curves described by similar graphs

Polyod of straight line, sand-glass curve, trisectrix of Delanges (Figure 497).
Antiversiera, pearformed quartic, pearl, quartic of Bonnet, top (Figure 498).
Agnesiana, duplicatrix cubic, pseudoversiera, secluded wave, versiera, visiera (Figure 499).
Atriphtaloid, conchoid of Külp (Figure 500).
Curve of Watt, hyperbolic lemniscate, lemniscate of Bernoulli, lemniscate of Booth, lemniscate of Gerono, lemniscate of Montferrier, logarithmic lemniscate, spiril curve, virtual parabola (Figure 501).
Cartesian folium, cissoid of second-order curves, cubic of Tschirnhausen, curve of Cramer, oblique strophoid, oblique versiera, ophiuride, parabolic folium, strophoid, trifolium, trisectrix of Maclaurin (Figure 502).
Curve of Baum, curve of Clairaut, generalized kampyle, kampyle (Figure 503).
Conchal, conchoid of Nicomedes, conchoids of Slüse (Figure 504).
Conchoid of Dürer, oblique conchoid, oblique pancappa (Figure 505).
Curve of Kiepert, three-leafed rose (Figure 506).
Right trifolium, trifolium of Cramer (Figure 507).
Besace, polyod of circle (Figure 508).
Bullet-nose curve, generalized hyperbola (Figure 509).
Cassinian oval, elliptic lemniscate, lemniscate of Booth, spiril curve (Figure 510).
Curve of Legendre, ellipse, folium, oval of Münger (Figure 511).

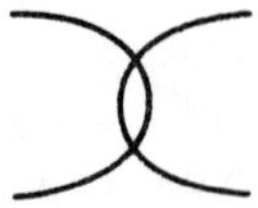

Figure 497.

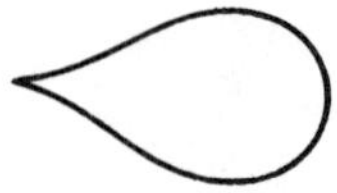

Figure 498.

Figure 499.

Figure 500.

Figure 501.

Figure 502.

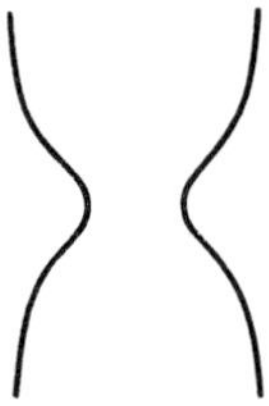

Figure 503.

Figure 504.

Figure 505.

Figure 506.

Figure 507.

Figure 508.

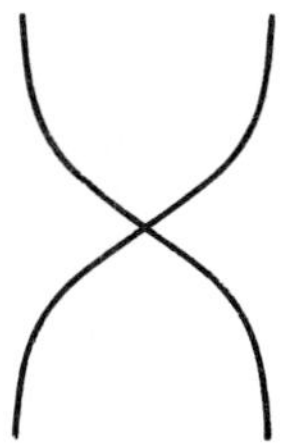

Figure 509.

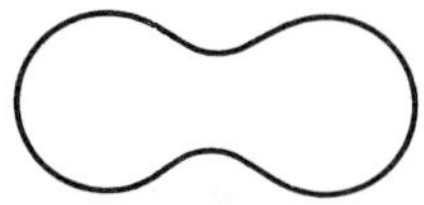

Figure 510.

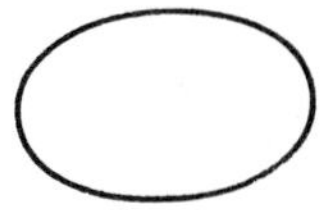

Figure 511.

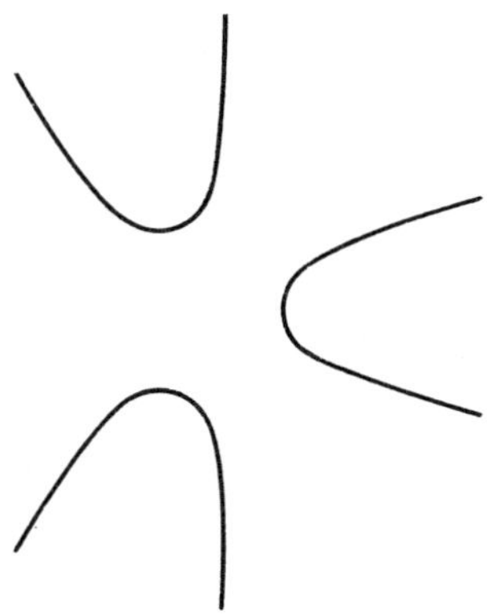

Figure 512.

Figure 513.

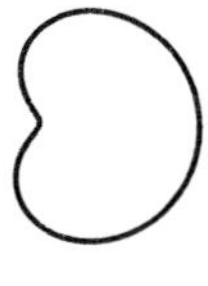

Figure 514.

Figure 515.

Cissoid of second-order curves, Côtes' spiral, trifolium, trisectrix of Longchamps (Figure 512).
Cissoid, parabolic curve, tractrix (Figure 513).
Conchoid of higher-order circle, limaçon (Figure 514).
Cartesian oval, conchoid of higher-order circle, limaçon (Figure 515).

Chapter 3

Space Curves

In this chapter we present the basics of the differential geometric approach to investigations of curve properties in three-dimensional space and describe several interesting classes of the space curves.

3.1 Main definitions and facts

In this section we expand and elucidate the principal concepts — notions, formulae and facts — of the smooth space curve theory unveiling the potential of calculus in the geometric problems.

3.1.1 Parametric equations of space curves

A set γ of points $M(x, y, z)$ whose coordinates x, y and z are defined by formulae

$$x = x(t), \quad y = y(t), \quad z = z(t), \quad a \leq t \leq b \qquad (*)$$

where functions $x(t)$, $y(t)$ and $z(t)$ are continuous on segment $[a, b]$, or in vector form,

$$\mathbf{r} = \mathbf{r}(t), \quad a \leq t \leq b \qquad (*)$$

where $\mathbf{r}(t) = (x(t), y(t), z(t))$, is called *a parametrically defined (parametrized) space curve.*

The relations $(*)$ are called *parametric equations of space curve* γ, or simply *a parametrization of curve* γ.

Point $M(x, y, z)$ whose coordinates x, y and z are calculated by formulae $(*)$ is called *a point of curve* γ.

Point $M(x(t), y(t), z(t))$, or $M(t)$, is called *the point of curve* γ *corresponding to parameter value* t. In this case, curve γ is said *to go through point* M. Point $A(x(a), y(a), z(a))$ is called *the initial point*, and point $B(x(b), y(b), z(b))$ — *the terminal point* of the curve. Points A and B are named *end-points* of curve γ; curve γ *connects* point A with point B.

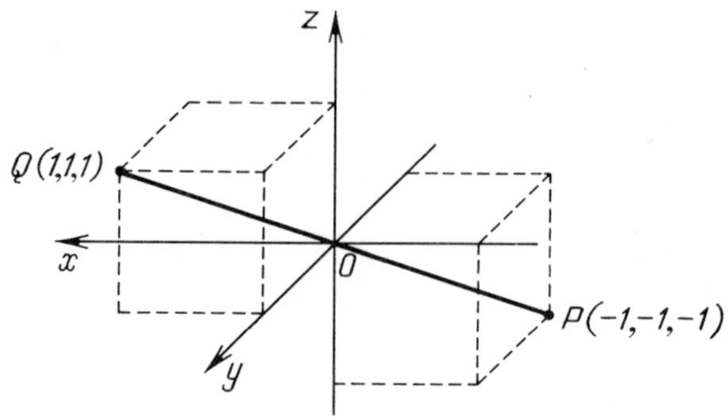

Figure 1.

Curve γ is called a *simple*, or *Jordan* curve when all three equalities

$$x(t_1) = x(t_2), \quad y(t_1) = y(t_2), \quad z(t_1) = z(t_2)$$

hold, if and only if

$$t_1 = t_2$$

A simple curve has no self-intersection points.

A curve is named *a closed curve*, or *a contour*, if its initial and terminal points coincide.

Closed curve γ is named *a simple closed* curve when all three equalities

$$x(t_1) = x(t_2), \quad y(t_1) = y(t_2), \quad z(t_1) = z(t_2)$$

hold, if and only if

$$t_1 = t_2$$

or

$$t_1 = a, \quad t_2 = b$$

Important note. The relations $(*)$ define not only a space point-set, but indicate the order of its points under changes of parameter as well. It is most clearly seen when parameter t is interpreted to be a physical quantity — *time.* We may then consider a parametrically defined curve to be a trajectory of running point $M(x, y, z)$ moving according to rule $(*)$.

Example. Relations

$$\mathbf{r}_+ = (t, t, t), \quad -1 \le t \le 1$$

and

$$\mathbf{r}_- = (-t, -t, -t), \quad -1 \le t \le 1$$

describe the same spatial point-set — the straight-line segment forming equal angles with the coordinate axes. However, under changes of parameter t from -1 to 1, running point $M(t)$ moves from position P to position

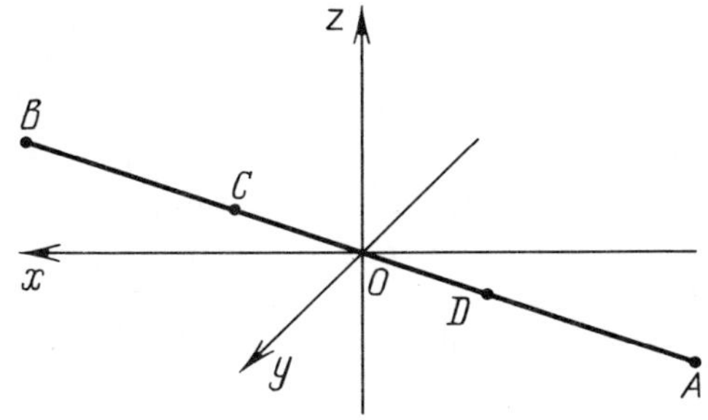

Figure 2.

Q in the first case, and in the opposite direction, from point Q to point P, in the second (Figure 1).

It is significant that the same spatial point-sets may be described by various parametric equations. As this takes place, the curve orientation may coincide or be opposite.

Let us consider one more example.

Example. Each of the following vector functions describes one spatial point-set — the straight-line segment connecting points $A(-1,-1,-1)$ and $B(1,1,1)$ (Figure 2):

$$\mathbf{r}(t) = (t^5, t^5, t^5), \quad -1 \le t \le 1 \tag{a}$$

$$\mathbf{r}(t) = (a(t), a(t), a(t)) \text{ where } a(t) = \begin{cases} t^3, & -1 \le t < 0 \\ 8t^3, & 0 \le t \le 1/2 \end{cases} \tag{b}$$

$$\mathbf{r}(t) = \left(\frac{1}{6}(t^3 - t), \frac{1}{6}(t^3 - t), \frac{1}{6}(t^3 - t)\right), \quad -2 \le t \le 2 \tag{c}$$

$$\mathbf{r} = (b(t), b(t), b(t)) \text{ where } b(t) = \begin{cases} (t+1)^3, & 2 \le t < -1 \\ 0, & -1 \le t \le 1 \\ (t-1)^3, & 1 < t \le 2 \end{cases} \tag{d}$$

In cases (a) and (b), running point $M(t)$ moves along each curve from position A to position B with parameter t increasing from -1 to 1, even though the velocities are different in each case. In case (c), running point $M(t)$ moves along the curve from position A to position D when parameter t increases from -2 to $-1/\sqrt{3}$, then in the opposite direction from position D to position C when parameter t increases from $-1/\sqrt{3}$ to $1/\sqrt{3}$, and finally when parameter t increases from $1/\sqrt{3}$ to 2, the running point once more changes direction of movement and completes it at point B (line segment CD is the superposition section of the curve). In case (d), running point $M(t)$ moves from position A to position B without changing the direction of the motion, but point $M(t)$ rests at origin $O(0,0,0)$ when parameter t lies within segment $[-1,1]$.

Thus, in all cases discussed above we have the same spatial point-sets. But can we think that we have only one curve?

There is more than a single answer to this question. But, ultimately we consider a curve in this handbook, nevertheless, to be a point-set. For this reason, all parametric equations defining the same set of points may be considered, under certain conditions (described below), as *various* parametrizations of the same curve.

Parametric equations of space curves:

in the vector form

$$\mathbf{r} = \mathbf{r}(t), \quad a \le t \le b$$

in Cartesian coordinates

$$x = x(t), \quad y = y(t), \quad z = z(t), \quad a \le t \le b$$

in cylindrical polar coordinates

$$\rho = \rho(t), \quad \theta = \theta(t), \quad z = z(t), \quad a \le t \le b$$

in spherical polar coordinates

$$r = r(t), \quad \varphi = \varphi(t), \quad \psi = \psi(t), \quad a \le t \le b$$

3.1.2 Reparametrization

Let us consider two parametrized curves: γ with vector equation

$$\mathbf{r} = \mathbf{r}(t), \quad a \le t \le b \tag{$*$}$$

and $\tilde{\gamma}$ with vector equation

$$\mathbf{r} = \tilde{\mathbf{r}}(\tau), \quad \alpha \le \tau \le \beta \tag{$\tilde{*}$}$$

Let

$$\tau = \tau(t)$$

be a continuous monotonically increasing function with segment $[a, b]$ as a range of definition and with segment $[\alpha, \beta]$ as a range of values, so that

$$\mathbf{r}(t) = \tilde{\mathbf{r}}(\tau(t))$$

for all t from segment $[a, b]$ (Figure 3).

In this case, the set of points defined by equation $(*)$ coincides with the set of points defined by equation $(\tilde{*})$,

$$\gamma = \tilde{\gamma}$$

Their velocity vectors point in the same direction but are distinguished from one another.

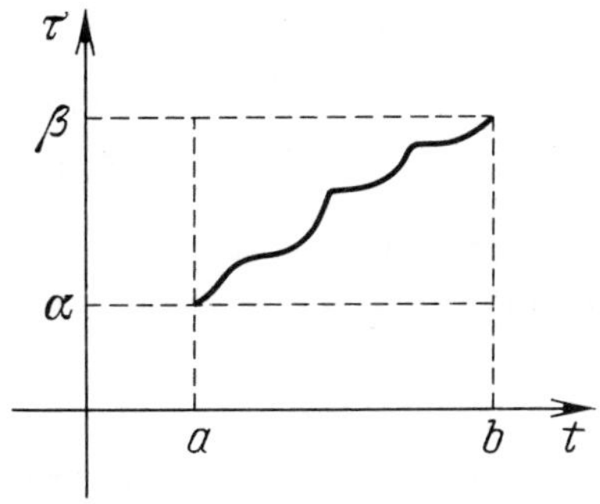

Figure 3.

Relations $(*)$ and $(\tilde{*})$ are *distinct parametrizations* of curve γ.

We say that

$$\tau = \tau(t), \quad a \leq t \leq b$$

is *a continuous transformation of the parameter* on curve γ if

a) $\tau(t)$ is a continuous function on line segment $[a, b]$,

b) $\tau(t)$ is a strictly increasing function on line segment $[a, b]$, and

c) its range of values is segment $[\alpha, \beta]$.

The vector equation

$$\mathbf{r} = \tilde{\mathbf{r}}(\tau), \quad \alpha \leq \tau \leq \beta$$

is *another parametrization* of curve γ.

It is easy to see that any curve may be parametrized in many ways.

Example. The curve defined by equation

$$\mathbf{r}(t) = (t, t, t), \quad -1 \leq t \leq 1$$

has the following parametrizations:

$$\mathbf{r}(\tau) = (\ln\tau, \ln\tau, \ln\tau), \quad e^{-1} \leq \tau \leq e; \tag{a}$$

$$\mathbf{r}(\tau) = (\tau^3, \tau^3, \tau^3), \quad -1 \leq \tau \leq 1; \tag{b}$$

$$\mathbf{r}(\tau) = (\tau \mid \tau \mid, \tau \mid \tau \mid, \tau \mid \tau \mid), \quad -1 \leq \tau \leq 1; \tag{c}$$

$$\mathbf{r}(\tau) = (a(\tau), a(\tau), a(\tau)) \text{ where } a(\tau) = \begin{cases} \tau, & -1 \leq \tau < 0 \\ 2\tau, & 0 \leq \tau \leq 1/2 \end{cases} \tag{d}$$

etc.

3.1.3 Smooth space curves

To use the apparatus and the methods of mathematical analysis for examination of curve properties, it is necessary to place some additional restrictions on the functions through which the curves may be described. One of such restrictions, among other factors, is the smoothness of a curve.

We say that space curve γ is *an s-smooth curve with respect to given parametrization*

$$\mathbf{r} = \mathbf{r}(t), \quad a \le t \le b \tag{$*$}$$

if vector function $\mathbf{r}(t)$ is an s-smooth function on segment $[a, b]$ (see Appendix B). A space curve is called *an ∞-smooth curve with respect to parametrization* $(*)$ if vector function $\mathbf{r}(t)$ is ∞-smooth on segment $[a, b]$.

We say simply that curve γ and the associated parametrization $(*)$ are *smooth* when order s of smoothness is not essential.

Let γ be a closed curve. We can construct the periodic extension of vector function $\mathbf{r}(t)$ onto all t-axes (with the period equal to $b - a$).

Closed curve γ is called *s-smooth* (*∞-smooth*, *analytic*), if $(b - a)$-periodic extension of vector function $\mathbf{r}(t)$ onto the whole t-axis is an s-smooth (∞-smooth, analytic) function.

The s-smoothness of a closed curve with vector equation

$$\mathbf{r} = \mathbf{r}(t), \quad a \le t \le b$$

is equivalent to the following conditions:

1) vector function $\mathbf{r}(t)$ is s-smooth on segment $[a, b]$, and

2) one-sided derivatives of the function at $t = a$ and at $t = b$ are respectively equal:

$$\mathbf{r}(a + 0) = \mathbf{r}(b - 0), \quad \mathbf{r}'(a + 0) = \mathbf{r}'(b - 0), \ldots, \mathbf{r}^{(s)}(a + 0) = \mathbf{r}^{(s)}(b - 0)$$

The next example shows that a space curve may have both smooth and nonsmooth parametrizations.

Example. We can parametrize the set-point in Figure 4 in several ways:

$$\mathbf{r}(t) = (a(t), a(t), a(t)) \text{ where } a(t) = \begin{cases} t, & -1 \le t < 0 \\ 2t, & 0 \le t \le 1/2 \end{cases} \tag{a}$$

(nonsmooth parametrization: derivative of vector function $\mathbf{r}(t)$ has discontinuity at $t = 0$);

$$\mathbf{r}(t) = (b(t), b(t), b(t)) \text{ where } b(t) = \begin{cases} t^3, & -1 \le t < 0 \\ 8t^3, & 0 \le t \le 1/2 \end{cases} \tag{b}$$

(2-smooth parametrization); and

$$\mathbf{r}(t) = (c(t), c(t), c(t)) \text{ where } c(t) = t, \quad -1 \le t \le 1/2 \tag{c}$$

(analytic parametrization).

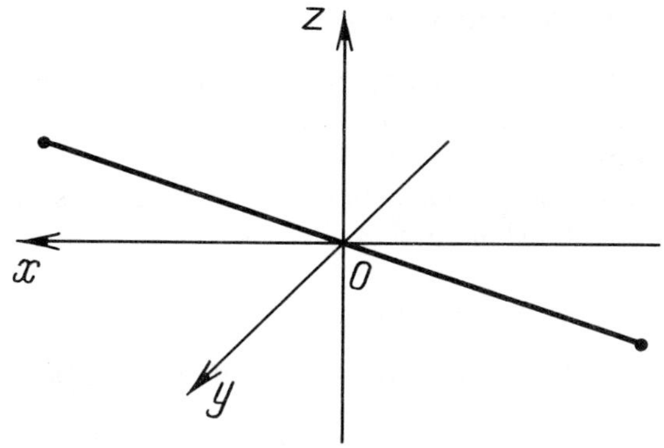

Figure 4.

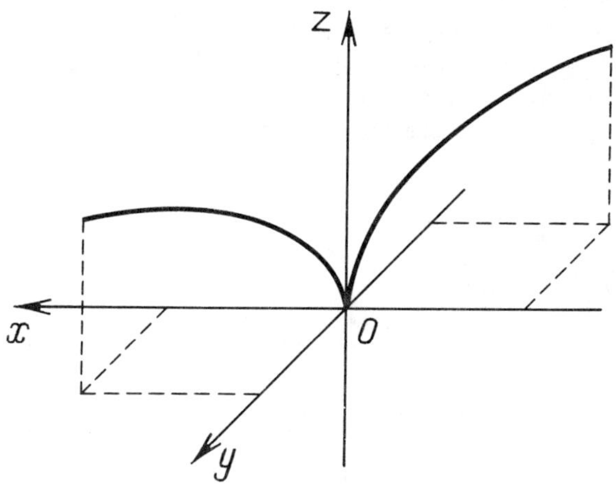

Figure 5.

Let us consider an example of a different type.

Example. Let γ be a curve defined by equation

$$\mathbf{r} = (t^6, t^6, t^4), \quad -1 \leq t \leq 1$$

With respect to this parametrization, curve γ is ∞-smooth. But the corresponding point-set (geometric image) described by the formula has a singularity (Figure 5). Note that derivative

$$\mathbf{r}'(t) = (6t^5, 6t^5, 4t^3)$$

is equal to zero at $t = 0$.

The last example shows that the requirements of smoothness on parametrization may not be sufficient for the absence of curve singularities.

Let γ be an s-smooth curve, and

$$\mathbf{r} = \mathbf{r}(t), \quad a \le t \le b$$

is the corresponding equation of the curve.

We say that point $M(t_0)$, $a < t_0 < b$, is *a singular point (with respect to given parametrization)* of curve γ, if

$$\mathbf{r}'(t_0) = \mathbf{0}$$

Point $M(t)$, $a < t < b$, of curve γ is called *regular (nonsingular)* if

$$\mathbf{r}'(t_0) \neq \mathbf{0}$$

Vector $\mathbf{r}'(t)$ is called *a velocity vector for curve γ with respect to given parametrization* $\mathbf{r} = \mathbf{r}(t)$. The vector equal to zero at a singular point differs from zero at any regular point.

Parametrization singularities may be of two kinds.

A. *Parametrization singularities (remediable)* disappear with the appropriate reparametrization of curve.

Example. It is easy to see that point $O(0,0,0)$ is a singular point of parametrized curve

$$\mathbf{r} = (t^3, t^3, t^3), \quad -1 \le t \le 1$$

Let τ be a new parameter for this curve so that

$$\tau = t^3$$

Then every point of parametrized curve

$$\tilde{\mathbf{r}} = (\tau, \tau, \tau), \quad -1 \le \tau \le 1$$

is nonsingular (regular).

B. *Geometric singularities (nonremediable)* are impossible to remedy by reparametrization (see the above example of curve (t^6, t^6, t^4)).

Ranges of the parameter may not only be segments, but also intervals (finite or infinite) and half-intervals. It is easy to reformulate the definitions above for each of these cases. There are only two notions (of a simple curve and of an analytic curve) which deserve special attention.

Let, for example, the range of definition be an interval.

1. Parametrized curve

$$\mathbf{r} = \mathbf{r}(t),$$

$$a < t < b$$

is said to be *a simple Jordan curve*, if equality

$$\mathbf{r}(t_1) = \mathbf{r}(t_2)$$

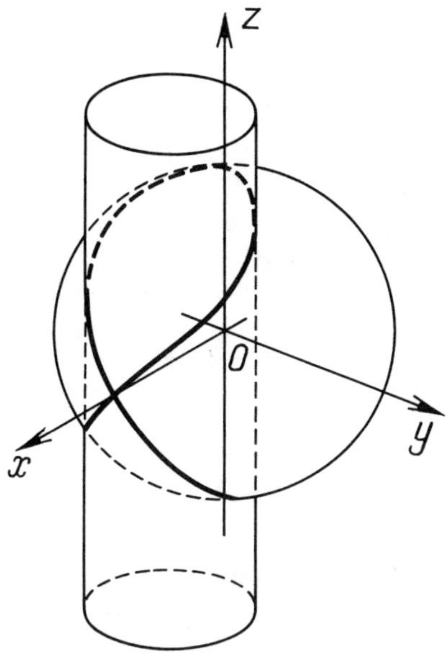

Figure 6.

holds for $t_1 = t_2$ only, and close points of the curve correspond to close parameter values.

Example. The necessity of the second requirement in the definition above may be explained with the help of the curve described by vector function

$$\mathbf{r} = (\cos^2 t, \cos t \sin t, \sin t),$$

$$0 < t < \frac{3\pi}{2}$$

(an arc of the Viviani curve).

To parameter values $t_1 = \varepsilon$ ($\varepsilon > 0$ is a small number) and $t_2 = \pi$ there correspond points $M_1(\cos^2 \varepsilon, \cos\varepsilon \sin\varepsilon, \sin\varepsilon)$ and $M_2(1, 0, 0)$ on the curve. The smaller is ε (consequently, the greater is the distinction between t_1 and t_2), the closer are points M_1 and M_2.

The curve on Figure 6 is not said to be a simple curve.

2. We say that a space curve is *an analytic curve with respect to parametrization*

$$\mathbf{r} = \mathbf{r}(t),$$

$$a < t < b$$

if vector function $\mathbf{r}(t)$ is analytic on interval (a, b) (see Appendix B).

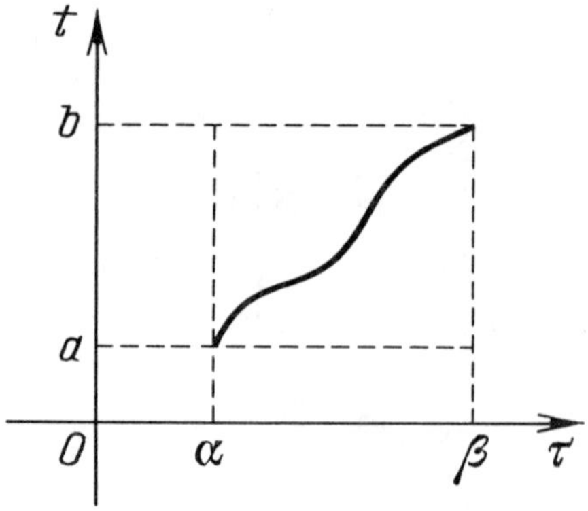

Figure 7.

3.1.4 Regular space curves

Space curve γ is named *an s-regular curve with respect to parametrization*

$$\mathbf{r} = \mathbf{r}(t), \quad a \le t \le b \tag{$*$}$$

if the following conditions are met:

1) this curve is s-smooth with respect to parametrization $(*)$, and
2) $\mathbf{r}'(t) \neq \mathbf{0}$ at every point of segment $[a, b]$.

The last condition implies that the velocity vector $\mathbf{r}'(t)$ is nonvanishing at every point of the definition. Further, we shall see how essential is the second condition.

If $\mathbf{r}(t) = (x(t), y(t), z(t))$, then the condition is equivalent to inequality

$$|\, \mathbf{r}'(t) \,| = \sqrt{[x'(t)]^2 + [y'(t)]^2 + [z'(t)]^2} > 0$$

Length $|\mathbf{r}'(t)|$ of the velocity vector of curve γ is said to be *the curve velocity* (with respect to the given parametrization) at point t, or at point $M(t)$ of the curve.

As the parameter t changes, running point $M(t)$ moves along regular curve γ without stopping and turning back because the curve velocity (with respect to given parametrization) is distinct from zero for any value of parameter.

Reparametrization of regular curves. Regular transformation of parameter. Consider function

$$t = t(\tau)$$

defined on segment $[\alpha, \beta]$ and possessing the following properties:

1) function $t = t(\tau)$ is s-smooth on segment $[\alpha, \beta]$,
2) its set of values coincides with segment $[a, b]$, and
3) $t'(\tau) > 0$ at every point of segment $[\alpha, \beta]$ (Figure 7).

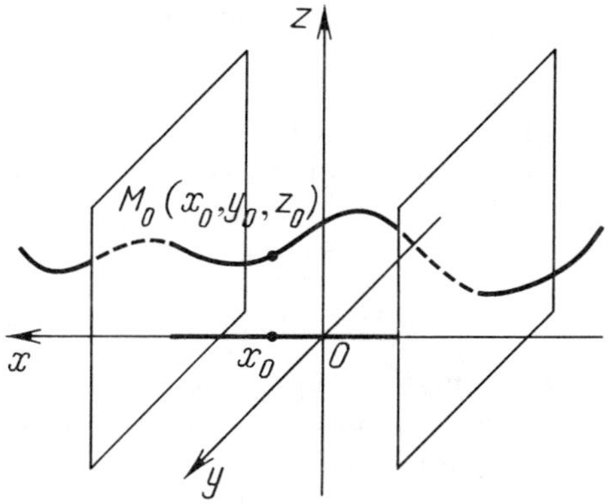

Figure 8.

Note. Under the stated conditions there exists function

$$\tau = \tau(t)$$

inverse to function $t = t(\tau)$, and with analogous properties:

1′) function $\tau(t)$ is s-smooth on segment $[a, b]$,

2′) the set of values of function $\tau(t)$ coincides with segment $[\alpha, \beta]$, and

3′) $\tau'(t) > 0$ at every point of segment $[a, b]$.

Let s-regular curve γ be defined by formula $\mathbf{r} = \mathbf{r}(t)$, $a \leq t \leq b$. By replacing parameter t with function $t(\tau)$ being subject to the conditions 1–3, we obtain vector relation $\mathbf{r} = \tilde{\mathbf{r}}(\tau) = \mathbf{r}(t(\tau))$ — a new vector equation of the curve. With respect to the new parametrization, curve γ is also s-regular: $|\tilde{\mathbf{r}}'(\tau)| = |\mathbf{r}'(t)| \cdot |t'(\tau)| > 0$.

The parameter transformation

$$t = t(\tau)$$

with properties 1–3 is called *s-regular* or simply *regular* (when the value of s is immaterial). Curve γ is said to be *s-regular* if there exists for it such a parametrization

$$\mathbf{r} = \mathbf{r}(t), \quad a \leq t \leq b$$

with respect to which curve γ is s-regular, that is, s-smooth, and the inequality $|\mathbf{r}'(t)| > 0$ holds at every point of segment $[a, b]$.

Advantages of regular parametrization. Let curve γ with parametric equations

$$\mathbf{r} = \mathbf{r}(t) = (x(t), y(t), z(t)), \quad a \leq t \leq b$$

be s-regular and $M_0(t_0)$ — a point on the curve (for the sake of definition, $x'(t_0) \neq 0$).

According to the well-known implicit-function theorem, equation

$$x = x(t)$$

is solvable with respect to t in the neighborhood of point t_0:

$$t = t(x)$$

and, thus, we may describe curve γ in the vicinity of point M_0 by equations

$$y = y(x),$$

$$z = z(x)$$

that is, the curve allows a one-to-one projection onto the x-axis (Figure 8).

In a similar manner, the notions of a regular curve and a regular transformation of parameter may be used in cases where domain T of the parameter variability is an interval or a half-interval.

Let us formulate, for example, the definitions of an ∞-regular curve and a regular analytic curve (in case T is an interval).

Curve γ is called an *∞-regular* curve (*regular analytic* curve, respectively) if there exists for it such a parametrization

$$\mathbf{r} = \mathbf{r}(t),$$

$$a < t < b$$

with respect to which the curve is ∞-smooth (analytic, respectively), and inequality $\mathbf{r}'(t) \neq \mathbf{0}$ holds at every point of interval (a, b).

Closed curve

$$\mathbf{r} = \mathbf{r}(t),$$

$$a \leq t \leq b$$

is said to be an *s-regular (∞-regular, regular analytic)* curve if the periodic expansion of vector function $\mathbf{r}(t)$ is an s-smooth (∞-smooth, analytic) function on the whole numerical axis and $\mathbf{r}'(t) \neq \mathbf{0}$.

3.1.5 Implicit definition of a space curve

Another abundant method for defining a space curve is the *implicit way* — when the curve is described by an equation system in the following form:

$$F(x, y, z) = 0, \quad G(x, y, z) = 0 \qquad (*)$$

It is apparent that functions $F(x, y, z)$ and $G(x, y, z)$ are nonarbitrary. To obtain a point-set corresponding to the intuitive view of a space curve, we must subject these functions $F(x, y, z)$ and $G(x, y, z)$ to definite restrictions. The requirements imposed on these restrictions must be such that we would have a reason to call the set of the points $M(x, y, z)$ whose coordinates x, y and z transform the equation $(*)$ to identity (or each connected component of this point-set) a spatial curve.

Let us point out an important special case which we shall use further.

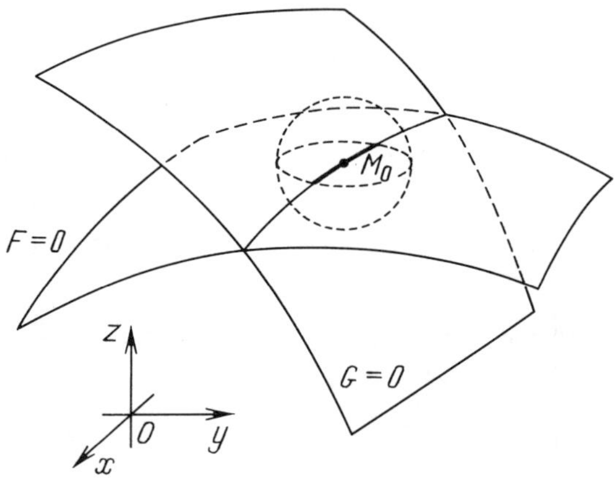

Figure 9.

Let $F(x, y, z)$ and $G(x, y, z)$ be smooth functions and at some point $M_0(x_0, y_0, z_0)$ the following conditions be met:

$$F(x_0, y_0, z_0) = 0, \quad G(x_0, y_0, z_0) = 0$$

$$\text{rang}\begin{pmatrix} F_x(x_0, y_0, z_0) & F_y(x_0, y_0, z_0) & F_z(x_0, y_0, z_0) \\ G_x(x_0, y_0, z_0) & G_y(x_0, y_0, z_0) & G_z(x_0, y_0, z_0) \end{pmatrix} = 2 \qquad (**)$$

The last relation is equivalent to inequality

$$\begin{vmatrix} F_x(x_0, y_0, z_0) & F_y(x_0, y_0, z_0) \\ G_x(x_0, y_0, z_0) & G_y(x_0, y_0, z_0) \end{vmatrix}^2 + \begin{vmatrix} F_y(x_0, y_0, z_0) & F_z(x_0, y_0, z_0) \\ G_y(x_0, y_0, z_0) & G_z(x_0, y_0, z_0) \end{vmatrix}^2$$

$$+ \begin{vmatrix} F_z(x_0, y_0, z_0) & F_x(x_0, y_0, z_0) \\ G_z(x_0, y_0, z_0) & G_x(x_0, y_0, z_0) \end{vmatrix}^2 > 0$$

It is well known (the implicit-function theorem) that the fulfillment of these conditions is sufficient for expressing any two variables as a function of the third. In particular, the fulfillment of inequality

$$\begin{vmatrix} F_x(x_0, y_0, z_0) & F_y(x_0, y_0, z_0) \\ G_x(x_0, y_0, z_0) & G_y(x_0, y_0, z_0) \end{vmatrix} \neq 0$$

is sufficient for the existence of a solution of system $(*)$ in the form

$$y = y(x), \quad z = z(x)$$

In this case, the system defines a simple Jordan curve in some neighborhood of point $M_0(x_0, y_0, z_0)$. Moreover, the curve is regular (Figure 9) and we may choose coordinate x as a parameter for this curve.

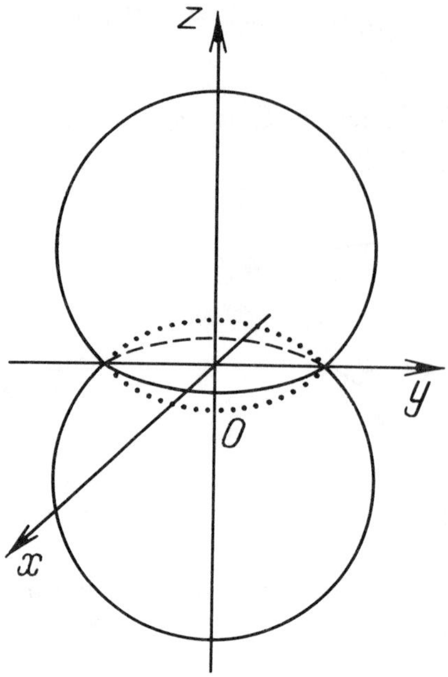

Figure 10.

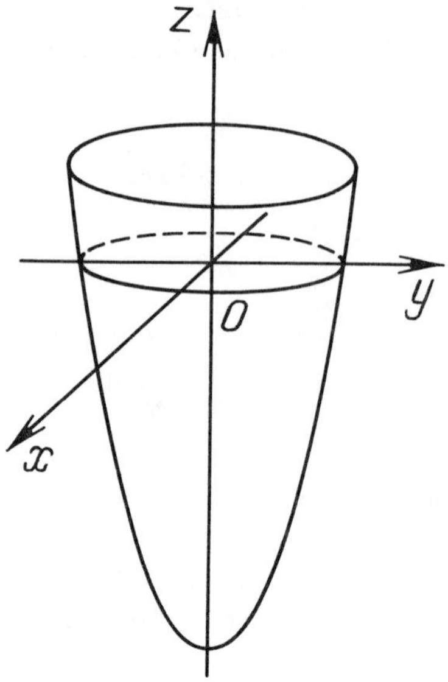

Figure 11.

It is clear that the same curve may be described by various systems of the form $(*)$.

Example. System

$$x^2 + y^2 + (z-1)^2 = 5, \quad x^2 + y^2 + (z+1)^2 = 5$$

(Figure 10) and

$$x^2 + y^2 - z - 4 = 0, \quad z = 0$$

(Figure 11) present the same circle with radius 2 and center at origin $O(0,0,0)$ lying on plane Oxy.

Point $M_0(x_0, y_0)$ of the implicitly defined curve is called *a simple (non-singular, regular) point* if conditions $(**)$ are met.

Point $M_0(x_0, y_0)$ of the implicitly defined curve is called its *singular* point if

$$F(x_0, y_0, z_0) = 0, \quad G(x_0, y_0, z_0) = 0,$$

$$\operatorname{rang} \begin{pmatrix} F_x(x_0, y_0, z_0) & F_y(x_0, y_0, z_0) & F_z(x_0, y_0, z_0) \\ G_x(x_0, y_0, z_0) & G_y(x_0, y_0, z_0) & G_z(x_0, y_0, z_0) \end{pmatrix} < 2$$

at the point.

3.1.6 Other definitions of a space curve

Curve as an equivalence class.

A continuous mapping of real axis segment $[a, b]$ into space $\mathbf{R}^n$

$$\gamma : [a, b] \to \mathbf{R}^n \tag{$*$}$$

is called *a path*. In other words, a path is a vector-valued function

$$r = r(t) = (x_1(t), x_2(t), \dots, x_n(t))$$

of real parameter t continuous at every point of segment $[a, b]$.

If mapping $(*)$ is bijective, then one says that the path is *simple*. Points $A = r(a)$ and $B = r(b)$ are called *the end-points* of path γ (A is *an initial* point and B — *a terminal* point).

Path γ is said to be *simple* when equality $r(t_1) = r(t_2)$ holds if, and only if, $t_1 = t_2$.

Path γ is said to be *closed*, if its end-points A and B coincide. The path is called *smooth* if its coordinate functions $x_1(t), x_2(t), \dots, x_n(t)$ are smooth. Smooth path γ is said to be *regular*, if $r'(t) \neq 0$ at every point of segment $[a, b]$.

Two regular paths

$$\gamma : \quad r = r(t), \quad a \le t \le b,$$

$$\tilde{\gamma} : \quad r = \tilde{r}(\tau), \quad \alpha \le \tau \le \beta,$$

are said to be *equivalent*, if there exists smooth function $\tau = \tau(t)$ satisfying the following conditions:

1) its domain of definition is segment $[a, b]$;

2) its converse domain is segment $[\alpha, \beta]$;

3) derivative $\tau'(t) > 0$ at every point of segment $[a, b]$

and such that $r(t) = \tilde{r}(\tau(t))$ for any $t \in [a, b]$.

An equivalence class of regular paths with respect to the above relation is called *a regular curve*.

An equivalence class of regular closed paths is respectively called *a regular closed curve*, and an equivalence class of regular simple paths is called *a regular simple curve*.

A regular curve is said to be *a (geometric) carrier* for any one of its generating paths.

Parametrized curve.

A continuous mapping

$$\gamma : G \to \mathbf{R}^n \tag{$*$}$$

where G is a one-dimensional manifold (interval or circle) is called *a parametrized curve*. When mapping $(*)$ is bijective we say that the curve is *simple*. In case G is a circle, the parametrized curve is said to be *closed*.

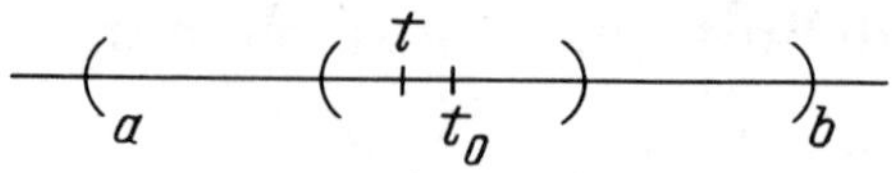

Figure 12.

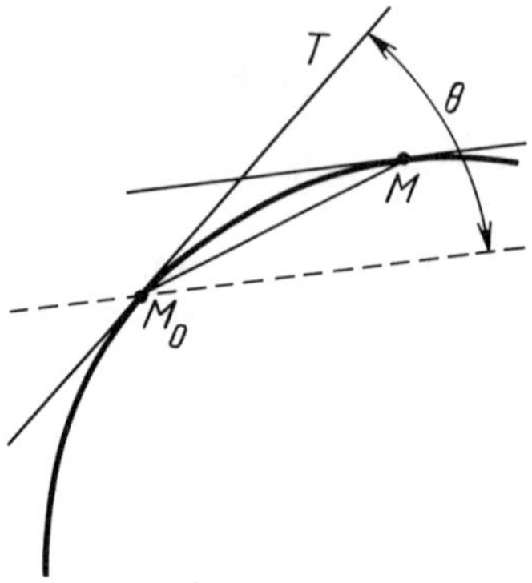

Figure 13.

Parametrized curve γ is said to be *smooth* if its coordinate functions $x_1(t), x_2(t), \ldots, x_n(t)$ are smooth (with respect to the local coordinate taken on manifold G). Smooth parametrized curve γ is said to be *regular*, if $r'(t) \neq 0$ for any $t \in G$. Two regular parametrized curves $\gamma : G \to \mathbf{R}^n$ and $\tilde{\gamma} : \tilde{G} \to \mathbf{R}^n$ are said to be *equivalent*, if there exists diffeomorphism $p : G \to \tilde{G}$ such that $\gamma = \tilde{\gamma} \circ p$.

An equivalence class of regular parametrized curves with respect to the above-described equivalence relation is called *a non parametrized* regular curve (simple or closed).

3.1.7 Tangents and normal planes of space curves

Let γ be a smooth curve defined by vector equation

$$\mathbf{r} = \mathbf{r}(t), \quad t \in T \qquad (*)$$

Denote by M_0 a point on the curve defined by parameter value t_0. Let t be a parameter value from some neighborhood of point t_0, and M — a corresponding point on the curve (Figure 12). It is obvious that $M \to M_0$ when $t \to t_0$. That is, the length of straight line segment MM_0 tends to zero.

Straight line M_0T is said to be *tangent* to curve γ at point M_0, if the lesser angle between this line and variable straight line M_0M (angle $\Delta\Theta$ in Figure 13) tends to zero when $M \to M_0$.

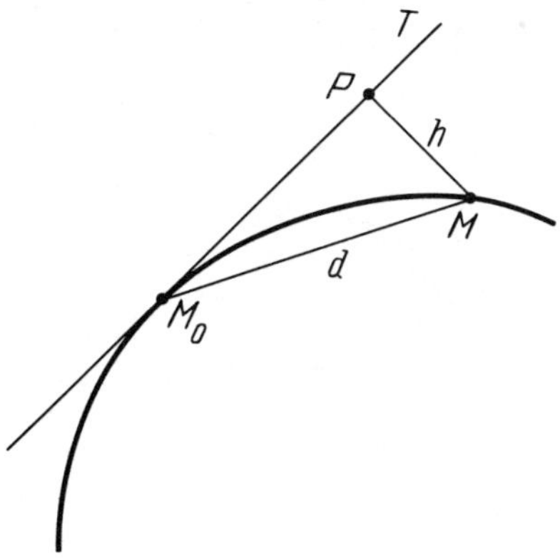

Figure 14.

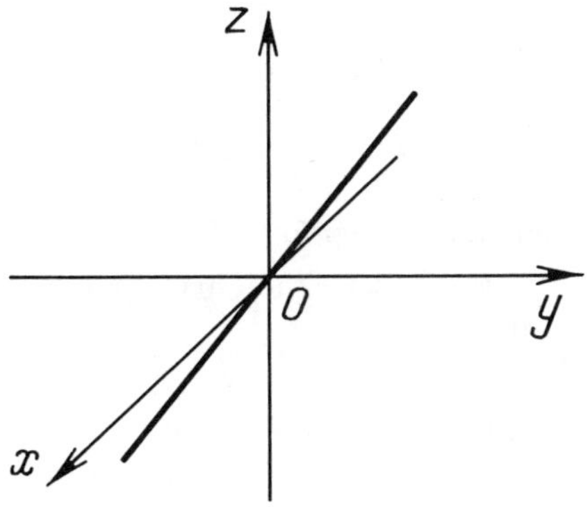

Figure 15.

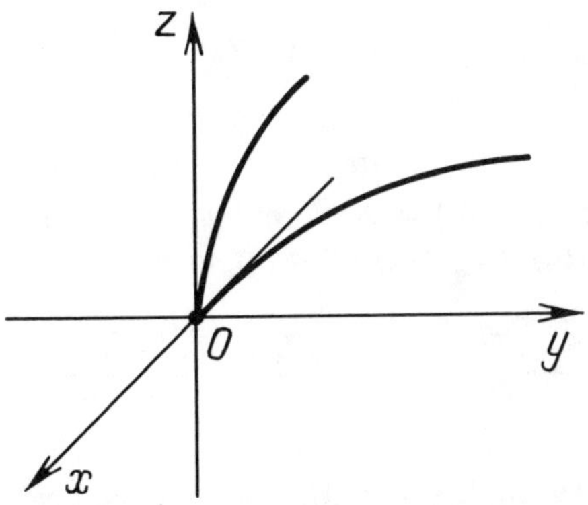

Figure 16.

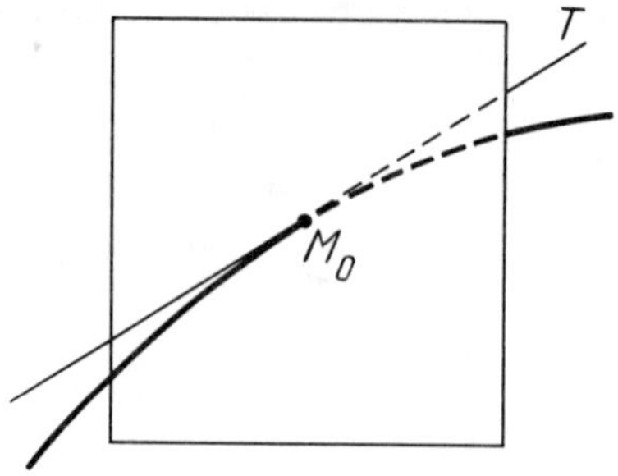

Figure 17.

Let P be an orthogonal projection of point M on line M_0T, and h — a length of line segment MP. Straight line M_0T is tangent to curve γ at point M_0, if

$$\frac{h}{d} \to 0$$

as $M \to M_0$ (Figure 14). If M_0 is a regular point of curve γ, that is, $\mathbf{r}'(t_0) \neq \mathbf{0}$, then there exists a tangent to the curve at this point and the tangent is uniquely determined by point M_0.

In case $\mathbf{r}'(t_0) = \mathbf{0}$ the tangent to the curve at point M_0 may exist, for example, for the curve in Figure 15:

$$\mathbf{r} = (t^5, t^5, t^5), \quad -1 \leq t \leq 1 \quad (t_0 = 0)$$

or it just as well might be nonexistent, for example, for the curve $\mathbf{r}(t) = (x(t), y(t), z(t))$ in Figure 16:

$$x(t) = \begin{cases} t^3 & -1 \leq t \leq 0 \\ 0 & 0 < t \leq 1 \end{cases} \quad y(t) = \begin{cases} 0 & -1 \leq t \leq 0 \\ t^3 & 0 < t \leq 1 \end{cases} \quad z(t) = \sin t^3$$

A regular curve has a unique tangent at each of its points.

If $\mathbf{r}'(t_0) = \ldots = \mathbf{r}^{(m-1)}(t_0) = \mathbf{0}$ and $\mathbf{r}^{(m)}(t_0) \neq \mathbf{0}$, then equations of the tangent at point $M_0(t_0)$ have the following form:

$$\frac{x - x(t_0)}{x^{(m)}(t_0)} = \frac{y - y(t_0)}{y^{(m)}(t_0)} = \frac{z - z(t_0)}{z^{(m)}(t_0)}$$

A plane passing through point M_0 on curve γ perpendicularly to its tangent M_0T at the point is called *a normal plane* of the curve at point M_0 (Figure 17).

Equations of a tangent and a normal plane for a space curve at regular point $M_0(x_0, y_0, z_0)$

If the curve is defined *in a vector form*

$$\mathbf{r} = \mathbf{r}(t), \quad \mathbf{r}_0 = \mathbf{r}(t_0), \mathbf{r}'(t_0) = \mathbf{0}$$

then an equation of the tangent has the following form:

$$\mathbf{r} = \mathbf{r}_0 + u\mathbf{r}'(t),$$

$$-\infty < u < \infty$$

and

$$(\mathbf{r} - \mathbf{r}_0, \mathbf{r}'(t_0)) = 0$$

is an equation of the normal plane.

If the curve is defined *in a parametric form*

$$x = x(t), \quad y = y(t), \quad z = z(t),$$

$$x_0 = x(t_0), \quad y_0 = y(t_0), \quad z_0 = z(t_0),$$

$$[x'(t_0)]^2 + [y'(t_0)]^2 + [z'(t_0)]^2 > 0$$

then equations of the tangent have the form

$$\frac{x - x_0}{x'(t_0)} = \frac{y - y_0}{y'(t_0)} = \frac{z - z_0}{z'(t_0)}$$

and

$$x'(t_0)(x - x_0) + y'(t_0)(y - y_0) + z'(t_0)(z - z_0) = 0$$

is an equation of the normal plane.

If the curve is defined *in an explicit form*

$$y = y(x), \quad z = z(x),$$

$$y_0 = y(x_0), \quad z_0 = z(x_0)$$

then equations of the tangent have the form

$$x - x_0 = \frac{y - y_0}{y'(x_0)} = \frac{z - z_0}{z'(x_0)}$$

and

$$x - x_0 + y'(x_0)(y - y_0) + z'(x_0)(z - z_0) = 0$$

is an equation of the normal plane.

If the curve is defined *in an implicit form*

$$F(x, y, z) = 0, \quad G(x, y, z) = 0,$$

$$F(x_0, y_0, z_0) = 0, \quad G(x_0, y_0, z_0) = 0,$$

$$\text{rang} = \begin{pmatrix} F_x & F_y & F_z \\ G_x & G_y & G_z \end{pmatrix}_{M_0} = 2$$

then equations of the tangent have the form

$$x = x_0 + u \begin{vmatrix} F_y & F_z \\ G_y & G_z \end{vmatrix}_{M_0}$$

$$y = y_0 + u \begin{vmatrix} F_z & F_x \\ G_z & G_x \end{vmatrix}_{M_0}$$

$$z = z_0 + u \begin{vmatrix} F_x & F_y \\ G_x & G_y \end{vmatrix}_{M_0}$$

where

$$-\infty < u < \infty$$

and

$$\begin{vmatrix} x - x_0 & y - y_0 & z - z_0 \\ F_x(x_0, y_0, z_0) & F_y(x_0, y_0, z_0) & F_z(x_0, y_0, z_0) \\ G_x(x_0, y_0, z_0) & G_y(x_0, y_0, z_0) & G_z(x_0, y_0, z_0) \end{vmatrix} = 0$$

is an equation of the normal plane.

3.1.8 Arc lengths of space curves

Let curve γ be defined by vector equation

$$\mathbf{r} = \mathbf{r}(t), \quad a \leq t \leq b$$

Divide segment $[a, b]$ into parts by points

$$t_0, t_1, \ldots, t_{n-1}, t_n$$

so that

$$a = t_0 < t_1 < \ldots < t_{n-1} < t_n = b$$

Mark points M_i on it, coresponding to parameter values t_i, and construct broken line $M_0 M_1 \ldots M_{n-1} M_n$ inscribed in curve γ (Figure 18).

Curve γ is said to be *rectifiable* if the set of lengths of all broken lines inscribed in curve γ as above is bounded. Least upper bound of this length set is called *the arc length of curve* γ, or simply *the length of curve* γ.

Notation: $s(\gamma)$, S.

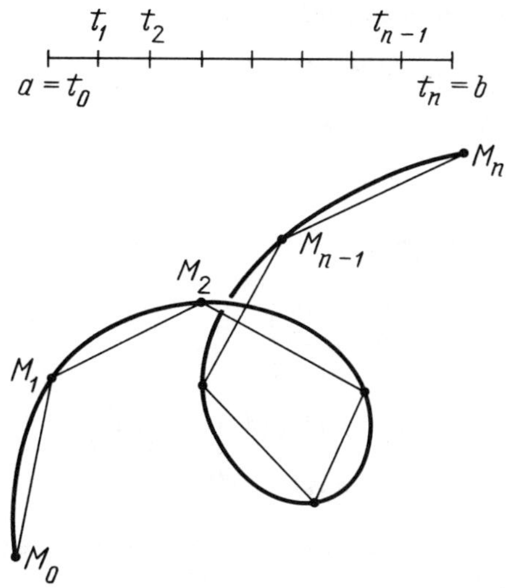

Figure 18.

Every smooth plane curve is rectifiable; its length may be calculated by one of the following formulae:

$$s = \int_a^b |\mathbf{r}'(t)|\, dt$$

for vector description ($\mathbf{r} = \mathbf{r}(t)$),

$$s = \int_a^b \sqrt{[x'(t)]^2 + [y'(t)]^2 + [z'(t)]^2}\, dt$$

for parametric description with respect to Cartesian coordinates ($x = x(t)$, $y = y(t)$, $z = z(t)$),

$$s = \int_a^b \sqrt{[r'(t)]^2 + r^2(t)[\varphi'(t)]^2 + [z'(t)]^2}\, dt$$

for parametric description with respect to cylindrical polar coordinates ($r = r(t)$, $\varphi = \varphi(t)$, $z = z(t)$),

$$s = \int_a^b \sqrt{[\rho'(t)]^2 + \rho^2(t)[\theta'(t)]^2 + \rho^2(t)\sin^2\theta(t)[\varphi'(t)]^2}\, dt$$

for parametric description with respect to spherical polar coordinates ($\rho = \rho(t)$, $\theta = \theta(t)$, $\varphi = \varphi(t)$).

3.1.9 Natural parametrization of regular space curves

Let γ be an s-regular curve with vector equation

$$\mathbf{r} = \mathbf{r}(t), \quad a \le t \le b$$

and S — a length of the curve.

A function

$$s(t) = \int_a^t |\mathbf{r}'(\xi)|\, d\xi$$

strictly increases on segment $[a, b]$ and its derivative

$$s'(t) = |\mathbf{r}'(t)|$$

is positive. In addition, function $s(t)$ is s-smooth on segment $[a, b]$ and its set of values coincides with segment $[0, S]$.

Thus, the parameter transformation

$$t = t(s)$$

is s-regular and we may take the arc length as a new *natural* parameter on curve γ.

A curve parametrization with an arc length as a parameter is called *a natural parametrization*. The natural parametrization of an s-regular curve is s-regular.

If

$$\mathbf{r} = \mathbf{r}(s), \quad 0 \le t \le S$$

is the natural parametrization of curve γ, then

$$|\mathbf{r}'(s)| = 1$$

For this reason, a naturally parametrized curve is said to be *a unit velocity curve*.

For the parametrization

$$\mathbf{r} = \mathbf{r}(t), \quad a \le t \le b$$

of a smooth curve to be natural, it is necessary and sufficient that

$$[x'(t)]^2 + [y'(t)]^2 + [z'(t)]^2 = 1$$

for every $t \in [a, b]$.

Let t_0 be an inner point of segment $[a, b]$. Then function

$$s(t) = \int_{t_0}^t |\mathbf{r}'(\xi)|\, d\xi$$

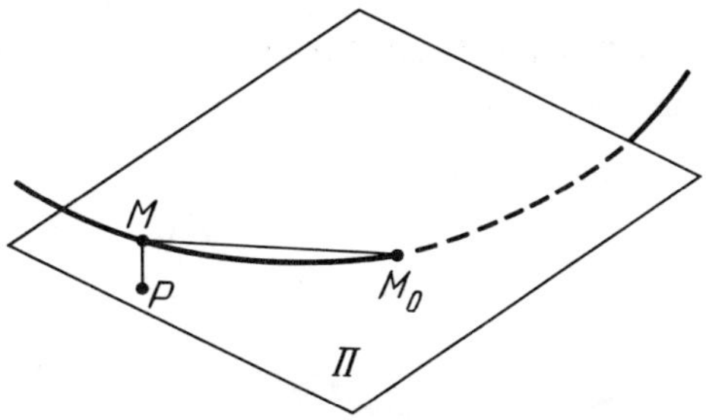

Figure 19.

takes both negative ($a \leq t < t_0$) and positive ($t_0 < t \leq b$) values. The function keeps the name of *a natural parameter.*

Note. The above permits the use of natural parametrization in cases where domain T of a parameter definition is an interval or a half-interval (finite or infinite).

The natural parametrization of an ∞-regular (regular analytic, respectively) curve is ∞-regular (regular analytic, respectively).

3.1.10 Osculating plane of a space curve

Let γ be a regular curve, M_0 — a point on the curve, and Π — a plane passing through the point. Then let M be a point on curve γ close to point M_0 and P — an orthogonal projection of point M on plane Π (Figure 19). Denote by h a length of segment MP, and by d — length of segment M_0M.

Plane Π is said to be *an osculating plane* of curve γ at point M_0 if

$$\frac{h}{d^2} \to 0$$

for $M \to M_0$. The osculating plane Π of curve γ is a limiting plane for variable plane Π_M passing through tangent M_0T to curve γ at point M_0 and running point M of the curve; that is, the least angle between planes Π and Π_M (angle Θ in Figure 20) or the least angle between the normals to these planes (equal to it) tends to zero when $M \to M_0$.

Let curve γ be defined by equation $\mathbf{r} = \mathbf{r}(t)$, and point M_0 of the curve be corresponding to parameter value t_0.

If vectors $\mathbf{r}'(t_0)$ and $\mathbf{r}''(t_0)$ are noncolinear, then an osculating plane of curve γ exists at the point, and the plane is uniquely determined.

A vector equation of an osculating plane for curve γ at point $M_0(t_0)$ has the following view:

$$(\mathbf{r} - \mathbf{r}(t_0), \mathbf{r}'(t_0), \mathbf{r}''(t_0)) = 0$$

In the case of the curve coordinate description

$$x = x(t), \quad y = y(t), \quad z = z(t)$$

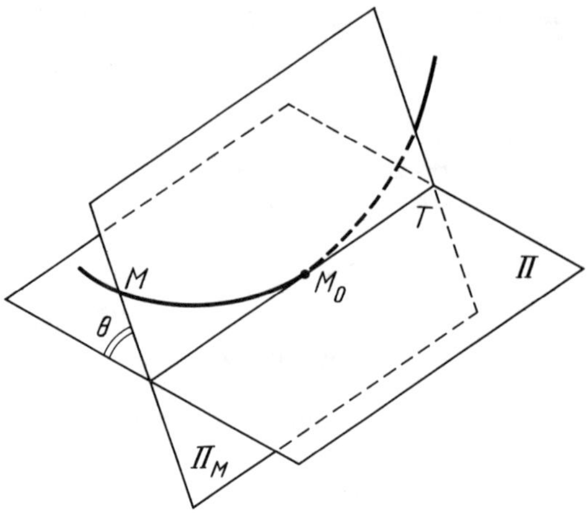

Figure 20.

an equation of the osculating plane may be written as follows:

$$\begin{vmatrix} x - x(t_0) & y - y(t_0) & z - z(t_0) \\ x'(t_0) & y'(t_0) & z'(t_0) \\ x''(t_0) & y''(t_0) & z''(t_0) \end{vmatrix} = 0$$

If vectors $\mathbf{r}'(t_0)$ and $\mathbf{r}''(t_0)$ are colinear, then any plane containing a tangent to the curve at point M_0 will be its osculating plane at point M_0.

If

$$(\mathbf{r}'(t_0), \mathbf{r}''(t_0), \mathbf{r}'''(t_0)) \neq 0$$

(a torsion of the curve differs from zero at point M_0), then an osculating plane intersects the curve at point M_0 (see 3.1.14 below regarding torsion of space curves).

An orthogonal projection of a curve on its osculating plane at point M_0 has the same curvature at the point as the projected curve (see 3.1.12 below regarding curvature of space curves).

3.1.11 Principal normal and binormal of a space curve. Rectifying plane

Let γ be a regular curve, and M_0 — a point on the curve. Any straight line passing through point M_0 perpendicularly to a tangent of curve γ at the point is called *a normal* of the curve at point M_0. All normals of the curve at point M_0 lie on its normal plane at the point. The normal of curve γ at point M_0 lying in an osculating plane of the curve at the point is said to be *the principal normal* of curve γ at point M_0. The normal of curve γ perpendicular to the osculating point of the curve at the point is called *a binormal* of curve γ at point M_0 (Figure 21).

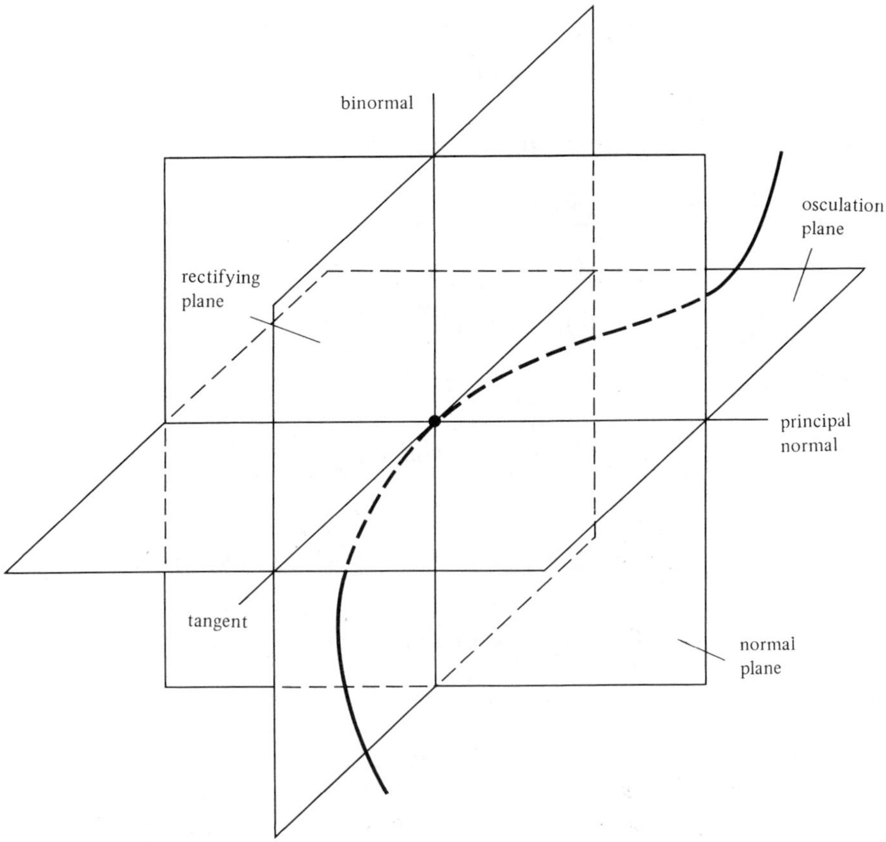

Figure 21.

A plane passing through a tangent and a binormal of curve γ at point M_0 is called *a rectifying plane* of curve γ at point M_0.

Equations of principal normal, binormal and rectifying plane.

If the curve is defined *in a vector form*

$$\mathbf{r} = \mathbf{r}(t), \quad \mathbf{r}_0 = \mathbf{r}(t_0), \quad \mathbf{r}'(t_0) = \mathbf{0}$$

then equations of the principal normal and binormal have the form

$$\mathbf{r} = \mathbf{r}_0 + u[\mathbf{r}'(t_0) \times \mathbf{r}''(t_0)] \times \mathbf{r}'(t_0), \quad -\infty < u < \infty$$

and

$$\mathbf{r} = \mathbf{r}_0 + u[\mathbf{r}'(t_0) \times \mathbf{r}''(t_0)], \quad -\infty < u < \infty$$

respectively, and

$$(\mathbf{r} - \mathbf{r}_0, \mathbf{r}'(t_0), [\mathbf{r}'(t_0) \times \mathbf{r}''(t_0)]) = 0$$

is an equation of the osculating plane.

If the curve is defined *in a parametric form*

$$x = x(t), \quad y = y(t), \quad z = z(t),$$

$$x_0 = x(t_0), \quad y_0 = y(t_0), \quad z_0 = z(t_0),$$

$$[x'(t_0)]^2 + [y'(t_0)]^2 + [z'(t_0)]^2 > 0$$

then equations of the principal normal and binormal have the form

$$x = x_0 + u\left(\begin{vmatrix} z'(t_0) & x'(t_0) \\ z''(t_0) & x''(t_0) \end{vmatrix} z'(t_0) - \begin{vmatrix} x'(t_0) & y'(t_0) \\ x''(t_0) & y''(t_0) \end{vmatrix} y'(t_0)\right),$$

$$y = y_0 + u\left(\begin{vmatrix} x'(t_0) & y'(t_0) \\ x''(t_0) & y''(t_0) \end{vmatrix} x'(t_0) - \begin{vmatrix} y'(t_0) & z'(t_0) \\ y''(t_0) & z''(t_0) \end{vmatrix} z'(t_0)\right),$$

$$z = z_0 + u\left(\begin{vmatrix} y'(t_0) & z'(t_0) \\ y''(t_0) & z''(t_0) \end{vmatrix} y'(t_0) - \begin{vmatrix} z'(t_0) & x'(t_0) \\ z''(t_0) & x''(t_0) \end{vmatrix} x'(t_0)\right),$$

and

$$\frac{x - x_0}{\begin{vmatrix} y'(t_0) & z'(t_0) \\ y''(t_0) & z''(t_0) \end{vmatrix}} = \frac{y - y_0}{\begin{vmatrix} z'(t_0) & x'(t_0) \\ z''(t_0) & x''(t_0) \end{vmatrix}} = \frac{z - z_0}{\begin{vmatrix} x'(t_0) & y'(t_0) \\ x''(t_0) & y''(t_0) \end{vmatrix}}$$

respectively, and

$$\begin{vmatrix} x - x_0 & y - y_0 & z - z_0 \\ x'(t_0) & y'(t_0) & z'(t_0) \\ \begin{vmatrix} y'(t_0) & z'(t_0) \\ y''(t_0) & z''(t_0) \end{vmatrix} & \begin{vmatrix} z'(t_0) & x'(t_0) \\ z''(t_0) & x''(t_0) \end{vmatrix} & \begin{vmatrix} x'(t_0) & y'(t_0) \\ x''(t_0) & y''(t_0) \end{vmatrix} \end{vmatrix} = 0$$

is an equation of the osculating plane.

In conclusion write the formulae for a unit principal normal vector

$$\mathbf{n} = \frac{[\mathbf{r}' \times \mathbf{r}''] \times \mathbf{r}'}{|\mathbf{r}' \times \mathbf{r}''| \cdot |\mathbf{r}'|}$$

and for a unit binormal vector

$$\mathbf{b} = \frac{\mathbf{r}' \times \mathbf{r}''}{|\mathbf{r}' \times \mathbf{r}''|}$$

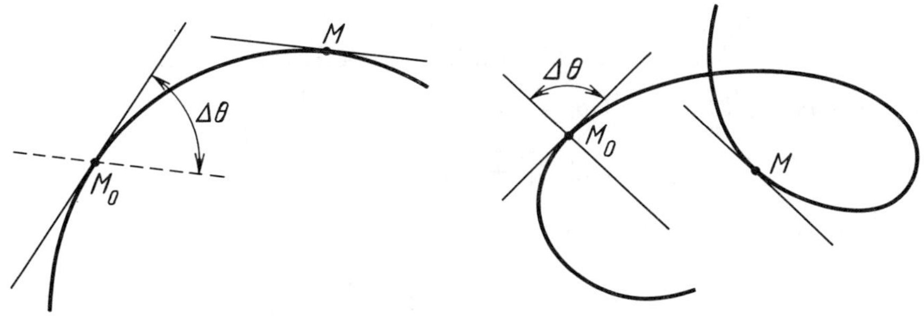

Figure 22.

3.1.12 Curvature of space curves

Let γ be a regular curve and M_0 — a point on it.

Curvature k_1 of the curve is defined as a limit of fraction

$$\frac{\Delta\Theta}{\Delta s}$$

as $M \to M_0$, where $\Delta\Theta$ is the lesser of the two angles between tangents to curve γ at points M and M_0, and Δs is a length of arc M_0M (Figure 22). A 2-regular curve has definite curvature k_1 at every point and the curvature is nonnegative by definition.

If

$$\mathbf{r} = \mathbf{r}(s)$$

is a natural parametrization of curve γ, then its curvature

$$k_1(s) = |\mathbf{r}''(s)|$$

Vector $\mathbf{r}''(s)$ is called *a curvature vector* of the curve. A length of the curvature vector is equal to the curvature of the curve at the given point.

For an arbitrary 2-smooth parametrization

$$\mathbf{r} = \mathbf{r}(t), \quad t \in T$$

the curvature of a 2-regular curve is calculated by formula

$$k_1(t) = \frac{|\mathbf{r}'(t) \times \mathbf{r}''(t)|}{|\mathbf{r}'(t)|^3}$$

(in general, vectors $\mathbf{r}'(t)$ and $\mathbf{r}''(t)$ are not orthogonal).

If the curve is defined by parametric equations

$$x = x(t), \quad y = y(t), \quad z = z(t)$$

then its curvature

$$k_1(t) = \frac{\sqrt{\begin{vmatrix} x'(t_0) & y'(t_0) \\ x''(t_0) & y''(t_0) \end{vmatrix}^2 + \begin{vmatrix} y'(t_0) & z'(t_0) \\ y''(t_0) & z''(t_0) \end{vmatrix}^2 + \begin{vmatrix} z'(t_0) & x'(t_0) \\ z''(t_0) & x''(t_0) \end{vmatrix}^2}}{([x'(t)]^2 + [y'(t)]^2 + [z'(t)]^2)^{3/2}}$$

If the curve is defined by explicit equations

$$y = y(x), \quad z = z(x)$$

then its curvature

$$k_1(x) = \frac{\sqrt{[y''(x)]^2 + [z''(x)]^2 + [y'(x)z''(x) - y''(x)z'(x)]^2}}{(1 + [y'(x)]^2 + [z'(x)]^2)^{3/2}}$$

In case the parametrization of curve γ is arbitrary,

$$\mathbf{r} = \mathbf{r}(t), \quad t \in T$$

its *curvature vector* $K(t)$ may be calculated via formula

$$K(t) = \frac{[\mathbf{r}'(t) \times \mathbf{r}''(t)] \times \mathbf{r}'(t)}{|\mathbf{r}'(t)|^4}$$

or

$$K(t) = \frac{(\mathbf{r}'(t), \mathbf{r}'(t))\mathbf{r}''(t) - (\mathbf{r}'(t), \mathbf{r}''(t))\mathbf{r}'(t)}{|\mathbf{r}'(t)|^4}$$

A curvature of a regular curve measures a rate of its deflection of the tangent to the curve. If the curvature of a curve is equal to zero at every point then this curve is a straight line or a connected part of a straight line.

3.1.13 Moving trihedral of space curves

Let γ be a 2-regular curve, and $\mathbf{r} = \mathbf{r}(s)$ — its natural parametrization. Denote by M_0 a point on the curve corresponding to value s_0 of the natural parameter. Let $\mathbf{t}_0 = \mathbf{r}'(s_0)$ be a unit tangent vector of curve γ at the point. If point M_0 is not a point of straightness ($k_1(s_0) \neq 0$), then unit principal normal vector

$$\mathbf{n}_0 = \frac{1}{k_1(s_0)}\mathbf{r}''(s_0)$$

is defined at the point. Vector product

$$\mathbf{b}_0 = \mathbf{t}_0 \times \mathbf{n}_0$$

is a unit binormal vector of curve γ.

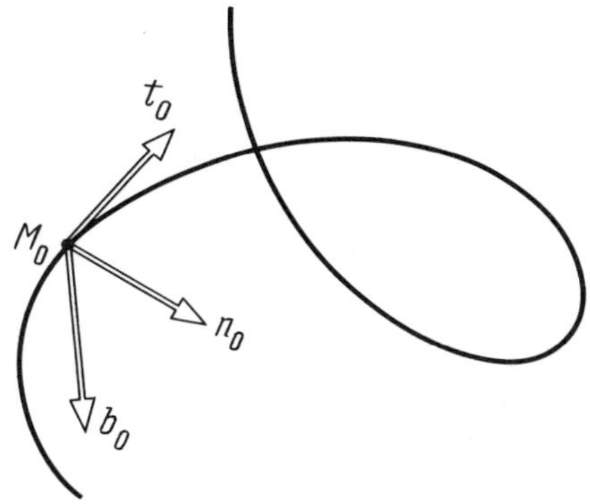

Figure 23.

A triple of vectors

$$\mathbf{t}_0 = \mathbf{t}(s_0), \quad \mathbf{n}_0 = \mathbf{n}(_0), \quad \mathbf{b}_0 = \mathbf{b}(t_0)$$

is named *a moving trihedral (a basic trihedral, a trihedral of Frénet, Frénet frame)* of curve γ at point M_0 (Figure 23).

If a curve is defined by the parametric vector equation of the form

$$\mathbf{r} = \mathbf{r}(t), \quad t \in T$$

then the vectors of a moving trihedral may be calculated by formulae

$$\mathbf{t}(t) = \frac{\mathbf{r}'(t)}{|\mathbf{r}'(t)|}, \quad \mathbf{n}(t) = \frac{[\mathbf{r}'(t) \times \mathbf{r}''(t)] \times \mathbf{r}'(t)}{|\mathbf{r}'(t) \times \mathbf{r}''(t)| \cdot |\mathbf{r}'(t)|}, \quad \mathbf{b}(t) = \frac{\mathbf{r}'(t) \times \mathbf{r}''(t)}{|\mathbf{r}'(t) \times \mathbf{r}''(t)|}$$

In case the curve is given in the coordinate form

$$x = x(t), \quad y = y(t), \quad z = z(t)$$

the formulae for these vectors may be written as follows:

$$\mathbf{t}(t) = \frac{(x'(t), y'(t), z'(t))}{\sqrt{[x'(t)]^2 + [y'(t)]^2 + [z'(t)]^2}},$$

$$\mathbf{n}(t) = \frac{\left(\begin{vmatrix} Y(t) & Z(t) \\ Y'(t) & Z'(t) \end{vmatrix}, \begin{vmatrix} Z(t) & X(t) \\ Z'(t) & X'(t) \end{vmatrix}, \begin{vmatrix} X(t) & Y(t) \\ X'(t) & Y'(t) \end{vmatrix} \right)}{\left(\begin{vmatrix} Y(t) & Z(t) \\ Y'(t) & Z'(t) \end{vmatrix}^2 + \begin{vmatrix} Z(t) & X(t) \\ Z'(t) & X'(t) \end{vmatrix}^2 + \begin{vmatrix} X(t) & Y(t) \\ X'(t) & Y'(t) \end{vmatrix}^2 \right)^{3/2}},$$

$$\mathbf{b}(t) = \frac{(X(t), Y(t), Z(t))}{([X'(t)]^2 + [Y'(t)]^2 + [Z'(t)]^2)^{3/2}}$$

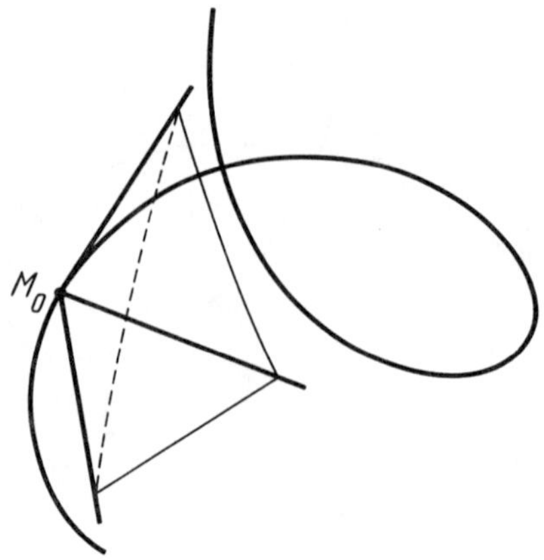

Figure 24.

where

$$X(t) = \begin{vmatrix} y'(t) & z'(t) \\ y''(t) & z''(t) \end{vmatrix}, \quad Y(t) = \begin{vmatrix} z'(t) & x'(t) \\ z''(t) & x''(t) \end{vmatrix},$$

$$Z(t) = \begin{vmatrix} x'(t) & y'(t) \\ x''(t) & y''(t) \end{vmatrix}$$

Let γ be a regular curve defined by explicit equations

$$y = y(x), \quad z = z(x)$$

Then

$$\mathbf{t}(x) = \frac{(1, y'(x), z'(x))}{\sqrt{1 + [y'(x)]^2 + [z'(x)]^2}},$$

$$\begin{aligned}\mathbf{n}(x) &= ([-z'(x)z''(x) - y'(x)y''(x)]^2 \\ &+[y''(x) - z'(x)(y'(x)z''(x) - y''(x)z'(x))]^2 + [z''(x) + y'(x)(y'(x)z''(x) \\ &\qquad -y''(x)z'(x))]^2)^{-1/2}(-z'(x)z''(x) - y'(x)y''(x), \\ &\qquad y''(x) - z'(x)(y'(x)z''(x) - y''(x)z'(x)), \\ &\qquad z''(x) + y'(x)(y'(x)z''(x) - y''(x)z'(x)))\end{aligned}$$

$$\mathbf{b}(x) = \frac{(y'(x)z''(x) - y''(x)z'(x), -z''(x), y''(x))}{([y'(x)z''(x) - y''(x)z'(x)]^2 + [z''(x)]^2 + [y''(x)]^2)^{3/2}}$$

Three beams (half-lines) outgoing from point M_0 in the directions determined by vectors $\mathbf{t}_0, \mathbf{n}_0, \mathbf{b}_0$ are the sides of a trihedral angle which is called *the natural trihedral angle (a moving trihedral (angle), a basic trihedral)* of curve γ at point M_0 (Figure 24).

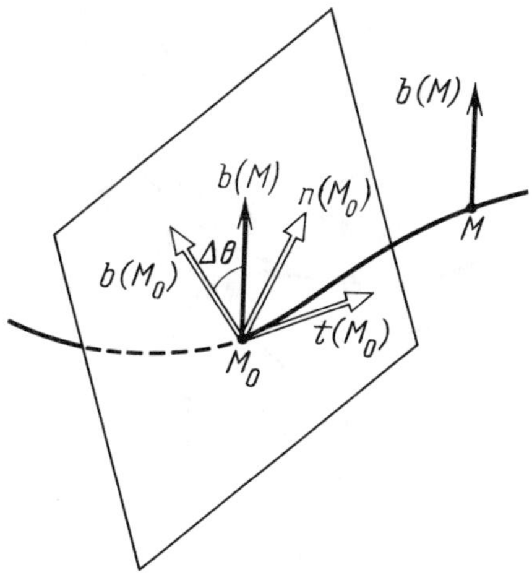

Figure 25.

3.1.14 Torsion of space curves

Let γ be a regular curve. Take two sufficiently close points M_0 and M on the curve. Suppose that point M_0 is not a straightness point of curve γ ($k_1(M_0) \neq 0$). Denote by $\Delta\Theta$ the lesser of the angles between osculating planes Π_0 and Π of the curve at points M_0 and M, respectively (the angle coincides with the lesser angle between binormals of curve γ at these points), and by Δs — a length of arc M_0M of the curve (Figure 25).

Torsion (second curvature) k_2 of the curve is defined as a limit of fraction

$$\frac{\Delta\Theta}{\Delta s}$$

as $M \to M_0$ completed with a sign in accordance with the following *sign-choosing rule*:

if vectors $\mathbf{b}'$ and $\mathbf{n}$ are unidirectional (these vector are always collinear), then sign "−" should be chosen (osculating plane rotates from vector $\mathbf{n}$ to vector $\mathbf{b}$), and if vectors $\mathbf{b}'$ and $\mathbf{n}$ are in the opposite directions, then sign "+" (osculating plane rotates from vector $\mathbf{b}$ to vector $\mathbf{n}$ (Figure 26).

Let γ be a 3-regular curve, and

$$\mathbf{r} = \mathbf{r}(t), \quad t \in T$$

its regular parametrization. If vectors $\mathbf{r}'(t_0)$ and $\mathbf{r}''(t_0)$ are noncollinear (the curve curvature $k_1(t_0) \neq 0$), then at point $M_0(t_0)$ the curve has torsion $k_2(t_0)$ which is calculated by formula

$$k_2(t_0) = \frac{(\mathbf{r}'(t_0), \mathbf{r}''(t_0), \mathbf{r}'''(t_0))}{(\mathbf{r}'(t_0) \times \mathbf{r}''(t_0))^2}$$

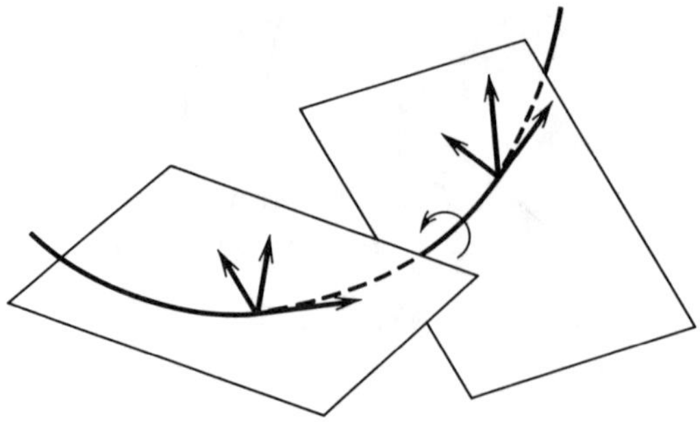

Figure 26.

or, in coordinate form,

$$k_2(t_0) = \frac{\begin{vmatrix} x'(t_0) & y'(t_0) & z'(t_0) \\ x''(t_0) & y''(t_0) & z''(t_0) \\ x'''(t_0) & y'''(t_0) & z'''(t_0) \end{vmatrix}}{\begin{vmatrix} y'(t_0) & z'(t_0) \\ y''(t_0) & z''(t_0) \end{vmatrix}^2 + \begin{vmatrix} z'(t_0) & x'(t_0) \\ z''(t_0) & x''(t_0) \end{vmatrix}^2 + \begin{vmatrix} x'(t_0) & y'(t_0) \\ x''(t_0) & y''(t_0) \end{vmatrix}^2}$$

in particular, for equations $y = y(x), z = z(x)$ we have

$$k_2(x_0) = \frac{y''(x_0)z'''(x_0) - y'''(x_0)z''(x_0)}{([y'(x_0)z''(x_0) - y''(x_0)z'(x_0)]^2 + [z''(x_0)]^2 + [y''(x_0)]^2)^2}$$

In case of natural parametrization, $\mathbf{r} = \mathbf{r}(s)$, the torsion of the curve may be found by formula

$$k_2(s_0) = \frac{(\mathbf{r}'(s_0), \mathbf{r}''(s_0), \mathbf{r}'''(s_0))}{|\mathbf{r}''(s_0)|^2}$$

A torsion of a regular curve measures a rate of its deflection from the osculating plane. If the torsion of a curve is equal to zero at every point, then this curve is a plane curve. The opposite is also true: a torsion of a plane curve is equal to zero.

Vector $\mathbf{w} = k_2\mathbf{t} + k_1\mathbf{b}$ (*a Darboux vector of the curve*) is a instantaneous velocity vector of Frénet frame when a point moves along the curve with a unit speed.

If, at point M_0, vectors $\mathbf{r}'$ and $\mathbf{r}''$ are collinear and vectors $\mathbf{r}'$ and $\mathbf{r}'''$ are noncollinear, then a torsion of curve γ may be calculated by formula

$$k_2(t_0) = \frac{(\mathbf{r}'(t_0), \mathbf{r}''(t_0), \mathbf{r}^{\text{iv}}(t_0))}{[\mathbf{r}'(t_0) \times \mathbf{r}'''(t_0)]^2}$$

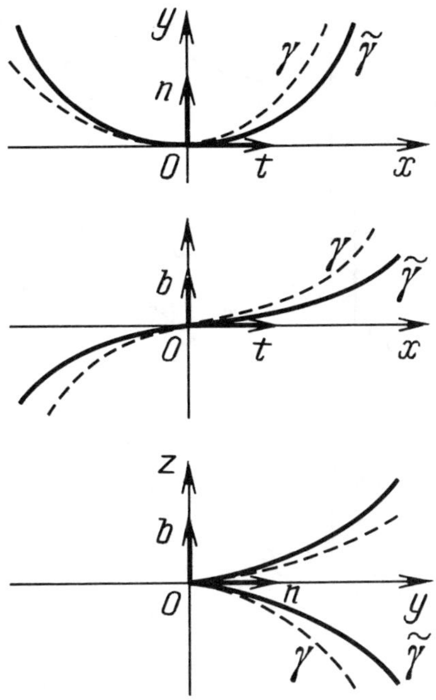

Figure 27.

3.1.15 Frénet formulae for space curves

Let γ be a 3-regular curve, and $\mathbf{r} = \mathbf{r}(s)$ — its natural parametrization.

The unit vectors of tangent $\mathbf{t}(s)$, principal normal $\mathbf{n}(s)$ and binormal $\mathbf{b}(s)$ of the curve, its curvature $k_1(s)$ and torsion $k_2(s)$ are connected at every point by the following relations:

$$\frac{d\mathbf{t}}{ds} = k_1\mathbf{n}, \quad \frac{d\mathbf{n}}{ds} = -k_1\mathbf{t} + k_2\mathbf{n}, \quad \frac{d\mathbf{b}}{ds} = k_2\mathbf{n}$$

The relations are called *Frénet (Serret–Frénet) formulae.*

A view of a space curve close to a given point. Let γ be a regular curve, and $\mathbf{r} = \mathbf{r}(s)$ — its regular parametrization.

Choose coordinate system $Oxyz$ in a space so that its origin $O(0,0,0)$ coincides with point M_0 on curve γ corresponding to value $s_0 = 0$ of the natural parameter, $\mathbf{r}(s_0) = \mathbf{r}(0) = \mathbf{0}$, and let vectors $\mathbf{t}(s_0)$, $\mathbf{n}(s_0)$ and $\mathbf{b}(s_0)$ of the curve basic frame be unit vectors of coordinate axes Ox, Oy and Oz, respectively,

$$\mathbf{i} = \mathbf{t}(s_0), \quad \mathbf{j} = \mathbf{n}(s_0), \quad \mathbf{k} = \mathbf{b}(s_0)$$

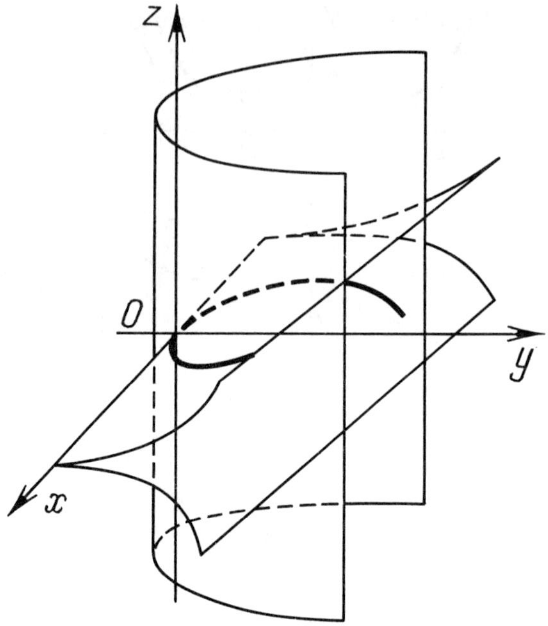

Figure 28.

Then the following formula is true:

$$\mathbf{r}(s) = \left(s - \frac{k_1^3(0)}{6}s^3 + o(s^3)\right)\mathbf{i}$$

$$+\left(\frac{k_1(0)}{2}s^2 + \frac{k_1'(0)}{6}s^3 + o(s^3)\right)\mathbf{j} + \left(\frac{k_1(0)k_2(0)}{6}s^3 + o(s^3)\right)\mathbf{k}$$

Retain the principal parts only in each summand and then obtain equality

$$\tilde{\mathbf{r}}(s) = s\mathbf{i} + \frac{k_1(0)}{2}s^2\mathbf{j} + \frac{k_1(0)k_2(0)}{6}s^3\mathbf{k}$$

of curve $\tilde{\gamma}$ close to curve γ.

Write the last vector relation in the coordinate form

$$x = s, \quad y = \frac{k_1(0)}{2}s^2, \quad z = \frac{k_1(0)k_2(0)}{6}s^3$$

Under the assumption that $k_1(0) > 0$, $k_2(0) > 0$, the projections of curve γ on the coordinate planes have the view outlined in Figure 27. The general view of curve $\tilde{\gamma}$ is shown in Figure 28.

3.1.16 Natural equations of space curves

Let $k_1(s)$ and $k_2(s)$ be arbitrary smooth functions, and $k_1(s) > 0$. Then there exists a single (up to moving) curve, the curvature of which coincides with function $k_1(s)$, and torsion — with function $k_2(s)$. Equalities of the form

$$k_1 = k_1(s), \quad k_2 = k_2(s)$$

are called *the natural equations* of the curve.

The natural equations (uniquely up to the position in space) define a space curve. The latter implies that, for a given right triple of mutually orthogonal unit vectors $\mathbf{t}_0$, $\mathbf{n}_0$ and $\mathbf{b}_0$, there exists a curve with given curvature k_1 and torsion k_2 passing through point M_0, so that vectors $\mathbf{t}_0$, $\mathbf{n}_0$ and $\mathbf{b}_0$ are a unit tangent, a unit principal normal and a unit binormal vectors, respectively.

When natural equations are given, the searching process of a curve involves integration of the Frénet equation system.

Sometimes a space curve is called *a curve of double curvature*: the first curvature is a curvature proper, and the second — a torsion of the curve.

3.1.17 Asymptotes of space curves

We say that curve γ defined by equations

$$x = x(t), \quad y = y(t), \quad z = z(t) \quad a < t < b$$

goes to infinity, if

$$x^2(t) + y^2(t) + z^2(t) \to \infty$$

when $t \to a + 0$ and/or $t \to b - 0$.

This property is independent of parametrization.

Straight line A is called *an asymptote* of curve γ going to infinity if distance $d(t)$ from curve point $M(x(t), y(t), z(t))$ to the line tends to zero when $t \to a + 0$ and/or $t \to b - 0$.

Let curve γ go to infinity (for the sake of definition when $t \to a + 0$).

The existence criterion for asymptotes of space curves. In order for straight line $y = B + Mx, z = C + Nx$ to be an asymptote of curve γ with reference to a given Cartesian coordinate system, it is necessary and sufficient that the following four limits

$$\lim_{t \to a+0} \frac{y(t)}{x(t)} = M, \quad \lim_{t \to a+0} \frac{z(t)}{x(t)} = N,$$

$$\lim_{t \to a+0} (y(t) - Mx(t)) = B, \quad \lim_{t \to a+0} (z(t) - Nx(t)) = C$$

exist concurrently.

An asymptote of the space curve lies in plane $x = D$ if

$$\lim_{\to a+0} x(t) = D, \quad \text{and} \quad y^2(t) + z^2(t) \to \infty \quad \text{as} \quad t \to a + 0$$

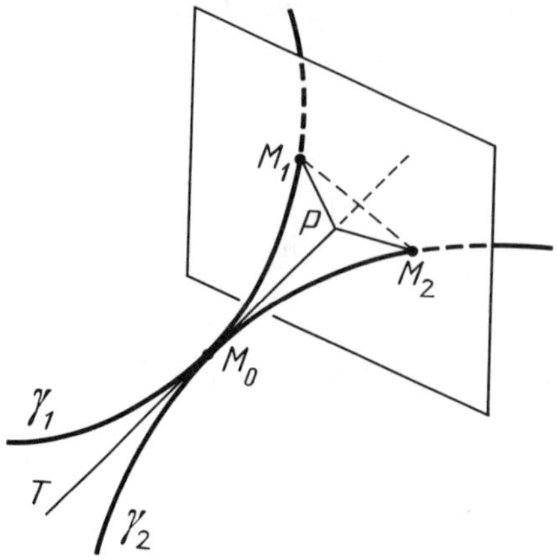

Figure 29.

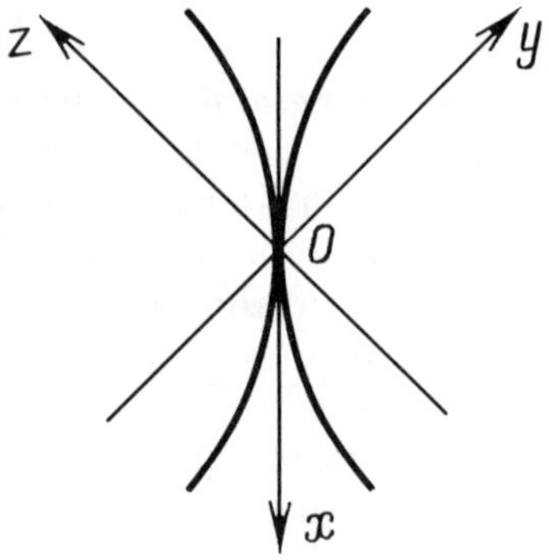

Figure 30.

3.1.18 Osculation of space curves

Two curves γ_1 and γ_2 are said *to be tangent to one another at common point* M_0, or *to be contacting curves*, if both curves have tangents at the point and the tangents coincide.

Let γ_1 and γ_2 be regular curves, M_0 — their common point, and M_0T — their common tangent at the point. Let us take point P on tangent M_0T close to point M_0 and pass plane Π through point P perpendicular to straight line M_0T. The plane intersects curves γ_1 and γ_2 at points M_1 and M_2, respectively. Denote the length of line segment M_0P by d and the length of line segment M_1M_2 by h (Figure 29).

We say that *an order of osculation (an order of contact)* of curves γ_1 and γ_2 at common point M_0 is not below $n \in \mathbf{N}$, if

$$\lim_{d\to 0} \frac{h}{d^n} = 0 \tag{$*$}$$

If, in addition,

$$\lim_{d\to 0} \frac{h}{d^n} \neq 0$$

or a ratio

$$\frac{h}{d^{n+1}}$$

has no limit under $d \to 0$, then *the order of osculation (the order of contact)* of curves γ_1 and γ_2 at point M_0 is said *to be equal to* n.

If the limiting relation $(*)$ holds for every natural number n, curves γ_1 and γ_2 have at common point M_0 *an infinite order of osculation (contact)* by definition.

Let regular curves γ_1 and γ_2 have a common tangent M_0T at point M_0. Let, for the sake of simplicity, straight line M_0T coincide with the x-axis, and $M_0 \equiv O$. Let P be a common projection of points $M_1 \in \gamma_1$, and $M_2 \in \gamma_2$ on tangent M_0T. Choose an abscissa of the projection as a parameter on each of these curves. Then the parametric equations of curves γ_1 and γ_2 may be written in the following form

$$\mathbf{r} = \mathbf{r}_1(x), \quad \mathbf{r} = \mathbf{r}_2(x)$$

respectively (Figure 30). If

$$\mathbf{r}_1'(0) = \mathbf{r}_2'(0), \ldots, \mathbf{r}_1^n(0) = \mathbf{r}_2^n(0)$$

then, at point $M_0 = O$, curves γ_1 and γ_2 have the order of osculation (the order of contact) not below n. If, in addition, $\mathbf{r}_1^{n+1}(0) \neq \mathbf{r}_2^{n+1}(0)$, then the order of osculation (the order of contact) is equal to n.

Curves

$$\begin{aligned} \gamma : \quad & y = y(x), \quad z = z(x) \\ \Gamma : \quad & Y = Y(x), \quad Z = Z(x) \end{aligned}$$

have a tangency of order $\geq n$ at common point $M_0(x_0, y_0, z_0)$ if

$$\begin{aligned} y'(x_0) &= Y'(x_0), & z'(x_0) &= Z'(x_0) \\ y''(x_0) &= Y''(x_0), & z''(x_0) &= Z''(x_0) \\ &\ldots & &\ldots \\ y^n(x_0) &= Y^n(x_0), & z^n(x_0) &= Z^n(x_0) \end{aligned}$$

Let M_0 be a common point of regular curves

$$\begin{aligned} \gamma : \quad & x = x(t), \quad y = y(t), \quad z = z(t) \\ \Gamma : \quad & X = X(\tau), \quad Y = Y(\tau), \quad Z = Z(\tau) \end{aligned}$$

corresponding to parameter values t_0 and τ_0, respectively:

$$x_0 = x(t_0) = X(\tau_0), \quad y_0 = y(t_0) = Y(\tau_0), \quad z_0 = z(t_0) = Z(\tau_0 1)$$

Let (for the sake of definiteness) $x'(t_0) \neq 0$ and $X'(\tau_0) \neq 0$. This allows equations $x = x(t)$ and $X = X(\tau)$ to be solved with respect to t and τ, respectively: $t = t(x)$, $\tau = \tau(x)$, and obtain the following parametrizations

$$\begin{aligned} \gamma : \quad & x = x, \quad y = y(t(x)), \quad z = z(t(x)) \\ \Gamma : \quad & x = x, \quad Y = Y(\tau(x)), \quad Z = Z(\tau(x)) \end{aligned}$$

Curves γ and Γ have the osculation order no less than 2 at point M_0, if

$$\frac{y'(t_0)}{x'(t_0)} = \frac{Y'(\tau_0)}{X'(\tau_0)}, \quad \frac{z'(t_0)}{x'(t_0)} = \frac{Z'(\tau_0)}{X'(\tau_0)}$$

$$\frac{y''(t_0)x'(t_0) - y'(t_0)x''(t_0)}{[x'(t_0)]^3} = \frac{Y''(\tau_0)X'(\tau_0) - Y'(\tau_0)X''(\tau_0)}{[X'(\tau_0)]^3}$$

$$\frac{z''(t_0)x'(t_0) - z'(t_0)x''(t_0)}{[x'(t_0)]^3} = \frac{Z''(\tau_0)X'(\tau_0) - Z'(\tau_0)X''(\tau_0)}{[X'(\tau_0)]^3}$$

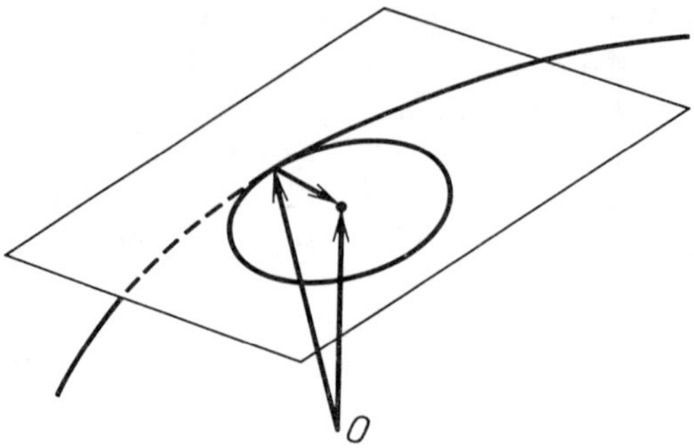

Figure 31.

3.1.19 Osculating circle of a space curve

Let γ be a regular curve, and M_0 — a point on the curve. A circle passing through point M_0 of curve γ is said to be *an osculating circle* of the curve at point M_0 if their order of osculation is no less than 2 (Figure 31).

If the curvature of curve γ differs from zero at point M_0, $k_1(t_0) > 0$ (point M_0 is not a straightness point of the curve), then there exists a unique osculating circle of curve γ at point M_0. The circle lies in an osculating plane of curve γ at point M_0, and its radius R is defined by equality

$$R = \frac{1}{k_1(M_0)}$$

Let $\mathbf{r} = \mathbf{r}(t), t \in T$ be a vector equation of curve γ and $k_1(t_0) > 0$. Then the center of an osculating circle is defined by vector

$$\mathbf{r}(t_0) + \frac{1}{k(t_0)}\mathbf{n}(t_0)$$

where $\mathbf{n}(t_0)$ is a unit principal normal vector of the curve at point M_0.

The center of an osculating circle is called *a center of curvature* of the curve at a given point, and its radius — *a radius of curvature* at the point.

An osculating plane of a curve intersects an osculating sphere (see 3.1.22) of the curve in an osculating circle.

An osculating circle is often called *a circle of curvature.*

3.1.20 Evolute and evolvent of a space curve

Let γ be a regular plane curve, and M_0 — a point on it. Let a flexible, nonstreching thread be superimposed on the given curve and one of its ends be fixed at point M_0. Leading the stretched thread from the curve as though it were a template, the other end-point M describes curve γ^* — *an evolvent* of curve γ (Figure 32). Curve γ is called *an evolute* of curve γ^*.

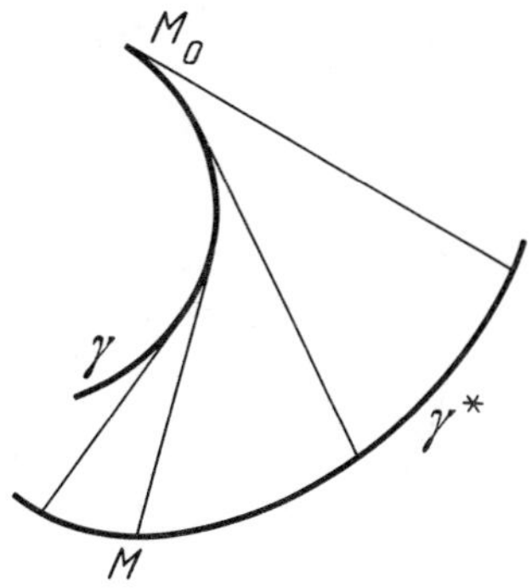

Figure 32.

Let

$$\mathbf{r} = \mathbf{r}(s)$$

be a natural parametrization of curve γ. Then a vector equation of the evolvent has the following form:

$$\rho(s) = \mathbf{r}(s) + (c - s)\mathbf{t}(s)$$

where $\mathbf{t}(s)$ is a unit tangent vector of the curve, and c — an arbitrary constant.

A regular (plane or space) curve has an infinite number of evolvents.

The evolutes of curve γ depend on a numerical parameter and are described by the following formula:

$$\rho = \mathbf{r}(s) + \frac{1}{k_1(s)}\mathbf{n}(s) + \cot\left(\int k_2(\sigma)\,d\sigma + c\right) \cdot \mathbf{b}(s)$$

An angle at which two evolutes of curve γ are viewed from point $\mathbf{r} = \mathbf{r}(s)$ of the curve is independent of the point.

If curve γ is plane ($k_2 \equiv 0$), then its evolutes are described by equation

$$\rho = \mathbf{r}(s) + \frac{1}{k_1(s)}\mathbf{n}(s) + \frac{c}{k_1(s)}\mathbf{b}(s)$$

where $\mathbf{n}$(s) and $\mathbf{b}$(s) are a unit principal normal and binormal vectors of curve γ.

A plane curve has a single plane evolute ($c = 0$). The remaining evolutes are *slope lines*, that is, the curves whose tangents are inclined at a constant angle to the plane of the curve.

3.1.21 Spherical evolute and evolvent

Let γ be a spherical curve. By definition, *spherical evolvent* γ^* of curve γ is a set generated by end-points of the great circle arcs tangent to the

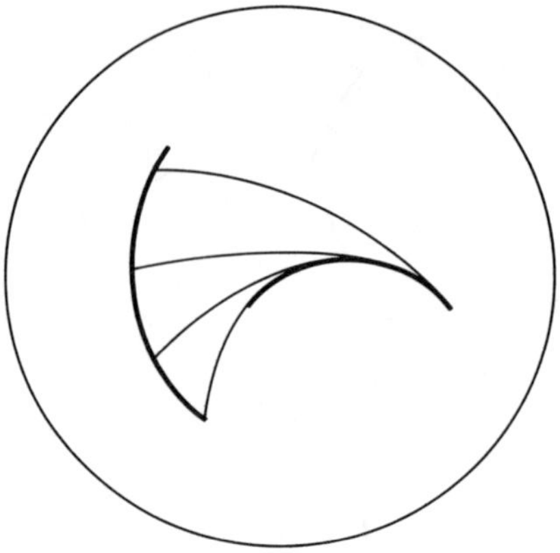

Figure 33.

curve under the condition that their lengths decrease in accuracy with increment of the curve arcs. Curve γ is called *the spherical evolute* of curve γ^* (Figure 33).

1. Every spherical curve has an infinitely large number of spherical evolvents. Through each point of an arbitrary great circle tangent to an evolute, only one spherical evolvent passes.

2. A spherical evolvent is an orthogonal trajectory of the great circles tangent to an evolute. The opposite is also true: every orthogonal trajectory of the great circle family tangent to curve γ is an evolvent of the curve.

3. Every spherical curve has a single spherical evolute (consisting of two symmetric branches).

4. A spherical evolvent of the spherical indicatrix of binormals (see 3.1.24) to regular curve γ and the curve have correspondingly parallel tangents, principal normals and binormals.

5. A spherical indicatrix of binormals to a spherical curve is one of its spherical evolute branches.

3.1.22 Osculating sphere of a space curve

Let regular curve γ and sphere $\mathcal{S}$ have common point M_0. Take point M on the curve close to point M_0 and connect the point with the center of sphere $\mathcal{S}$ — point O. Denote by P an intersection point of straight line MO and sphere $\mathcal{S}$ nearest to point M (Figure 34). Let h be a length of segment MP, and d — a length of chord M_0P.

We say that curve γ has *an order of osculation (an order of contact)* $\leq n$ with sphere $\mathcal{S}$, if

$$\frac{h}{d^n} \to 0$$

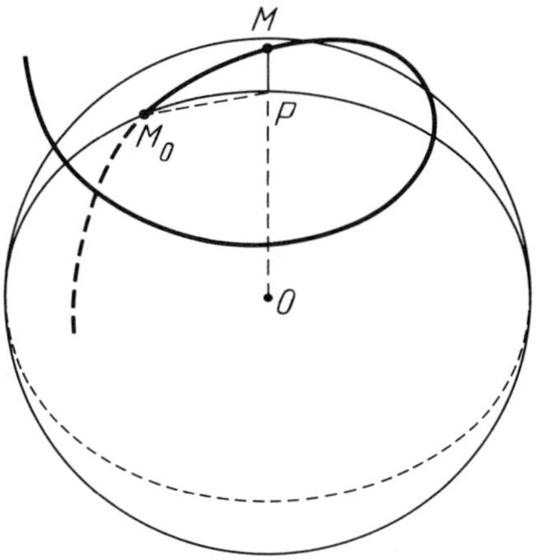

Figure 34.

as $d \to 0$. If, in addition, a ratio

$$\frac{h}{d^{n+1}}$$

has a nonvanishing limit or has no limit at all as $d \to 0$, then we shall say that *the order of osculation (the order of contact) is equal to n.*

A sphere whose osculation order with curve γ at point M_0 is no less than 3 is called *the osculating sphere* of the curve at point M_0.

If curvature k_1 and torsion k_2 of curve γ at point M_0 differ from zero, then there exists an osculating sphere of the curve at the point. The center of the sphere lies in a normal plane of curve γ at point M_0.

Let $\mathbf{r} = \mathbf{r}(t)$ be a parametrization of curve γ. Then a center of the osculating sphere is defined by vector

$$\mathbf{r}(t_0) + \frac{1}{k_1(t_0)}\mathbf{n}(t_0) - \frac{k_1'(t_0)}{k_1^2(t_0)k_2(t_0)}\mathbf{b}(t_0)$$

and its radius R may be found by formula

$$R = \sqrt{\left(\frac{1}{k_1(t_0)}\right)^2 + \left(\frac{k_1'(t_0)}{k_1^2(t_0)k_2(t_0)}\right)^2}$$

In order for sphere $(x-a)^2+(y-b)^2+(z-c)^2 = R^2$ to be an osculating sphere of curve γ with $x = x(t), y = y(t), z = z(t)$ at regular point M_0 corresponding to parameter value t_0, it is necessary and sufficient that the following conditions be met:

$$\varphi(t_0) = 0, \quad \varphi'(t_0) = 0, \quad \varphi''(t_0) = 0, \quad \varphi'''(t_0) = 0$$

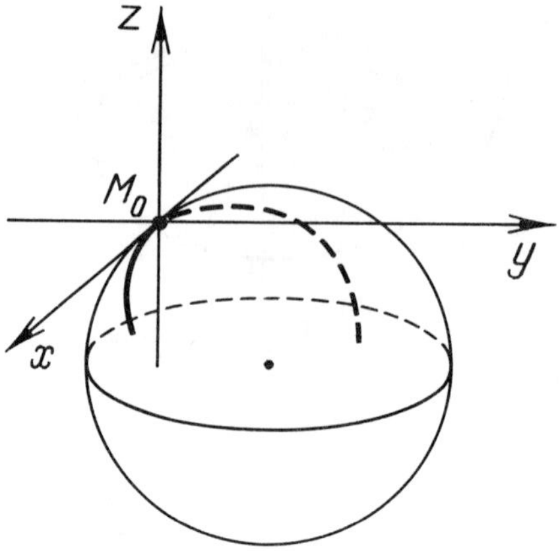

Figure 35.

where

$$\varphi(t) = [x(t) - a]^2 + [y(t) - b]^2 + [z(t) - c]^2 - R^2$$

If a coordinate system in space is chosen so that its origin coincides with point M_0, x-axis — with a tangent, y-axis — with a principal normal, and z-axis — with binormal of the curve at point M_0 (Figure 35), then

$$a = x(t_0) = 0, \quad b = y(t_0) + \frac{1}{k_1(t_0)} = \frac{1}{k_1(t_0)}$$

$$c = z(t_0) - \frac{k_1'(t_0)}{k_1^2(t_0)k_2(t_0)} = -\frac{k_1'(t_0)}{k_1^2(t_0)k_2(t_0)}$$

When a radius of an osculating sphere of curve γ is constant, the curve is either a circle or has a constant curvature.

An osculating plane of curve γ intersects an osculating sphere in an osculating circle (see 3.1.19).

3.1.23 Singular points of space curves

The implicit description. Let plane curve γ be defined by equations

$$F(x, y, z) = 0, \quad G(x, y, z) = 0$$

where $F(x, y, z)$ and $G(x, y, z)$ are smooth functions of variables x, y and z.

Point $M_0(x_0, y_0, z_0)$ is called *a singular point* of curve γ, if the following three conditions

$$F(x_0, y_0, z_0) = 0, \quad G(x_0, y_0, z_0) = 0,$$

$$\operatorname{rang}\begin{pmatrix} F_x(x_0,y_0,z_0) & F_y(x_0,y_0,z_0) & F_z(x_0,y_0,z_0) \\ G_x(x_0,y_0,z_0) & G_y(x_0,y_0,z_0) & G_z(x_0,y_0,z_0) \end{pmatrix} < 2$$

are met concurrently.

The last inequality may be written as follows:

$$\begin{vmatrix} F_x(x_0,y_0,z_0) & F_y(x_0,y_0,z_0) \\ G_x(x_0,y_0,z_0) & G_y(x_0,y_0,z_0) \end{vmatrix} = \begin{vmatrix} F_y(x_0,y_0,z_0) & F_z(x_0,y_0,z_0) \\ G_y(x_0,y_0,z_0) & G_z(x_0,y_0,z_0) \end{vmatrix}$$

$$= \begin{vmatrix} F_z(x_0,y_0,z_0) & F_x(x_0,y_0,z_0) \\ G_z(x_0,y_0,z_0) & G_x(x_0,y_0,z_0) \end{vmatrix} = 0$$

However, a character of the smooth space curve singularities is best (and easiest) to investigate in the case of the parametric description.

Parametric description. Let smooth curve γ be defined by equation $\mathbf{r} = \mathbf{r}(t)$. Point $M_0(t_0)$ of curve γ where $\mathbf{r}'(t_0) = 0$ is called *a singular point* of curve γ with respect to the given parametrization.

On examination of the smooth curve construction in vicinity of its singular point, it is useful to employ the expasion of vector function $\mathbf{r}(t)$ via Taylor formula

$$\mathbf{r}(t) = \mathbf{r}(t_0) + \frac{1}{2}\mathbf{r}''(t_0)(t-t_0)^2 + \frac{1}{6}\mathbf{r}'''(t_0)(t-t_0)^3 + \dots$$

$$+\frac{1}{k!}\mathbf{r}^{(k)}(t_0)(t-t_0)^k + \varepsilon \cdot (tt_0)^k$$

where vector function ε tends to zero as $t \to t_0$.

If $\mathbf{r}''(t_0) \neq \mathbf{0}$ and $\mathbf{r}''(t_0) \times \mathbf{r}'''(t_0) \neq \mathbf{0}$, then the singular point is *a cuspidal point of the first order*. This singularity is *geometric* and *unremediable* by parameter transformations.

If $\mathbf{r}'(t_0) = \mathbf{0}$, $\mathbf{r}''(t_0) = \mathbf{0}$ and $\mathbf{r}'''(t_0) \neq \mathbf{0}$ at the examined point, then the singlarity is nongeometric and *remediable*. By changing the parameter to $\tau = (t-t_0)^3$, it may disappear. But such a transformation may lower the smoothness class of the curve.

Let a given curve be analytic (of class C^A), that is, vector function $\mathbf{r} = \mathbf{r}(t), a < t < b$ defining the curve is a sum of its Taylor series

$$\mathbf{r}(t) = \mathbf{r}(t_0) + \frac{1}{6}\mathbf{r}'''(t_0)(t-t_0)^3 + \dots$$

at the neighborhood of point t_0 (recall that $\mathbf{r}'(t_0) = \mathbf{0}$, $\mathbf{r}''(t_0) = \mathbf{0}$). If the Taylor expansion contains terms of the form $(t-t_0)^{3p}$ only (the terms with powers multiple to 3), then, after changing $\tau = (t-t_0)^3$, the curve remains analytic. In case there exists a nonzero term whose power is not divided by 3 in the Taylor expansion, the curve smoothness will be of a finite order, that is, the curve will cease to be analytic.

If $\mathbf{r}'(t_0) = \mathbf{0}$, $\mathbf{r}''(t_0) = \mathbf{0}$, $\mathbf{r}'''(t_0) = \mathbf{0}$ and $\mathbf{r}^{\text{iv}}(t_0) \neq \mathbf{0}$, then we have *a cuspidal point* again, but with the higher-order osculation of the curve branches at point M_0.

Let $\mathbf{r}^{(p)}(t_0)$ be the least-order derivative of analytic vector function $\mathbf{r}(t)$ nonvanishing at point t_0, and $\mathbf{r}^{(q)}(t_0)$ — the least-order derivative of the function noncollinear to derivative $\mathbf{r}^{(p)}(t_0)$ at point t_0. Then the following cases are possible.

Point M_0 corresponding to value t_0 of the parameter may be:

1) a regular point (if p is odd, and q is even),

2) a regular point (if p is odd, and q is odd),

3) a cuspidal point of the first kind (if p is even, and q is odd), or

4) a cuspidal point of the second kind (if both p and q are even).

Note. In cases 1 and 2, singularities may disappear with reparametrization $\tau = (t - t_0)^p$. In some cases, it is possible for a curve to become regular after reparametrization, but often under a lesser smoothness class.

In conclusion, it is useful to note that the notions of singular points for the curves defined implicitly and for those defined explicitly are *different*. For example, in case of the parametric description, the self-intersection points are ordinary points of the curve.

3.1.24 Spherical indicatrices of space curves

Let γ be a regular curve, and $\mathbf{r} = \mathbf{r}(s)$ — its natural parametrization. A curve defined by vector equation $\mathbf{t} = \mathbf{t}(s)$, where $\mathbf{t}(s)$ is a unit tangent vector of curve γ, lies on a unit sphere and is said to be *a (spherical) indicatrix of tangents* to the curve, or *its tangent spherical image*.

A curve defined by vector equation $\mathbf{n} = \mathbf{n}(s)$, where $\mathbf{n}(s)$ is a unit principal normal vector of curve γ, lies on a unit sphere and is said to be *a (spherical) indicatrix of principal normals* to the curve, or *its principal normal spherical image*.

A curve defined by vector equation $\mathbf{b} = \mathbf{b}(s)$, where $\mathbf{b}(s)$ is a unit binormal vector of curve γ, lies on a unit sphere and is said to be *a (spherical) indicatrix of binormals* to the curve, or *its binormal spherical image*.

All three indicatrices lie on the unit sphere, and, hence, are spherical curves.

A spherical indicatrix of tangents of plane curve γ is a great circle arc parallel to the plane where the curve lies. An osculating plane of a curve is parallel to a plane of the great circle which is tangent to the spherical indicatrix of tangents. A tangent vector of the spherical indicatrix of tangents of curve γ and the unit principal normal vector of the curve have the same direction; that is, the spherical indicatrix of principal normals to the curve is at the same time a spherical indicatrix of tangents to the spherical indicatrix of tangents to curve γ. A spherical indicatrix of tangents to curve γ degenerates into a point if, and only if, the curve is a straight line. A spherical indicatrix of principal normals to a curve can not degenerate into a point. A spherical indicatrix of binormals to curve γ degenerates into a point if, and only if, the curve is plane.

Let

$$s^* = \int_0^s |\mathbf{t}'(\sigma)|\, d\sigma$$

be the arc length of the spherical indicatrix of tangents to curve γ. Then, for a curvature of the curve, the following formula

$$k_1 = \frac{ds^*}{ds}$$

is true.

Let

$$s^* = \int_0^s |\mathbf{b}'(\sigma)|\, d\sigma$$

be the arc length of the spherical indicatrix of principal normals to curve γ. Then, for a torsion of the curve, the following formula

$$|k_2| = \frac{ds^*}{ds}$$

is true.

Let

$$s^* = \int_0^s |\mathbf{h}'(\sigma)|\, d\sigma$$

be the arc length of the spherical indicatrix of binormals to curve γ. Then

$$\frac{ds^*}{ds} = \sqrt{k_1^2 + k_2^2}$$

The Jacobi theorem. A spherical indicatrix of principal normals to a closed curve divides the sphere into two parts of equal areas (under the conditions that the indicatrices of tangents and principal normals to the curve are continuous and nonconcurrent).

If the spherical indicatrix of tangents to curve γ lies in a plane passing through the center of the sphere, then the curve is plane. A spherical indicatrix of tangents to curve γ is a circle arc if, and only if, the curve is a screwline (a cylindrical helix). For a spherical indicatrix of tangents to a closed curve, there does not exist an open hemisphere nor a closed hemisphere containing the indicatrix (the exception for the latter is when the curve is a great circle).

A closed curve lying in a unit sphere is a spherical indicatrix of tangents to a closed curve if, and only if, no hemisphere contains the given curve.

3.1.25 Conditions for a space curve to be plane

1. If all tangents of curve γ are parallel to the same plane, then the curve is plane.

2. If all tangent planes of curve γ are parallel to the same straight line, then the curve is plane.

3. If all osculation planes of curve γ have a common point, then the curve is plane.

4. If all osculation planes of curve γ are parallel to the same straight line, then the curve is plane; in particular, γ is a straight line.

5. If all osculation planes of curve γ have a constant inclination, then the curve is plane.

6. If a torsion of curve γ is equal to zero at each of its points, then the curve is plane.

7. If curve γ is defined by equation

$$\mathbf{r} = \mathbf{r}(t)$$

and equality

$$(\mathbf{r}'(t), \mathbf{r}''(t), \mathbf{r}'''(t)) = 0$$

holds at every of its points, then the curve is plane.

8. If for closed regular curve γ

$$\int_{\gamma} k_1(s)\, ds = 2\pi$$

where $k_1(s)$ is its curvature, then the curve is plane.

9. If a spherical indicatrix of binormals of regular curve γ degenerates into a point, then the curve is plane.

10. If a spherical indicatrix of tangents of regular curve γ lies in a plane passing through a center of the sphere, then the curve is plane.

3.1.26 Envelope of a family of space curves

Let a family of space curves be described by smooth vector function

$$\mathbf{r} = \mathbf{r}(u, v), \quad A < u < B, \quad C < v < D \tag{*}$$

Regular curve

$$\rho = \rho(t), \quad a < t < b \tag{**}$$

is called *(a portion of) envelope* of the space curve family (*) if there exists a continuous *attachment law* (for the concrete curves of the family to the envelope points), that is, there exists a pair of functions

$$u = u(t), \quad v = v(t)$$

with the following properties:

1) $\mathbf{r}(u(t), v(t)) = \rho(t)$;

2) for every value t_1 subject to inequality $a < t_1 < b$, the curve of family $(*)$ described by equation $\mathbf{r}(u, v(t_1))$ is tangent to the curve at point $\rho(t_1)$;

3) function $v(t)$ differs from constant on any segment of the parameter variability.

The necessary criterion for the envelope existence. If family $(*)$ has an envelope $(**)$, then vector equality

$$\mathbf{r}_u \times \mathbf{r}_v = \mathbf{0}$$

holds at the points (u, v) corresponding to values $u - u(t), v = v(t)$.

Note. The system of scalar equations

$$\begin{vmatrix} y_u & y_v \\ z_u & z_v \end{vmatrix} = 0, \quad \begin{vmatrix} z_u & z_v \\ x_u & x_v \end{vmatrix} = 0, \quad \begin{vmatrix} x_u & x_v \\ y_u & y_v \end{vmatrix} = 0$$

corresponding to the last vector relation is overdeterminated: it has three equations and only two variables. By this is meant that the availability of an envelope for a one-parameter family of space curves is rather an exception than the rule.

Nevertheless, every curve is an envelope of the family of its tangents.

3.2 Some classes of space curves

3.2.1 Space curves with constant curvature

$k_1 =$ const

A. Let γ be a space curve with constant curvature $k_1 > 0$ (for $k_1 = 0$ curve γ is a line segment), and γ^* — a set of its curvature centers. Then curve γ in its turn is a set of the curvature centers of curve γ^*, and both curves are of the same curvature.

B. The general equation of a curve with constant curvature k_1 has the view

$$\mathbf{r}(s) = \frac{1}{k_1} \int_0^s \mathbf{t}(\sigma)\, d\sigma$$

where $\mathbf{t} = \mathbf{t}(\sigma)$ is a vector equation of an arbitrary curve on a unit sphere.

C. If curve γ has a constant curvature, then an osculating sphere and an osculating circle of the curve have constant curvature.

D. Every curve with constant curvature $k_1 > 0$ different from an arc of a circle and connecting two points separated by distance

$$d < \frac{2}{k_1}$$

is either longer than the greater arc or shorter than the lesser arc of the circle with radius $R = 1/k_1$ tightened by a chord of length d.

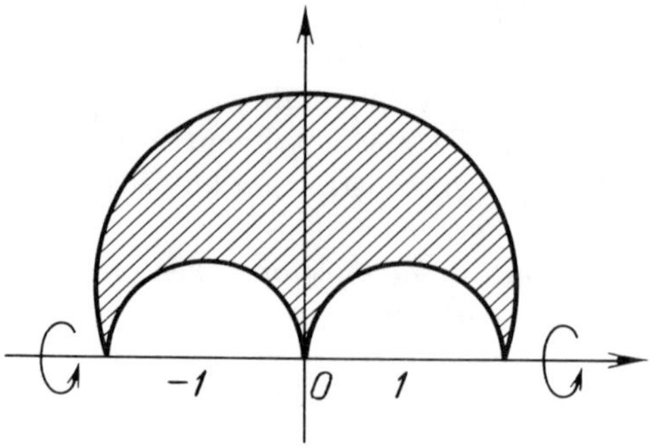

Figure 36.

E. All the curves with curvature 1 outgoing from point O in direction l whose lengths are no more than π lie in the revolution solid formed by rotation of the plane figure with boundary consisting of two semi-circles with radius 1 and their evolvent arcs (Figure 36).

3.2.2 Space curves with constant torsion

$k_2 =$ const

A. If $k_2 \equiv 0$, then the curve is plane.

B. The general equation of a curve with constant torsion $k_2 \neq 0$ has the following form:

$$\mathbf{r}(s) = k_2 \int_0^s \mathbf{b}(\sigma) \frac{d\mathbf{b}}{ds}(\sigma)\, d\sigma$$

where $\mathbf{b} = \mathbf{b}(\sigma)$ is a vector equation of an arbitrary curve on a unit sphere.

3.2.3 Slope lines. Generalized helical curves

The curves with constant curvature-to-torsion ratio are called *generalized helical curves* or *slope lines* (Figure 37).

A. A generalized helical curve is the curve whose tangents form a constant angle with a given direction l. The reverse is also true.

B. The fulfilment of one of the following four conditions:

1) principal normals of a curve are perpendicular to a fixed direction,
2) binormals form a constant angle with a fixed direction;
3) $(\mathbf{r}''_{ss}, \mathbf{r}'''_{sss}, \mathbf{r}^{\text{iv}}_{ssss}) = 0$;
4) vector

$$\frac{1}{\sqrt{\frac{1}{k_1^2} + \frac{1}{k_2^2}}} \left(\frac{\mathbf{t}}{k_1} + \frac{\mathbf{b}}{k_2} \right) = \frac{k_2 \mathbf{t} + k_1 \mathbf{b}}{\sqrt{k_1^2 + k_2^2}} = \text{const}$$

is necessary and sufficient for the curve to be a generalized helical curve.

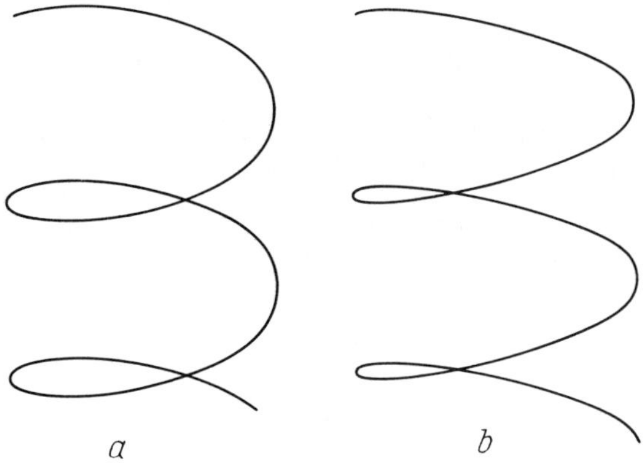

a *b*

Figure 37.

C. Let a cylindrical surface pass through a slope line so that its generators are parallel to a given direction l. Then the slope line intersects the generators at a constant angle.

D. A standard helical curve stands out from the helical curves of the general type because of its cylinder C being circular.

3.2.4 Spherical curves

A curve lying on a sphere is said to be *spherical.*

Let $\mathcal{S}$ be a sphere of radius R.

A. The circles with

$$k_1 = \text{const} \geq \frac{1}{R}, \quad k_2 = 0$$

form the simplest subclass of the spherical curves.

The circles with $k_1 = 1/R$ are *great circles* of sphere $\mathcal{S}$.

B. Every curve of a nonconstant curvature where natural equation (that is, the equation connecting curvature k_1 and torsion k_2)

$$\left(\frac{d}{ds}\left(\frac{1}{k_1}\right)\right)^2 = \frac{1}{k_2^2}\left(R^2 - \frac{1}{k_1^2}\right)$$

or

$$\left(k_2\frac{d}{ds}\left(\frac{1}{k_1}\right)\right)^2 + \frac{1}{k_1^2} = R^2$$

holds lies on a sphere with radius R.

C. A natural equation describing all spherical curves has the following view:

$$\frac{1}{k_1} + \frac{1}{k_2}\frac{d}{ds}\left(k_2\frac{d}{ds}\left(\frac{1}{k_1}\right)\right) = 0$$

or

$$\frac{k_2}{k_1} + \frac{d}{ds}\left(k_2 \frac{d}{ds}\left(\frac{1}{k_1}\right)\right) = 0$$

D. A curve with nonvanishing curvature k_1 and nonvanishing torsion k_2 is a spherical curve if, and only if, the relation

$$\frac{k_2}{k_1} = \frac{d}{ds}\left(\frac{1}{k_2 k_1^2}\frac{d}{ds}k_1\right)$$

holds.

Let

$$\rho = \frac{1}{k_1}, \quad \sigma = \int k_2\, ds$$

Then the last equality may be written as

$$\frac{d^2\rho}{d\sigma^2} + \rho = 0$$

E. A regular curve lies on a sphere if, and only if, there exist numbers A and B such that

$$k_1\left(A\cos\left(\int_0^s k_2(\sigma)\, d\sigma\right) + B\sin\left(\int_0^s k_2(\sigma)\, d\sigma\right)\right) = 1$$

F. A closed spherical curve is contained in an open hemisphere with radius R if its length S is either less than $2\pi R$, or equal to $2\pi R$, but in the latter case the curve is not a union of two great half-circles.

G. If γ is a regular closed spherical curve with curvature $k_1 > 0$ and torsion k_2, then

$$\int_\gamma k_2(s)\, ds = 0, \quad \int_\gamma \frac{k_2(s)}{k_1(s)}\, ds = 0$$

H. A spherical curve intersecting every meridian at a given angle is called *a loxodrome (a loxodromic curve)* (Figure 38).

Let sphere $\mathcal{S}$ be defined by equation

$$x^2 + y^2 + z^2 = R^2$$

and θ — a given angle ($0 < \theta \leq \pi/2$). Then the equations of loxodrome have the following form:

$$\begin{aligned} x &= R\cos\left[\tan\theta \ln\tan\left(\tfrac{\pi}{4} + \tfrac{t}{2}\right)\right]\cos t, \\ y &= R\sin\left[\tan\theta \ln\tan\left(\tfrac{\pi}{4} + \tfrac{t}{2}\right)\right]\sin t, \\ z &= R\sin t \end{aligned}$$

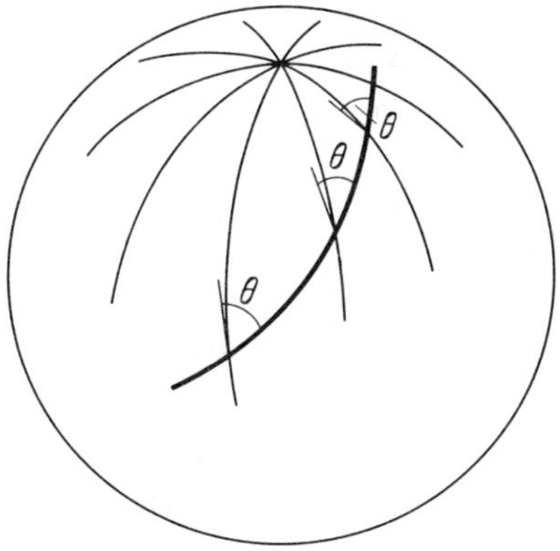

Figure 38.

Curvature k_1 and torsion k_2 of the loxodrome are equal to

$$k_1 = \frac{\cos\theta}{R}\sqrt{1 + \frac{\tan^2\theta}{\cos^2 t}}, \quad k_2 = \frac{\tan\theta}{R(\cos^2 t + \tan^2\theta)}$$

respectively.

I. If all normal planes of a curve pass through a fixed point, then the curve is spherical.

J. If a closed spherical curve has no intersection points, and no more than two osculating planes of the curve pass through the center of the sphere, then the curve lies in a hemisphere.

K. Let γ be a closed spherical curve with length $S < 2\pi R$. Then on the sphere there exists point A such that a spherical distance between the point and every point M of the curve (a length of the lesser arc of great circle passing through points A and M) is no more than $S/4$.

3.2.5 Bertrand curves

Curves γ and $\tilde{\gamma}$ are called *Bertrand curves* if there exists a one-to-one correspondence between them such that their normals coincide at the corresponding points.

Properties of the Bertrand curves

A. The distance between the corresponding points of curves γ and γ^* is constant.

B. Tangents to curves γ and γ^* at the corresponding points form a constant angle.

C. Curvature k_1 and torsion k_2 of either of the Bertrand curves are connected by relation

$$a(k_1 \sin\theta - k_2 \cos\theta) = \sin\theta \qquad (*)$$

where a is a distance between the corresponding points, and θ — an angle formed by tangents at the points.

D. If a curvature and a torsion of a curve are connected by relation (*), then the curve is *a curve of Bertrand*, that is, a curve whose principal normals coincide with principal normals of other curve (different from it).

A curve defined by equation

$$\mathbf{r} = a \int \mathbf{e}(t)\,dt + b \int \mathbf{e}(t) \times \mathbf{e}'(t)\,dt \qquad (**)$$

where $|\mathbf{e}(t)| = 1$, $|\mathbf{e}'(t)| = 1$, $a = \text{const}$, $b = \text{const}$ is a curve of Bertrand. The reverse is also true: every curve of Bertrand may be described by equation (**).

The existence of constant numbers λ and μ such that

$$\lambda k_1 + \mu k_2 = 1$$

is necessary and sufficient for a space curve to be a curve of Bertrand.

3.2.6 Closed space curves

A. If

$$0 \leq k_1 \leq a$$

where k_1 is a curvature of a space curve, then its length S satisfies the condition

$$S \geq \frac{2\pi}{a}$$

B. For any regular closed space curve γ the following inequality

$$\int_\gamma k_1(\sigma)\,d\sigma \neq 2\pi$$

is true. The equality takes place if, and only if, curve γ is plane and convex (theorem of Fenchel).

C. Let A be an arbitrary real number. There exists regular closed curve γ such that

$$\int_\gamma k_2(s)\,ds = A$$

D. Let γ be a regular closed curve of length S lying in a ball with radius R. Then

$$S \int_\gamma k_2^2(s)\,ds \leq R^2 \Big(\int_\gamma k_1^2(s)\,ds \cdot \int_\gamma k_2^2(s)\,ds - \Big(\int_\gamma k_1(s) k_2(s)\,ds \Big)^2 \Big)$$

$$S\frac{16}{S}\int\limits_{\gamma} k_2^2(s)\,ds \leq \int\limits_{\gamma} k_1^2(s)\,ds \cdot \int\limits_{\gamma} k_2^2(s)\,ds - \left(\int\limits_{\gamma} k_1(s)k_2(s)\,ds\right)^2$$

$$S \leq R\int\limits_{\gamma} k_1(s)\,ds$$

E. Let simple closed regular curve γ possess the following property: through every point of the curve one may draw a plane which has no other common points with the curve. Then the curve has no less than four common points with the steady-state osculation plane.

F. Let an arc of a plane curve and its tightening chord form a boundary of a convex domain. Then under any twisting of the curve (that is, under the transformations holding its length and curvature) the length of its tightening chord increases.

G. For any knotted closed curve the following formula

$$\int\limits_{\gamma} k_1(s)\,ds \geq 4\pi$$

is true (formula of Fairy–Milnor).

H. Let arcs of a unit circle $\alpha < \pi$, $\beta = 2\pi - \alpha$ be tightened by a chord MN of length $d < 2$. Then every other curve with curvature 1 connecting points M and N has length S such that either $S < \alpha$ or $S > \beta$.

I. Let γ be a closed curve of length $S < 2\pi$ lying on a unit sphere. Then there exists point A on the sphere such that a distance between the point and any point of curve γ is no more than $S/4$.

3.2.7 Torus knots

Let $\mathcal{K}$ be a circle of radius a and a center at point $A(0, l)$, where $a < l$, on plane Oxy (Figure 39). A surface formed by rotation of the circle about the x-axis is called *a torus* (Figure 40). The parametric equations of the torus have the following view

$$x = a\sin u, \quad y = (l + a\cos u)\cos v, \quad z = (l + a\cos u)\sin v$$

Two circles are naturally connected with the torus: an axial circle (a trajectory of point A under the rotation) and a null meridian (given circle $\mathcal{K}$).

Let m and n be coprime numbers. A curve defined by equations

$$\begin{aligned} x &= a\sin 2\pi mt, \\ y &= (l + a\cos 2\pi mt)\cos 2\pi nt, \\ z &= (l + a\cos 2\pi mt)\sin 2\pi nt \end{aligned}$$

is called *a torus knot of type* (m, n). The curve makes n turns about the torus in the axial circle direction and m turns in a direction of the meridian. The torus knots of types (1,1), (5,2) and (4,3) are pictured in Figure 41.

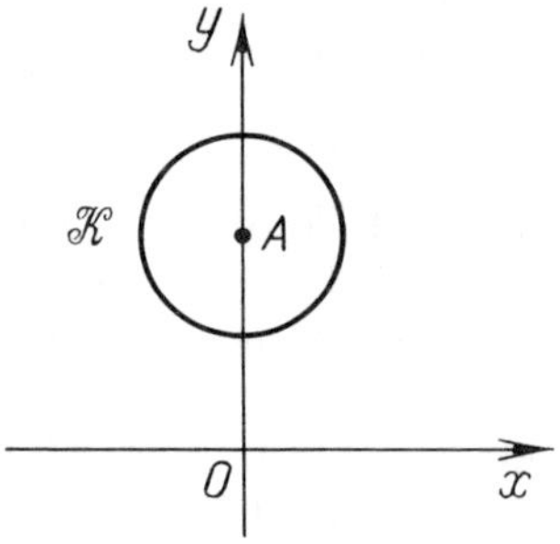

Figure 39.

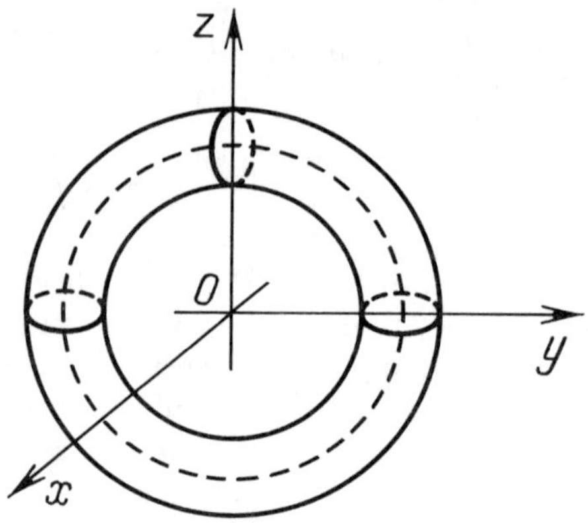

Figure 40.

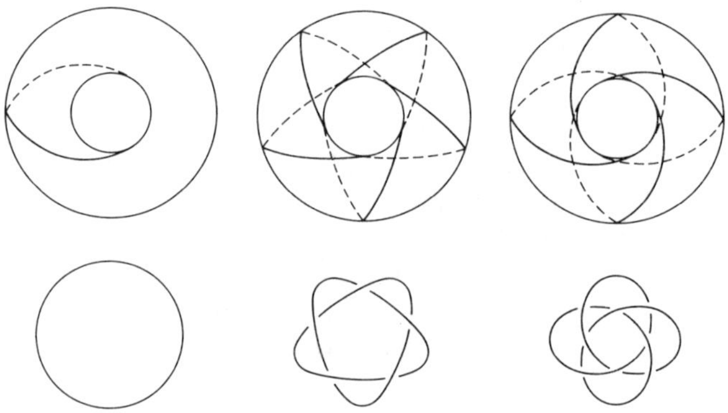

Figure 41.

3.2.8 Bicylindrical curves and Viviani curve

An intersection line of two cylinders whose axes intersect at the right angle is called *a bicylindrical curve* or *a bicylindric*. The parametric equations of the bicylindrical curve have the following form:

$$x = a\cos t, \quad y \pm \sqrt{b^2 - a^2\sin^2 t}, \quad z = a\sin t$$

where a and $b > a$ are radii of the cylinders. For $a = b$ the intersection line is a pair of equal ellipses.

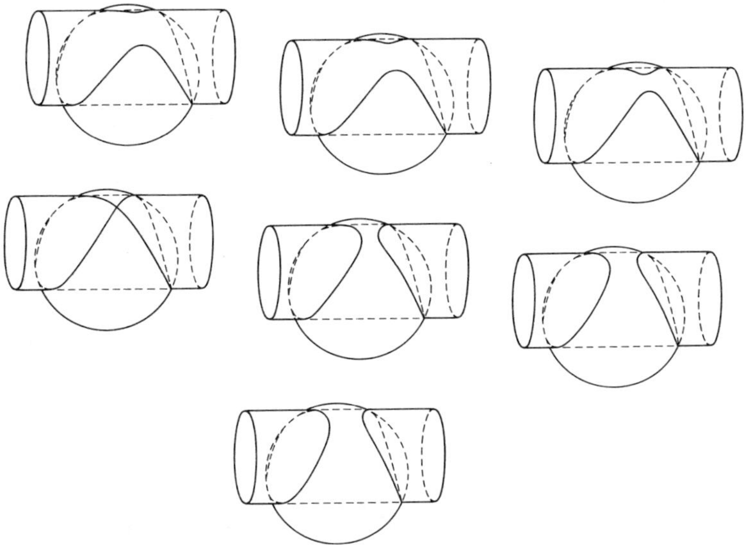

Figure 42.

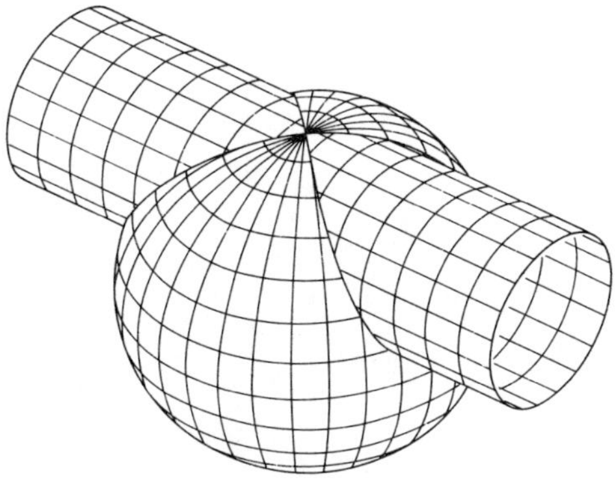

Figure 43.

The intersection lines of a circular cylinder and a sphere (under varying mutual arrangement) are shown in Figure 42. In the special case that radius of the sphere is two times greater than radius of the cylinder and the cylinder generator passes through a center of the sphere we have *a curve of Viviani* (Figure 43).

Chapter 4

Some Applications of Curves

Many of the plane curves from the Atlas possess interesting geometric, kinematic and dynamic properties. The investigation of the properties clarifies a natural origination of some curve classes, unveils reasons for unremitting attention to the others and discloses the possibilities of the curve applications to various technical problems.

The elements of the geometric spline theory and simple algorithms for constructing some spline-curves also are of practical significance.

4.1 Dynamic properties of plane curves

This section describes the noncoordinate ways for constructing certain of the curve classes from the Atlas. The presentation follows a unified scheme: first we describe the simplest geometric objects (points, straight lines, circles and their parts) required for construction of a curve, and then present the method for constructing points on the curve and corresponding drawing.

Given: straight line $\mathcal{L}_1$, circle $\mathcal{K}$ of radius a tangent to the line at point A, and straight line $\mathcal{L}_2$ parallel to line $\mathcal{L}_1$ and located at distance $b > a$ from it. It is suggested that the circle and line $\mathcal{L}_2$ are located in the same half-plane defined by straight line $\mathcal{L}_1$.

Denote by M and N the intersection points of variable beam AM with line $\mathcal{L}_2$ and circle $\mathcal{K}$ respectively, and pass straight line MP perpendicular and straight line NP parallel to line $\mathcal{L}_1$ (Figure 1). Point P of the intersection of these two lines lies on *an agnesiana*.

Given: straight lines $\mathcal{L}_1$ and $\mathcal{L}_2$ intersecting at right angle at point O, point A lying on line $\mathcal{L}_1$ at distance $2a$ from point O, and natural number n.

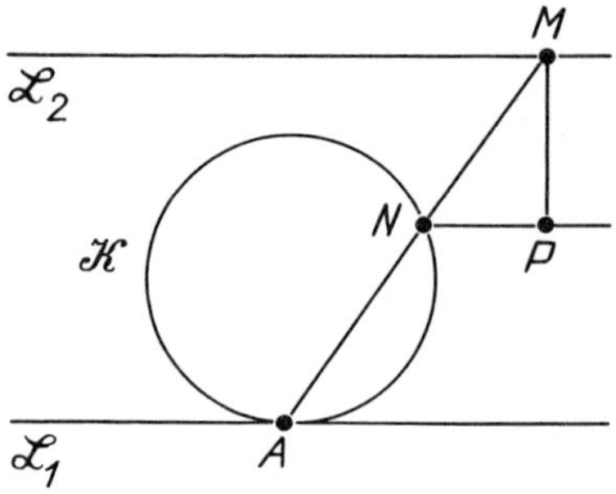

Figure 1.

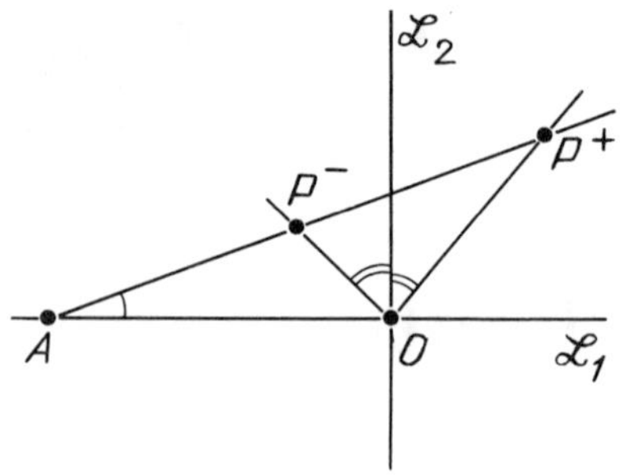

Figure 2.

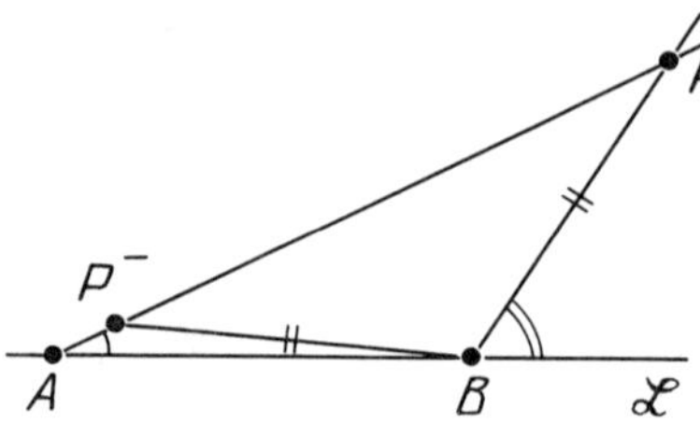

Figure 3.

Let M be an intersection point of variable beam AM and line $\mathcal{L}_2$. Construct angles $\angle MOP^+$ and $\angle MOP^-$ with common side OM such that

$$\angle MOP^+ = \angle MOP^- = n\angle OAM$$

(Figure 2). Intersection points P^+ and P^- of beam AM and the sides of the angles lie on *an arachnida* (A).

Given: straight line $\mathcal{L}$, points A and B on the line separated by distance a, and natural number $n \geq 3$.

Plot an angle with vertex B from line $\mathcal{L}$ being n times greater than an angle which variable beam AM forms with the line (Figure 3). Intersection point P^+ of the angle side and beam AM and point P^- of the beam such that $P^-B = P^+B$ lies on *arachnida* (B).

Given: straight lines $\mathcal{L}_1$ and $\mathcal{L}_2$ intersecting at right angle at point O, and positive number l.

Denote by MN a variable segment with end-points M and N lying on lines $\mathcal{L}_1$ and $\mathcal{L}_2$, respectively. Pass straight line MQ parallel to line $\mathcal{L}_1$, and NQ — to line $\mathcal{L}_2$. Foot P of perpendicular QP dropped from point Q to segment MN lies on *an astroid* (Figure 4).

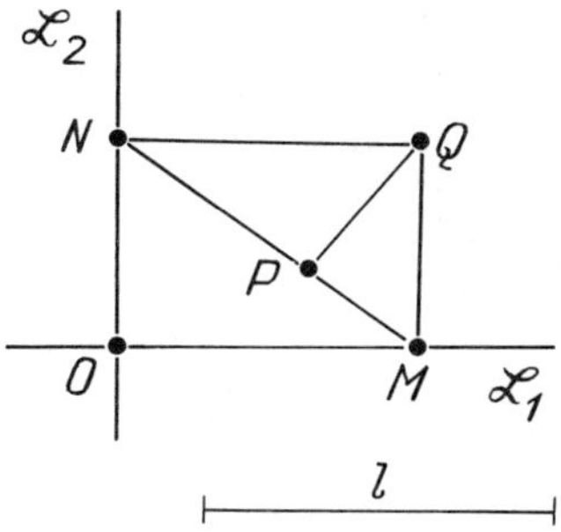

Figure 4.

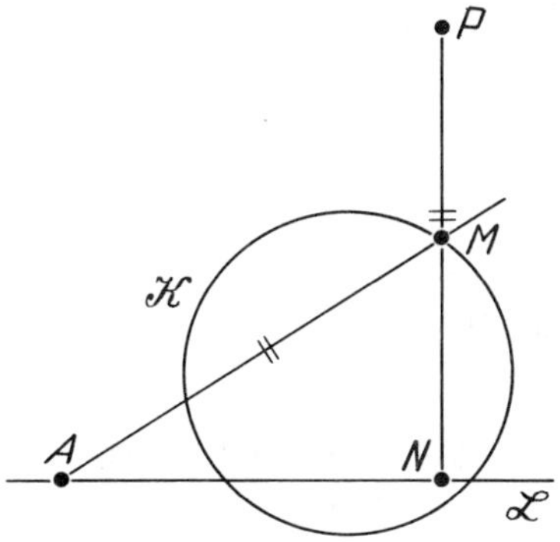

Figure 5.

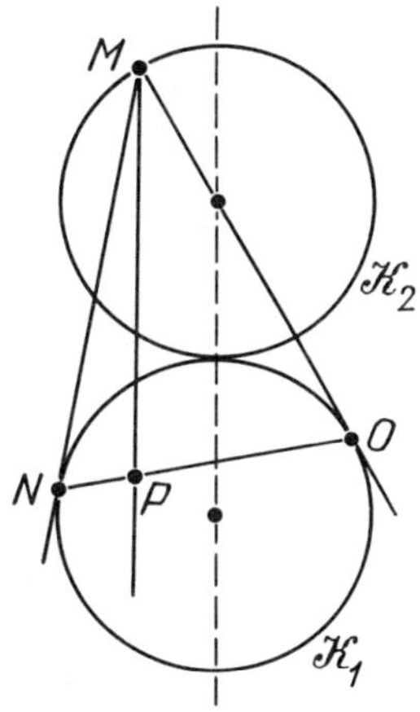

Figure 6.

Given: straight line $\mathcal{L}$, circle $\mathcal{K}$ intersecting the line, and point A on line $\mathcal{L}$ but outside of the circle.

Let M be an intersection point of line $\mathcal{L}$ and circle $\mathcal{K}$. Drop perpendicular MN to line $\mathcal{L}$ from point M and plot segment NP on the perpendicular from intersection point N of lines NM and $\mathcal{L}$ equal to segment AM (Figure 5). End-point P of the segment lies on *a besace*.

Given: circles $\mathcal{K}_1$ and $\mathcal{K}_2$ of radius r with an exterior tangency.

Pass tangent lines MN and MQ to circle $\mathcal{K}_1$ through running point M of circle $\mathcal{K}_2$. Intersection point P of segment NQ and straight line MP parallel to the center line of circles $\mathcal{K}_1$ and $\mathcal{K}_2$ lies on *a bicorn curve* (Figure 6).

Given: straight line $\mathcal{L}$, variable circle $\mathcal{K}$ with a center on line $\mathcal{L}$ and, fixed point A lying on the line outside the circle.

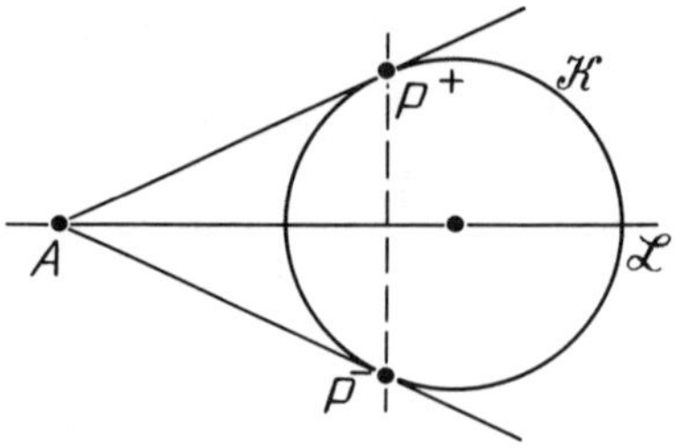

Figure 7.

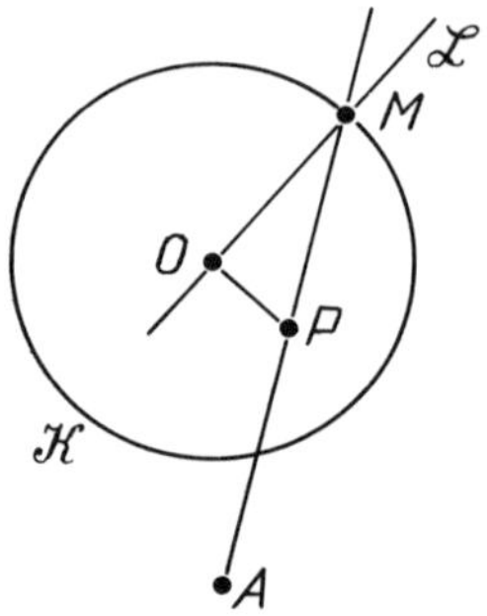

Figure 8.

Pass tangent lines AP^+ and AP^- from point A to variable circle $\mathcal{K}$ (Figure 7). Tangency points P^+ and P^- lie on *a cappa*.

Given: circle $\mathcal{K}$ of radius c, and point A lying outside of the circle at distance $g > c$ from its center O.

Pass straight line $\mathcal{L}$ through intersection point M of variable beam AM and circle $\mathcal{K}$ and center O of the circle, and erect perpendicular OP to the straight line from point O. Intersection point P of the perpendicular and beam AM lies on *a capricornoid* (Figure 8).

Given: circles $\mathcal{K}_1$ and $\mathcal{K}_2$ of radius r with exterior tangency, and point P on circle $\mathcal{K}_2$.

When circle $\mathcal{K}_2$ rolls along circle $\mathcal{K}_1$ without slipping point P traces out *a cardioid* (Figure 9).

Given: non-intersected circles $\mathcal{K}_1$ and $\mathcal{K}_2$ of radii r_1 and r_2 with centers O_1 and O_2, respectively, length $a > r_1 + r_2$ of segment O_1O_2, point A lying on the center line outside these circles and the segment O_1O_2 at distance $b > r_1$ from point O_1.

Variable beam AM intersects circles $\mathcal{K}_1$ and $\mathcal{K}_2$ at four points N_1, Q_1, N_2 and Q_2. Intersection points P_1, P_2, P_3, P_4 of straight lines O_1N_1 and

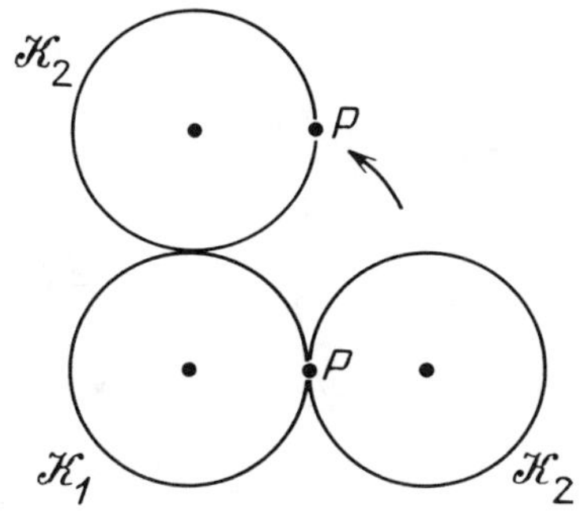

Figure 9.

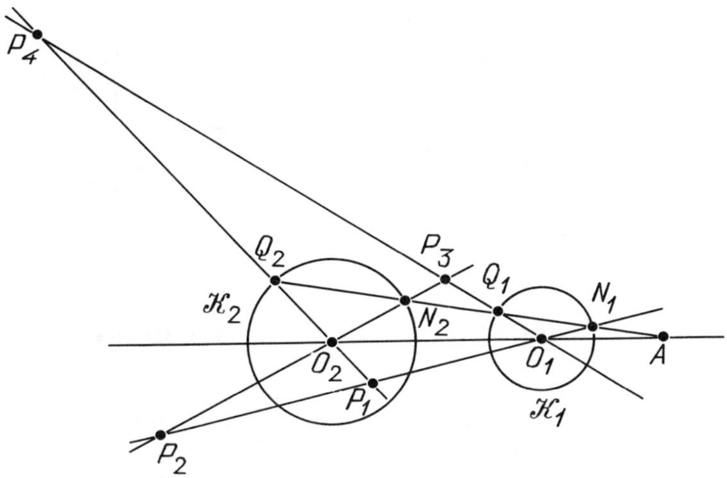

Figure 10.

O_2N_2, O_1N_1 and O_2Q_2, O_1Q_1 and O_2N_2, O_1Q_1 and O_2Q_2, respectively, lie on *a Cartesian oval* (Figure 10).

Given: straight line $\mathcal{L}$, circle $\mathcal{K}$ of radius r with center O on line $\mathcal{L}$, and point A lying on the line at distance a from point O.

Let B and C be the intersection points of line $\mathcal{L}$ and circle $\mathcal{K}$. Pass a variable straight line through point A and denote by M and N the intersection points of the line and circle $\mathcal{K}$. Draw two circles with centers at points B and C and radii equal to lengths of segments AM and AN, respectively. Their intersection point P lies on *a Cassinian oval* (Figure 11).

Given: straight line $\mathcal{L}$, circle $\mathcal{K}$ of radius a touching line $\mathcal{L}$, and point A on the circle opposite to the point of tangency.

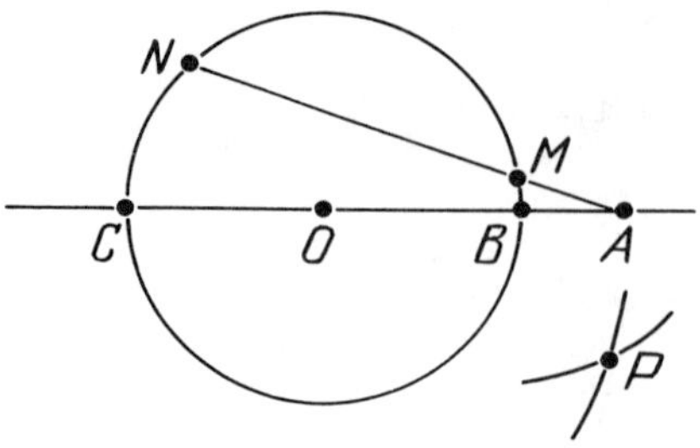

Figure 11.

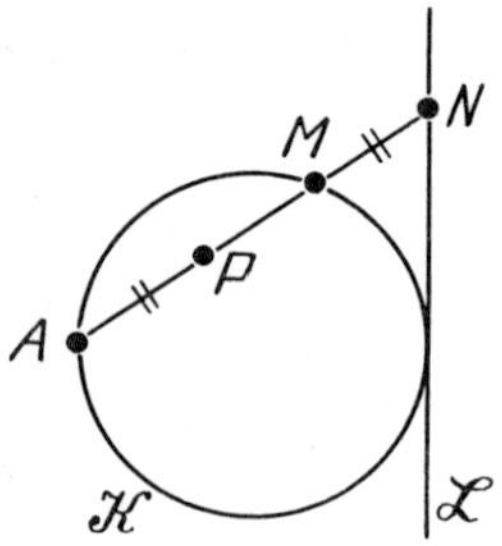

Figure 12.

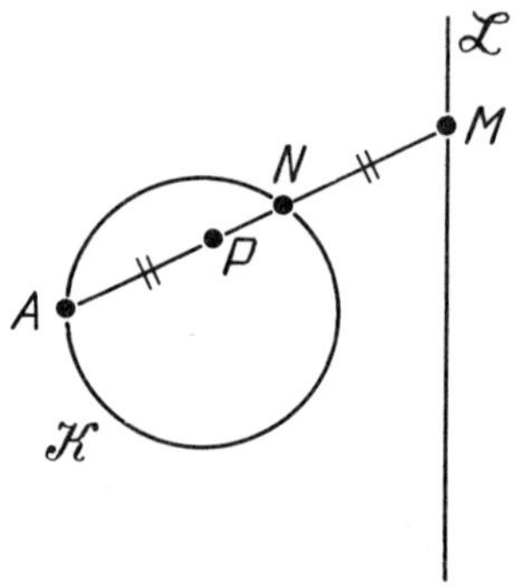

Figure 13.

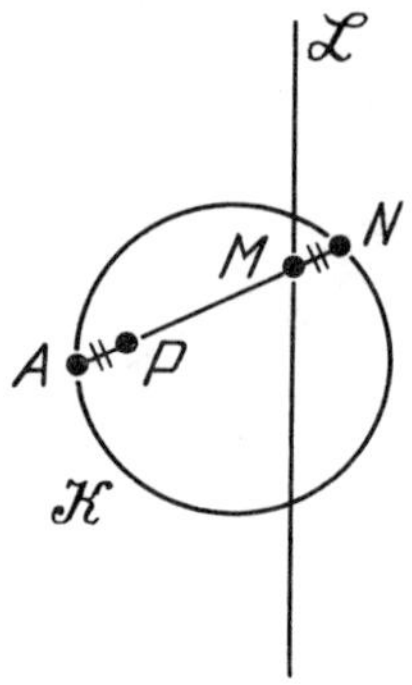

Figure 14.

Denote by M and N the intersection points of variable beam AM and lines $\mathcal{L}$ and $\mathcal{K}$, respectively, and plot segment AP on the beam from point A equal to segment MN (Figure 12). End-point P of the segment lies on *a cissoid*.

Given: straight line $\mathcal{L}$, circle $\mathcal{K}$ of radius a, and point A placed on the circle at maximal distance l from line $\mathcal{L}$.

Denote by M and N the intersection points of variable beam AM and lines $\mathcal{L}$ and $\mathcal{K}$, respectively, and plot segment AP on the beam from point A equal to segment MN. End-point P of the segment lies on *a prolate cissoid* ($l > 2a$) (Figure 13) or on *a curtate cissoid* ($l < 2a$) (Figure 14).

Given: straight lines $\mathcal{L}_1$ and $\mathcal{L}_2$ intersecting at right angle at point O, point A lying on line $\mathcal{L}_1$ at distance $2a$ from point O, and number k.

Pass straight line $\mathcal{L}$ through variable point M of line $\mathcal{L}_1$ parallel to line $\mathcal{L}_1$. Point P of line $\mathcal{L}$ such that $MP \cdot PA = ka$ lies on *a conchal* (Figure 15).

Given: straight lines $\mathcal{L}_1$ and $\mathcal{L}_2$ intersecting at right angle at point O, point A lying on line $\mathcal{L}_1$ at distance a from point O, and number b.

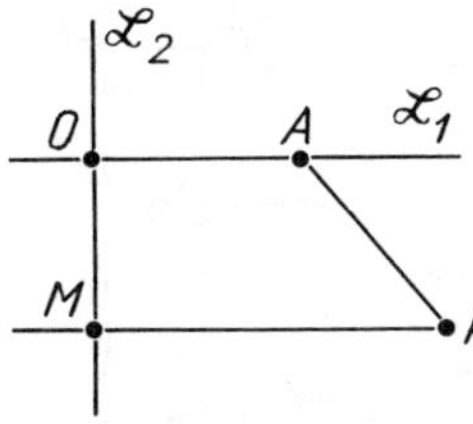

Figure 15.

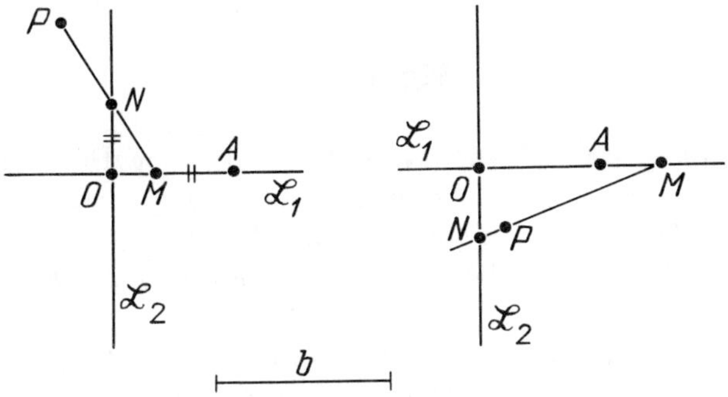

Figure 16.

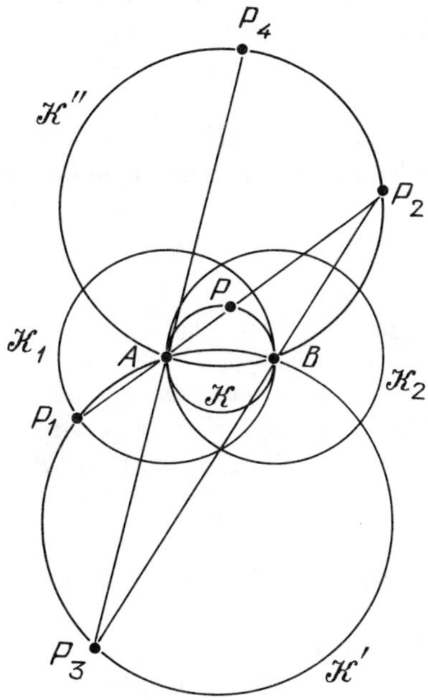

Figure 17.

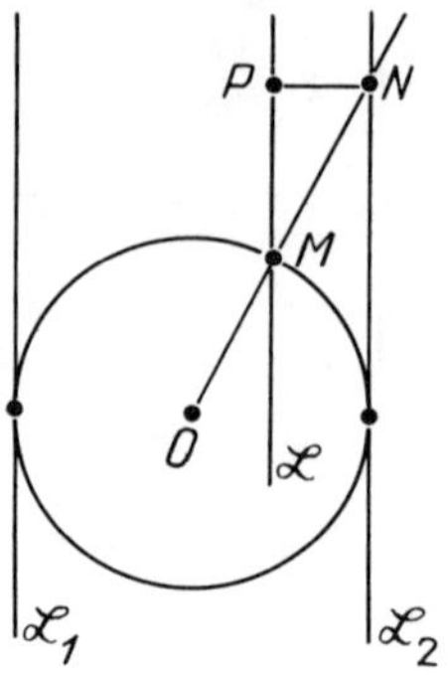

Figure 18.

Plot variable segments AM and ON equal in length on line $\mathcal{L}_1$ from point A and on line $\mathcal{L}_2$ from point O respectively. Pass straight line $\mathcal{L}$ through points M and N, and plot segment MP of length b on it from point M (Figure 16). End-point P of the segment lies on *a conchoid of Dürer.*

Given: circle $\mathcal{K}$ of diameter d, and two equal circles $\mathcal{K}_1$ and $\mathcal{K}_2$ of radius d which have inner tangency with circle $\mathcal{K}$ at diametrically opposite points A and B.

Denote by M an intersection point of variable straight line AM and circle $\mathcal{K}$, and plot segment MP_2 equal to segment MP_1 where P_1 is the intersection point of lines AM and $\mathcal{K}_1$ on the line. Circumscribe circle $\mathcal{K}'$ about triangle ΔABP_1, pass a straight line through points P_2 and B and denote by P_3 an intersection point of these circle and line. Next circumscribe circle $\mathcal{K}''$ about triangle ΔABP_2, pass a straight line through points P_3 and B, and denote by P_4 an intersection point of these circle and line, etc. Points $P_1, P_2, P_3, P_4, \ldots$ lie on *a conchoid of higher order circle* (Figure 17).

Given: parallel straight lines $\mathcal{L}_1$ and $\mathcal{L}_2$ and circle $\mathcal{K}$ of radius a with center at point O tangent to them.

Pass straight line $\mathcal{L}$ through intertsection point M of variable beam OM and circle $\mathcal{K}$ parallel to line $\mathcal{L}_1$, and drop perpendicular NP to line $\mathcal{L}$ from intersection point N of beam OM and line $\mathcal{L}_2$ (Figure 18). Point P lies on *a conchoid of Külp.*

Given: straight line $\mathcal{L}$, point A lying at distance a from the line, and positive number l.

Denote by M the intersection point of variable beam AM and line $\mathcal{L}$, and plot segments MP^+ and MP^- of length l on the beam on each side of line $\mathcal{L}$ (Figure 19). End-points P^+ and P^- of these segments lie on *a conchoid of Nicomedes.*

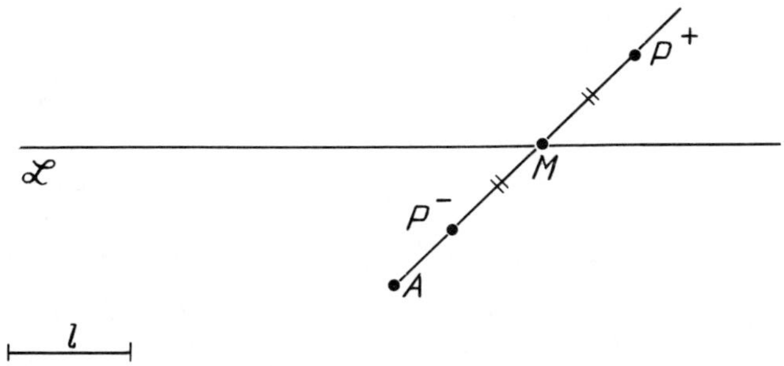

Figure 19.

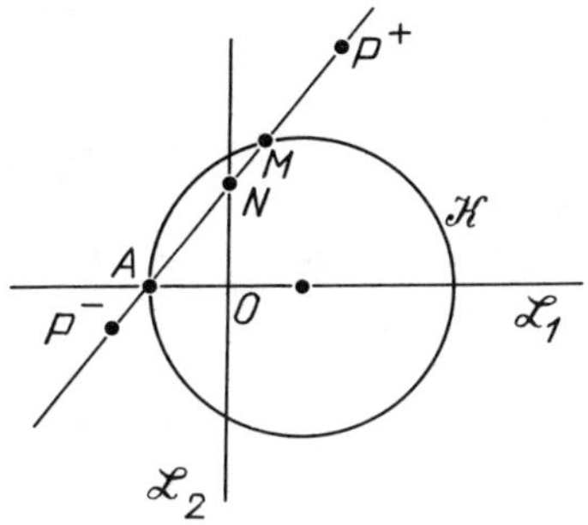

Figure 20.

Given: straight lines $\mathcal{L}_1$ and $\mathcal{L}_2$ intersecting at right angle at point O, circle $\mathcal{K}$ of radius R center of which lies on line $\mathcal{L}_1$ at distance $R - a$ from point O ($k^2 = 2aR$), and intersection point A of line $\mathcal{L}_1$ and the circle nearest to point O.

Denote by N the intersection point of variable straight line AM and line $\mathcal{L}_2$, and plot segments NP^+ and NP^- on line AM from the point equal to segment AM (here M is an intersection point of line AM and circle $\mathcal{K}$) (Figure 20). End-points P^+ and P^- of these segments lie on *conchoids of Slüse*.

Given: circle $\mathcal{K}$ of radius r and straight line $\mathcal{L}$ passing through its center O.

Pass tangent line $\mathcal{L}'$ to the circle through end-point M of variable chord MN parallel to line $\mathcal{L}$. Foot P of perpendicular NP dropped to line $\mathcal{L}$ from the second end-point of the chord lies on *a cornoid* (Figure 21).

Given: concentric circles $\mathcal{K}_1$ and $\mathcal{K}_2$ of radii c and $d > c$, respectively, with common center O, point A lying at distance $g < c$ from point O, and number l.

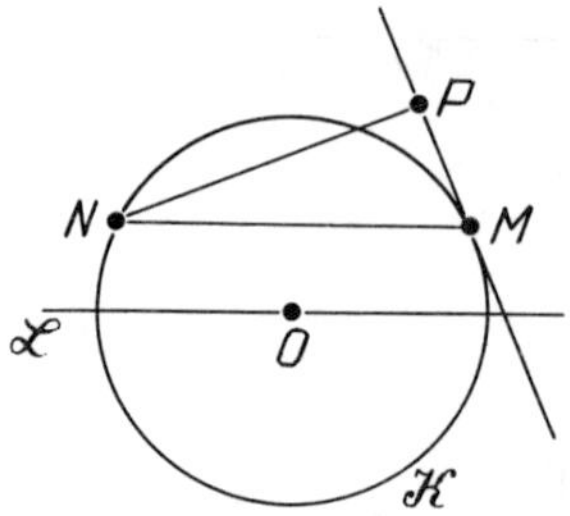

Figure 21.

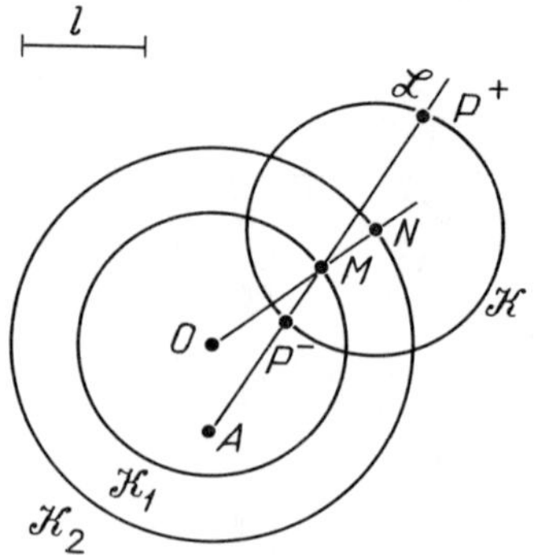

Figure 22.

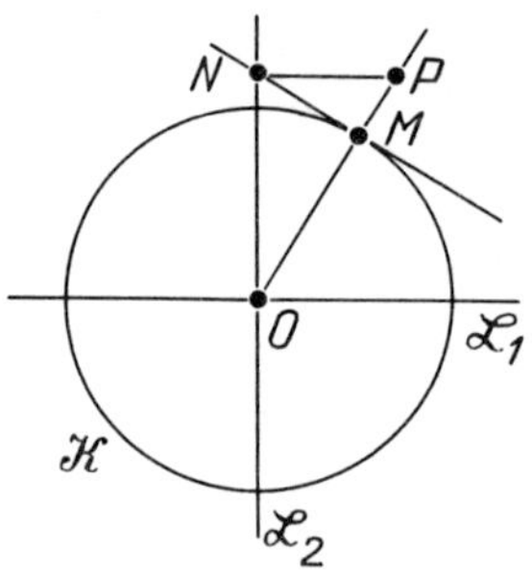

Figure 23.

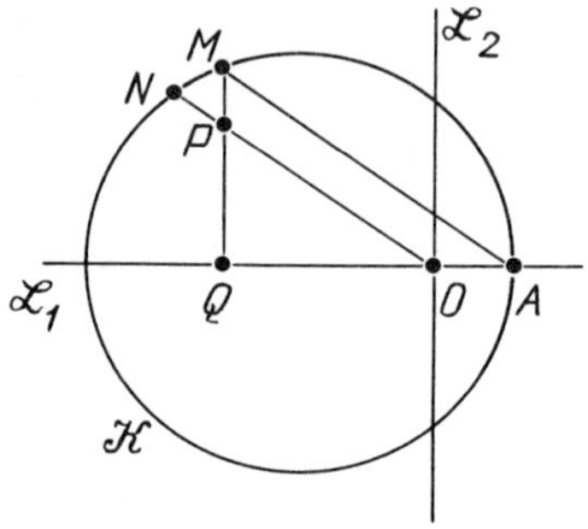

Figure 24.

Pass straight line $\mathcal{L}$ through intersection point M of variable beam OM and circle $\mathcal{K}_1$ and the point A, and describe circle $\mathcal{K}$ of radius l with intersection point N of the beam and circle $\mathcal{K}_2$ as a center (Figure 22). Intersection points P^+ and P^- of circle $\mathcal{K}$ and line $\mathcal{L}$ lie on *a cranioid.*

Given: straight lines $\mathcal{L}_1$ and $\mathcal{L}_2$ intersecting at right angle at point O, and circle $\mathcal{K}$ with radius a and center at point O.

Let M be an intersection point of circle $\mathcal{K}$ and variable beam OM. Pass a tangent line to the circle through the point and erect a perpendicular to line $\mathcal{L}_2$ from an intersection point of the line and the tangent (Figure 23). Point P of the intersection of the perpendicular and beam OM lies on *a curve of Clairaut.*

Given: straight lines $\mathcal{L}_1$ and $\mathcal{L}_2$ intersecting at right angle at point O, circle $\mathcal{K}$ with radius r center of which lies on line $\mathcal{L}_1$ at distance $l < r$ from point O, and intersection point A of line $\mathcal{L}_1$ and the circle lying at distance $r - l$ from point O.

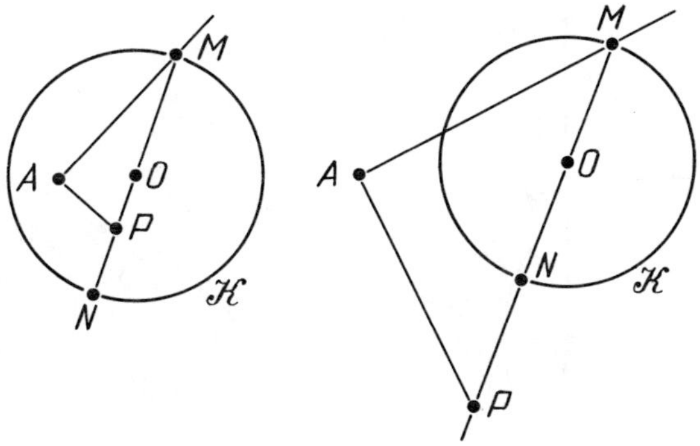

Figure 25.

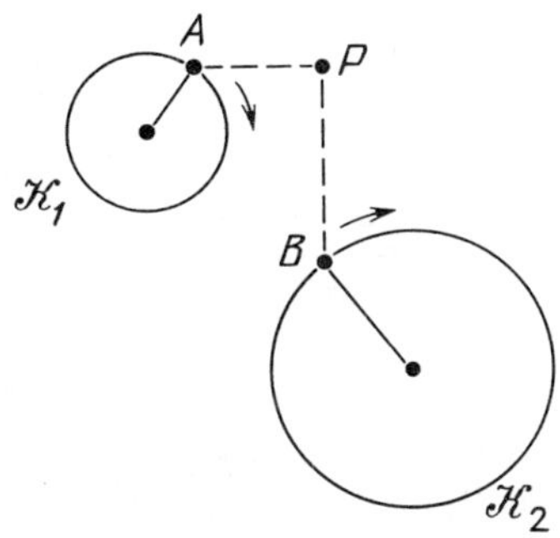

Figure 26.

Denote by NO a segment parallel to variable chord AM of circle $\mathcal{K}$ (Figure 24). Intersection point P of the segment and perpendicular MQ to line $\mathcal{L}_1$ lies on *a curve of Cramer.*

Given: circle $\mathcal{K}$ of radius r with center at point O, and point A at distance $a \neq r$ from point O.

Pass straight line MN through intersection point M of variable beam AM and circle $\mathcal{K}_1$ and the center O of the circle. Intersection point P of perpendicular AP erected to beam AM and line MN lies on *a curve of Jeřabek* (Figure 25).

Given: circle $\mathcal{K}_1$ of radius a, point A on circle $\mathcal{K}_1$, circle $\mathcal{K}_2$ of radius b and point B on circle $\mathcal{K}_2$.

Let points A and B move with constant velocities along circles $\mathcal{K}_1$ and $\mathcal{K}_2$, respectively (Figure 26). Intersection point P of variable horizontal straight line AP and variable vertical straight line BP traces out *a curve of Lissajous.*

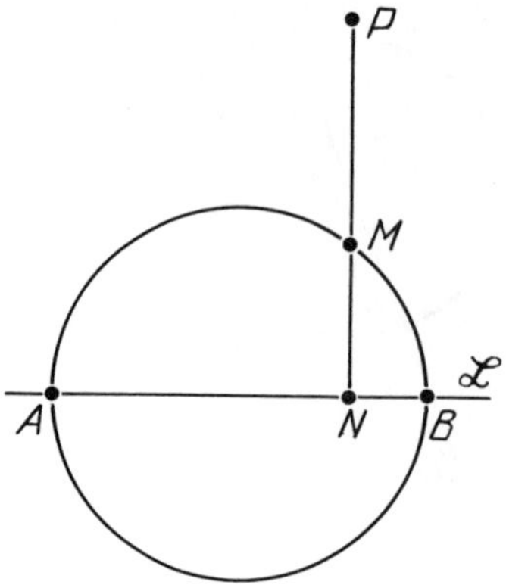

Figure 27.

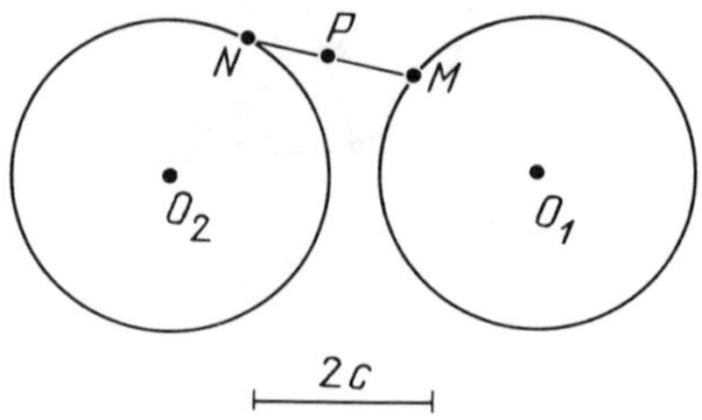

Figure 28.

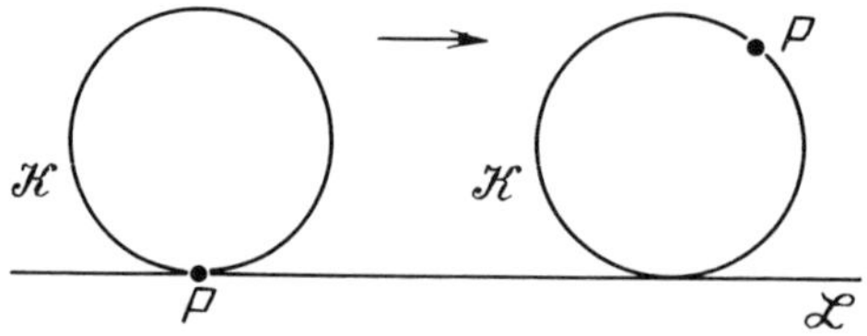

Figure 29.

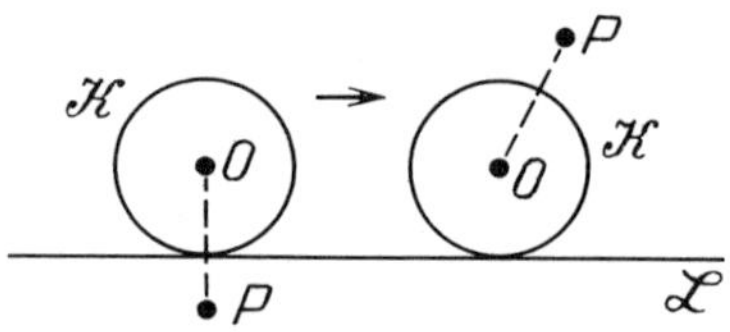

Figure 30.

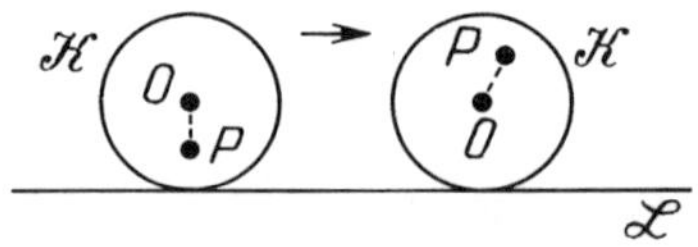

Figure 31.

Given: straight line $\mathcal{L}$, circle $\mathcal{K}$ of radius a with center on line $\mathcal{L}$, and one of their common points — point A.

Drop perpendicular MN to line $\mathcal{L}$ from variable point M of circle $\mathcal{K}$ and plot segment PN on the perpendicular from its intersection point with line $\mathcal{L}$ whose length is equal to the length of circle arc AM (Figure 27). End-point P of the segment lies on *a curve of Ozanam*.

Given: nonintersecting circles $\mathcal{K}_1$ and $\mathcal{K}_2$ of radius b with centers O_1 and O_2, respectively, length $2a$ of segment O_1O_2 and number c.

Midpoint P of variable segment MN of length $2c$ whose end-points move along circles $\mathcal{K}_1$ and $\mathcal{K}_2$ lies on *a curve of Watt* (Figure 28).

Given: straight line $\mathcal{L}$, circle $\mathcal{K}$ of radius r touching line $\mathcal{L}$, and point P mounted rigidly to the circle.

When circle $\mathcal{K}$ rolls along line $\mathcal{L}$ without slipping point P traces out *a cycloid* (Figure 29).

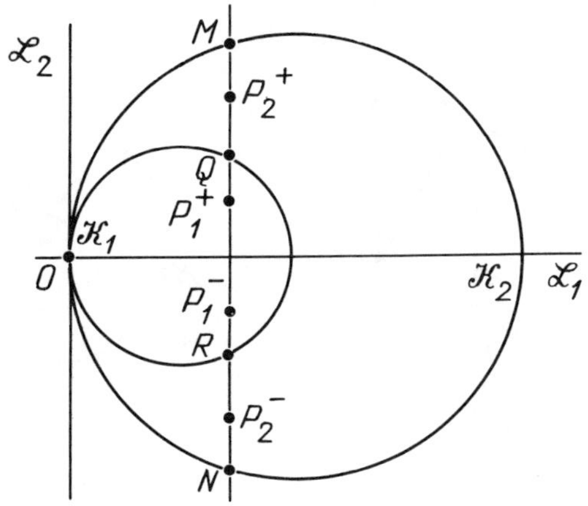

Figure 32.

Given: straight line $\mathcal{L}$, circle $\mathcal{K}$ of radius r with center at point O touching line $\mathcal{L}$, and segment OP of length $d \neq r$ mounted rigidly to the circle.

When circle $\mathcal{K}$ rolls along circle $\mathcal{K}_0$ without slipping end-point P traces out *a prolate cycloid* ($d > r$) (Figure 30) or *a curtate cycloid* ($d < r$) (Figure 31).

Given: straight lines $\mathcal{L}_1$ and $\mathcal{L}_2$ intersecting at right angle at point O, circles $\mathcal{K}_1$ and $\mathcal{K}_2$ of radii a and $2a$, respectively, with centers on line $\mathcal{L}_1$, and point O of inner tangency.

Pass variable straight line MN perpendicularly to line $\mathcal{L}_1$. Let M, N, Q and R be points of its intersection with circles $\mathcal{K}_1$ $\mathcal{K}_2$, respectively. End-points P_1^+ and P_1^- of segment $P_1^+P_1^-$ lying on line MN symmetrically about line $\mathcal{L}_1$ and such that $P_1^+P_1^- = MQ = NR$, and end-points P_2^+ and P_2^- of segment $P_2^+P_2^-$ lying on line MN in a similar way and such that $P_2^+P_2^- = MR = NQ$ lie on *a double-heart curve* (Figure 32).

Given: straight lines $\mathcal{L}_1$ and $\mathcal{L}_2$ intersecting at right angle at point O, and point A lying on line $\mathcal{L}_1$ at distance a from point O.

Erect perpendicular MN from intersection point M of variable beam AM and line $\mathcal{L}_2$ to the beam, and pass straight line NP parallel to line $\mathcal{L}_2$ through intersection point N of the perpendicular and line $\mathcal{L}_1$ (Figure 33). Point of intersection of line NP and beam AM lies on *a duplicatrix*.

Given: circle $\mathcal{K}$ of radius a with center O, and point A inside the circle at distance $b > a$ from point O.

Pass variable straight line through point A and denote by M and N its intersection points with the circle (Figure 34). End-point P of segment

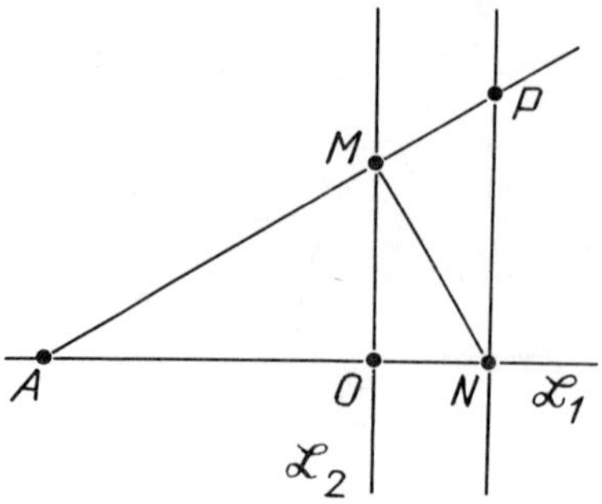

Figure 33.

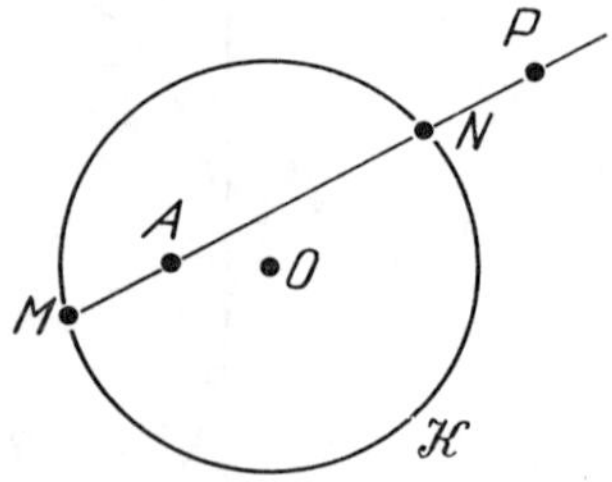

Figure 34.

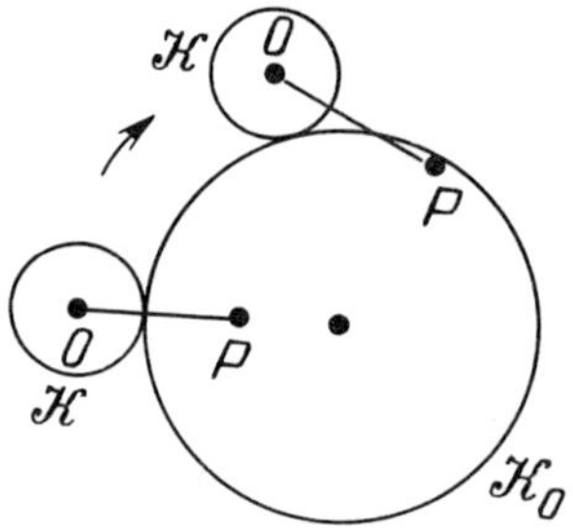

Figure 35.

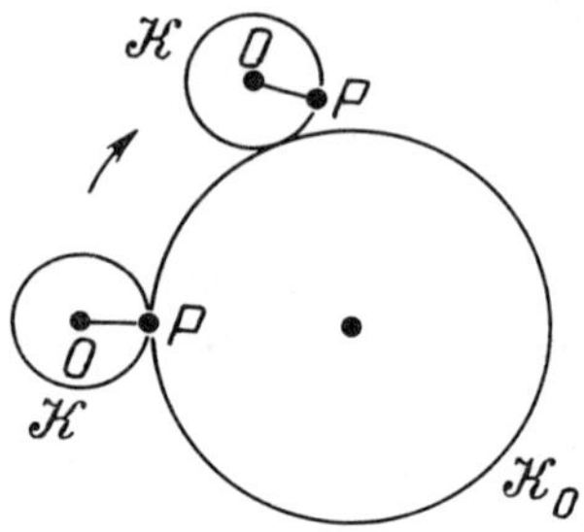

Figure 36.

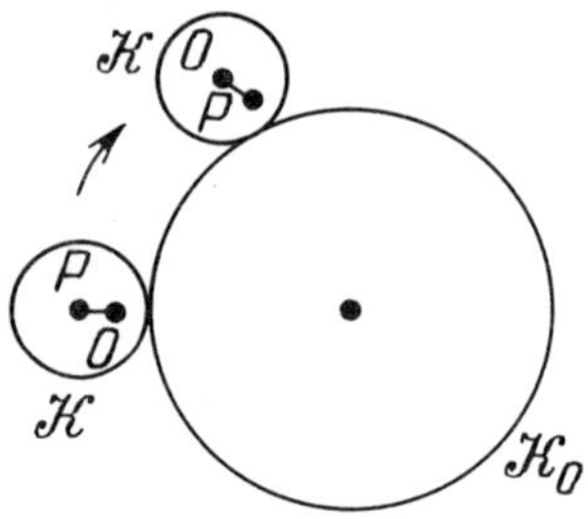

Figure 37.

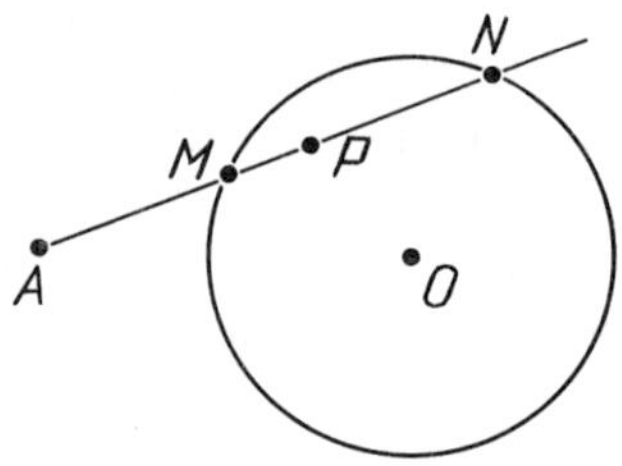

Figure 38.

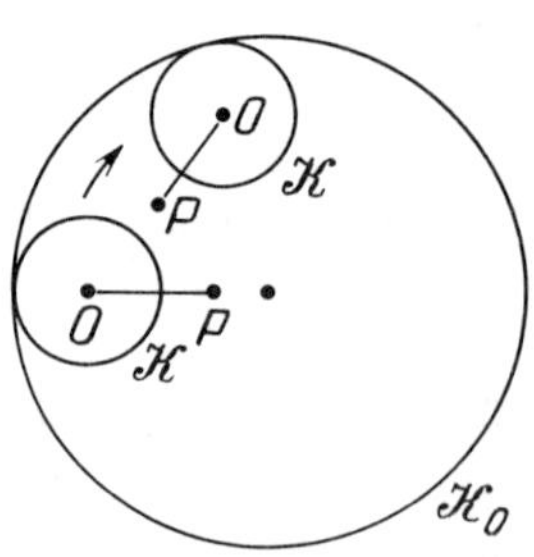

Figure 39.

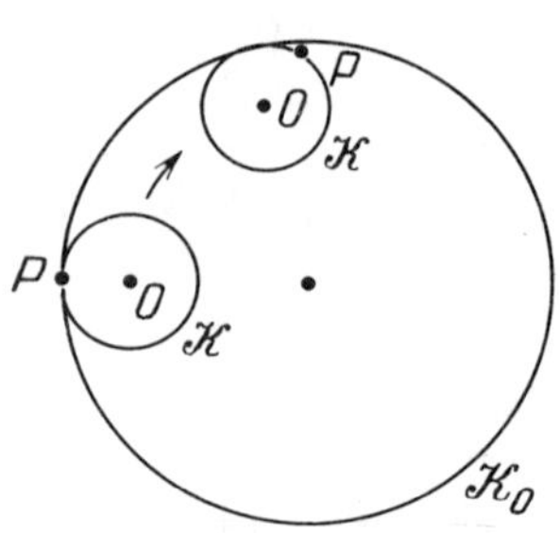

Figure 40.

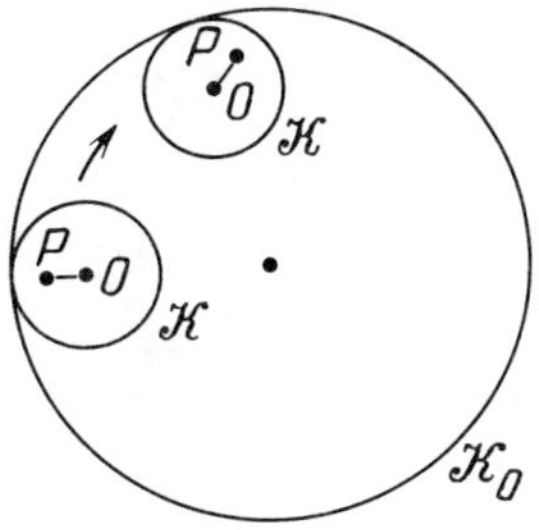

Figure 41.

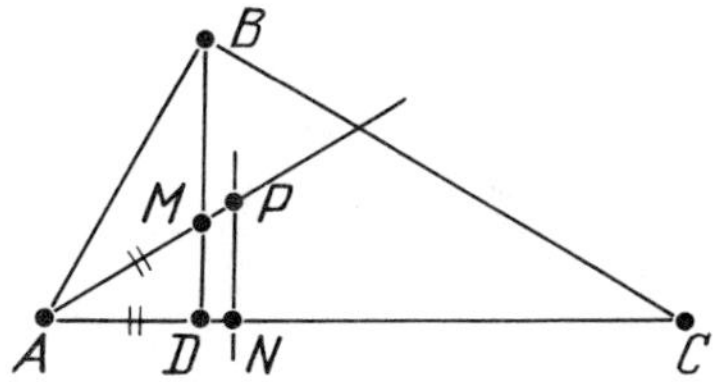

Figure 42.

AP on the variable line whose length is equal to the length of chord MN lies on *an elliptic lemniscate.*

Given: externally tangent circles $\mathcal{K}_0$ with radius R and $\mathcal{K}$ with radius r and center O, and segment OP of length h mounted rigidly to the latter circle.

When circle $\mathcal{K}$ rolls along circle $\mathcal{K}_0$ without slipping end-point P of segment OP traces out *a prolate epitrochoid* ($h > r$) (Figure 35), or *an epicycloid* ($h = r$) (Figure 36), or *a curtate epitrochoid* ($h < r$) (Figure 37).

Given: circle $\mathcal{K}$ with radius a and center at point O, and point A lying outside the circle at distance $b > a$ from point O.

Pass a variable straight line through point A, and denote by MN a chord intercepted on the line by circle $\mathcal{K}$. End-point P of segment AP of straight line MN equal to chord MN lies on *a hyperbolic lemniscate* (Figure 38).

Given: circle $\mathcal{K}_0$ of radius R, circle $\mathcal{K}$ of radius r with center O, and segment OP of length h.

Let circle $\mathcal{K}$ roll on the inside of circle $\mathcal{K}_0$ without slipping. End-point P of segment OP mounted rigidly to circle $\mathcal{K}$ traces out: *a curtate hypotrochoid* (for $h < r$) (Figure 39), or *a hypocycloid* (for $h = r$) (Figure 40), or *a prolate hypotrochoid* (for $h > r$) (Figure 41).

Given: right-angle triangle ΔABC with hypotenuse $AC = b$ and other side $BC = a$.

Let variable beam AM intersect height BD of the triangle at point M. Plot segment AN equal to segment AM on hypotenuse AC from point A. Intersection point P of a straight line parallel to height BD and beam AM lies on *a kampyle* (Figure 42).

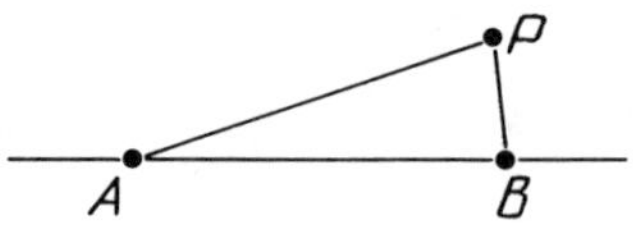

Figure 43.

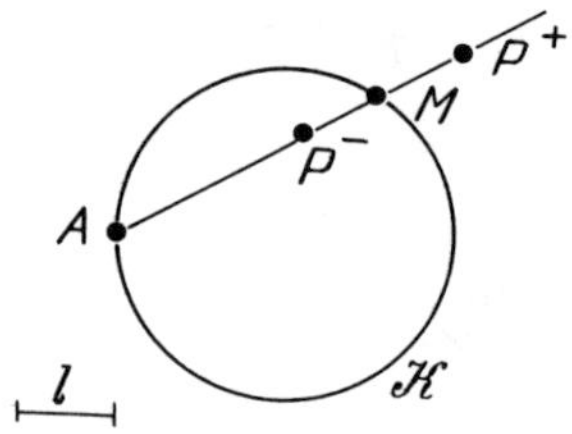

Figure 44.

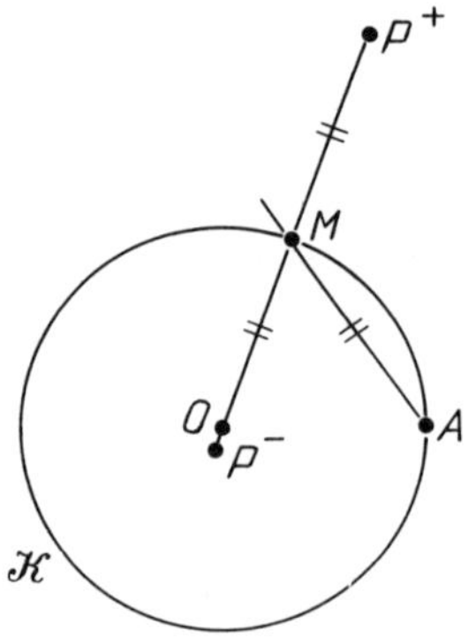

Figure 45.

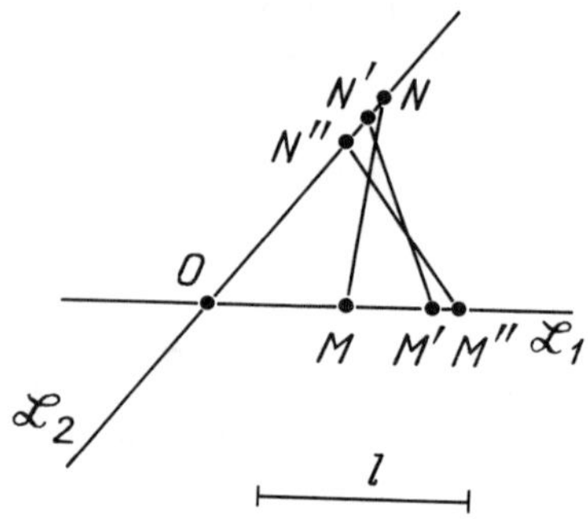

Figure 46.

Given: points A and B separated by distance $2a$.

Point P such that $AP \cdot BP = a^2$ lies on *a lemniscate of Bernoulli* (Figure 43).

Given: circle $\mathcal{K}$ of radius a, point A on the circle and number l.

Denote by M an intersection point of variable beam AM and the circle, and plot two segments MP^+ and MP^- of length l on the beam from the point (Figure 44). End-points P^+ and P^- of the segments lie on *a limaçon*.

Given: circle $\mathcal{K}$ with radius r and center at point O, and point A on the circle.

Let M be an intersection point of variable beam AM and circle $\mathcal{K}$. Pass a straight line through the point and center O of the circle, and plot segments MP^+ and MP^- equal to segment AM on the line from point M (Figure 45). Points P^+ and P^- lie on *a nephroid*.

Given: straight lines $\mathcal{L}_1$ and $\mathcal{L}_2$ intersecting at angle α at point O, and number l.

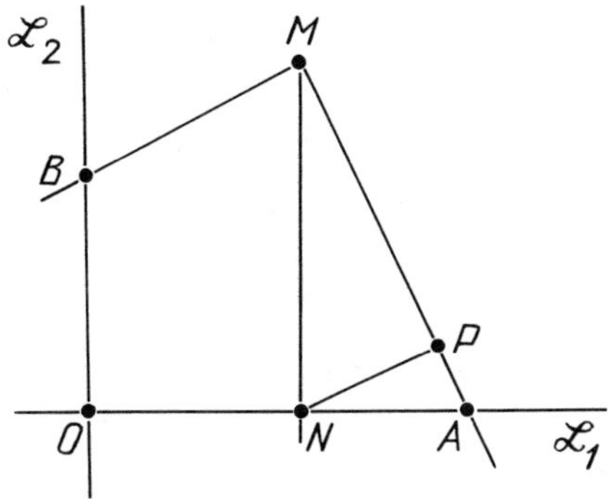

Figure 47.

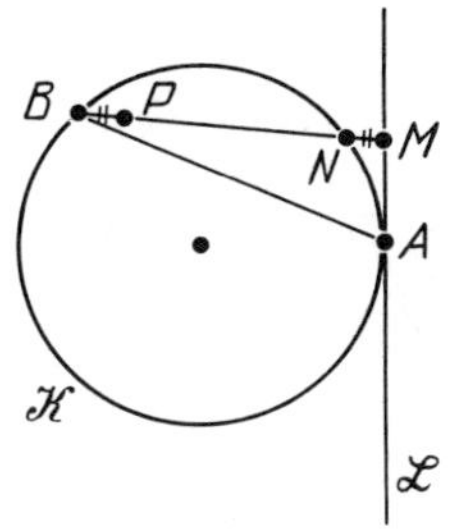

Figure 48.

An oblique astroid is an envelope of variable segments MN of length l whose end-points M and N lie on given straight lines $\mathcal{L}_1$ and $\mathcal{L}_2$ (Figure 46).

Given: straight lines $\mathcal{L}_1$ and $\mathcal{L}_2$ intersecting at right angle at point O, point A lying on line $\mathcal{L}_1$ at distance a from point O, and point B lying on line $\mathcal{L}_2$ at distance b from point O.

Drop a perpendicular to line $\mathcal{L}_1$ from a vertex of variable right angle $\angle AMB$, and pass straight line NP through intersection point N of the perpendicular and line $\mathcal{L}_1$ parallel to line BM. Point P of intersection of lines NP and AM lies on *an oblique bifolium* (Figure 47).

Given: straight line $\mathcal{L}$, circle $\mathcal{K}$ of radius a tangent to the line at point A, and chord AB of the circle forming an angle

$$\frac{\pi}{2}\alpha$$

with line $\mathcal{L}$.

Let variable beam BM with initial point B on circle $\mathcal{K}$ intersect line $\mathcal{L}$ at point M and the circle at point N. Plot segment BP equal to segment

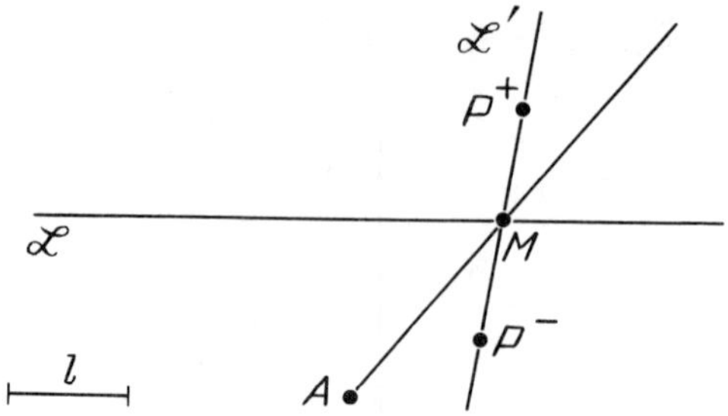

Figure 49.

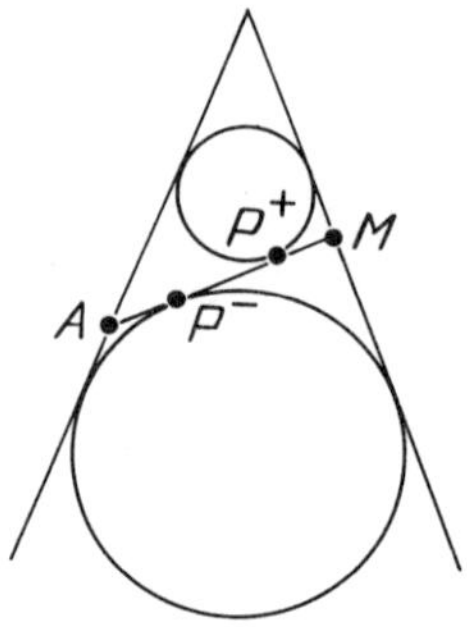

Figure 50.

MN on the beam from initial point B (Figure 48). Point P lies on *an oblique cissoid.*

Given: straight line $\mathcal{L}$, point A lying at distance a from the line, and numbers l and α.

Let N be an intersection point of variable beam AM and line $\mathcal{L}$. Pass straight line $\mathcal{L}'$ through the point forming angle α with beam AM. Plot segments MP^+ and MP^- of length l on line $\mathcal{L}'$ from point N on each side of it (Figure 49). End-points P^+ and P^- of the segments lie on *an oblique conchoid.*

Given: angle α and point A lying on one of its sides at distance

$$\alpha \csc \frac{\alpha}{2}$$

from the vertex of the angle.

Construct two circles tangent to the sides of the angle and variable segment AM (Figure 50). Tangency points P^+ and P^- of these circles and segment AM lie on *an oblique strophoid.*

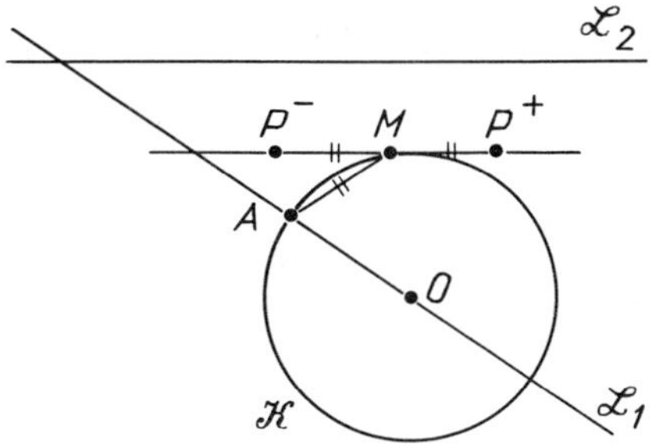

Figure 51.

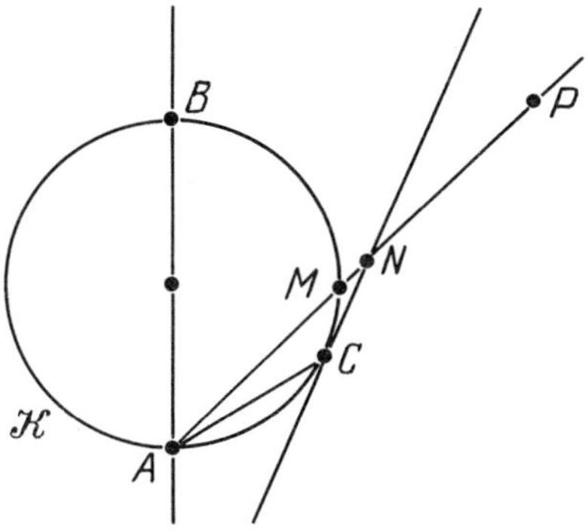

Figure 52.

Given: circle $\mathcal{K}$ of radius a with center O, straight line $\mathcal{L}_1$ passing through its center O and intersecting the circle at point A, and straight line $\mathcal{L}_2$ lying outside the circle and forming angle α with line $\mathcal{L}_1$.

Let AM be a variable chord of circle $\mathcal{K}$. Pass a straight line through its end-point M parallel to line $\mathcal{L}_2$, and plot segments MP^+ and MP^- with lenghts being equal to the length of chord AM on the line on each side of point M (Figure 51). Points P^+ and P^- lie on *an oblique trifolium.*

Given: circle $\mathcal{K}$ of radius r, diametrically opposite points A and B of the circle, and chord AC forming angle α with diameter AB.

Let N be an intersection point of variable beam AM and tangent CM to circle $\mathcal{K}$. Plot segment NP equal to segment AM (here M is an intersection point of beam AM and circle $\mathcal{K}$) on the beam from point N (beyond point N) (Figure 52). End-point P of the segment lies on *an oblique versiera.*

Given: straight lines $\mathcal{L}_1$ and $\mathcal{L}_2$ intersecting at right angle at point O, and point A lying at distance b from line $\mathcal{L}_1$ and at distance a from line $\mathcal{L}_2$.

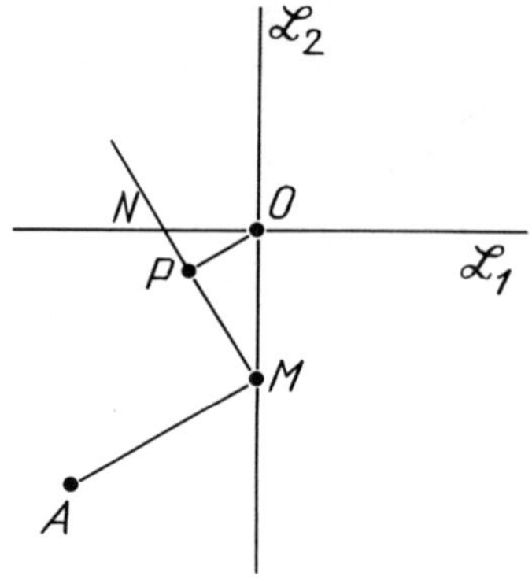

Figure 53.

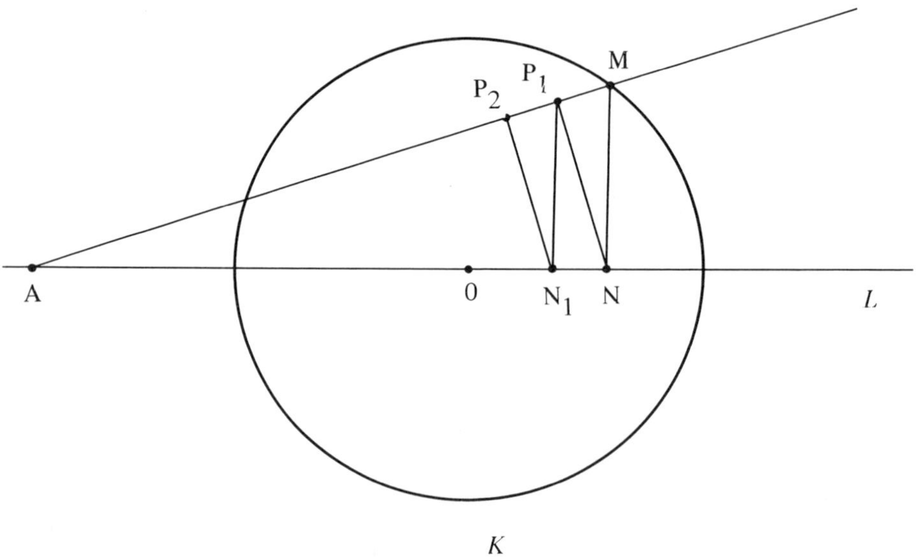

Figure 54.

Denote by M an intersection point of variable beam AM and line $\mathcal{L}_2$, and erect perpendicular MN to the beam at the point. Foot P of perpendicular OP dropped to line MN lies on *an ophiuride* (Figure 53).

Given: straight line $\mathcal{L}$, circle $\mathcal{K}$ with radius r and center O on line $\mathcal{L}$, and point A on the line at distance $d > r$ from center O.

Drop perpendicular MN to line $\mathcal{L}$ from intersection point M of variable beam AM and circle $\mathcal{K}$, and denote by N a foot of the perpendicular (Figure 54). Then drop perpendicular NP_1 from the point to the beam, and denote by P_1 a foot of the perpendicular. Point P_1 lies on *a sixth order oval*

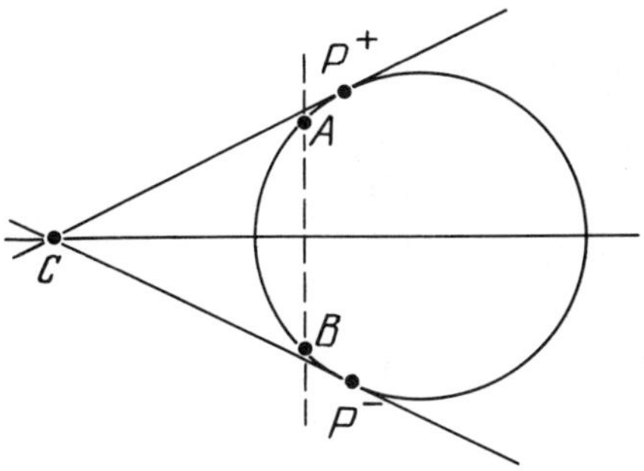

Figure 55.

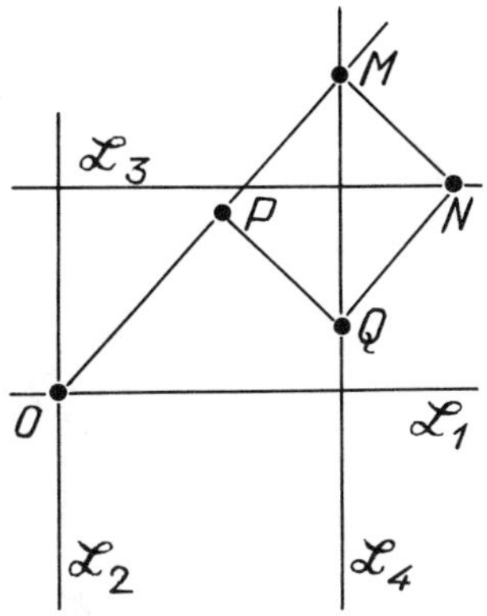

Figure 56.

of Münger. In a similar way do the next step: drop perpendicular P_1N_1 from point P_1 to line $\mathcal{L}$ and then from its foot N_1 — perpendicular N_1P_2 to beam AM. Intersection point P_2 of the beam and line N_1P_2 lies on *a tenth order oval of Münger.* In a similar manner we may construct points P_3, P_4,..., P_n lying on ovals of the corresponding orders.

Given: points A and B, length $2k$ of the segment AB, and point C on midperpendicular of the segment at distance g from it.

Draw tangent lines CP^+ and CP^- to a variable circle from point C passing through points A and B (Figure 55). Points P^+ and P^- of tangency lie on *a panstrophoid.*

Given: straight lines $\mathcal{L}_1$ and $\mathcal{L}_2$ intersecting at right angle at point O, straight line $\mathcal{L}_3$ parallel to line $\mathcal{L}_1$ and lying at distance a from it, and straight line $\mathcal{L}_4$ parallel to line $\mathcal{L}_2$ and lying at distance b from it.

Denote by M an intersection point of variable beam OM and line $\mathcal{L}_4$, and erect perpendicular MN to the beam at the point. Pass straight line NQ through intersection point N of the perpendicular and line $\mathcal{L}_3$ parallel to beam OM, and drop perpendicular QP to the beam from intersection point Q of lines NQ and $\mathcal{L}_4$ (Figure 56). A foot of the perpendicular lies on *a parabolic folium.*

Given: straight lines $\mathcal{L}_1$ and $\mathcal{L}_2$ intersecting at right angle at point O, circle $\mathcal{K}_1$ of radius R tangent to the lines, and $\mathcal{K}_2$ of the same radius with a center at point O.

Pass tangents MN and MQ to circle $\mathcal{K}_2$ from variable point M of line $\mathcal{L}_2$. Denote by RT a chord intercepted on a straight line passing through tangent-points N and Q by circle $\mathcal{K}_1$, and erect perpendiculars to the line at points R and T. Intersection points P^+ and P^- of these two perpendic-

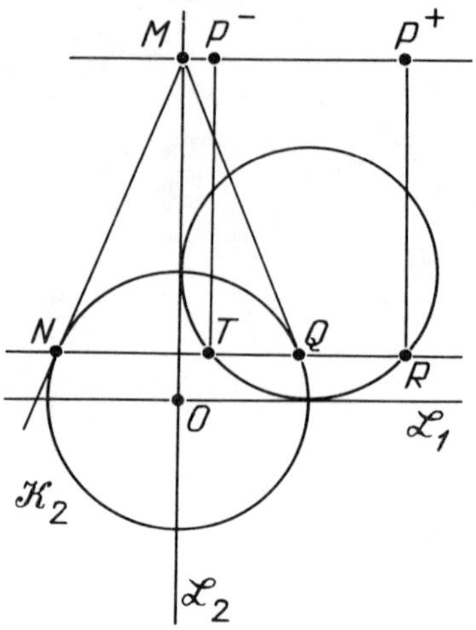

Figure 57.

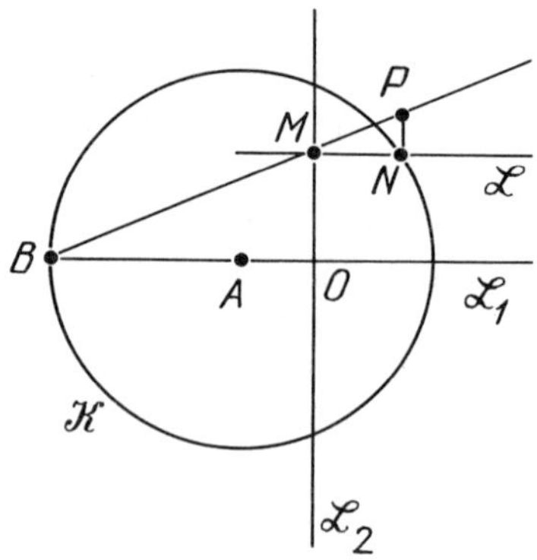

Figure 58.

ulars and a straight line passing through variable point M parallel to line $\mathcal{L}_1$ lie on *a pear* (Figure 57).

Given: straight lines $\mathcal{L}_1$ and $\mathcal{L}_2$ intersecting at right angle at point O, circle $\mathcal{K}_1$ of radius a with center A lying on line $\mathcal{L}_1$ at distance b from point O, and intersection point B of circle $\mathcal{K}_2$ and line $\mathcal{L}_1$ lying at distance $a+b$ from point O.

Pass straight line $\mathcal{L}$ through intersection point M of line $\mathcal{L}_2$ and variable beam BM parallel to line $\mathcal{L}_1$, and erect a perpendicular to line $\mathcal{L}$ at intersection point N of the line and circle $\mathcal{K}$. Point P of intersection of the perpendicular and beam BM lies on *a pearformed quartic* (Figure 58).

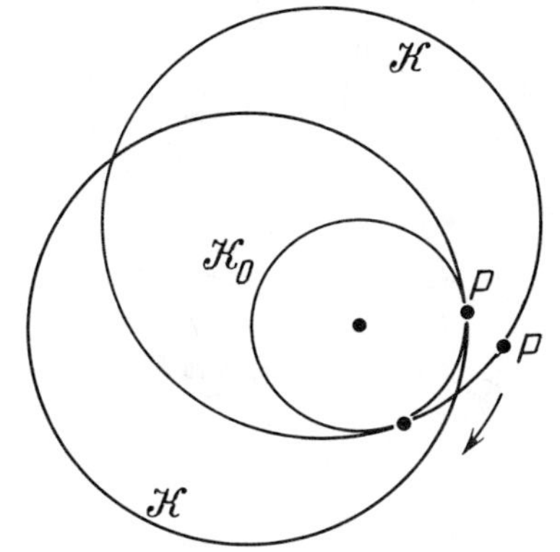

Figure 59.

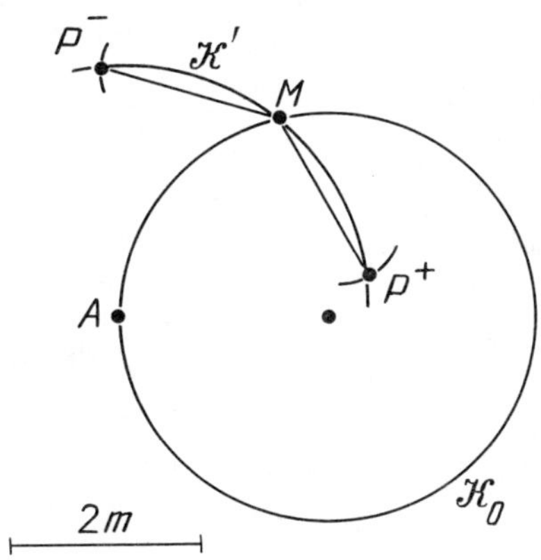

Figure 60.

Given: externally tangent circles $\mathcal{K}_0$ and $\mathcal{K}$ of radii r and $r > R$, respectively, and point P on circle $\mathcal{K}$.

When circle $\mathcal{K}$ rolls along circle $\mathcal{K}_0$ without slipping point P traces out *a pericycloid* (Figure 59).

Given: circle $\mathcal{K}_0$ of radius a, point A on the circle and number $2m$.

Let M be an intersection point of variable circle $\mathcal{K}$ with a center at point A and circle $\mathcal{K}_0$. Draw circle $\mathcal{K}'$ of radius $2m$ with a center at the point (Figure 60). Intersection points P^+ and P^- of circles $\mathcal{K}$ and $\mathcal{K}'$ lie on *a polyod of circle*.

Given: straight line $\mathcal{L}$, point O and number $2m$.

Let M be an intersection point of variable circle $\mathcal{K}$ and line $\mathcal{L}$. Draw circle $\mathcal{K}'$ of radius $2m$ with a center at the point (Figure 61). Inter-

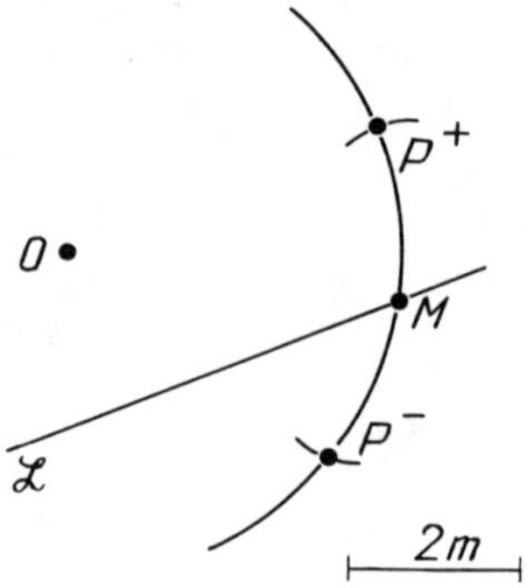

Figure 61.

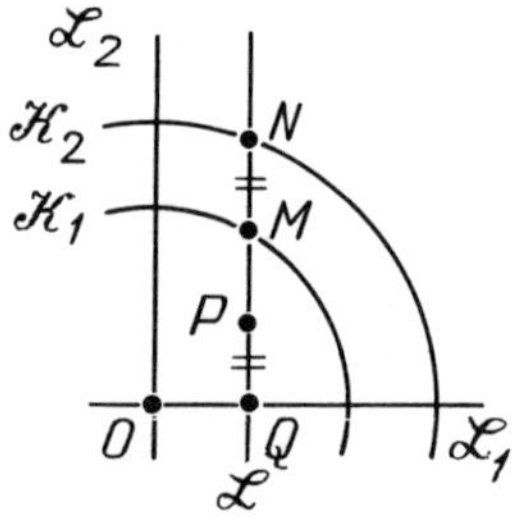

Figure 62.

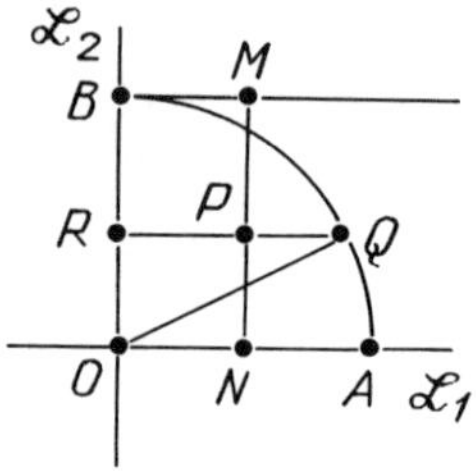

Figure 63.

section points P^+ and P^- of circles $\mathcal{K}$ and $\mathcal{K}'$ lie on *a polyod of straight line.*

Given: straight lines $\mathcal{L}_1$ and $\mathcal{L}_2$ intersecting at right angle at point O, and concentric circles $\mathcal{K}_1$ and $\mathcal{K}_2$ of radii a and $a\sqrt{2}$, respectively, with common center O.

Let $\mathcal{L}$ be a variable straight line parallel to line $\mathcal{L}_2$. Denote by MN a segment intercepted on the line by these circles, and by Q — an intersection point of lines $\mathcal{L}_1$ and $\mathcal{L}$. Plot segment PQ equal to segment MN on the latter line from point Q (Figure 62). Point P lies on *a polyzomal curve of Bernoulli.*

Given: straight lines $\mathcal{L}_1$ and $\mathcal{L}_2$ intersecting at right angle at point O, and a quater of circle $\mathcal{K}$ of radius a with center at point O whose end-points A and B lie on lines $\mathcal{L}_1$ and $\mathcal{L}_2$, respectively.

Let vertical segment MN move with a constant velocity along line $\mathcal{L}_1$ (from line $\mathcal{L}_2$), and one of the end-points of variable horizontal segment QR proceed with a constant velocity along curve $\mathcal{K}$ (Figure 63). Intersection point P of these two segments lies on *a quadratrix of Tschirnhausen.*

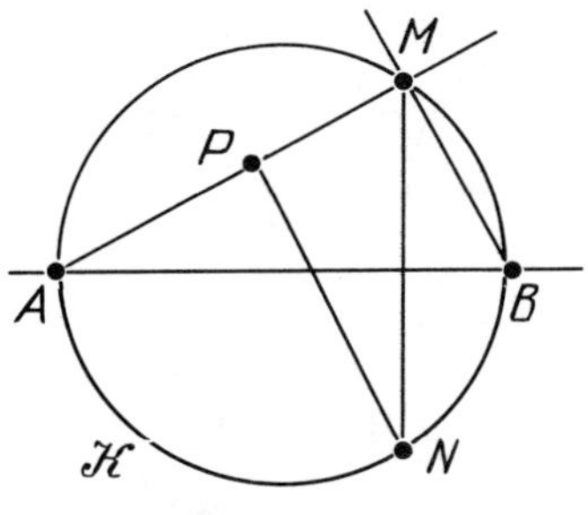

Figure 64.

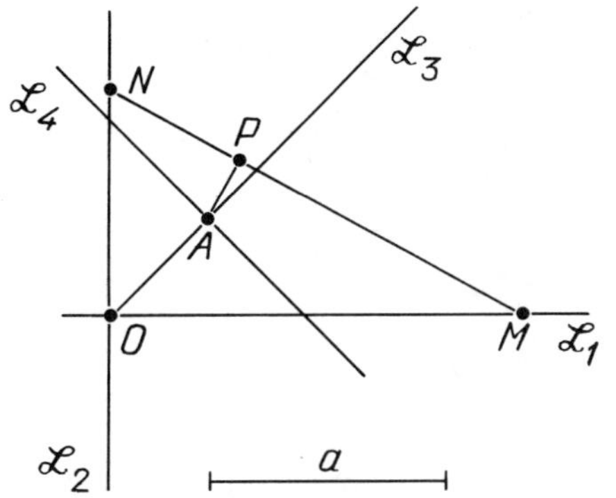

Figure 65.

Given: circle $\mathcal{K}$ of diameter d, and diametrically opposite points A and B on the circle.

Let $\angle AMB$ be a variable right angle resting upon diameter AB. Drop perpendicular MN to the diameter from vertex M, denote by N an intersection point of the perpendicular and circle $\mathcal{K}$, and pass straight line NP through the point parallel to beam BM (Figure 64). Intersection point P of the line and beam AM lies on *a right trifolium*.

Given: straight lines $\mathcal{L}_1$ and $\mathcal{L}_2$ intersecting at right angle at point O, bisectrix $\mathcal{L}_3$ of the angle between these lines, straight line $\mathcal{L}_4$ intersecting the bisectrix at point A (at distance c from point O), and number a.

Foot P of a perpendicular dropped to variable segment MN of length $2a$ with end-points M and N on lines $\mathcal{L}_1$ and $\mathcal{L}_2$, respectively, lies on *a scarabaeus* (Figure 65).

Given: straight line $\mathcal{L}$ and circle $\mathcal{K}$ of radius a with center O on line $\mathcal{L}$.

Let P be an intersection point of variable beam OP and circle $\mathcal{K}$. Pass straight line $\mathcal{L}_1$ through point P so that a length of segment PM_1 (here M_1

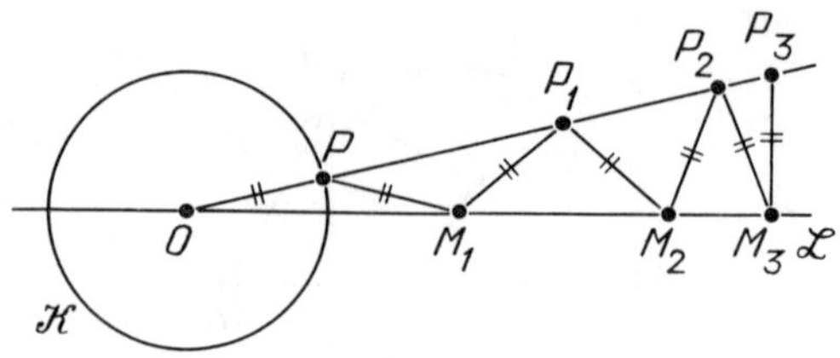

Figure 66.

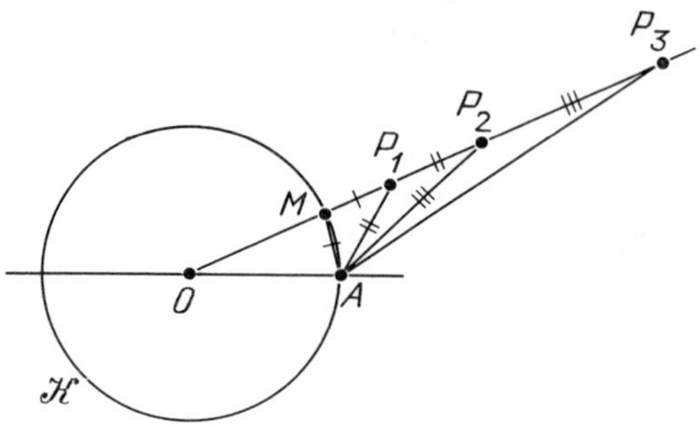

Figure 67.

is an intersection point of lines $\mathcal{L}$ and $\mathcal{L}_1$) be equal to a (Figure 66). Next pass straight line $\mathcal{L}_2$ through point M_1 so that a length of segment M_1P_1 (here P_1 is an intersection point of the line $\mathcal{L}_1$ and beam OM) be equal to a. In the manner described above, points P_2, P_3, ... are constructed. All the points lie on *a sectrix of Ceva.*

Given: circle $\mathcal{K}_0$ of radius a with a center at point O, and point A on the circle.

Let M be an intersection point of variable beam OM and circle $\mathcal{K}$. Plot segment MP_1 equal to segment AM on the beam from point M (Figure 67). Then plot segment P_1P_2 equal to segment AP_1 on beam OM from point P_1, etc. Points P_1, P_2, ... lie on *a sectrix of Coppola.*

Given: circle $\mathcal{K}_0$ of radius a with a center at point O, and point A on the circle.

Let M be an intersection point of variable beam AM and circle $\mathcal{K}$. Plot segment MP_1 equal to segment OM on the beam from point M (Figure 68). Then plot segment P_1P_2 equal to segment OP_1 on beam AM from point P_1, etc. Points P_1, P_2, ... lie on *a sectrix of Kempe.*

Given: straight line $\mathcal{L}$, points A and A' on the line separated by distance $2a$, and two natural numbers n and $n' > n$.

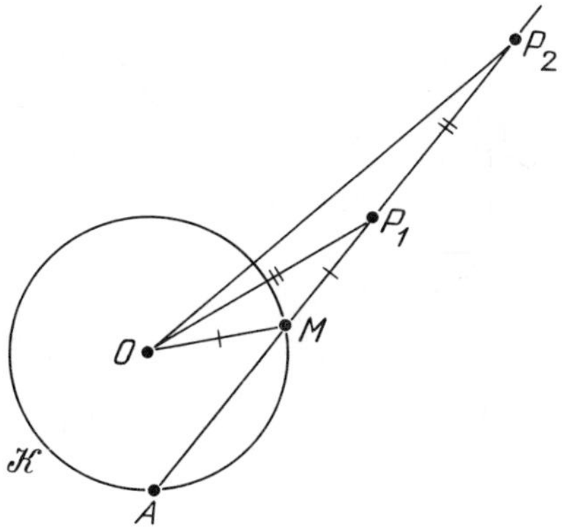

Figure 68.

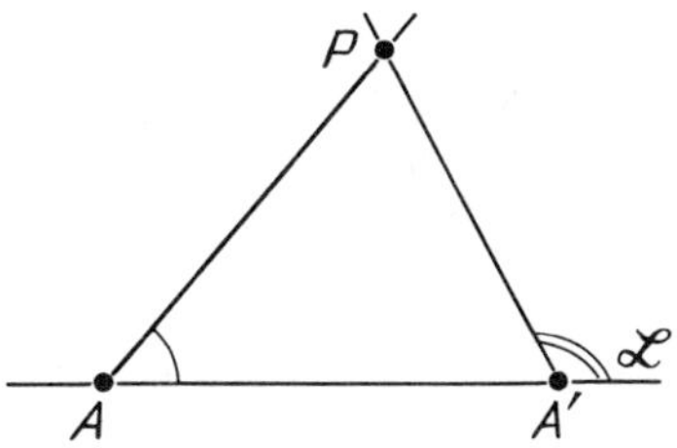

Figure 69.

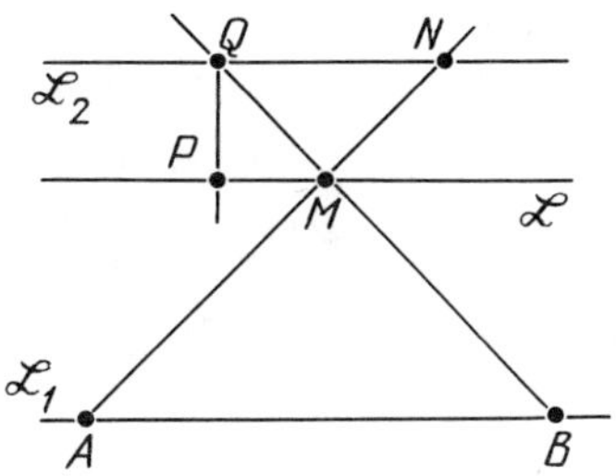

Figure 70.

Denote by $\angle A'AP$ and $\angle AA'P$ variable angles with vertices at points A and A', respectively, connected by equality

$$\frac{1}{n}\angle A'AP = \frac{1}{n'}(\pi - \angle AA'P)$$

(Figure 69). Intersection point P of angle sides AP and $A'P$ lies on *a sectrix of Plateau*.

Given: parallel straight lines $\mathcal{L}_1$ and $\mathcal{L}_2$ separated by distance a, and two points A and B of line $\mathcal{L}_1$ separated by distance c.

Let M be an intersection point of two mutually perpendicular straight lines AN and BQ. Pass straight line $\mathcal{L}$ through the point parallel to line $\mathcal{L}_1$, and drop a perpendicular to line $\mathcal{L}$ from intersection point Q of lines BQ and $\mathcal{L}_2$ (Figure 70). Point P of intersection of the perpendicular and line $\mathcal{L}$ lies on *a serpentine of Newton*.

Given: straight lines $\mathcal{L}_1$ and $\mathcal{L}_2$ intersecting at right angle at point O, and point A lying on line $\mathcal{L}_1$ at distance a from point O.

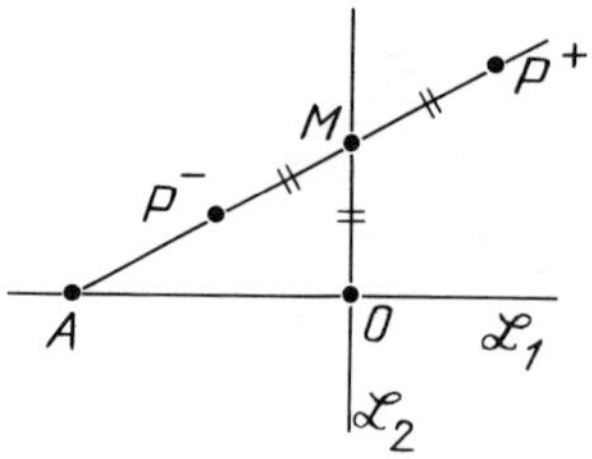

Figure 71.

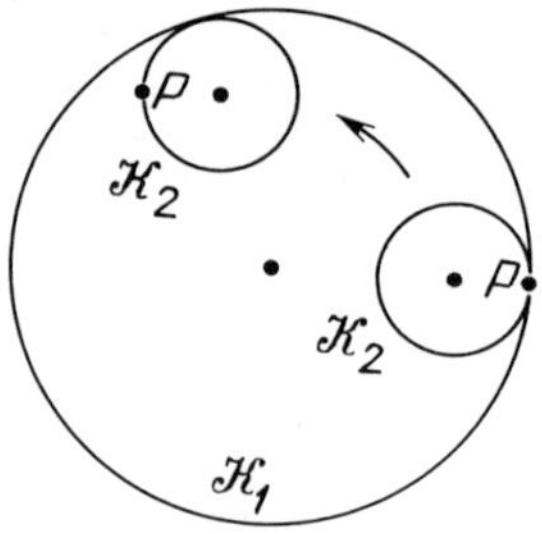

Figure 72.

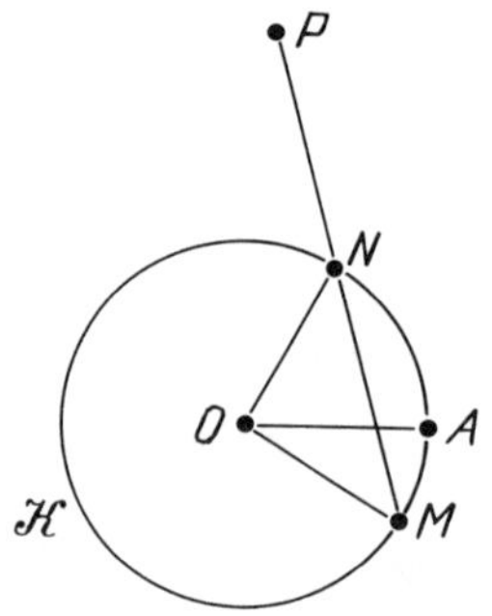

Figure 73.

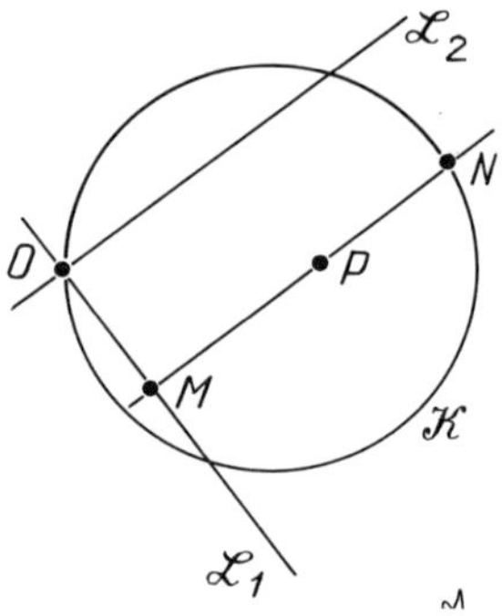

Figure 74.

Let M be an intersection point of variable beam AM and line $\mathcal{L}_2$. Plot segments MP^+ and MP^- equal to segment OM on the beam on each side of point M (Figure 71). End-points P^+ and P^- of these segments lie on *a strophoid.*

Given: circle $\mathcal{K}_1$ of radius $3r$, circle $\mathcal{K}_2$ of radius r, and point P on circle $\mathcal{K}_2$.

When circle $\mathcal{K}_2$ rolls on the inside of circle $\mathcal{K}_1$ without slipping point P traces out *a three-cuspid hypocycloid* (Figure 72).

Given: circle $\mathcal{K}$ of radius r with a center at point O and point A on the circle.

Plot arcs AM and AN on circle $\mathcal{K}$ from point A (on each side of the point) so that associated central angles $\angle AOM$ and $\angle AON$ be connected by equality $\angle AON = 2\angle AOM$. Denote by NP a segment of straight line MN equal to segment MN (Figure 73). End-point P of the segment lies on *a trefoil.*

Given: straight lines $\mathcal{L}_1$ and $\mathcal{L}_2$ intersecting at right angle at point O, and transversal circle $\mathcal{K}$ of radius $a\sqrt{2}$ passing through point O.

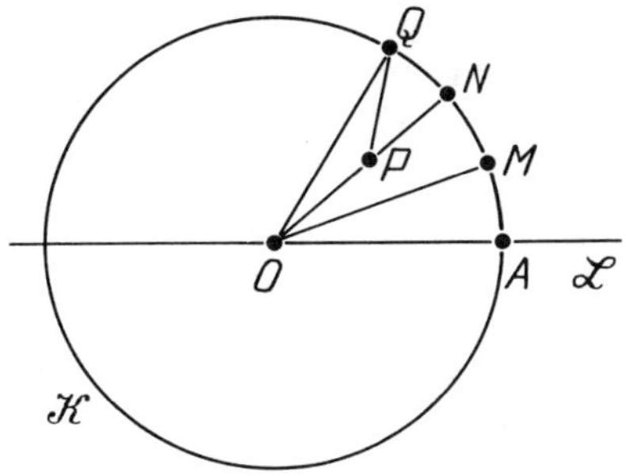

Figure 75.

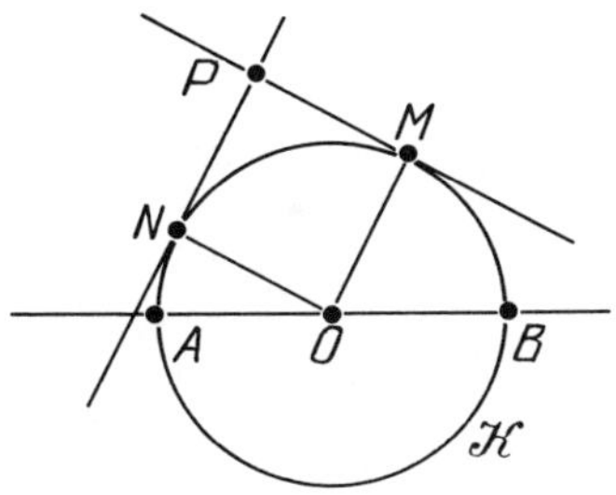

Figure 76.

Let M and N be intersection points of variable straight line $\mathcal{L}$ parallel to line $\mathcal{L}_2$ and line $\mathcal{L}_1$ and circle $\mathcal{K}$, respectively (Figure 74). Point P on line $\mathcal{L}$ such that $MP^2 = OM \cdot OP$ lies on *a trifolium of Cramer.*

Given: circle $\mathcal{K}$ of radius a, straight line $\mathcal{L}$, and intersection point A of these two lines.

Construct three variable equal angles $\angle AOM$, $\angle MON$ and $\angle NOQ$, and plot angle $\angle OQP$ equal to the above-constructed from intersection point Q of beam OQ and the circle (Figure 75). Intersection point P of lines ON and NQ lies on *a trisectrix of Delanges.*

Given: circle $\mathcal{K}$ of radius a, and diametrically opposite points A and B on the circle.

Plot arcs AN and BM on the circle from points A and B so that points N and M lie on one side of straight line AB and corresponding central angles $\angle AON$ and $\angle BOM$ be connected by equality $\angle BOM = 2\angle AON$ (Figure 76). Intersection point P of the tangents to circle $\mathcal{K}$ at points A and B respectively lie on *a trisectrix of Longchamps.*

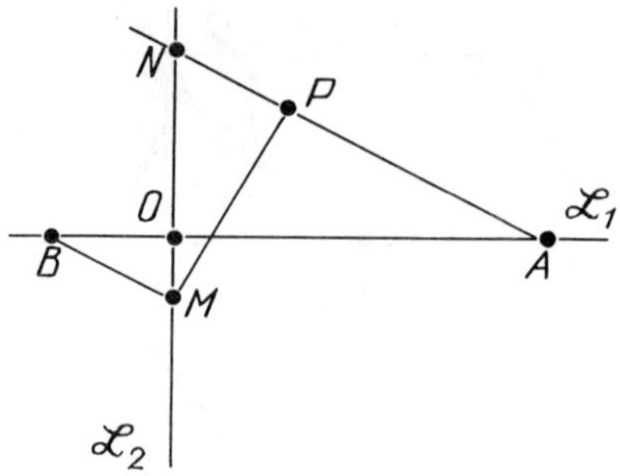

Figure 77.

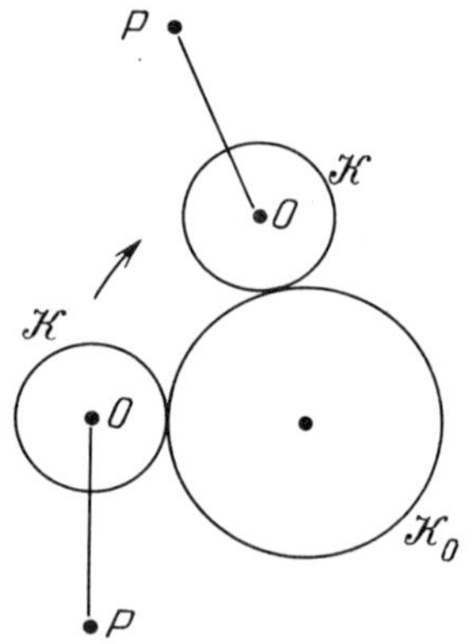

Figure 78.

Given: straight lines $\mathcal{L}_1$ and $\mathcal{L}_2$ intersecting at right angle at point O, and points A and B lying on line $\mathcal{L}_1$ at distances $3a$ and a from point O so that the point is an inner point of segment AB.

Let AN and BM be variable parallel beams, and N and M — intersection points of the beams and line $\mathcal{L}_2$ (Figure 77). Drop perpendicular MP to beam AN from point M. Foot P of the perpendicular lies on *a trisectrix of Maclaurin*.

Given: externally tangent circles $\mathcal{K}_0$ of radius R and $\mathcal{K}$ of radius r with center O, and segment OP of length $R + r$ mounted rigidly to the latter circle.

When circle $\mathcal{K}$ rolls along circle $\mathcal{K}_0$ without slipping end-point P traces out *a trochoidal rose* (Figure 78).

Given: parallel straight lines $\mathcal{L}_1$ and $\mathcal{L}_2$, and circle $\mathcal{K}$ of diameter a tangent to these lines at points A and B, respectively.

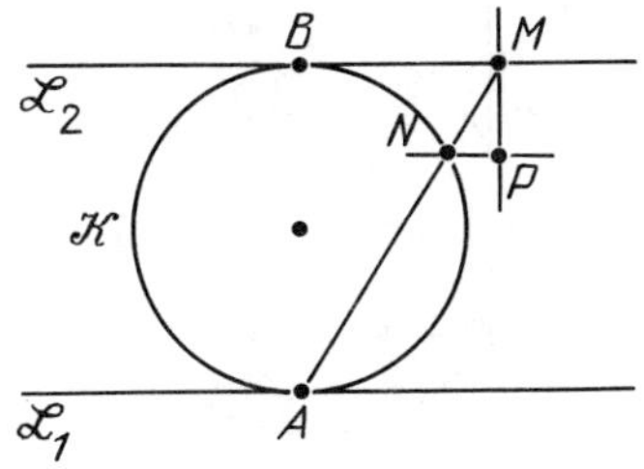

Figure 79.

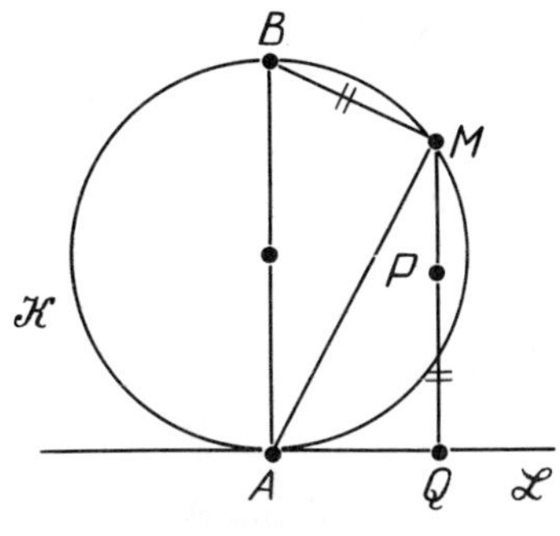

Figure 80.

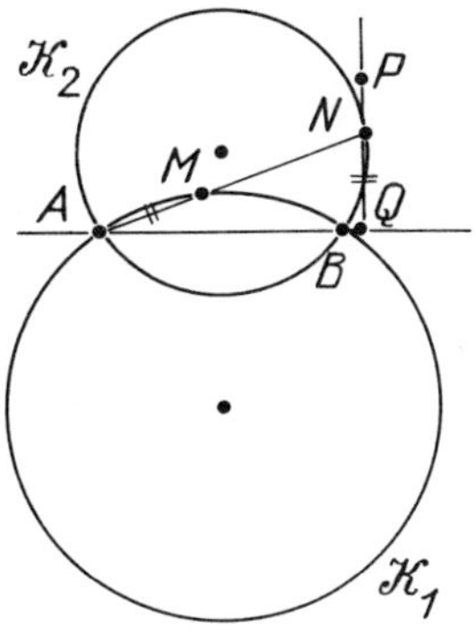

Figure 81.

Denote by M and N intersection points of variable beam AM with line $\mathcal{L}_2$ and circle $\mathcal{K}$, respectively, and pass straight line MP parallel to line $\mathcal{L}_2$ and straight line NP perpendicular to it (Figure 79). Point P of intersection of these two lines lies on *a versiera.*

Given: straight line $\mathcal{L}$, circle $\mathcal{K}$ of diameter d tangent to the line at point A, and its diametrically opposite point B.

Construct variable right angle $\angle AMB$ rested on diameter AB of circle $\mathcal{K}$, and plot segment QP equal to segment MB on perpendicular MN to line $\mathcal{L}$ from intersection point Q of these two lines (Figure 80). Point P lies on *a virtual parabola* (A).

Given: circles $\mathcal{K}_1$ and $\mathcal{K}_2$ of radii a and b, respectively, intersecting at points A and B.

Let variable beam AM intersect $\mathcal{K}_1$ and $\mathcal{K}_2$ at points M and N, respectively. Plot two lines segment QP equal to segment AM on perpendicular NQ to straight line AB from intersection point Q of these two lines (Figure 81). Point P lies on *a virtual parabola* (B).

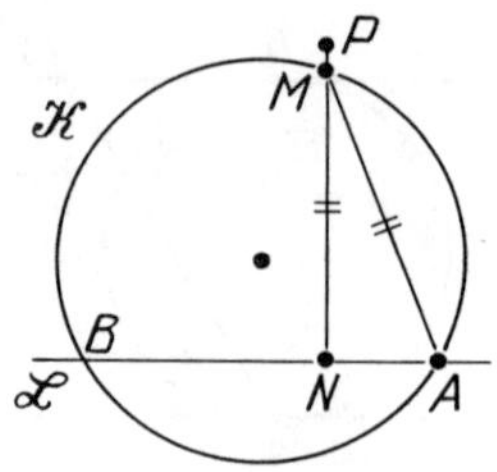

Figure 82.

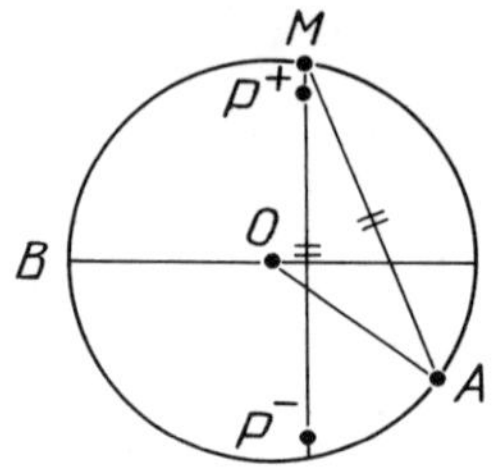

Figure 83.

Given: circle $\mathcal{K}$ of radius r, and straight line $\mathcal{L}$ intersecting the circle at points A and B.

Drop a perpendicular to line $\mathcal{L}$ from point M of intersection of circle $\mathcal{K}$ and variable beam AM, and plot segment NP equal to segment MN from intersection point N of the perpendicular and line $\mathcal{L}$ (Figure 82). Its end-point P lies on *a virtual parabola* (C).

Given: circle $\mathcal{K}$ of radius r with a center at point O and angle $\angle AOB$ of span α.

Drop a perpendicular to straight line OB from end-point M of variable chord AM. End-points P^+ and P^- of a segment of the perpendicular symmetric about line OB and equal to segment AM lie on *a virtual parabola* (D) (Figure 83).

Given: concentric circles $\mathcal{K}_1$ and $\mathcal{K}_2$ of radii r and R, respectively, with common center O, and point A on the exterior circle.

Drop a perpendicular to straight line AO from intersection point M of variable straight line AM and circle $\mathcal{K}_2$, and plot segments NP^+ and NP^- equal to segments MQ and MR, respectively (here Q and R are intersection points of line AM and circle $\mathcal{K}_1$), on the perpendicular from point N of its intersection with line AO (Figure 84). Points P^+ and P^- lie on *a virtual parabola* (E).

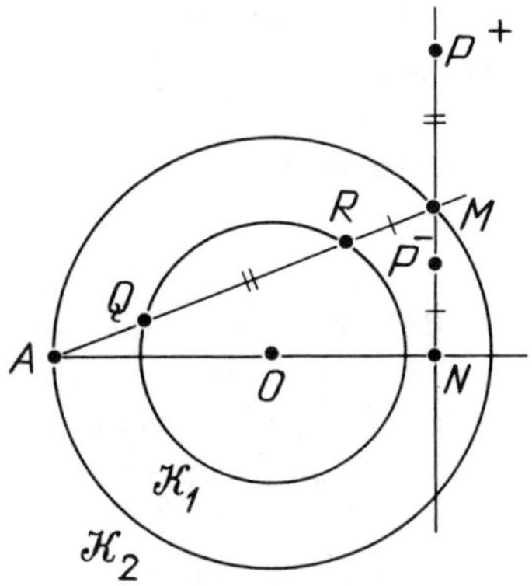

Figure 84.

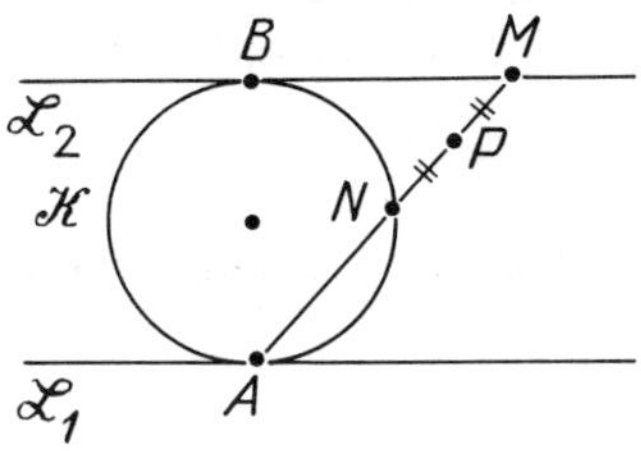

Figure 85.

Given: parallel straight lines $\mathcal{L}_1$ and $\mathcal{L}_2$, and circle $\mathcal{K}$ of diameter a tangent to these lines at points A and B, respectively.

Denote by M and N intersection points of variable beam AM with line $\mathcal{L}_2$ and circle $\mathcal{K}$, respectively (Figure 85). Midpoint P of segment MN lies on *a visiera*.

4.2 Plane curves and mechanisms

There exist many ways for solving problems of concrete curve construction, among which are widely known computer-aided methods. Using simply a personal computer (IBM PC AT), all graphics in the Atlas of curves were obtained.

There exist alternate approaches too.

Here we describe several of the numerous mechanisms by means of which the various plane curves — ellipses and hyperbolae, cissoids and trifoliums, strophoids and virtual parabolae, spirals and other curves — can be reproduced. It may be useful in the search for new solutions in engineering design. Detailed information on the reproduced mechanisms is contained in the voluminous handbook for engineers, designers and inventors, *Mechanisms in Modern Engineering Design* by Ivan I. Artobolevsky, Mir Publishers, 1975–1980.

In this section some simple mechanisms for constructing plane curves are given.

VYATKIN LINK-GEAR MECHANISM FOR TRACING HYPERBOLIC SPIRAL. Link 1 turns about fixed axis A and is connected by turning pair B to slider 4 which, in turn, moves along the axis of link 5. Link 3 is connected by turning pair K to link 5 and slides in guide a of link 1. The axis of guide a is perpendicular to the line AB. Roller 2 has a sharp-edged rim and rotates about axis $b - b$. When link 1 turns about axis A, roller 2, cutting into the surface of the drawing with its sharp edge, moves during each instant along the straight line BK. The envelope of the successive positions of straight line BK is *a hyperbolic spiral* with the polar equation, with respect to pole A,

$$\rho\varphi = a$$

where $\rho = AK$, $a = AB$ and φ is angle of rotation of radius-vector ρ (Figure 86).

FOUR-BAR SLIDING-LINK MECHANISM FOR TRACING LOGARITHMIC SPIRAL. Slider 1 turns about fixed axis A. Link 3 moves in slider 1. Roller 2 rotates about axis $D - D$. When slider 1 turns about axis A, roller 2, cutting into the surface of the drawing with its sharp edge, moves during each instant along a straight line coinciding with the plane of the roller. The envelope of the successive positions of this straight line is *a logarithmic spiral* with the polar equation, with respect to pole A,

$$\rho = ae^{\varphi/\tan\alpha}$$

where $\rho = AK$, φ is angle of rotation of radius-vector ρ, $a = AB$ is an initial value of the radius-vector, e — a base of natural logarithms, and α — an angle between axis AK and the plane of roller 2 (Figure 87).

PONCELET LINK-GEAR MECHANISM FOR TRACING CAPRICORNOID. Link 1, having the form of a bent lever with angle $\angle BOf$ equal to 90°, turns about fixed axis O. Link 1 is connected by turning pair B to slider 3 which moves along axis Ad of link 4. Link 4 turns about fixed axis A. Sliders 2 and 5 are connected together by turning pair D and move along axes Ad and Of of links 4 and 1. When link 1 turns about axis O, point D of slider 2 describes *capricornoid* $q - q$ with the equation

$$\rho_D = \frac{ab\sin\varphi}{a + b\cos\varphi}$$

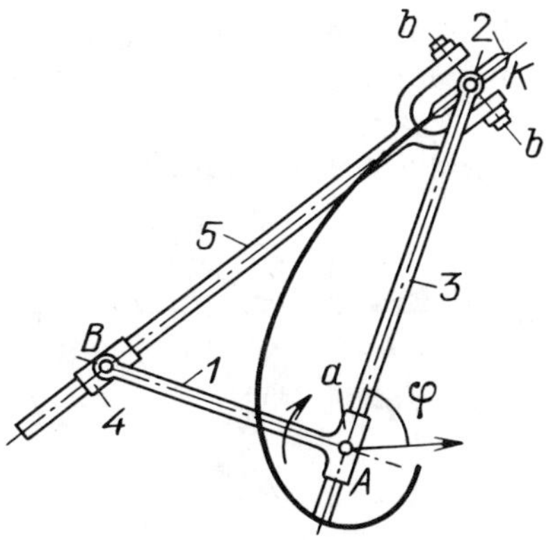

Figure 86.

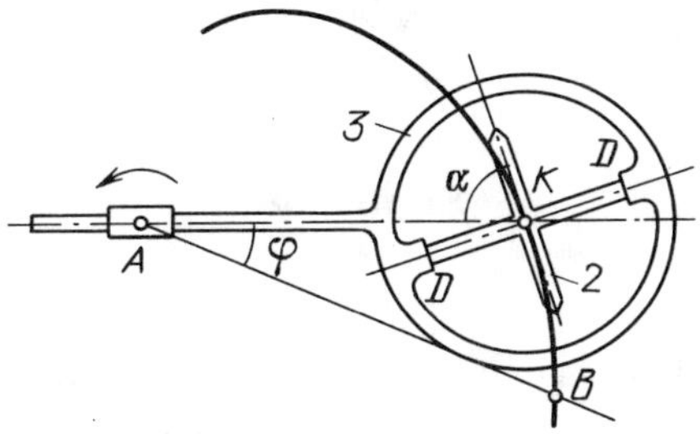

Figure 87.

or

$$b^2x^2(x^2+y^2) = a^2(by - x^2 - y^2)^2$$

where $a = OB$, $b = OA$ are constant dimensions of the mechanism and φ is polar angle between radius-vector ρ_D and polar axis Ox (Figure 88).

LINK-GEAR MECHANISM FOR TRACING CRANIOID. The lengths of the links comply with the conditions:

$$AD = CF = EA', \quad CF > CE$$

Link 1 turns about fixed axis C. Slotted link 2 turns about fixed axis E. Slider 3, moving along the slot of link 2, is connected by turning pair F to link 1. Link 4 is connected by turning pairs D and A to link 1 and slider 5 which moves along the slot of link 2. When link 1 turns about axis C, points A and A' describe *a cranioid* consisting of branches a and b with the polar equation

$$\rho = d\cos\varphi + b\sqrt{1 - m^2\sin^2\varphi} + c\sqrt{1 - k^2\sin^2\varphi}$$

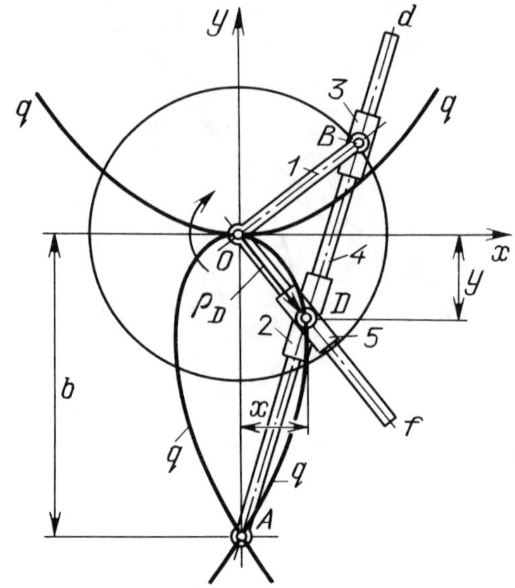

Figure 88.

where

$$\rho = EA, \quad d = EC, \quad b = CD, \quad c = AD = CF,$$
$$m = \frac{EC}{FC}, \quad k = m\frac{FD}{AD}$$

and φ is the angle between radius-vector ρ and axis Ey.

In the indicated position of link AD, point A describes the first branch a of the cranioid. If link 4 is put into position 4′ (shown by broken lines), point A' describes the second branch b of the cranioid which is self-intersecting at point E (Figure 89).

Artobolevsky link-gear mechanism for tracing virtual parabola of Vincentio. The lengths of the links comply with the conditions:

$$AB = OA = a, \quad OC = CE = d, \quad \tan\alpha = \frac{b}{a}$$

where α is fixed angle $\angle COf$ and $b = \sqrt{d^2 - a^2}$. Link 1 has the form of a bent lever with angle $\angle COf$ equal to α. Link 1 turns about fixed axis O and is connected by turning pair C to link 5 and by a sliding pair to slider 3. Slider 3 is connected by turning pair B to link 4 which turns about fixed axis A. Slider 3 is also connected by turning pair B to link 7 which moves in cross-shaped slider 2 whose guides are perpendicular to each other. Link 5 is connected by turning pair E to slider 6 which moves along fixed guides

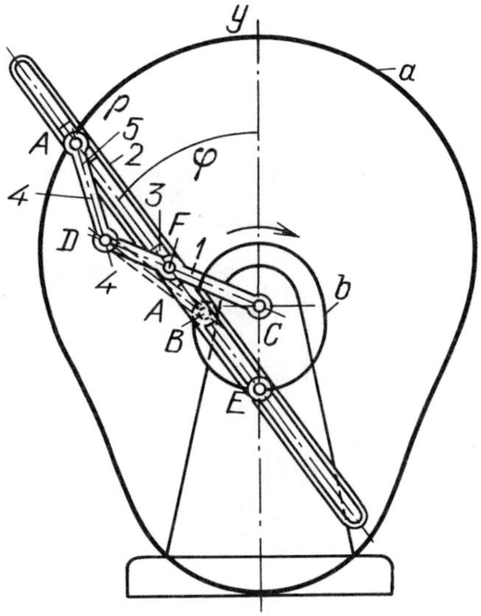

Figure 89.

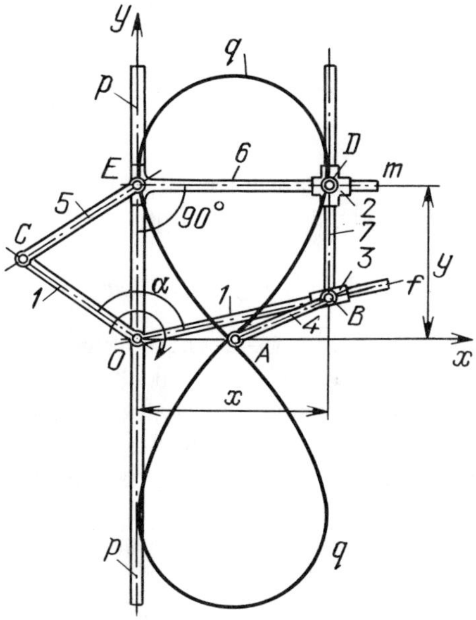

Figure 90.

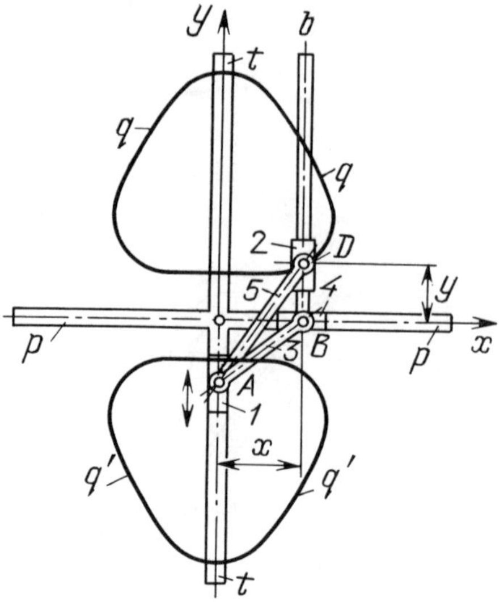

Figure 91.

$p-p$ whose axis coincides with axis Oy. When link 1 turns about point O, point D of slider 2 describes *virtual parabola of Vincentio* $q-q$ with the equation

$$y = \sqrt{2ax} - \sqrt{4b^2 - \frac{2b^2}{a}x}$$

(Figure 90).

Artobolevsky link-gear mechanism for tracing polyzomal curve of Bernoulli. The lengths of the links comply with the conditions:

$$AD = a\sqrt{2}, \quad AB = a$$

Slider 1 moves along fixed guides $t-t$ whose axis coincides with axis Oy. Slider 1 is connected by turning pair A to links 3 and 5. Link 3 is connected by turning pair B to slider 4 which moves along fixed guides $p-p$ whose axis coincides with axis Ox. Link 5 is connected by turning pair D to slider 2 which moves along cross-piece Bb of slider 4. The axis of cross-piece Bb is parallel to axis Oy. When slider 1 moves along guides $t-t$, point D of slider 2 describes the upper branch $q-q$ of *a polyzomal curve of Bernoulli* with the equation

$$y = \sqrt{2a^2 - x^2} + \sqrt{a^2 - x^2}$$

To trace the lower branch $q'-q'$ of this curve, the links of the mechanism are reassembled symmetric to the axis Ox (Figure 91).

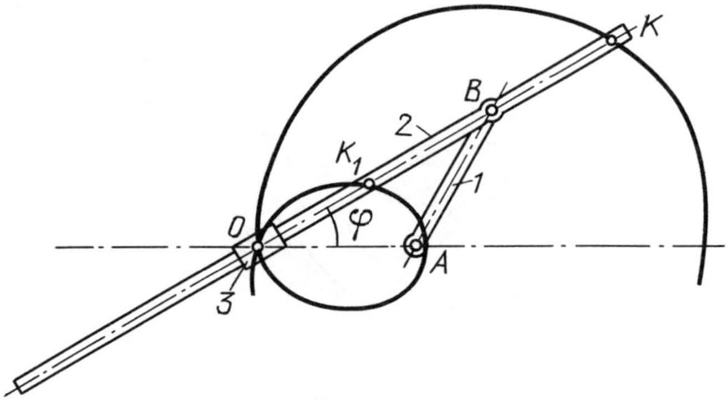

Figure 92.

FOUR-BAR SLIDING-LINK MECHANISM FOR TRACING LIMAÇON. Crank 1, turning about axis A, is connected by turning pair B to link 2 which moves in slider 3. Slider 3 turns about fixed axis O. When crank 1 turns about axis A, point K (and K_1) describes *a limaçon* whose polar equation with respect to center O is of the form

$$\rho_K = 2r\cos\varphi \pm b$$

where

$$b = KB = BK_1, \quad r = OA$$

ρ_K is a radius-vector drawn from center O to point $K(K_1)$ and φ — a polar angle (Figure 92).

FOUR-BAR SLIDING-LINK MECHANISM FOR TRACING CARDIOID. When bent sliding link 1 slides in guides 2 and 3, which turn about fixed axes A and B, point K of the link describes *a cardioid* with the equation

$$\rho = b - 2a\frac{\sin(\alpha+\varphi)}{\sin\alpha}$$

where

$$a = \frac{AB}{2}, \quad b = CK, \quad \rho = BK$$

φ is polar angle and $\alpha = \angle ACB$ of bent sliding lever 1.

If $\alpha = \pi/2$ and $a = b$ then the equation of the cardioid is

$$\rho = 2a(1 - \cos\varphi)$$

(Figure 93).

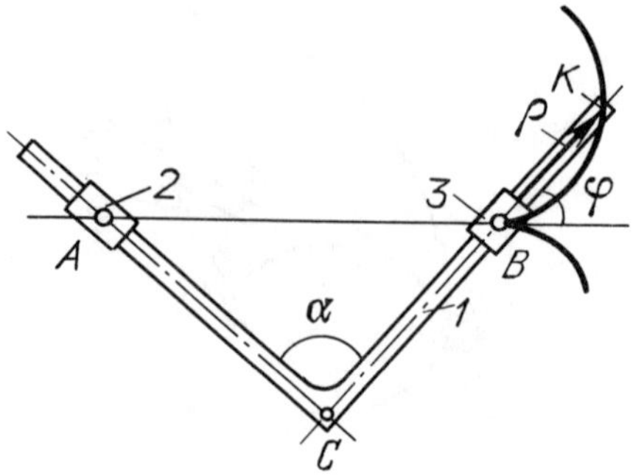

Figure 93.

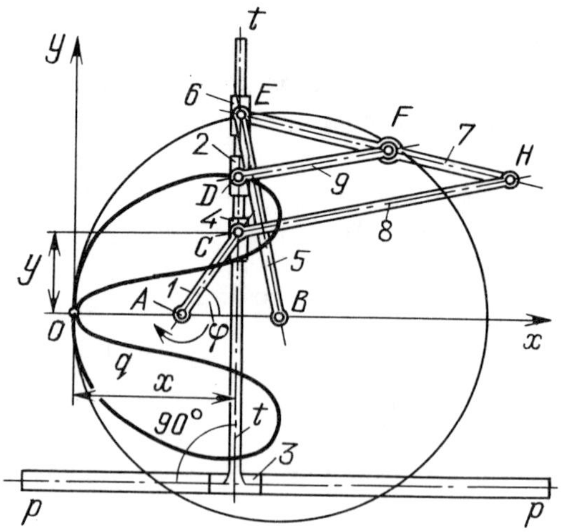

Figure 94.

ARTOBOLEVSKY LINK-GEAR MECHANISM FOR TRACING VIRTUAL PARABOLA OF CRAMER. The lengths of the links comply with the conditions:

$$AC = AB = a, \quad BE = 2a, \quad CH : DF = HE : FE = 2$$

Link 1, rotating about fixed axis A, is connected by turning pair C to slider 4 which moves along cross-piece $t-t$ of slider 3. Slider 3 moves along fixed guides $p-p$ whose axis is parallel to axis Ox. Link 8 is connected by turning pairs C and H to slider 4 and link 7. Link 7 is connected by turning pairs E and F to slider 6 and link 9. Link 9 is connected by turning pair D to slider 2 which moves along cross-piece $t-t$ of slider 3. Link 5, rotating about fixed axis B, is connected by turning pair E to slider 6 which also moves along cross-piece $t-t$ of slider 3. When link 1 rotates about axis

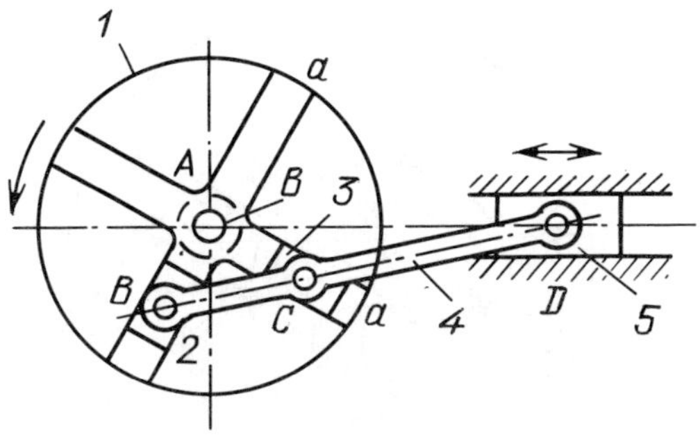

Figure 95.

A, point D of slider 2 describes *virtual parabola of Cramer* $q-q$ with the equation

$$2y = \sqrt{x(2a-x)} + \sqrt{x(4a-x)}$$

(Figure 94).

LINK-GEAR INVERTED ELLIPSOGRAPH MECHANISM. Disk 1 has two mutually perpendicular diametral grooves a. Sliders 2 and 3 move along these grooves. Link 4 is connected by turning pair D to slider 5 which moves along fixed guides. If slider 5 is removed and disk 1 is held stationary, the mechanism is converted into *an ellipsograph* (elliptic trammel) in which all points of link 4 describe ellipses. In one revolution of sharf A, piston 5 has two double strokes (back and forth) (Figure 95).

LEONARDO DA VINCI'S SLOTTED-LINK ELLIPSOGRAPH. Ring 2 is secured by screws b to upright 1 which has slots a for this purpose. Ring 2 can be set to any position along slots a to obtain required coordinate x which is equal to AO where O is the center of ring 2. Rotating about fixed axis A is slider 3 which moves along groove $d-d$ of disk 5. Slider 4, designed as a ring encircling fixed ring 2, has extensions c that fit in groove $f-f$ of disk 5. The axes of grooves $d-d$ and $f-f$ are perpendicular to each other. When slider 3 rotates about axis A, a pencil clamped on upright 1 describes *an ellipse* on disk 5 which rotates about axis A and slides along the axes of its grooves $d-d$ and $f-f$. The size of the ellipse depends upon the preset distance x (Figure 96).

NEWTON'S FOUR-BAR LINK-GEAR MECHANISM FOR TRACING CISSOID OF DIOCLES. Slider 2 moves along fixed guides $p-p$. Link 1 is connected by turning pair A to slider 2 and by a sliding pair to slider 3. Slider 3 turns about fixed axis C. When slider 2 moves along guides $p-p$, point K of

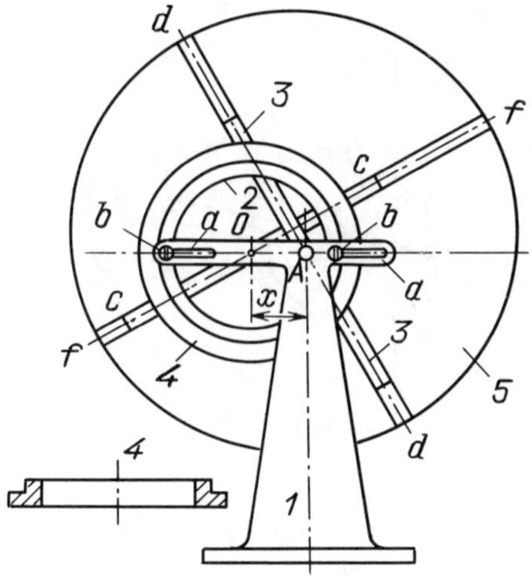

Figure 96.

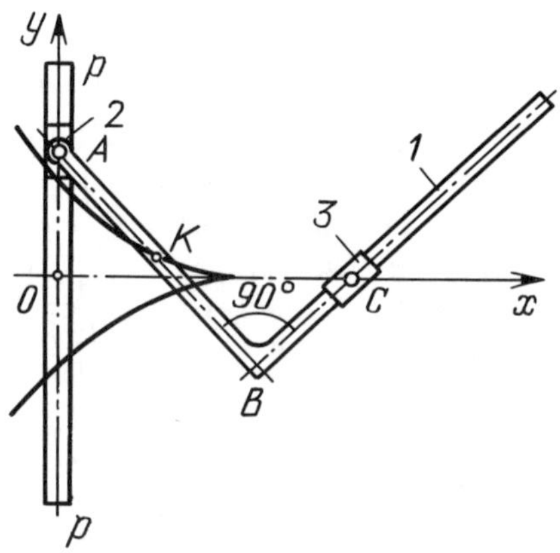

Figure 97.

sliding link 1 describes *a cissoid* with the equation

$$y^2 = \frac{(d-x)^3}{d+x}$$

where

$$d = AK = KB, \quad OC = 2d$$

(Figure 97).

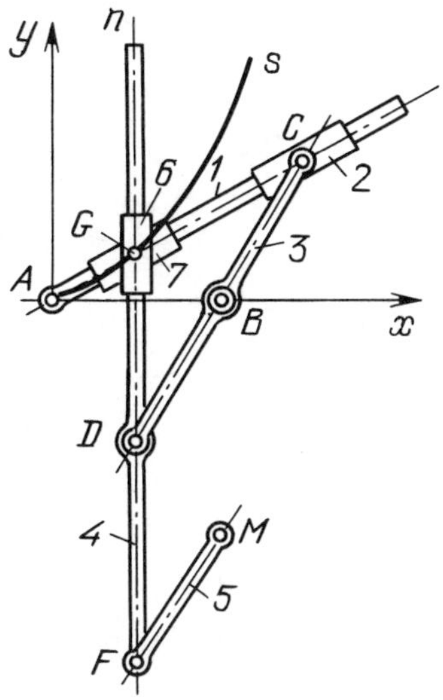

Figure 98.

ARTOBOLEVSKY FOUR-BAR LINK-GEAR MECHANISM FOR TRACING CISSOID OF DIOCLES. The lengths of the links comply with the conditions:

$$BC = BD = MF = AB = \frac{a}{2}, \quad DF = BM$$

Figure $BDFM$ is a parallel-crank linkage. Link 1, turning about fixed axis A, is connected by sliding pairs to sliders 7 and 2. Link 3, turning about fixed axis B, is connected by turning pairs C and D to slider 2 and link 4. Link 4 is connected by turning pair F to link 5 which turns about fixed axis M. Slider 6 is connected to slider 7 by turning pair G and moves along axis Fn of link 4. When link 1 turns about axis A, point G describes *cissoid s of Diocles* with the equation

$$y^2 = \frac{x^2}{a - x}$$

(Figure 98).

4.3 Geometric splines

In practice, the next problem is widely met. For a given finite set of points on plane or in three-dimensional space it is required to construct a smooth curve closely approximating the set. In this section we describe a solution of the problem with the help of simple but natural geometric considerations and cubical splines.

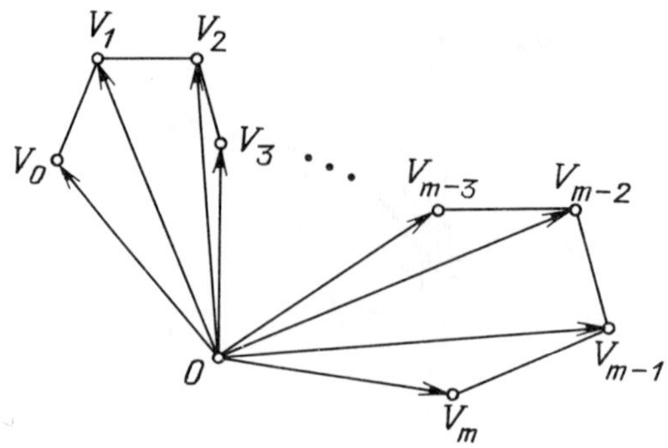

Figure 99.

In the construction of a composite regular curve the conditions of geometric continuity of the comprising regular partial curves at the points of contact are of great importance.

Let 2-regular curves γ_1 and γ_2 be given by parametric equations $\mathbf{r} = \mathbf{r}_1(t)$, $0 \le t \le 1$ and $\mathbf{r} = \mathbf{r}_2(t)$, $0 \le t \le 1$, respectively. The curves will be considered to have a common point

$$\mathbf{r}_1(1) = \mathbf{r}_2(0), \tag{1}$$

For a composite curve $\gamma = \gamma_1 \bigcup \gamma_2$ to be a 2-regular curve it is necessary that the unit tangent vectors and the vectors of curvature of the comprising curves γ_1 and γ_2 be equal at the common point:

$$\frac{\mathbf{r}_1'(1)}{|\,\mathbf{r}_1'(1)\,|} = \frac{\mathbf{r}_2'(0)}{|\,\mathbf{r}_2'(0)\,|}, \tag{2}$$

$$\frac{[\mathbf{r}_1'(1) \times \mathbf{r}_1''(1)] \times \mathbf{r}_1'(1)}{|\,\mathbf{r}_1'(1)\,|^4} = \frac{[\mathbf{r}_2'(0) \times \mathbf{r}_2''(0)] \times \mathbf{r}_2'(0)}{|\,\mathbf{r}_2(0)\,|^4}, \tag{3}$$

It is easy to verify that if radii vectors of the curves γ_1 and γ_2 are connected by *conditions of geometric continuity*

$$\begin{aligned} \mathbf{r}_2(0) &= \mathbf{r}_1(1), \\ \mathbf{r}_2'(0) &= \beta_1 \mathbf{r}_1(1), \\ \mathbf{r}_2''(0) &= \beta_1^2 \mathbf{r}_1''(1) + \mathbf{r}_1'(1) \end{aligned} \tag{4}$$

where $\beta_1 > 0$, $\beta_2 \ge 0$ are numerical parameters, then all three requirements (1)–(3) are fulfilled.

Let us take a set of $m+1$ points

$$\{\mathbf{V}_0, \mathbf{V}_1, \ldots, \mathbf{V}_{m-1}, \mathbf{V}_m\}$$

defined by their radii vectors (Figure 99). *A control broken line* sequentially joining these points plays an important role in the construction of the composite curve.

Let us seek an approximating composite 3-regular curve γ with the help of the partial curves γ_i described by equations of the form

$$\mathbf{r}_i(t) = \sum_{j=-2}^{1} b_j(t)\mathbf{V}_{i+j}, \quad 0 \le t \le 1, \tag{5}$$

where

$$b_j(t) = \sum_{k=0}^{3} c_{kj}(\beta_1, \beta_2)t^k, \quad j = -2, -1, 0, 1, \tag{6}$$

are the weight coefficients independent of t.

Let us require for vectors $\mathbf{r}_i(t)$ and $\mathbf{r}_{i+1}(t)$ fulfillment of the conditions (4) at a common point. With the formulae (5) the conditions of geometric continuity can be written in the following way:

$$\sum_{j=-2}^{1} b_j(0)\mathbf{V}_{i+1+j} = \sum_{j=-2}^{1} b_j(1)\mathbf{V}_{i+j},$$

$$\sum_{j=-2}^{1} b_j'(0)\mathbf{V}_{i+1+j} = \beta_1 \sum_{j=-2}^{1} b_j'(1)\mathbf{V}_{i+j}, \tag{7}$$

$$\sum_{j=-2}^{1} b_j''(0)\mathbf{V}_{i+1+j} = \beta_1^2 \sum_{j=-2}^{1} b_j''(1)\mathbf{V}_{i+j} + \beta_2 \sum_{j=-2}^{1} b_j'(1)\mathbf{V}_{i+j}$$

To find the weight coefficients c_{kj} let us rewrite the first of the relations (7) in detail

$$b_{-2}(0)\mathbf{V}_{i-1} + b_{-1}(0)\mathbf{V}_i + b_0(0)\mathbf{V}_{i+1} + b_1(0)\mathbf{V}_{i+2}$$

$$= b_{-2}(1)\mathbf{V}_{i-2} + b_{-1}(1)\mathbf{V}_{i-1} + b_0(1)\mathbf{V}_i + b_1(1)\mathbf{V}_{i+1}$$

and then equate coefficients of the same vectors. As a result we have

$$0 = b_{-2}(1), \quad b_{-2}(0) = b_{-1}(1), \quad b_{-1}(0) = b_0(1),$$

$$b_0(0) = b_1(1), \quad b_1(0) = 0$$

In a similar manner, equations for the first and second derivatives of the weight functions $b_j(t)$ can be obtained.

Considering formulae (6), we finally arrive at the linear algebraic system for the coefficients c_{kj} with positive determinant

$$\delta = 2\beta_1^3 + 4\beta_1^2 + 4\beta_1 + \beta_2 + 2 > 0$$

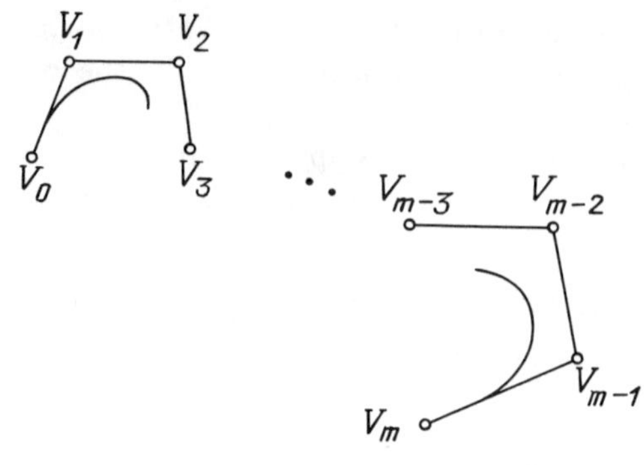

Figure 100.

Calculate coefficients c_{kj} and insert their values into formulae (6). The found formulae for the weight functions

$$b_{-2}(t) = \frac{2\beta_1^3}{\delta}(1-t)^3,$$

$$b_{-1}(t) = \frac{1}{\delta}[2\beta_1^3 t(t^2 - 3t + 3) + 2\beta_1^2(t^3 - 3t^2 + 2)$$

$$+2\beta_1(t^3 - 3t + 2) + \beta_2(2t^3 - 3t^2 + 1)],$$

$$b_0(t) = \frac{1}{\delta}[2\beta_1^2 t^2(-t+3) + 2\beta_1 t(-t^2+3) + \beta_2 t^2(-2t+3) + 2(-t^3+1)],$$

$$b_1(t) = \frac{2t^3}{\delta}$$

are suitable for the whole construction.

For a composite curve to pass through vertices $\mathbf{V}_0$ and $\mathbf{V}_m$ and to be tangent to segments $\mathbf{V}_0\mathbf{V}_1$ and $\mathbf{V}_{m-1}\mathbf{V}_m$ (Figure 100) it is necessary to add to the constructed set four more vector functions

$$\mathbf{r}_0(t) = \left(1 - \frac{2t^3}{\delta}\right)\mathbf{V}_0 + \frac{2t^3}{\delta}\mathbf{V}_1,$$

$$\mathbf{r}_1(t) = (b_{-2}(t) + b_{-1}(t))\mathbf{V}_0 + b_0(t)\mathbf{V}_1 + b_1(t)\mathbf{V}_2,$$

$$\mathbf{r}_m(t) = b_{-2}(t)\mathbf{V}_{m-1} + b_{-1}(t)\mathbf{V}_m + (b_0(t) + b_1(t))\mathbf{V}_m,$$

$$\mathbf{r}_{m+1}(t) = \frac{2\beta_1^3}{\delta}(1-t)^3\mathbf{V}_{m-1} + \left(1 - \frac{2\beta_1^3}{\delta}(1-t)^3\right)\mathbf{V}_m$$

The desired composite curve is constructed.

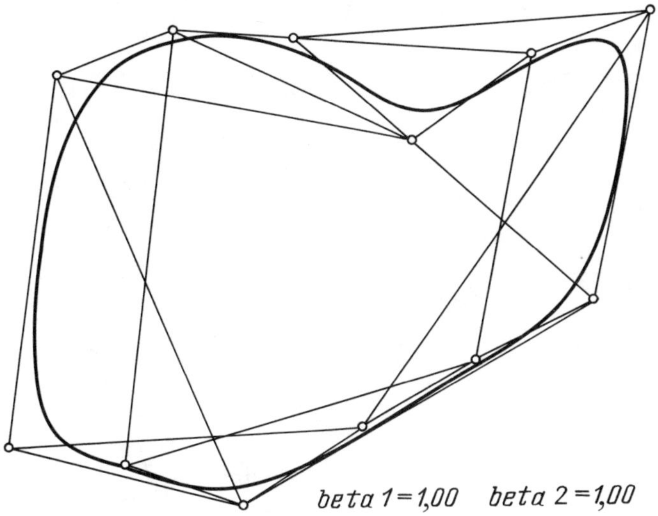

Figure 101.

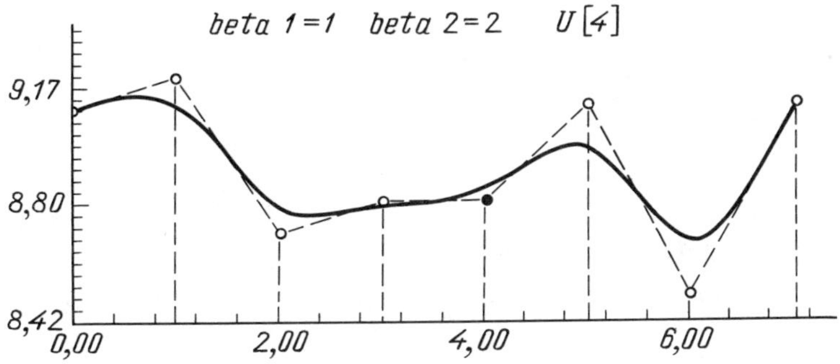

Figure 102.

Remark 1. The weight functions $b_j(t)$ are subject to the conditions

$$\sum_{j=-2}^{1} b_j(t) = 1, \quad b_j(t) \geq 0$$

where $0 \leq t \leq 1$. By this is meant that the whole curve lies in a unity of the convex hulls constructed for all sequentially taken quadruple vertices (Figure 101).

Remark 2. The outlined method for constructing a composite curve possesses *a local propety*: every vertex $\mathbf{V}_i$ takes part in construction of no more than four partial vertices. As one vertex changes, the composite curve changes only in its zone of influence (Figures 102–103).

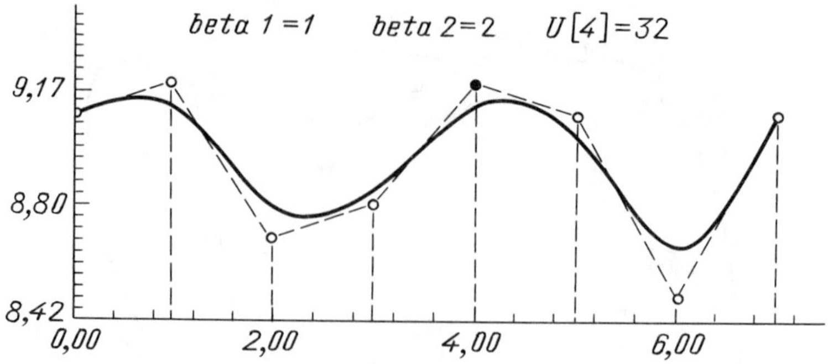

Figure 103.

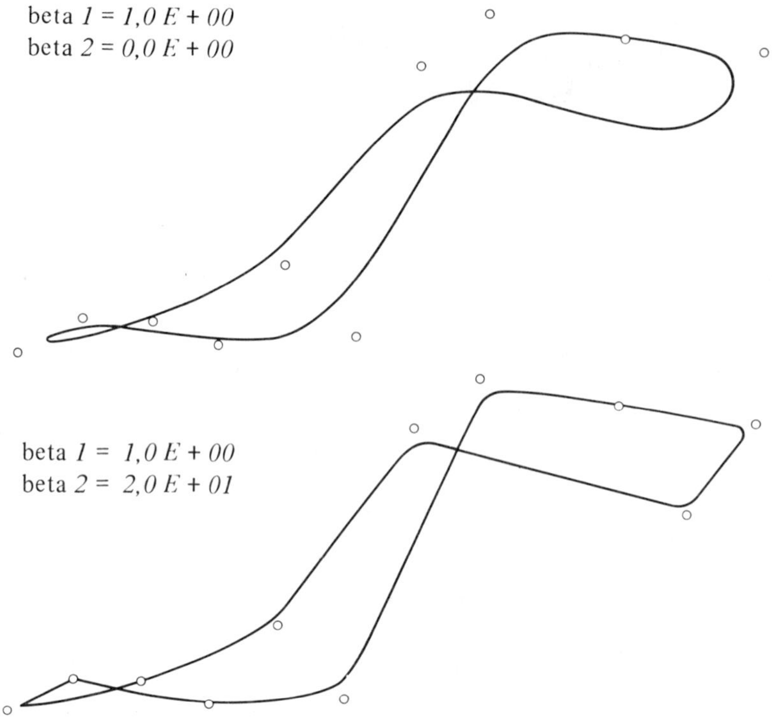

Figure 104.

Remark 3. By choosing different values of the parameters β_1 and β_2 we can change the general view of the curve (Figure 104).

For those who wish to read more in particular subjects it may be well to use the following two books:

Computational Geometry for Design and Manufacture by Ivor D. Faux and Michael J. Pratt, Ellis Horwood, Chichester, Sussex, 1979, and

Computer Graphics and Geometric Modeling Using Beta Splines by Brian A. Barsky, Springer, Berlin, 1988.

Appendix A

Smooth Functions of One and Two Variables

A real-valued function defined on interval (a, b) is called *a smooth function of class* C^n $(n \in \mathbf{N})$ or *n-smooth on interval* (a, b) if all derivatives of the function up to and including order n exist on the interval and are continuous. In case the function has on interval (a, b) continuous derivatives of any order we say that the function is *infinitely differentiable*, or *infinitely smooth*, or *of class* C^∞. A function is named *an analytical function* on interval (a, b) or *of class* C^A if for every point t_0 from the interval there exists a neighborhood of the point such that the given function may be represented as a sum of a convergent power series

$$\sum_{k=0}^{\infty} a_k (t - t_0)^k$$

A real-valued function defined on segment $[a, b]$ is called *a smooth function of class* C^n $(n \in \mathbf{N})$ or *an n-smooth on segment* $[a, b]$ if all derivatives of the function up to and including order n exist on interval (a, b) and are continuous, and at points $t = a$ and $t = b$ there exist continuous one-sided derivatives of the same orders (right-hand derivatives at point $t = a$ and left-hand derivatives at point $t = b$). In a similar manner *an n-smooth function on half-closed interval* $(a, b]$ (half-closed interval $[a, b)$) is defined. A vector function $\mathbf{r}(t)$ is called *n-smooth* on interval (a, b) if its coordinate functions are n-smooth on the interval. In a similar way *infinitely smooth* and *analytical* vector functions are defined.

A real-valued function defined in a plane domain D is called *n-smooth*, or *n-differentiable*, or *of class* C^n in domain D if in the domain all partial derivatives of the function up to and including order n exist and are continuous. In case the given function has in the domain continuous derivatives of any order we say that the function is *infinitely differentiable*, or *infinitely smooth*, or *of class* C^∞. A function is named *an analytical function* in domain D or *of class* C^A if for every point from the domain there exists a neighborhood of the point such that the given function may be represented as a sum of a convergent power series.

Let $F = F(x, y) \in C^A$ and $(x_0, y_0) \in D$. Then

$$F(x,y) = F(x_0, y_0) + \sum_{n=0}^{\infty} \frac{1}{n!} \sum_{l+k=n} C_n^k F^{(n)}_{x^k y^l}(x_0, y_0)(x - x_0)^k (y - y_0)^l$$

Appendix B

Vector Functions of Scalar Argument

Let $\mathbf{r} = \mathbf{r}(t)$, $t \in T$, be *a vector function* defined on set T of real axis. By values of the vector function are meant free vectors or bound (fixed) vectors depending on studied problems. Let application points of all vectors $\mathbf{r}(t)$, $t \in T$ (initial points) be fixed at origin O. A set of their end-points is said to be *a hodograph* of the vector function and each vector $\mathbf{r}(t)$ *a radius vector*.

Let a three-dimensional Euclidean space be provided with a Cartesian rectangular coordinate system $Oxyz$. Then to every vector $\mathbf{r}$ corresponds an ordered triple of real numbers x, y, z — its coordinates, and vice versa to every ordered triple of real numbers x, y, z corresponds a vector whose coordinates coincide with these numbers. Hence, presentation of vector function $\mathbf{r} = \mathbf{r}(t)$ is equivalent to presentation of three scalar coordinate functions $x = x(t), y = y(t), z = z(t)$:

$$\mathbf{r}(t) = \big(x(t), y(t), z(t)\big), \quad t \in T$$

If an equality $z(t) = 0$ holds for each $t \in T$, then the vector function is described by two functions $x(t), y(t)$ only:

$$\mathbf{r}(t) = \big(x(t), y(t)\big), \quad t \in T$$

Vector $\mathbf{a}$ is called *a limit* of vector function $\mathbf{r} = \mathbf{r}(t)$ for $t \to t_0$ if a length of difference $\mathbf{r} - \mathbf{a}$ tends to zero for $t \to t_0$:

$$\lim_{t \to t_0} \mid \mathbf{r}(t) - \mathbf{a} \mid = 0$$

Designation:

$$\lim_{t \to t_0} \mathbf{r}(t) = \mathbf{a}$$

If $\mathbf{a} = (l, m, n)$, then the vector limit relation

$$\lim_{t \to t_0} \mathbf{r}(t) = \mathbf{a}$$

is equivalent to concurrent fulfilment of the following three scalar relations:

$$\lim_{t \to t_0} x(t) = l, \quad \lim_{t \to t_0} y(t) = m, \quad \lim_{t \to t_0} z(t) = n$$

If $t_0 \in T$ is a finite point (number) and

$$\lim_{t \to t_0} \mathbf{r}(t) = \mathbf{r}(t_0)$$

for the vector function $\mathbf{r}(t), t \in T$, then the vector function is said to be *continuous at point* t_0. Vector function $\mathbf{r}(t), t \in T$, is called *continuous on set* T if the function is continuous at every (finite) point of the set. Vector function $\mathbf{r}(t), t \in T$, is continuous on set T if and only if each of coordinate functions $x(t), y(t), z(t)$ is continuous on the set.

Limiting properties of vector functions.

1^+. If $\lim_{t \to t_0} \mathbf{r}(t) = \mathbf{a}$ then $\lim_{t \to t_0} \mid \mathbf{r}(t) \mid = \mid \mathbf{a} \mid$.

2^+. $\lim_{t \to t_0} \mathbf{r}_1(t) + \lim_{t \to t_0} \mathbf{r}_2(t) = \lim_{t \to t_0}(\mathbf{r}_1(t) + \mathbf{r}_2(t))$.

3^+. $\lim_{t \to t_0} \alpha(t) \cdot \lim_{t \to t_0} \mathbf{r}(t) = \lim_{t \to t_0}[\alpha(t)\mathbf{r}(t)]$, $\alpha(t)$ is a scalar function.

4^+. $(\lim_{t \to t_0} \mathbf{r}_1(t), \lim_{t \to t_0} \mathbf{r}_2(t)) = \lim_{t \to t_0}(\mathbf{r}_1(t), \mathbf{r}_2(t))$.

5^+. $\lim_{t \to t_0} \mathbf{r}_1(t) \times \lim_{t \to t_0} \mathbf{r}_2(t) = \lim_{t \to t_0}(\mathbf{r}_1(t) \times \mathbf{r}_2(t))$.

Vector

$$\lim_{t \to t_0} \frac{\mathbf{r}(t) - \mathbf{r}(t_0)}{t - t_0}$$

is called *a derivative* of vector function $\mathbf{r}(t)$ at point t_0.
Designations:

$$\mathbf{r}'(t_0), \quad \dot{\mathbf{r}}(t_0), \quad \frac{d\mathbf{r}}{dt}(t_0)$$

Geometric meaning of derivative of vector function.

If $\mathbf{r}'(t_0) \neq \mathbf{0}$, then there exists a tangent line to hodograph of vector function $\mathbf{r}(t)$ at point M_0 (corresponding to value t_0 of parameter t) and vector $\mathbf{r}'(t_0)$ is parallel to the tangent line. Vector function $\mathbf{r}(t) = (x(t), y(t), z(t))$ has a derivative at point t_0 if, and only if, each of functions $x(t), y(t), z(t)$ has a derivative at the point; as this takes place, the equalities

$$\mathbf{r}'(t_0) = (x'(t_0), y'(t_0), z'(t_0))$$

$$| \mathbf{r}'(t_0) | = \sqrt{[x'(t_0)]^2 + [y'(t_0)]^2 + [z'(t_0)]^2}$$

are true.

Derivative $\mathbf{r}'(t)$ of vector function $\mathbf{r}(t)$ is said to be *a rate of change* of vector $\mathbf{r}(t)$ with respect to parameter t. A vector function is called *differentiable* at point t_0 if the function has a continuous derivative at the point.

Differential properties of vector functions.

1^+. $\mathbf{r}_1'(t) + \mathbf{r}_2'(t) = (\mathbf{r}_1(t) + \mathbf{r}_2(t))'$.

2^+. $\alpha'(t)\mathbf{r}(t) + (\alpha(t)\mathbf{r}(t))' = [\alpha(t)\mathbf{r}(t)]'$, $\quad \alpha(t)$ is a scalar function.

3^+. $(\mathbf{r}_1'(t)\mathbf{r}_2(t)) + (\mathbf{r}_1(t)\mathbf{r}_2'(t)) = (\mathbf{r}_1(t)\mathbf{r}_2(t))'$

4^+. $\mathbf{r}_1'(t) \times \mathbf{r}_2(t) + \mathbf{r}_1(t) \times \mathbf{r}_2'(t) = [\mathbf{r}_1(t) \times \mathbf{r}_2(t)]'$

Second derivative $\mathbf{r}''(t)$ of vector function $\mathbf{r}(t)$ is, by definition, a derivative of derivative $\mathbf{r}'(t)$, and so on. And the formula

$$\mathbf{r}^{(k)}(t) = (x^{(k)}(t), y^{(k)}(t), z^{(k)}(t))$$

holds.

If a length of a vector function is constant, then the vector and its first derivative are mutually orthogonal.

If vector function $\mathbf{r}(t)$ has the n-th derivative in a neighborhood of point t_0, then *the Taylor formula*

$$\mathbf{r}(t) = \mathbf{r}(t_0) + \mathbf{r}'(t_0)(t - t_0) + \frac{1}{2!}\mathbf{r}''(t_0)(t - t_0)^2 + \dots$$

$$+\frac{1}{n!}\mathbf{r}^{(n)}(t_0)(t - t_0)^n + \varepsilon(t) \quad (t = t_0)^n$$

where $\varepsilon \to \mathbf{0}$ as $t \to t_0$ is true.

If vector function $\mathbf{r}(t)$ is continuous on segment $[a, b]$, then the function is integrable on it and

$$\int_a^b \mathbf{r}(t)\, dt = \left(\int_a^b x(t)\, dt, \int_a^b y(t), \int_a^b z(t)\, dt \right)$$

is true.

Properties of integrable vector functions.

1^+. $\int_a^b \mathbf{r}_1(t)\, dt + \int_a^b \mathbf{r}_2(t)\, dt = \int_a^b (\mathbf{r}_1(t) + \mathbf{r}_1(t))\, dt$.

2^+. $\int_a^b \alpha\mathbf{r}(t)\, dt = \alpha \int_a^b \mathbf{r}(t)\, dt$, $\quad \alpha = \text{const}$.

3^+. $\int_a^b (\mathbf{a}_0 \cdot \mathbf{r}(t))\, dt = \mathbf{a}_0 \cdot \int_a^b \mathbf{r}(t)\, dt, \quad \mathbf{a}_0 = \text{const.}$

4^+. $\int_a^b (\mathbf{a}_0 \times \mathbf{r}(t))\, dt = \mathbf{a}_0 \times \int_a^b \mathbf{r}(t)\, dt, \quad \mathbf{a}_0 = \text{const.}$

5^+. $\frac{d}{dt}\left(\int_a^t \mathbf{r}(\tau)\, d\tau\right) = \mathbf{r}(t), \quad a < t < b$

Appendix C

Coordinate Systems. Coordinate Transformations

In the sections of the handbook devoted to the plane curves two coordinate systems are used: Cartesian rectangular coordinates (x, y) and polar coordinates (ρ, φ) as a rule in conformity with them:

$$x = \rho \cos \varphi, \quad y = \rho \sin \varphi$$

For investigation of a concrete plane curve the following simple coordinate transformations ever so often turn out to be useful.

1. *A parallel translation* of the coordinate system (origin O is transferred to given point (a, b)):

$$X = x - a, \quad Y = y - b$$

2. *A rotation* of coordinate axes through given angle α:

$$X = x \cos \alpha + y \sin \alpha, \quad Y = -x \sin \alpha + y \cos \alpha$$

3. *A contraction (dilatation)*:

a) along x-axis

$$X = \lambda x, \quad Y = y$$

b) along y-axis

$$X = x, \quad Y = \mu y$$

4. *A uniform contraction (dilatation)*:

a) in Cartesian coordinates

$$X = \gamma x, \quad Y = \gamma y$$

b) in polar coordinates

$$R = \gamma \rho, \quad \Phi = \varphi$$

Appendix D

Search of Some Geometric Characteristics

A METHOD FOR SEARCHING ASYMPTOTES OF UNBOUNDED ALGEBRAIC CURVE

Let a plane curve be given by an equation of the form

$$P_n(x,y) + P_{n-1}(x,y) + \ldots + P_1(x,y) + P_0 = 0, \qquad (1)$$

where

$$P_n(x,y) = \sum_{p+q=n} a_{pq}x^p y^q, \quad P_{n-1}(x,y) = \sum_{r+s=n-1} b_{rs}x^r y^s, \quad \ldots$$

Asymptotes of the form $y = kx + b$ (inclined and horizontal) may be found in the following way.

Step 1. Let $y = kx$ in equation (1), divide both sides of it by x^n, and tend x to infinity. As a consequence of this, the equation

$$P_n(1,k) = \sum_{p+q=n} a_{pq}k^q = 0, \qquad (2)$$

is obtained.

If equation (2) has no real solutions, then the given curve has no asymptotes of the form $y = kx + b$.

Let k_0 be a real solution of equation (2). In this case the next step is in order.

Step 2. Let $y = k_0x+b$ in equation (1). In view of equality $P_n(1,k_0) = 0$ the resulting relation can be written in the form

$$Q_M(b)x^m + Q_{m-1}(b)x^{m-1} + \ldots + Q_1(b)x + Q_0(b) = 0, \qquad (3)$$

where $m \leq n-1$.

Divide both sides of equation (3) by x^m, tend x to infinity and finally secure equation

$$Q_m(b) = 0, \tag{4}$$

If equation (4) has no real solutions, then the given curve has no asymptotes of the form $y = kx + b$.

Let b_0 be a real solution of equation (4). In the case $y = k_0 x + b_0$ is an equation of the desired asymptote.

Similarly, the asymptotes of the form $x = ly + c$ (inclined and vertical) may be found.

Example 1.
Given: an equation of oblique versiera

$$(x^2 + y^2)(x \cos\alpha + y \sin\alpha - 4r\cos^2\alpha) + 2ry^2 = 0,$$

$$r > 0, \quad 0 < \alpha < \frac{\pi}{2}$$

Required: asymptotes of the versiera.

Let $y = kx$ in the equation of oblique versiera:

$$x^2(1 + k^2)(x(\cos\alpha + k\sin\alpha) - 4r\cos^2\alpha) + 2rk^2x^2 = 0$$

Divide both sides of the last equation by x^3 and tend x to infinity. A limiting equality has the form

$$\cos\alpha + k\sin\alpha = 0$$

Hence, it follows that

$$k_0 = -\cot\alpha$$

Let $y = -x\cot\alpha + b$ in the equation of oblique versiera:

$$(x^2 + (-x\cot\alpha + b)^2)(-4r\cos^{\alpha} + b\sin\alpha) + 2r(-x\cot\alpha + b)^2 = 0$$

Divide both sides of the last equation by x^2 and tend x to infinity. A limiting equality has the form

$$-2r\cos^2\alpha + b\sin\alpha = 0$$

Thus, it follows that

$$b_0 = 2r\frac{\cos^2\alpha}{\sin\alpha}$$

the oblique versiera has a single inclined asymptote and

$$x\cos\alpha + y\sin\alpha - 2r\cos^2\alpha = 0$$

is an equation of the line.

A METHOD FOR SEARCHING TANGENTS AT NODES OF ALGEBRAIC CURVES

Let (a, b) be a node of a plane curve described by equation (1). Replacing coordinates x and y with formulae

$$x = X + a, \quad y = Y + b$$

transforms the point to an origin and the equation to an equation of the form

$$Q_n(X, Y) + \ldots + Q_{m+1}(X, Y) + Q_m(X, Y) = 0, \tag{5}$$

where

$$Q_m(X, Y) = \sum_{p+q=m} c_{pq} X^p Y^q, \quad Q_{m+1}(X, Y) = \sum_{r+s=m+1} d_{rs} X^r Y^s, \ldots$$

$$m \geq 1$$

Let $Y = kX$ in equation (5) and tend X to zero. Thus, it follows that

$$Q_m(1, k) = \sum_{p+q=m} c_{pq} k^q = 0, \tag{6}$$

Let k_0 be a real solution of equation (6). In this case

$$y - b = k_0(x - a)$$

is an equation of a tangent to the curve at point (a, b).

Example 2.

Given: an equation of oblique versiera

$$(x^2 + y^2)(x \cos \alpha + y \sin \alpha - 4r \cos^2 \alpha) + 2ry^2 = 0,$$

$$r > 0, \quad 0 < \alpha < \frac{\pi}{2}$$

Required: tangents to the versiera at its node $(0, 0)$.

Let $y = kx$ in the equation of oblique versiera:

$$x^2(1 + k^2)(x(\cos \alpha + k \sin \alpha) - 4r \cos^2 \alpha) + 2rk^2x^2 = 0$$

Divide both sides of the last equation by x^2 and tend x to zero. A limiting equality has the following form:

$$-(1 + k^2)4r \cos^2 \alpha + 2rk^2 = 0$$

Hence, it follows that

$$k_0^2 = \frac{2 \cos^2 \alpha}{1 - 2 \cos^2 \alpha}$$

Thus,

$$\pm x\sqrt{2} \cos \alpha = y\sqrt{- \cos 2\alpha}$$

are equations of the tangents to the curve at its node (a, b).

Appendix E

Greek Roots of Names for Some Curves

Astroid (from *ástron* – star and *éidos* – form)
Atriphtaloid (from *atritos* – without friction and *thálassa* – derivative)
Brachistochrone (from *brachistos* – shortest and *chronos* –time)
Cardioid (from *kardia* – heart and *éidos* – form)
Cissoid (from *kissós* – ivy and *éidos* – form)
Cochleoid (from *kochlos* – cóchlea and *éidos* – form)
Conchoid (from *kónche* – shell and *éidos* – form)
Cranioid (from *kránion* – cranium and *éidos* – form)
Cycloid (from *kýklos* – circle and *éidos* – form)
Epi ... (*epi* – over)
Hypo ... (*hypo* – under)
Isochrone (from *isos* – similar and *chronos* – time)
Kampyle (from *kampylos* – curved)
Lemniscate (from *lemniskos* –bandage)
Neoid (from *neos* – novel and *éidos* – form)
Nephroid (from *nephróros* – kidney and *éidos* – form)
Ophiuride (from *ófis* – snake and *oira* – tail)
Pan ... (*pan* – enveloping)
Pericycloid (from *peri* – around and *kýklos* – circle)
Poly ... (*poly* – plenty of)
Polyod (from *poly* – plenty of and *hodós* – track)
Polytrop (from *poly* – plenty of and *trópos* – turning)
Polyzomal (from *poly* – plenty of and *tó soma* – belt)
Strophoid (from *stróphos* – twisted band and *éidos* – form)
Trochoid (from *trochós* – wheel and *éidos* – form)

Appendix F

English-German-French Index

Here the names of curve classes from the Atlas of curves in English, German and French are given. Some of the classes have several names in each of the languages. Widely used names are placed in the index. The names of the class in the same language are separated by commas, the names in different languages are separated by semicolons.

For example,

Astroid (asteroid, four-cuspid hypocycloid, tetracuspid; Asteroide, Astroide, Hypozykloide mit vier Ruckkehrpunkten, kubische Zykloide, Sternkurve, vierspitzige Hypozykloide; astroïde, cubo–cycloïde, quatre rebroissements)

denotes that the curve class disposed in the Atlas by basic name *astroid* possesses also the next names:
in English — asteroid, four-cuspid hypocycloid, tetracuspid;
in German — Asteroide, Astroide, Hypozykloide mit vier Ruckkehrpunkten, kubische Zykloide, Sternkurve, vierspitzige Hypozykloide;
in French — astroïde, cubo–cycloïde, quatre rebroissements.

Acceleration curve (acceleration; Accelerationskurve, Akzelerationskurve; courbe d'accélération curve)
Agnesiana (witch of Agnesi; Agnesiana; agnésienne)
Antiversiera (Antiversiera; antiversiere)
Arachnida (araneidan; Aranaide, Spinne; arachnide, araignée)
Arc cosinusoid (Arkuskosinusoide; arc cosinusoïde)
Arc cotangentoid (Arkuskotangensoide, Arkuskotangentoide; arc cotangentoïde)
Arc sinusoid (Arkussinusoide; arc sinusoïde)
Arc tangentoid (Arkustangensoide, Arkustangentoide; arc tangentoïde)

Archimedean spiral (Archimedean screw, Archimedes' spiral, spiral of Archimedes; Archimedische Spirale, Conons Spirale; spiral archimédienne, spirale d'Archimède, spirale de Conon)
Area-hyperbolic cosinusoid (Areakosinusoide; cosinusoïde hyperbolique inverse)
Area-hyperbolic cotangentoid (Areakotagentoide; cotangentoïde hyperbolique inverse)
Area-hyperbolic sinusoid (Areasinusoide; sinusoïde hyperbolique inverse)
Area-hyperbolic tangentoid (Areatangentoide; tangentoïde hyperbolique inverse)
Astroid (asteroid, four-cuspid hypocycloid,tetracuspid; Asteroide, Astroide, Hypozykloide mit vier Ruckkehrpunkten, kubische Zykloide, Sternkurve, vierspitzige Hypozykloide; astroïde, cubo–cycloïde, quatre rebroissements)
Atriphtaloid (atriphtothalassic curve; Atriftaloide; atriphtaloïde)

Bellshaped curve (error curve; Fehlerkurve, Glokenkurve; courbe en cloche, courbe d'erreurs)
Besace (Besace, Quersachkurve, Quersachkurve Cramer; bésace)
Bicorn curve (cocked hat; Kremphutkurve, Zweihorn, Zweihornkurve; bicorne, courbe bicorne, chapeau à cornes)
Bifolium (bifolia, double folium, right double folium; Bifolium, Doppelblatt, doppelte Blattkurve, gerades Zweiblatt; bifolium, feuill double droite, folium double droit)
Biquadratic cubic (parabola) (biquadratische kubische Parabel, biquadratische–kubische Parabel von Schooten; parabole biquadratique cubique)
Bullet-nose curve (Kohlenspitzenkurve; Kohlenspitzenkurve)

Cappa (cappa curve, kappa curve; Kappa-Kurve; kappa)
Capricornoid (Capricornoide; capricornoïde)
Cardioid (cardioid curve, cardioid line, heart cotour, heart line, heart shape; Herzkurve, Herzlinie, Kardioide; cardioïde)
Cartesian folium (Descartes' folium, folium of Desartes; blatt des Descartes, Cartesisches Blatt, Descartesisches Blatt, Kartesisches Blatt, Folium Cartesii, Jasminkurve; feuille cartésienne, feuille de Descartes, fleur de jasmin, folium de Descartes, gland, nœud de ruban)
Cartesian oval (Cartesische Kurve, Cartesisches Oval, Descartesisches Oval, Kartesische Kurve, Kartesisches Oval, Oval von Descartes; ligne aplanétique, ovale cartésienne, ovale de Descartes, quartique bicirculaire cartésienne)
Cassinian oval (Cassinian ellipse, oval of Cassini; Cassinische Ellipse, Cassinische Kurve, Cassinische Linie, Cassinisches Oval, Cassinoide, Kassinoide; cassinienne, cassinoïde, courbe de Cassini, ellipse cassienne, ellipse de Cassini, ligne cassinienne, ligne de Cassini, ovale de Cassini)
Catenary curve (catenary; chainette, funiculare curve, funicular line; Katenoide, Kettenlinie, Segelkurve, Seilkurve, Velaria; arc de chaînette, caté-

naire, chaînette, courbe de chaînette, courbe funiculaire, ligne funiculaire, vélaire)
Catenary of equal resistance (Kettenlinie des gleichen Widerstandes; chaînette d'égale résistance)
Circle (Kreis; cercle)
Circular tractrix (Kreistraktrix, zirkuläre Traktrix; tractrice circulaire)
Cissoid (cissoid of Diocles, Diokles'cissoid; Efeublattkurve, Efeulinie, gerade Kissoide, Kissoide des Diokles, Kissoide Dioklea, Zissoide des Diokles, Zissoide Dioklea; cissoïde, cissoïde de Diocle, cissoïde de Dioclès, cissoïde de la droite, cissoïde de la droite cuspidale, courbe de Diocle, dioclëa)
Cissoid of the fourth order (Kissoide der viertan Ordnung, Zissoide der viertan Ordnung; cissoïde d'ordre quatrième)
Cissoid of second-order curves (Kissoide der Kurven der zweiter Ordnung, Zissoide der Kurven der zweiter Ordnung; cissoïde des courbes de ordre deuxième, cissoïde des coniques)
Cochleoid (Kochleoide; cochleoïde)
Conchal (Konchale; conchale)
Conchoid of Dürer (Dürers Muschellinie, Muschellinie von Dürer; conchoïde de Dürer)
Conchoid of higher-order circle (Kreiskonchoide von hoherer Ordnung; conchoïde de cercles d'ordre supérieur)
Conchoid of hyperbolic spiral (Konchoide der hyperbolischen Spirale; conchoïde de la spirale hyperbolique)
Conchoid of Külp (Külpsche Konchoide; conchoïde de Külp)
Conchoid of Nicomedes (cochloid, conchoid of the line, Nicomedian conchoid; Konchoide von einer Geraden, Konchoide von Nicomedes, Muscellinie des Nicomedes; conchoïde de Nicomède, conchoïde de Nicomedes, conchoïde de la droite, conchoïde première)
Conchoid of Slüse (Slüsesche Konchoide; conchoïde de Slüse)
Cornoid (Kornoide; cornoïde)
Cosecantoid (Kosekantoide; cosecantoïde)
Cosinusoid (Kosinusoide; cosinusoïde)
Cotangentoid (Kotangensoide; cotangensoïde)
Côtes' spiral (Ährenkurve, Côtessche Spirale, Kurve von Stammer; spirale de Côtes)
Cranioid (Kranioide; cranioïde)
Cruciform curve (cross curve; Kreuzkurve; courbe cruciforme)
Cubic of Tschirnhausen (trisectrix of Catalan; l'Hospital Kubic, Trisektrix von Catalan, Tschirnhausensche Kubik; cubique de Tschirnhausen, trisectrice de Catalan)
Curtate arachnida (gestreckte Araneide, verkürzte Araneide; araignée raccourcie)
Curtate cissoid (gestreckte Kissoide, verkürzte Kissoide; cissoïde raccourcie)

Curtate cycloid (gestreckte Zykloide, verkürzte Zykloide; cycloïde raccourcie)
Curtate epicycloid (gestreckte Epizykloide, verkürzte Epizykloide; epicycloïde raccourcie)
Curtate epitrochoid (gestreckte Epitrochoide, verkürzte Epitrochoide; epitrochoïde raccourcie)
Curtate hypocycloid (gestreckte Hypozykloide, verkürzte Hypozykloide; hypocycloïde raccourcie)
Curtate hypotrochoid (gestreckte Hypotrochoide, verkürzte Hypotrochoide; hypotrochoïde raccourcie)
Curtate quadratrix (gestreckte Quadratrix, verkürzte Quadratrix; quadratrice raccourcie)
Curve of Baum (Kurve von Baum; courbe de Baum)
Curve of Caraccioli (Kurve von Caraccioli; courbe de Caraccioli)
Curve of Clairaut (curve of Eudoxus, kampyle of Eudoxus; Clairauts Mediatrix, Clairauts Multiolikatrix, Clairautsche Kurve, Mediatrix von Clairaut; courbe des médianes, courbe de Clairaut, courbe multiplicatrice de Clairaut, kampule d'Eudoxe)
Curve of Cramer (Kurve von Cramer; courbe de Cramer)
Curve of damped oscillations (Kurve der erlöschenden Schwingungen; courbe de oscillations solitaire)
Curve of Debeaune (Debeaunesche Kurve; courbe de Debeaune)
Curve of Gutschoven (Gutschovensche Kurve; courbe de Gutschoven)
Curve of Jeřabek (Jeřabeksche Kurve; courbe de Jeřabek)
Curve of Kiepert (Kiepertsche Kurve; courbe de Kiepert)
Curve of Lamet (Lametsche Kurve; courbe de Lamet)
Curve of Legendre (Legendre's curve of the sixth order, Legendre's sextic; Legendresche Kurve sechster Ordnung; courbe sextique de Legendre)
Curve of Lissajous (Bowditch curve; Lissajoussche Kurve; courbe de Lissajous)
Curve of Mascheroni (Kurve von Mascheroni, Mascheronische Kurve; courbe de Mascheroni)
Curve of Ozanam (Ozanamsche Kurve; courbe de Ozanam)
Curve of Rolle (Rolle's curve of the third order; Rolles Kurve, Rollesche Kurve; courbe de Rolle)
Curve of Slüse (lemniscate of Slüse; Slüsesche Lemniskate; lemniscate de Slüse)
Curve of Watt (lemniscatoid, lemniscoid; Inflexionskurve, Lemniskoide, Wattsche Kurve; courbe à longue inflexion, courbe de Watt, lemniscoïde)
Curve with convexities (Kurve mit Bäuchen; courbe à ventres)
Cycloid (common cycloid; gemeine Zykloide,gespitzte Zykloide, gewöhnliche Zykloide, Radkurve, Radlinie, Zykloide; cycloïde, cycloïde common)

Double-heart curve (Doppel-Herz-Kurve; courbe double cœur, ligne double cœur)

Double sinusoid (Doppelsinuskurve, Doppelsinuskurve von de Longchamps; double sinusoïde)
Duplicatrix cubic (duplicatrix; Duplikatrix, Toxoide, verdopplungskurve; courbe de duplication, cubique duplicatrice, duplicatrice)

Ellipse (Ellipse; ellipse)
Elliptic lemniscate (elliptische Lemniskate; lemniscate elliptique)
Epicycloid (Epizykloide; epicycloïde)
Epitrochoid (Epitrochoide; epitrochoïde)
Exponential curve (exponent; Exponentialkurve, Exponentiallinie; courbe exponentielle, exponentielle)

Fermat's spiral (parabolic spiral; Fermatsche Spirale; spirale de Fermat)
First pearl (erste Perlkurve; première perle)
Folium (simple folium; Einblatt, einfaches Blatt; Unifolium; folium, folium simple)

Galileo's spiral (Galilesche Spirale; spirale de Galilée)
Generalized hyperbola (verallgemeinerte Hyperbola; hyperbole généralisée)
Generalized kampyle (verallgemeinerte Kampyla; kampule généralisée)
Geometric petal (botanic curve; botanische Kurve, geometrisches Blatt; courbe botanique, pétale géometrique)

Hyperbola (Hyperbel, Hyperbola; hyperbole)
Hyperbolic cosinusoid (Hyperbelkosinusoide; cosinusoïde hyperbolique)
Hyperbolic cotangentoid (Hyperbelkotangensoide; cotangensoïde hyperbolique)
Hyperbolic curve (hyperbolische Kurve; courbe hyperbolique)
Hyperbolic lemniscate (hyperbolische Lemniskate; lemniscate hyperbolique)
Hyperbolic sinusoid (Hyperbelsinusoide; sinusoïde hyperbolique)
Hyperbolic spiral (reciprocal spiral; hyperbolische Spirale; spirale hyperbolique)
Hyperbolic tangentoid (Hyperbeltangensoide; tangensoïde hyperbolique)
Hypocycloid (Hypozykloide; hypocycloïde)
Hypotrochoid (Hypotrochoide; hypotrochoïde)

Isochrone of Varignon (Isochrone von Varignon; isochrone de Varignon)
Inverse logarithmic curve (reciprocal logarithmic curve; inverse logarithmische Linie, reziproke logarithmische Linie; courbe logarithmique inverse, courbe logarithmique réciproque)

Kampyle (Kampyla; kampule)

Lemniscate of Bernoulli (lemniscate; Bernoullische hyperbolische Lemniskate, Bernoullische Lemniskate; lemniscate, lemniscate de Bernoulli)
Lemniscate of Booth (Lemniskate von Booth; lemniscate de Booth)

Lemniscate of Gerono (eight curve; Geronos Lemniskate, Lemniskate von Gerono; lemniscate de Gerono)
Lemniscate of Montferrier (Montferriersche Lemniskate; lemniscate de Montferrier)
Lemniscatrix of Ökinghaus (Lemniskatrix von Ökinghaus; lemniscatrice de Ökinghaus)
Limaçon (Pascal's limaçon; Pascalsche limaçon; limaçon de Pascal)
Lituus (Krummstab, Lituus, Lituus von Côtes; lituus)
Logarithmic curve (logarithmische Kurve; courbe logarithmique)
Logarithmic lemniscate (Logarithmische Lemniskate; lemniscate logarithmique)
Logarithmic sinusoid (logarithmic sine; logarithmische Sinus-Kurve, logarithmische Sinus-Linie, logarithmische Sinusoide; sinusoïde logarithmique)
Logarithmic spiral (equiangular spiral, logistic spiral logistique; gleichwinklige Spirale, logarithmische Spirale; spirale équiangulaire, spirale logarithmique)

Nephroid (nephroid of Freeth; Nephroide, Nierenkurve von Freeths; nephroïde)
Neoid (stream-line, stream-line rankine, swan-like line; Neoide, Stromlinie; neoïde, cycnoïde)
Nodal curve (Knotenkurve; coubre nodale)

Oblique astroid (schiefe Astroide, Sternkurve; astroïde oblique)
Oblique bifolium (schiefes Zweiblatt; bifolium oblique)
Oblique cappa (schiefe Kappa-Kurve; cappa oblique, kappa oblique)
Oblique cissoid (schiefe Kissoide; cissoïde oblique)
Oblique conchoid (schiefe Konchoide, conchoïde oblique)
Oblique pancappa (schiefe Pankappa; pancappa oblique, pankappa oblique)
Oblique parabolic folium (schiefes parabolisches Blatt; folium parabolique oblique)
Oblique strophoid (strophoid; Fokale mit Knoten, Fokale Quetelets, Knotenfokale, schiefe Strophoide, schräge Strophoide, Strophoide; focale à nœud, focale de Quetelett, strophoïde oblique, strophoïde)
Oblique trifolium (oblique three-leafed rose; schiefes Dreiblatt, schiefes Kleeblatt; rose à trois branches oblique, rose à trois feuilles oblique, tréfle oblique, trifolium oblique)
Oblique versiera (schiefe Versiera; versiere oblique)
Ophiuride (Ophiuride, Schlangenschwanzkurve, Schlangenschwanzlinie; ophiuride)
Orthoconchoid (Orthokonchoide; orthoconchoïde)
Oval of Münger (Oval Münger, einformige Kurve; ovale de Münger)

Pancappa (Pankappa; pancappa, pankappa)
Panstrophoid (Panstrophoide; panstrophoïde)

angulare Hypozykloide; courbe tricuspid ale, courbe quartique tricuspidale, hypocycloïde de Cremona, hypocycloïde steinérienne, hypocycloïde steinérienne à trois rebroussements, hypocycloïde steinérienne triangulairte)
Three-leafed rose (dreiblätterige Rosenkurve, Kleeblatt, regelmässiges Dreiblatt, reguläres Dreiblatt; rose à trois branches, rose à trois feuilles)
Top (Kreiselkurve; toupie, toton)
Tractrix (Schleppkurve, Traktorie von Huygens, Tractrix; tractoire de Huygens et Leibnitz, tractoria de Huygens et Leibnitz, tractrice)
Trefoil (trifolium; Kleeblatt; feuille de trèfle)
Trident of Newton (Cartesian parabola, Descartes' parabola, parabola of Descartes; Cartesische Parabel, Descartesische Parabel, Dreisack, Dreisackkurve, Kartesische Parabel, Newtons Dreisackkurve, Newtons Trident, Trident; conchoïde parabolique, parabole cartésienne, parabole de Descartes, trident, trident newtonien)
Trifolium (three-leafed figure; Dreiblatt, Dreiblattkurve, Kleeblattlinie, Trifolium; trèfle, trifolium)
Trifolium of Cramer (Cramersches Dreiblatt; trifolium de Cramer)
Trisectrix of Delanges (Trisekante Delanges, Trisektrix von Delanges; trisectrice de Delanges)
Trisectrix of Longchamps (Longchampssche Trisektrix, Trisektrix von Longchamps; trisectrice de Longchamps)
Trisectrix of Maclaurin (Trisectrix von Maclaurin; trisectrice de Maclaurin)
Trochoidal rose (trochoidal rhodonea; trochoidale Rhodonee; rose trochoïdale)

Versiera (Agnesi's versiera, witch, witch of Agnesi; Agnesische Kurve, Locke der Agnesi, Versiera, Versiera der Agnesi, Versiera von Guido Grandi; agnéesienne, courbe agnéesienne, courbe d'Agnesi, versiere d'Agnesi, versiere cubique d'Agnesi)
Virtual parabola (virtuelle Parabel; parabole virtuelle)
Visiera (aguinea, cissoidal of cubic, visiera Peano; begleitende Kissoide, Visiera; anguinea, anguinée, cissoïdale)

Windmill (Windmühle; moulin à vent)

Bibliography

Fladt, K., and A. Baur, *Analytische Geometrie Spezieller Flächen und Raumkurven*, Sammlung Vieweg 136, Friedr. Vieweg & Solm, Braunschweig, 1975.

Lawrence, J.D., *A Catalog of Special Plane Curves*, Dover Publication, New York, 1972.

Lockwood, E.H., *A Book of Curves*, Cambridge University Press, Cambridge, 1961.

Loria, G., *Curve Sghembe Speciale*, Nicola Zanichelli, Bologna, 1910.

Loria, G., *Spezielle algebraische and transcendente ebene Kurven. Theorie und Geschichte*, Leipzig, 1911.

Primrose, E.J.F., *Plane Algebraic Curves*, MacMillan, London, 1955.

von Seggern, D.H., *CRC Standard Curves and Surfaces*, CRC Press, Boca Raton, 1993.

Walker, A., *Algebraic Curves*, Princeton University Press, Princeton, NJ, 1950.

Some research papers published in recent years and devoted to curves

(1987 – 1992)

Benedetti, R., and M. Dedo, A geometric inequality for total curvature of plane curves, *Geom. dedic., 22*, 105, 1987.

Cieslak, W., and W. Mozgawa, On rosettes and almost rosettes, *Geom. dedic., 24*, 221, 1987.

Farran, H.R., and S.A. Robertson, Parallel immersions in euclidean space, *J. London Math. Soc. ser. 2, 35,* 527, 1987.

Epstein, C., and M. Gage, The curve shortening flow, in Wave motion: theory, modelling, and computation, *Proc. of a Conf. in honor of the 60th birthday of Peter D.Lax*, MSRI Publications, Springer-Verlag, 1987.

Grayson, M., The heat equation shrinks embedded plane curves to round points, *J. Diff. Geom., 26*, 285, 1987.

Pinkall, U., On the four vertex theorem, *Aequationes Math., 34*, 221, 1987.

Weiner, J.L., A spherical Fabricius-Bjerre formula with applications to closed space curves, *Math. Scand., 61, 2,* 286, 1987.

Zemyan, S.M., On the total torsion of certain non-closed sphere curves, *Bull. Austral. Math. Soc., 36, 1,* 39, 1987.

Costa, S.I.R., On closed twisted curves, preprint, Relatorio Tecnico N 14 IMECC-Unicamp, 1988.

Romero Fuster, M.C., Convexly generic curves in R^3, *Geom. dedic., 28, 1,* 7, 1988.

Šobr, V., Isoperimetrische Ungleichungen fur geschlossene Kurven, *Cas. pestov. mat., 113, 4,* 403, 1988.

Benedetti, R., and M. Dedo, Global inequalities for curves and surfaces in three-space, *Ann. mat. pura ed appl., 155*, 213, 1989.

Grayson, M., Shortening embedded curves, *Ann. of Math., 129*, 71, 1989.

Guggenheimer, H., Computing frames along a trajectory, *Comput. Aided Geom. Des., 6, 1,* 77, 1989.

Kosters, M., Curvature-dependent parametrisation of curves and surfaces, *Comput. Sci. Notes/Univ. Groningen. dep. Math. and Comput. Sci., N 8915,* 1, 1989.

Linner, A., Some properties of the curve straightening flow in the plane, *Trans. Amer. Math. Soc., 314, 2,* 605, 1989.

Romero Fuster, M.C., Costa, S.I.R., and J.J. Nuno Ballesteros, Some global properties of closed space curves, *Diff. Geom. Peniscola 1988, Lecture Notes in Math.*, 1410 Springer-Verlag, 286, 1989.

Stone, N.C., and T.D. DeRose, A geometric characterization of parametric cubic curves, *ACM Trans. Graph., 8, 3,* 147, 1989.

Wegner, B., Some global properties and constructions for closed curves in the plane, *Geom. dedic., 29, 3*, 317, 1989.

Cairns, G., and R.W. Sharpe, On the inversive differential geometry of plane curves, *L'Enseignement Math., 36*, 175, 1990.

Gage, M.E., Curve shortening on surfaces, *Ann. sci. Ec. norm. super., 23, 2*, 229, 1990.

Costa, S.I.R., On closed twisted curves, *Proc. Amer. Math. Soc., 109, 1*, 205, 1990.

Góźdź, S., An antipodal set of a periodic function, *J. Math. Anal. and Appl., 148, 1*, 11, 1990.

Góźdź, S., Barbier type theorems for plane curves, *Matematiche., 45, 2*, 369, 1990.

Grunbaum, B., and G.C. Shephard, Rotation and winding numbers for planar polygons and curves, *Trans. Amer. Math. Soc., 332, 1,* 169, 1990.

Kupitz, Y.S., and M.A. Perles, A condition for flatness of curves in R^n, *Amer. Math. Mon., 97, 5*, 401, 1990.

Nuño Ballesteros, J.J., and M.C. Romero Fuster, Curves with no tritangent planes in space and their convex envelopes, *J. Geom., 39, 1–2*, 120, 1990.

Stower-Grote, R., Isometrische kubische Bezierkurvensegmente, *J. Geom., 39, 1–2*, 158, 1990.

Baker, J.A., Plane curves, polar coordinates and winding numbers, *Math. Mag., 64, 2*, 75, 1991.

D'Andrea, A.C., and G. Romani, An extension of the Fenchel theorem, *Geom. dedic., 40, 1*, 59, 1991.

Fuclede, B., Bonnesen's inequality for the isoperimetric deficiency of closed curves in the plane, *Geom. dedic., 38, 3*, 283, 1991.

Michelacci, G., Inscribed polygons and fixed points of homeomorphisms of circle, *Geom. dedic., 40, 1*, 103, 1991.

Góźdź, S., and R. Smarzewski, Blaschke-Süss theorems for closed plane curves, *J.Geom., 44, 1–2*, 45, 1992.

Müller, V.H.R., Geschlossene Raumkurven mit Normalrissen von verschwindendem Flächeninhalt, *Abh. Braunschw. wiss. Ges., 18*, 7, 1992.

Pech, P., Chords and arclength of a closed space curve, *Czechosl. Math. J., 42, 3*, 471, 1992.

Strantzen, J., and J. Brooks, A chord-stretching map of a convex loop is isometry, *Geom. dedic., 41, 1*, 51, 1992.

Teufel, E., The isoperimetric inequality and total absolute curvature of closed curves in spheres, *Manuscr. math., 75, 1*, 43, 1992.

Index

Q

R

S